The Geometric Theory of Complex Variables

Peter V. Dovbush • Steven G. Krantz

The Geometric Theory of Complex Variables

Springer

Peter V. Dovbush
Institute of Mathematics and Computer Science
Moldova State University
Kishinev, Moldova

Steven G. Krantz
Mathematics, CB 1146
Washington University in St. Louis
Saint Louis, MO, USA

ISBN 978-3-031-77203-0 ISBN 978-3-031-77204-7 (eBook)
https://doi.org/10.1007/978-3-031-77204-7

Mathematics Subject Classification: 14H55, 30L05, 32D26, 32G15, 32Q02, 32M17, 30C62, 30F60

This Springer imprint is published by the registered company Springer Nature Switzerland AG
The registered company address is: Gewerbestrasse 11, 6330 Cham, Switzerland

Preface

The geometric theory of complex functions goes back at least to the mid-nineteenth century, with the efforts of Riemann. For he created Riemann surfaces, the Riemann mapping theorem, and Riemannian geometry. Another great milestone in the geometrization of complex analysis is Ahlfors's 1938 proof of the Schwarz lemma using curvature.

Since that time, geometry has blossomed in complex analysis. The geometric theory of Riemann surfaces is rather fully developed. Geometric techniques in one variable, such as harmonic measure, have achieved some maturity. In several complex variables, the study of complex manifolds is booming. Even such a basic question such as which complex manifolds support bounded holomorphic functions are unsolved and generating considerable interest. The classical question of whether there is a complex structure on the six-sphere is still open. There are thousands of papers and many open problems.

A book of the type we are producing here cannot possibly cover all aspects of what has been described above. Instead we must pick and choose. We have selected topics from one complex variable, from several complex variables, and from manifold theory that we feel are representative, and that will give the reader a feeling for this approach to our subject.

Thanks to efforts of Chern, Yau, Greene/Wu, Cheeger/Gromoll, Harvey/Lawson, Greene/Krantz, and dozens of other mathematicians, the subject of complex manifolds has exploded in the past 50 years. One could easily devote several volumes to these developments. Our choice in this book is different. We concentrate our efforts on classical geometric function theory of one and several complex variables. We do briefly treat complex mamifolds, but our discussion is only an introduction.

A few of the topics that we treat here have the flavor of hard analysis. Others are pure geometry. Still others are soft analysis or functional analysis. These aspects only serve to illustrate the diversity and variety of the subject matter.

Some of our topics, such as the boundary behavior of holomorphic functions, are subtly geometric. The geometry is in the shape of the boundary approach regions. In other topics, such as hyperbolic manifolds, the geometry is more explicit.

While some of the topics that we treat—such as Riemann surfaces—are classical, others—such as circle packing—are more modern. Hyperbolic manifolds are fairly modern. Ahlfors's proof of the Schwarz lemma borders on the classical.

Anyone knowledgeable in geometric complex analysis may lament some of our choices and some of our omissions. Having exercised our authors' prerogative, we can only welcome all comments and suggestions.

Kishinev, Moldova Peter V. Dovbush
Saint Louis, MO, USA Steven G. Krantz

Contents

Chapter 1
Introduction

Most of the theorems and definitions in this book may be found in well-known textbooks or articles in mathematical journals.

Geometric function theory of one variable is already a highly developed branch of mathematics, and it is not one in which an easily formulated classical problem awaits its solution. On the contrary it is a field in which the formulation of essential problems is almost as important as their solution. It is a subject in which methods and principles are all important, while an isolated result, however pretty and however difficult to prove, carries little weight.

Nothing could be more false than to say that classical function theory has solved its problems and has therefore outlived itself. Even without the introduction of completely new ideas, the classical problem of modules, vague as it is, and—to mention a more recent example—the investigation of the true role played by Teichmüller's extremal quasiconformal mappings are questions that can keep generations busy. The interaction of topology and function theory is likewise a field that has only been scratched on the surface. Above all, it has happened before that the whole outlook on function theory has changed abruptly, and it will happen again. "The spirit of Riemann will move future generations as it has moved us." See Ahlfors [13, p. 13].

The contents of this book can be summarized as follows. The central block of the book consists of four chapters. Chapters 2–6 treat the Riemann mapping theorem and related topics. Most readers undoubtedly have heard the name of this theorem. In fact it is a fundamental result, and we may even say that it serves as the basis of the geometric function theory of one and several complex variables. We give several proofs of this theorem including Koebe-Andreev-Thurston's Theorem. In addition to the many ordinary proofs, we provide a constructive proof of the Riemann mapping theorem. This appears in Chap. 2. It is based primarily on the Schwarz-Christoffel formula and Carathéodory's theorem.

The theme of Chap. 6 is the theory of quasiconformal mappings. Quasiconformal mappings are generalizations of conformal mappings. They can be considered not

P. V. Dovbush, S. G. Krantz, *The Geometric Theory of Complex Variables*,
https://doi.org/10.1007/978-3-031-77204-7_1

only on Riemann surfaces, but also on Riemannian manifolds in all dimensions, and even on arbitrary metric spaces. Quasiconformal mappings occur naturally in various mathematical and often a priori unrelated contexts. Of central importance in the theory of quasiconformal mappings in two dimensions is the measurable Riemann mapping theorem, proved by Lars Ahlfors and Lipman Bers.

In Chap. 7 we begin with the definition of manifolds and the fundamental ideas connected with them, such as tangent vectors, tangent spaces, etc. The description, although minimal, should be sufficient for understanding the rest of this book.

The concept of a manifold is central to many parts of geometry and modern mathematical physics because it allows complicated structures to be described in terms of well-understood topological properties of simpler spaces. Manifolds naturally arise as solution sets of systems of equations and as graphs of functions.

In Chap. 8 we explain Riemann surface theory, which is one of the highest summits of modern geometry, and we may even say that it serves as the basis of development of new mathematics. Algebraic and general topology, the theory of function fields, algebraic geometry, the theory of discrete groups, the theory of groups representation, the theory of (ordinary) differential equations with singularities, problems of the calculus of variations, potential theory, the theory of several complex variables, etc., owe their creation and development to the idea of a Riemann surface.

The theory of Riemann surfaces is undoubtedly the core of mathematics; the requirements of that theory led to the creation of general and algebraic topology, the theory of discontinuous groups of transformations, and the theory of function fields. The theory of Riemann surfaces shows much more clearly than any other theory that the "mathematical universe is inhabited not only by important species but also by interesting individuals" (Siegel).

In Chap. 9 we provide a fundamental result of Riemann's theory: the uniformization theorem (Koebe-Poincaré).

In Chap. 10 we present the basic theorem of Bun Wong [533] and Jean-Pierre Rosay [441] that was further developed in the work [201] of Greene and Krantz. The basic result is that the geometry of a boundary orbit accumulation point determines the global characteristics of the domain.

In Chap. 11 we prove the famous theorem of Daniel M. Burns and Steven G. Krantz that is in the nature of a boundary Schwarz lemma.

In Chap. 12 we prove the Schwarz-Pick lemma and give Ahlfors's version of Schwarz's lemma.

The idea to construct metrics invariant for suitable maps goes back to Bernhard Riemann. Among many other ideas that came to determine the subsequent development of the geometric approach to mathematics, this idea was explicated in his famous lecture "Über die Hypothesen, welche der Geometrie zu Grunde liegen" (1854) [435]. In a more precise form, it is formulated in the so-called Erlanger program of Felix Klein in the latter's inaugural lecture "Vergleichende Betrachtungen über neuere geometrische Forschungen" (1893) [287], where invariants of various transformation groups are exhibited.

Henri Poincaré, in his study of automorphic functions, constructed a model of non-Euclidean geometry on the unit disc $\mathbb{D}$ in the complex plane, where an important role is played by the metric

$$\rho\left(\zeta', \zeta''\right) = \frac{1}{2} \ln \frac{1 + \left|\frac{\zeta' - \zeta''}{1 - \overline{\zeta'}\zeta''}\right|}{1 - \left|\frac{\zeta' - \zeta''}{1 - \overline{\zeta'}\zeta''}\right|},$$

which is invariant for Möbious motions, i.e., conformal automorphisms of the disc. This metric is now called the Poincaré metric.

Chapter 13 is devoted to the metrics that are invariant with respect to biholomorphic maps of complex manifolds.

About the strongest statement one can make about a complex manifold from the standpoint of holomorphic mappings is that it is hyperbolic or complete hyperbolic. These concepts were introduced by Kobayashi, and their implications are discussed in his book [297]. Hyperbolicity is a natural generalization of the idea of boundedness. Chapter 14 is devoted to such questions.

In Chap. 14 Hurwitz's automorphisms theorem bounds the order of the group of automorphisms, via orientation-preserving conformal mappings, of a compact Riemann surface of genus $g > 1$. The result states that the number of such automorphisms cannot exceed $84(g - 1)$. A group for which the maximum is achieved is called a Hurwitz group, and the corresponding Riemann surface a Hurwitz surface.

In Chap. 15 we develop a new technique for studying the boundary limiting behavior of a holomorphic function on a domain—in both one and several complex variables. The approach involves two new localized maximal functions.

As a result of this methodology, theorems of Calderón type about local boundary behavior on a set of positive measures may be proved in a new and more natural way.

We also study the question of nontangential boundedness (on a set of positive measures) versus admissible boundedness. Under suitable hypotheses, these two conditions are shown to be equivalent.

In Chap. 16 we present a remarkable characterization of the unit ball in terms of its automorphism group.

In Chap. 17 we present the theorem of Greene and Krantz that is a substitute for the Riemann mapping theorem; the statement and proof are more in the spirit of Fefferman's celebrated mapping theorem.

The $\overline{\partial}$-Neumann problem was proposed by D. C. Spencer in the early 1950s as an approach to studying the inhomogeneous Cauchy-Riemann equations. The $\overline{\partial}$-Neumann problem is probably the most important and natural example of a non-elliptic boundary value problem, arising as it does from the Cauchy-Riemann equations. It has been known for some time, going back to the work of Kohn, how to determine solvability and prove regularity of solutions in the study of this problem. In Chap. 18 we present some results of Khenkin.

Chapter 19 includes a brief introduction to harmonic measure. We prove, among others, the celebrated Hadamard's three-circles theorem.

One of the most basic properties of a holomorphic (or harmonic) function f is its mean value property:

$$f(a) = A(D(a, r))^{-1} \int_{\mathbb{D}(a,r)} f dA. \tag{1.1}$$

By dA we denote the area element on $\mathbb{R}^2 \cong \mathbb{C}$.

In Chap. 20 we consider several extensions of equality (1.1).

Chapter 21 is an exposition of Teichmüller Theory based on Tromba's work.

The moduli problem. Riemann suggests that the set (space) $\mathcal{M}_p$ of isomorphy classes of Riemann surfaces of genus $p > 1$ could be parametrized by $3(p-1)$ complex parameters (or $6(p-1)$ real ones) which he called moduli.

This problem generated a huge theory: the Teichmüller theory. Teichmüller constructed a $3(p-1)$-dimensional unified covering $\mathcal{T}_p$ of the moduli space $\mathcal{M}_p, \mathcal{M}_p = \mathcal{T}_p/\Gamma_p$, with Γ_p being the Teichmüller modular group.

The manifold $\mathcal{T}_p$ has beautiful geometry: It is Stein, can be equipped with a Kodaira metric of negative curvature, etc.

There are many methods of proving these marvelous theorems (Ahlfors, Bers, Wolpert), but the most natural one is the "purely Riemannian method" by Tromba (and later Wolf and Jost): One starts with construction of an energy integral on $\mathcal{T}_p$ and shows that it is strictly convex and exhausting, whence $\mathcal{T}_p$ is Stein. Its Hessian gives a Kähler metric, the so-called Petersson-Weil metric. Here almost all of the previous material on manifolds will be used.

The moduli problem was the impulse for the huge theory of deformations of complex structures (Teichmüller, Kodaira, Spencer, Kuranishi, Grauert, and others) This was the first moduli problem. Nowadays we have moduli problems for complex surfaces, Kähler submanifolds, stable vector bundles, and Yang-Mills connections. N. Hitchin predicted that we are at the beginning of the moduli problems era.

In a recent article about the distinguished mathematician Nessim Sibony, Steven R. Bell (another distinguished complex analyst) said the following:

> I was young when I heard Nessim Sibony say "It's just analysis!" Now that I have witnessed his life's work and his prodigious contributions to geometric analysis and this venerable journal, I wonder if maybe he meant that the best analysis is geometric analysis. Otherwise it's just analysis.

Certainly Bell's remarks are very much in the spirit of the present book and serve in part as our inspiration.

Those readers who want to know the details of geometric function theory, Riemann surfaces, Teichmüller theory, invariant distances, hyperbolic manifolds that are briefly explained in this book are invited to consult textbooks on those subjects. If this book leads the reader to such a study, or if it awakens an interest in deeper theories, the authors will feel that they have accomplished their goals, and will be very happy indeed.

Chapter 2
The Riemann Mapping Theorem

We are interested in the following question: Given a domain in the complex plane, can we find a conformal map from this domain onto some simple domain? This is the uniformization problem in its simplest form. The first result in this direction is the famous Riemann mapping theorem that states that any simply connected planar domain can be conformally mapped onto the complex plane $\mathbb{C}$ or the unit disc $\mathbb{D}$.

The Riemann mapping theorem was first stated by Riemann [436] in 1851; as Weierstrass pointed out, the proof given there is insufficient. The complete proof evolved in more than half a century, together with the modern concepts of Riemann surface and of n-dimensional manifold, through the work of many people. The most important ones are Poincaré [416] (who introduced new powerful methods and gave a partial proof of the existence of the universal covering map), Osgood [405] (who proved the theorem for plane domains), Hilbert [234] (who rigorously proved the Dirichlet principle, a main tool), again Poincaré [416], and Koebe [300] (who in 1907 proved the complete statement), and Weyl [526] (who put the result in today's perspective). A direct proof of the Riemann mapping theorem can be found, e.g., in [447].

The paper of Jeremy Gray [187] traces the history of two closely related theorems in complex function theory: the Riemann mapping theorem and the uniformization theorem. They are taken from their origins in the work of Riemann, Poincaré, and Klein to their first rigorous proofs, in a slew of papers by several mathematicians around 1910, among whom Koebe was perhaps the most influential.

For Riemann's geometric ideas and their role in mathematics and physics, see Maurin's book [375].

2.1 The Riemann Mapping Theorem

The Riemann mapping theorem, even today, commands a nontrivial proof. Here we will sketch one style of proof with some details omitted.

P. V. Dovbush, S. G. Krantz, *The Geometric Theory of Complex Variables*,
https://doi.org/10.1007/978-3-031-77204-7_2

In this book, we often work with the "standard" domains $\widehat{\mathbb{C}}$, $\mathbb{C}$, $\mathbb{D}$, and $\mathbb{H} = \{z : \operatorname{Im} z > 0\}$. The first three cannot be mapped conformally onto each other, but it is easy to construct a conformal map between the last two. Indeed, it is easy to see that $\widehat{\mathbb{C}}$ cannot be equivalent to $\mathbb{C}$ or $\mathbb{D}$ since they are not even homeomorphic. To show that $\mathbb{C}$ and $\mathbb{D}$ are not equivalent, we assume the contrary, that is, that there is a univalent map from $\mathbb{C}$ onto $\mathbb{D}$. This function is a bounded entire function. By Liouville's theorem, this function must be constant, which contradicts our assumption that it is univalent.

The conformal equivalence of open sets in $\mathbb{C}$ is generally not very easy to compute or to verify, even in the special case of sets bounded by (generalized) circular arcs. For instance, it is the case (as will be verified soon) that the square $\{z = x + iy : |x| < 1, |y| < 1\}$ is conformally equivalent to the unit disc $\mathbb{D} = \{z \in \mathbb{C} : |z| < 1\}$. But it is quite difficult to write down explicitly the conformal equivalence. Even when a biholomorphic mapping is known (from other considerations) to exist, it is generally quite difficult to find it. Thus there is a priori interest in demonstrating the existence of the conformal equivalence of certain open sets by abstract, nonconstructive methods. In this section we will be concerned with the question of when an open set $U \subseteq \mathbb{C}$ is conformally equivalent to the unit disc.

Several restrictions must obviously be imposed on U. For instance, U obviously cannot be $\mathbb{C}$ because every holomorphic function $f : U \to \mathbb{D}$ would then be constant (Liouville's theorem). Also, U must be topologically equivalent (i.e., homeomorphic) to the unit disc $\mathbb{D}$ since we are in fact demanding much more in asking that U be conformally equivalent. For instance, $\mathbb{D}$ could not be conformally equivalent to $\{z \in \mathbb{C} : 1 < |z| < 2\}$ since it is not topologically (i.e., homeomorphically) equivalent. (Although at the moment we have no precise proof that the two are not homeomorphic, this assertion is at least intuitively plausible.)

Surprisingly, the two restrictions just indicated (that U not be $\mathbb{C}$ and that U be topologically equivalent to $\mathbb{D}$) are not only necessary, but they are sufficient to guarantee that U be conformally equivalent to the disc $\mathbb{D}$.

To give this notion a more precise formulation, we first define formally what we want to mean by topological (homeomorphic) equivalence:

Definition 2.1.1 Two open sets U and V in $\mathbb{C}$ are *homeomorphic* if there is a one-to-one, onto, continuous function $f : U \to V$ with $f^{-1} : V \to U$ also continuous. Such a function f is called a *homeomorphism* from U to V.

(It is a fact that the inverse of a continuous, one-to-one, onto function from one open subset of $\mathbb{R}^2$ to another is always continuous. Thus the condition in the definition of homeomorphism that f^{-1} be continuous is actually redundant. But this "automatic continuity" for the inverse function is rather difficult to prove and plays no role in our later work. Thus we shall skip the proof: The interested reader may consult an algebraic topology text.)

Theorem 2.1.2 (The Riemann Mapping Theorem) *If U is an open subset of $\mathbb{C}$, $U \neq \mathbb{C}$, and if U is homeomorphic to $\mathbb{D}$, then U is conformally equivalent to $\mathbb{D}$.*

We can find the Riemann mapping theorem in almost all textbooks on complex analysis (cf. also references in [341]), with all corresponding aspects such as the connection with the Green's function of U, construction of the Riemann mapping for polygonal domains U (Schwarz-Christoffel formula), construction with orthogonal expansions, etc. We can find the boundary behavior of the Riemann mapping, e.g., in [422, 423]. For numerical procedures, cf. [163] and [520] with many references.

The proof of the Riemann mapping theorem consists of two parts. First, one shows that if a domain U is homeomorphic to the disc, then each holomorphic function on U has an antiderivative (recall that this property is called "holomorphic simple connectivity"). The argument to prove this assertion is essentially topological: and indeed the conclusion will be deduced from a topological condition called *simple connectivity* that appears to be weaker than homeomorphism to the disc, though it is in fact equivalent to it for open sets in $\mathbb{C}$. The second step uses the first to construct a one-to-one holomorphic mapping from U into the unit disc. We then construct a mapping from U to $\mathbb{D}$ that is both one-to-one and onto by solving an "extremal problem"; that is, we shall find a function that maximizes some specific quantity (details will be provided momentarily).

The second part of the proof of the Riemann mapping theorem will occupy us for the rest of this chapter (the analytic form of the Riemann mapping theorem). The first, essentially topological, part of the proof will be given separately later on. Before presenting the (second part of the) proof, let us discuss the significance of the Riemann mapping theorem and of some of the issues that it raises. First, it is important to realize that homeomorphism is, in general, quite far from implying biholomorphic equivalence:

Example 2.1.3 The entire plane $\mathbb{C}$ is topologically equivalent to the unit disc $\mathbb{D}$. An equivalence is given by

$$f(z) = \frac{z}{1 + |z|}.$$

However, as previously noted, $\mathbb{C}$ is not conformally equivalent to $\mathbb{D}$.

It is a fact that two annuli

$$\{z \in \mathbb{C} : r_1 < |z| < r_2\} \text{ and } \{z \in \mathbb{C} : s_1 < |z| < s_2\}$$

are conformally equivalent if and only if $r_2/r_1 = s_2/s_1$. Yet two such annuli are always homeomorphic (exercise). Thus we see that the Riemann mapping theorem is startling by comparison.

The idea of the proof of the Riemann mapping theorem comes from the Schwarz lemma. Namely, one can think of this lemma in the following form: Consider a holomorphic (not necessarily one-to-one and onto) function $f : \mathbb{D} \to \mathbb{D}$ with $f(0) = 0$. Then f is conformal (i.e., one-to-one and onto) if and only if $|f'(0)|$ is as large as possible, that is, if and only if

$$\left|f'(0)\right| = \sup\left\{\left|h'(0)\right| : h : \mathbb{D} \to \mathbb{D}, h(0) = 0, h \text{ holomorphic }\right\}.$$

The special attention paid to the point 0 here is inessential. In fact if $f : \mathbb{D} \to \mathbb{D}$ is a holomorphic function such that

$$\left|f'(P)\right| = \sup\left\{\left|h'(P)\right| : h : \mathbb{D} \to \mathbb{D}, h(P) = P, h \text{ holomorphic}\right\},$$

then f must be conformal. (**Exercise:** Prove this, and find an analogous statement to cover the case when f maps P to Q and $Q \neq P$.)

If U satisfies the hypotheses of the Riemann mapping theorem, then one might look for a conformal mapping of U to $\mathbb{D}$ by choosing a point $P \in U$ and then looking for a holomorphic function $f : U \to \mathbb{D}$ such that $f(P) = 0$ and $\left|f'(P)\right|$ is maximal. It turns out to be technically simpler to consider only functions f that are assumed in advance to be one-to-one, and we shall take advantage of the resulting simplification.

Several questions arise about our plan of attack. First, do there exist any one-to-one holomorphic functions $f : U \to \mathbb{D}$? Assuming that there are such functions, is the set of values $\left\{\left|f'(P)\right|\right\}$ bounded? Also, is the least upper bound attained? That is, is there a holomorphic function f from U to $\mathbb{D}$ such that $f(P) = 0$ and $\left|f'(P)\right|$ is as large as possible?

The latter question is of particular historical interest because Riemann in fact neglected in his proof to verify that a similar sort of maximum was assumed by some particular function. Thus his theorem was in doubt for some time until what has become known as "Dirichlet's principle" gave a method to fill the gap. The modern approach to the Riemann mapping theorem involves a convergence idea for holomorphic functions called "normal families," which we develop in the next section.

Note that the issue of the boundedness of the derivatives of f at P is easily dispatched. For let $f : U \to \mathbb{D}$, $f(P) = 0$. Choose $r > 0$ such that $\overline{\mathbb{D}}(P, r) \subseteq U$. Then, of course, $|f(z)| \leq 1$ for all z; hence the Cauchy estimates imply that $\left|f'(P)\right| \leq 1/r$. Thus we have an a priori upper bound on the size of derivatives of f at P. In the next section we turn our attention to establishing the existence of an f whose derivative $f'(P)$ has modulus that actually equals the least upper bound of the absolute values of such derivatives.

In [334] is given an elementary proof of this remarkable theorem that does not use extremal problems.

2.2 The Proof of the Riemann Mapping Theorem

It is convenient to work with the upper half plane $\mathbb{H}$ rather than the disc $\mathbb{D}$. Of course they are conformally equivalent via the *Cayley transform*

$$\varphi(\zeta) = i \cdot \frac{1-\zeta}{1+\zeta}.$$

We begin by treating the case that U is bounded and has C^2 boundary. Let C be a large circle that encloses U and osculates ∂U at a point P to order 2.

With a translation, a dilation, and a rotation, we may map C into $\mathbb{H}$ with a map τ in such a way that $\tau(C)$ at the point $\tau(P)$ osculates $\partial\mathbb{H}$ at 0 to order 2 and the point $0+i$ lies inside the image of U. This means that the image of ∂U at $\tau(P)$ osculates $\partial\mathbb{H}$ at 0 to order 2. Of course τ is a conformal mapping. Let $Q \in U$ be the point that gets mapped to $0+i$. Let α_j be the dilation of U given by $\zeta \to j\zeta$. It is important to notice here that if $\mathbb{D}(0+iR, R)$ is a large disc in the upper half plane $\mathbb{H}$, then, for j sufficiently large, the image of U under $\alpha_j \circ \tau$ will contain $\mathbb{D}(0+iR, R)$. This is the significance of the osculation of the image with $\partial\mathbb{H}$. It means that the images of U under $\alpha_j \circ \tau$ exhaust $\mathbb{H}$. Unfortunately, the maps $\alpha_j \circ \tau$ do not form a normal family. Hence, we must work a bit harder.

The images of U under $\mu_j \equiv \log\left(\alpha_j \circ \tau\right)$ exhaust $\mathbb{R} \times (0, \pi)$. Since $\alpha_j \circ \tau$ is nonvanishing and U is simply connected, the logarithm function makes sense.

Now of course the image of μ_j lies in $\mathbb{R} \times (0, \pi)$. Also $0 + ij$ is mapped to $(\log j) + i\pi/2$. And the uppermost point of $\alpha_j \circ \tau(U)$ is mapped to $\log(j+C) + i\gamma_j$ for some γ_j and some positive constant C. We conclude by letting $\rho_j(\zeta) = \zeta - \log(j+C)$. Now we consider $\lambda_j \equiv \rho_j \circ \mu_j$.

We notice that as j increases, the sets $W_j \equiv \lambda_j(U) \cap [(-\infty, 0) \times (0, \pi)]$ increase. All these images lie to the left of $\mathrm{Re}\,\zeta = 0$. And the W_j exhaust $(-\infty, 0) \times (0, \pi)$.

Certainly the family of functions $\log\left(\alpha_j \circ \tau\right)$ forms a normal family (see [47]) just because the point Q is always mapped to $[\log j - \log(j+C) + i(\pi/2) \approx -C/j + i(\pi/2)]$. So there is a normally convergent subsequence. And the limit function is one-to-one by Hurwitz's theorem. It is onto $(-\infty, 0) \times (0, \pi) \equiv \{x+iy : -\infty < x < 0, 0 < y < \pi\} \equiv \mathcal{O}$. Of course the latter set can be explicitly mapped to the unit disc by a conformal mapping.

So we have established the Riemann mapping theorem for a bounded domain with C^2 boundary.

Proof of the Full Riemann Mapping Theorem:
If now U is an arbitrary simply connected (possibly unbounded) domain unequal to $\mathbb{C}$, then we may exhaust U by an increasing sequence of simply connected domains U_j, each bounded and having C^2 boundary. Fix a point Q that lies in all the U_j. By the preceding result, there is a conformal map $\gamma_j : U_j \to \mathbb{D}$. Since $\mathbb{D}$ has transitive automorphism group, we may suppose that $\gamma_j(Q) = 0$ for all j.

Of course the γ_j form a normal family. And the uniform limit on compact subsets of a subsequence of $\{\gamma_j\}$, which exists as a nondegenerate mapping because $\gamma_j(Q) = 0$ for all j, will be a conformal mapping of U to $\mathbb{D}$.

That proves the Riemann mapping theorem in full generality.

There are two approaches to function theory on a multiply connected planar domain.

The first is to work directly on the domain and examine the analytic functions in their own customary setting. This approach is quite natural, but difficulties do appear when the functions must be analyzed on or near the boundary of the domain, due to the possible complicated topological nature of the boundary.

The second approach is to "lift" the function theory of the domain to the unit disc by means of the uniformizer (uniformize—to bring into uniformity). With this technique, the boundary is well behaved (what could be nicer than the unit circle?) but new difficulties crop up with the requirement that the functions must be invariant under the group of linear fractional transformations that fix the uniformizer.

We begin by proving that every planar domain can be "uniformized" by either the unit disc or the plane. This is, of course, the famous uniformization theorem of Koebe. This theorem plays the role for a multiply connected planar domain that the Riemann mapping theorem does for a simply connected planar domain. Namely, it allows the function theory on the multiply connected domain to be "lifted" to the unit disc.

Remark 2.2.1 The Riemann mapping function given by $w = F(z)$, $F(z_0) = 0$, $F'(z_0) > 0$, determines an analytic univalent inverse function

$$z = F^{-1}(w) = f(w) : \mathbb{D} \to U$$

such that $f(0) = z_0$, $f'(0) > 0$. Then

$$z = f(w) = z_0 + a_1 w + a_2 w^2 + \dots \quad |w| < 1$$

with $a_1 > 0$. Taking instead $|w| > 1$ and $z_0 = 0$, we see that

$$z = f\left(\frac{1}{w}\right) = \frac{a_1}{w} + \frac{a_2}{w^2} + \dots$$

maps $|w| > 1$ onto U, sending ∞ to 0.

Suppose now that U^* is a simply connected domain in $\widehat{\mathbb{C}}$ with $\infty \in U^*$ and such that $E = \widehat{\mathbb{C}} \setminus U^*$ contains more than one point, i.e., E is a continuum. Then the auxiliary transformation

$$z = \frac{1}{W - \alpha}, \qquad \alpha \in E$$

maps U^* onto a simply connected domain $U \subset \mathbb{C}$ sending ∞ to 0. Combining this with $f\left(\frac{1}{w}\right)$ yields

$$W = \frac{1}{z} + \alpha = \frac{1}{f\left(\frac{1}{w}\right)} + \alpha = bw + b_0 + \frac{b_1}{w} + \frac{b_2}{w^2} + \dots = G(w),$$

which is a univalent (meromorphic) mapping of $|w| > 1$ onto U^*, with $b > 0$, and $\infty \to \infty$. Finally,

$$W = H(w) = G\left(\frac{w}{b}\right) = w + c_0 + \frac{c_1}{w} + \frac{c_2}{w^2} + \dots$$

maps $|w| > b$ onto U^*.

Similarly, by mapping U^* onto $|w| > b$, and then effecting the transformation

$$\zeta = w + \frac{e^{2i\theta}}{w},$$

one obtains a univalent mapping $\zeta = \phi(W)$ of U^* onto the ζ-plane with a slit at inclination θ with respect to the real axis, such that $\infty \to \infty$ and

$$\phi(W) = W + \frac{d_1}{W} + \frac{d_2}{W^2} + \dots$$

in a neighborhood of ∞. The representations of $H(w)$ and $\phi(W)$ feature in the proof of Theorem 5.1.2 on parallel slit mappings.

2.3 A New Proof of the Riemann Mapping Theorem

In a lecture given in the 1980s, W. Thurston proposed the idea that conformality can be understood in terms of circle packing. This idea captured the imagination of a number of mathematicians, and a new subject was quickly born.

B. Rodin and D. Sullivan were fairly quickly able to create a circle-packing proof of the Riemann mapping theorem. This result really put Thurston's idea on the map and also placed into focus some of the key techniques that would be widely used in this new study.

In this section, we explain the proof of Rodin and Sullivan.

There are many proofs of the Riemann Mapping theorem. Some of them are quite analytical—see, for instance, [97, Chapter IX, p. 585]. Others are more geometric. We present one of those here.

In fact, this proof stems from a lecture given by W.P. Thurston in 1985. The basic idea is that we can internally approximate a simply connected planar domain with a circle packing and then produce a corresponding packing in the unit disc. The natural mapping of the nerves of the packings then gives an approximate conformal mapping. In the limit, the Riemann mapping is produced.

Circle packing has given rise to an entirely new approach to complex function theory. The elegant book [488] is an homage to these new developments. B. Rodin and D. Sullivan [438] determined how to make Thurston's ideas concrete, and how to provide a rigorous circle-packing proof of the Riemann mapping theorem. That is the proof that we present here.

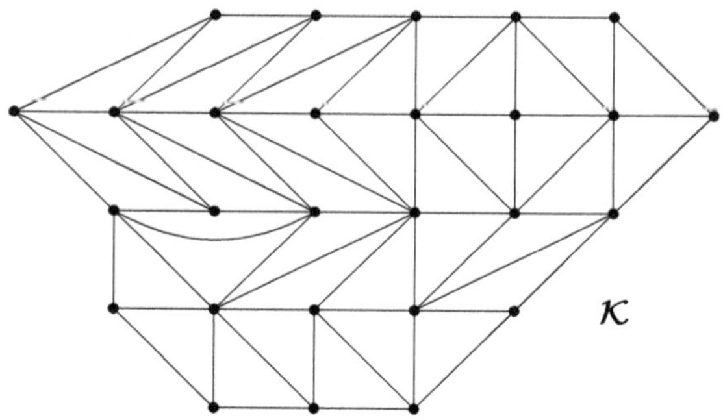

Fig. 2.1 Simple finite complex labeled $\mathcal{K}$

2.3.1 Some Definitions

We focus our attention on circle packings in the unit disc $\mathbb{D}$ with the Poincaré metric. Circle packing requires a surprisingly modest system of bookkeeping. The main objects and terms are as follows:

Let U be a planar domain.

Complex Combinatorics of a packing are encoded in an abstract simplicial 2-complex $\mathcal{K}$. The conditions are most easily described by requiring that $\mathcal{K}$ be a triangulation of an oriented topological surface. See Fig. 2.1.

Packing A *circle packing* in U is a collection of closed discs contained in U and having disjoint interiors. See Fig. 2.2 for a variety of circle packings.

Carrier The notation carr(P) denotes the concrete geometric complex in U formed by connecting the centers of tangent circles of a packing P with geodesic segments; this is simplicially equivalent to the abstract complex K.

It is easy to see that, in a circle packing of the unit disc $\mathbb{D}$, the carrier of the associated triangulation is a proper submanifold of the unit disc. See Fig. 2.3.

Chain A finite sequence of circles from a circle packing is called a *chain* if each circle except the last is tangent to its successor. The chain is a *cycle* if the first and last circles are tangent.

Nerve The *nerve* of the circle packing is the embedded 1-complex whose vertices are the centers of the discs and whose edges are the geodesic segments joining the centers of tangent discs and passing through the point of tangency.

We shall restrict our attention in this discussion to a circle packing whose nerve is the 1-skeleton of a triangulation of some open, connected subset in the plane or the Riemann sphere. In particular, a circle packing of the sphere is one whose associated triangulation is a triangulation of the sphere.

Now let c be a circle in a circle packing.

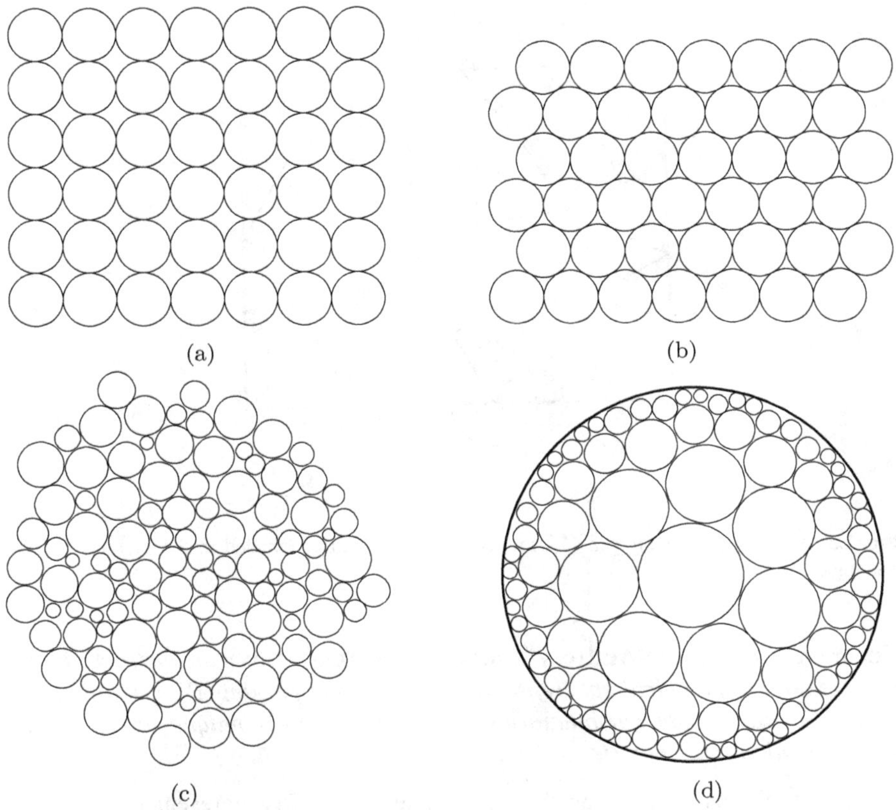

Fig. 2.2 Variety of circle packings. (**a**) A square circle packing. (**b**) A hexagonal circle packing. (**c**) Another circle packing. (**d**) A maximal circle packing in $\mathbb{D}$

Flower The *flower* centered at a circle c is the closed set consisting of c and its interior, together with all circles tangent to c and their interiors, and also the interiors of all the triangular interstices formed by these circles.

2.3.2 Discrete Analytic Function Theory

We shall not prove Theorem 2.3.1, but instead refer the reader to [27, 28, 506], [487, Chapter 19], and [438, Appendix 2]. This theorem is geometrically central to the theory. It tells us that any triangulation that we will encounter comes from a circle packing. It also gives an important uniqueness statement.

The theory of circle packing began on the sphere with this theorem, proven successively by P. Koebe [312], E.M. Andreev [27, 28], and W. Thurston [505]:

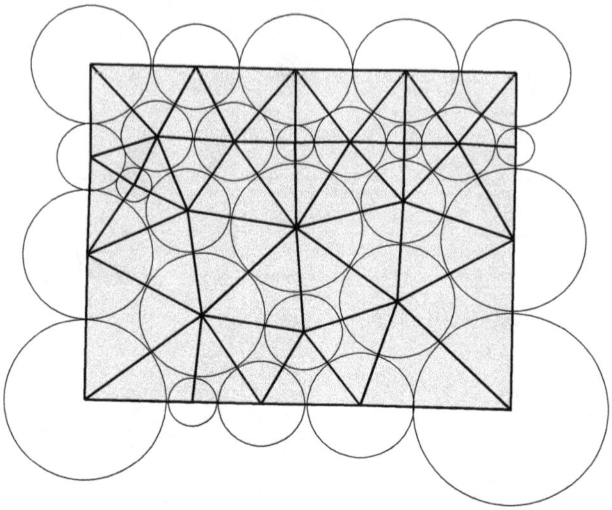

Fig. 2.3 A Euclidean packing in $\mathbb{C}$ arranged so that its carrier (shaded in the picture) forms a rectangle

Theorem 2.3.1 (Koebe-Andreev-Thurston Theorem) *Given an abstract triangulation $\mathcal{K}$ of a topological sphere, there exists a circle packing P on the Riemann sphere $\mathbb{S}^2$ having the combinatorics of $\mathcal{K}$. In fact P is unique up to Möbius transformations and inversions of $\mathbb{S}^2$.*

The conditions on P mean that each vertex v of $\mathcal{K}$ is represented by a circle $c_v \in P$, with c_v, c_u tangent whenever vertices v and u are connected by an edge in $\mathcal{K}$. The richness of circle packings may be traced to their dual natures: combinatorics in the pattern of prescribed tangencies and geometric in the realization by actual circles. We will use the phrase "has the combinatorics of" to describe this situation.

A topological annulus A in the plane has a *modulus* mod A that can be defined, without reference to conformal mapping, as the infimum of the L^2 norms of all Borel measurable functions ρ on the plane such that $1 \leq \int \rho(z)|dz|$ along all degree one curves in A.

This set of ideas is closely related to the idea of extremal length; see [17].

Definition 2.3.2 An orientation-preserving homeomorphism f between two planar domains is called *K-quasiconformal*, $1 \leq K < \infty$, if

$$K^{-1} \text{ mod } A \leq \text{mod}[f(A)] \leq K \text{ mod } A$$

for every annulus A in the domain of f.

Intuitively, a K-quasiconformal mapping does not distort the picture by more than a factor of K. Some useful facts about quasiconformal mappings are these:

(1) K-quasiconformality is a local property (see [16, p. 22]).

(2) A 1-quasiconformal map is conformal and conversely.

Further, we need the fact that a simplicial homeomorphism is K-quasiconformal for K depending only on the shapes of the triangles involved.

2.3.3 A Sketch of Thurston's Idea

Let U be a simply connected, proper subdomain of the plane. Now efficiently fill U with small circles from a regular, hexagonal circle packing. Surround these circles by a Jordan curve. Use the Koebe-Andreev-Thurston Theorem (Theorem 2.3.1) to produce a combinatorially equivalent packing of the unit disc, with the unit circle corresponding to the Jordan curve.

What we hope is that the correspondence between the circles of the two packings will approximate the Riemann mapping.

2.3.4 Details of the Proof

Lemma 2.3.3 (The Ring Lemma) *There is a constant r depending only on k such that, if k circles surround the unit disc (i.e., they form a cycle externally tangent to the unit disc), then each circle has radius at least r. See Fig. 2.4.*

Proof of the Lemma Fix k. There is a uniform lower bound for the radius of the largest outer circle c_1 (this occurs when all the k outer circles have equal radius). A circle c_2 adjacent to c_1 also has a uniform lower bound for its radius because, if c_2 were extremely small, then a chain of $k-1$ circles starting from c_2 could not escape from the crevice between c_1 and the unit circle. Repeat this reasoning for the circle c_3 adjacent to c_2, and so forth. □

Fig. 2.4 The ring lemma

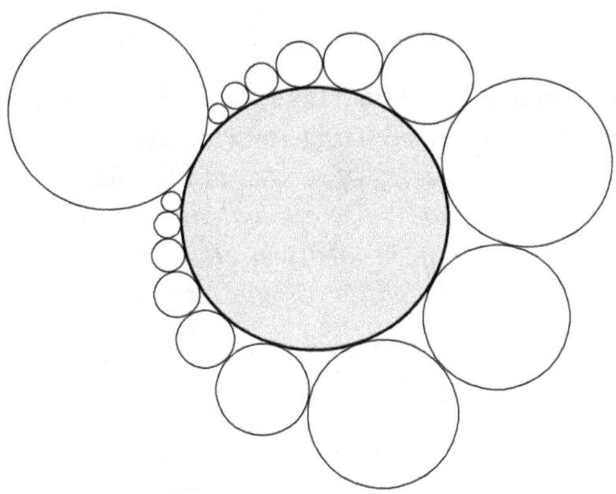

Lemma 2.3.4 (The Length-Area Lemma) *Let c be a circle in a circle packing of the unit disc. Let $S_1, S_2, \ldots, S_k$ be k disjoint chains that separate c from the origin and from a point of the boundary of the disc. Denote the combinatorial lengths of these chains by $n_1, n_2, \ldots, n_k$. Then*

$$\mathrm{radius}(c) \leq \left(n_1^{-1} + n_2^{-1} + \cdots n_k^{-1}\right)^{-1} .$$

Proof Suppose that the chain S_j consists of circles of radii $r_{j\ell}$. Then, by the Schwarz inequality,

$$\left(\sum_\ell r_{j\ell}\right)^2 \leq n_j \sum_\ell r_{j\ell}^2 .$$

Let $s_j = \sum_\ell r_{j\ell}$ be the geometric length of S_j. We find that

$$s_j^2 n_j^{-1} \leq 4 \sum_\ell r_{j\ell}^2$$

and

$$\sum_j s_j^2 n_j^{-1} \leq 4 \sum_{j,\ell} r_{j\ell}^2 \leq 4 .$$

Therefore

$$s \equiv \min\{s_1, s_2, \ldots, s_k\}$$

satisfies

$$s^2 \leq 4\left(n_1^{-1} + n_2^{-1} + \cdots + n_k^{-1}\right)^{-1} .$$

Since s is greater than the diameter of c, this last inequality proves the lemma. □

Lemma 2.3.5 (The Hexagonal Packing Lemma) *There is a sequence of real numbers s_j, decreasing to zero, with the following property. Let c_0 be a circle in a finite packing P of circles in the plane, and suppose that the packing P around c_0 is combinatorially equivalent to n generations of the regular hexagonal circle packing about one of its circles. Then the ratio of radii of any two circles in the flower around c differs from unity by less than s_n.*

Corollary 2.3.6 *A circle packing in the plane with the hexagonal pattern is the regular hexagonal packing.*

Proof *(Proof of Lemma 2.3.5)* Suppose that, for each $n = 1, 2, \ldots$, we have a circle packing P_n that is combinatorially equivalent to n generations of the regular hexagonal circle packing centered around the unit circle c_0. We apply the ring lemma to the P_n s. This shows that the radii of circles of generation k in $P_k, P_{k+1}, \ldots$ are uniformly bounded away from zero and infinity. Therefore we can select a subsequence of the P_j s so that the generation one circles converge geometrically. A further subsequence can be selected so that the generation two circles converge geometrically, and so forth.

In this fashion, a limit infinite packing of a planar domain is obtained. This packing has the combinatorics of the regular hexagonal packing, and the domain is an increasing sequence of discs and so is connected and simply connected.

If the lemma were false, then we could select the subsequences so that, in the limit packing, one of the six circles around c_0 would have radius different from 1. This contradicts the uniqueness of packings in the plane with the pattern of the hexagonal packing. □

Remark 2.3.7 It appears that we are using the corollary to prove the lemma, but an independent proof of the corollary can be given. See Appendix 1 in [438].

2.3.5 Convergence of Circle Packings to the Riemann Mapping

Let U be a simply connected, bounded domain in the plane with two distinguished points $z_0, z_1 \in U$. For a small $\epsilon > 0$, consider the regular hexagonal packing H_ϵ of the plane by circles of radius ϵ. Let c_0 be a circle whose flower contains z_0. Form all chains $\mathcal{C}_0, \mathcal{C}_1, \ldots, \mathcal{C}_k$ of circles from H_ϵ, starting with c_0, such that the flowers of the circles in the chain are contained in U. The circles that appear in such chains are called *inner circles*. The set of inner circles is denoted by I_ϵ.

The circles in H_ϵ that are not inner circles but which are tangent to inner circles will be called *border circles*. The set B_ϵ of border circles can be cyclically ordered to form a cycle called the *border*. The border has the property that the linear polygon obtained by joining the centers in order is a Jordan curve surrounding the inner circles. The inner circles I_ϵ and border circles B_ϵ form a packing C_ϵ whose nerve is the 1-skeleton of a triangulation T_ϵ.

Complete T_ϵ to a topological triangulation T_ϵ^* of the sphere by adding a vertex at ∞ plus disjoint (except for the point at ∞) Jordan arcs from ∞ to the centers of the border circles.

By Koebe-Andreev-Thurston's Theorem 2.3.1, there is a circle packing of the sphere whose associated triangulation is isomorphic to T_ϵ^* by an isomorphism that preserves the orientation of the sphere. This circle packing is unique up to Möbius transformations. We partially normalize this packing so that the exterior of the unit disc is the disc whose center corresponds to the vertex ∞ of T_ϵ^*. We then have a correspondence $c \to c'$ of circles c in $C_\epsilon \equiv I_\epsilon \cup B_\epsilon$ with circles c' in a circle packing C'_ϵ of the unit disc. We further normalize things by a Möbius transformation fixing the unit disc so that c'_0 is centered at the origin and c'_1, where c_1 is a circle whose flower contains z_1, is centered on the positive real axis. The correspondence

$c \to c'$ of C_ϵ to C'_ϵ defines an approximate mapping of H_ϵ into the unit disc $\mathbb{D}$ in the following fashion. Let z be any point in U. For $\epsilon > 0$ sufficiently small, z will be in the flower of some c in I_ϵ. As $\epsilon \to 0$, such a flower will be surrounded by more and more generations of cycles in I_ϵ. Therefore, by the length-area lemma and the divergence of the harmonic series, the radius of c' shrinks to 0. Thus c' determines an approximate position for the image of z in $\mathbb{D}$.

2.3.6 The Main Result

And now we have our main result:

Prelude This theorem is the heart of the matter. Now we see where the approximate conformal mapping comes from, and why it converges to the conformal mapping that we seek.

Caratheodory Convergence for Simply Connected Domains Let U be a simply connected domain in $\mathbb{C}$ with a marked point w, and let U_j be a sequence of simply connected domains with marked points $w_j \in U_j$. The classical notion of Carathéodory convergence is defined as follows.

Definition 2.3.8 A sequence of pointed domains (U_j, w_j) is said to converge to (U, w) in the *Carathéodory sense* if the following holds:

- $w_j \to w$.
- For any compact $K \subset U$, we have $K \subset U_j$ for every j sufficiently large.
- For any open connected set W containing w, if $W \subseteq U_j$ for infinitely many j, then $W \subseteq U$.

The pointed domains (U_j, w_j) converge to (U, w) in the Carathéodory sense if and only if the harmonic measures $\omega(U_j, w_j)$ converge weakly to the harmonic measure $\omega(U, w)$.

The classical result of Carathéodory [104, 421] states:

Theorem 2.3.9 (Carathéodory Convergence Theorem) *Let (U_j, w_j), and (U, w) be as above. Consider the conformal Riemann parametrizations $\phi_j : (\mathbb{D}, 0) \mapsto (U_j, w_j)$ with $\phi'_j(0) > 0$ and $\phi : (\mathbb{D}, 0) \mapsto (U, w)$ with $\phi'(0) > 0$. Then the Carathéodory convergence of (U_j, w_j) to (U, w) is equivalent to the uniform convergence of ϕ_j to ϕ on compact subsets of $\mathbb{D}$.*

It is worth noting that a similar statement holds for the Riemann mappings $f_n \equiv \phi_n^{-1}$:

Proposition 2.3.10 *Let $f_j = \phi_j^{-1}$ and $f \equiv \phi^{-1}$. The domains $(U_j, w_j) \to (U, w)$ in the Carathéodory sense if and only if for every compact subset $K \Subset U$ the following hold:*

- *For all j large enough, $K \Subset U_j$.*[1]
- $f_j \rightrightarrows f$ *on* K.

Proof To prove the "if" direction, let $D_m = \mathbb{D}(0, 1 - 1/m)$, and let $K_m = \phi(\mathbb{D}_m)$. Let γ be a boundary of a Jordan subdomain U' with $K_{m-1} \Subset U' \Subset K_m$. Then the winding number of $f(\gamma)$ around every $z \in D_{m-1}$ is 1. By the uniform convergence on γ, the same is true for $j \geq j_0$ large enough. Thus for these j, $f_j(K_m) \supset f_j(U') \supset D_{m-1}$. By Cauchy's Theorem, for such j, the maps ϕ_j are equicontinuous on D_{m-1}. By the Ascoli-Arzela theorem, it follows that $\phi_j \rightrightarrows \phi$ on D_{m-1}. Since every compact in $\mathbb{D}$ is a subset of D_m for some m, the claim follows.

The other direction is easier: (i) is the part of Definition 2.3.8, while (ii) again follows from equicontinuity of bounded family f_j on a compact K. □

Theorem 2.3.11 *The isomorphism $C_\epsilon \to C'_\epsilon$ of circle packings determines an approximate mapping that, as $\epsilon \to 0$, converges to a conformal homeomorphism of U with the unit disc $\mathbb{D}$.*

Proof As described in our preceding discussion, we have a circle packing C_ϵ in U and an isomorphic circle packing C'_ϵ of the unit disc $\mathbb{D}$. The associated isomorphic triangulations are denoted by T_ϵ and T'_ϵ. Let R_ϵ and R'_ϵ be the carriers of T_ϵ and T'_ϵ.

Let $f_\epsilon : R_\epsilon \to R'_\epsilon$ be the simplicial mapping determined by the correspondence of the vertices of T_ϵ and T'_ϵ. We may assume that f_ϵ is orientation preserving.

The construction makes it clear that R_ϵ converges to U in the sense that U is the union of the R_ϵ, and any compact subset of U is contained in all R_ϵ with sufficiently small positive ϵ (this is a special case of Carathéodory domain convergence). It is also true that R'_ϵ converges to the disc $\mathbb{D}$ in the same sense; this follows from the length-area lemma that shows that the radii of the border circles of C'_ϵ tend uniformly to 0 as $\epsilon \to 0$ (each border circle is separated from the origin of $\mathbb{D}$ by many disjoint chains of combinatorial length $\leq 6, 12, 18, \ldots$).

The ring lemma shows that the angles of the triangles in T'_ϵ are bounded away from 0 independently of ϵ. The ring lemma, in fact, shows this directly for the inner triangles of T'_ϵ. The proof of the ring lemma can also be applied to border circles to show that the ratio of the radius of a border circle to the radius of any adjacent circle in C'_ϵ is bounded above. It is also bounded away from zero. Hence the maps $f_\epsilon : R_\epsilon \to R'_\epsilon$ are uniformly K-quasiconformal since they map equilateral triangles to triangles of uniformly bounded distortion.

Since the f_ϵ are K-quasiconformal, it follows that they are equicontinuous on compact subsets of U (the same is true for the family f_ϵ^{-1} on compact subsets of $\mathbb{D}$). This standard fact can be seen, for example, by observing that, if z, z' vary in a compact subset of U and $|z - z'|$ becomes arbitrarily small, then $|f_\epsilon(z) - f_\epsilon(z')|$ cannot remain bounded away from 0 for a sequence of ϵ tending to 0. Indeed, one can surround z, z' with annuli A of arbitrarily large modulus;[2] hence $f_\epsilon(A)$

[1] We shall write $G \Subset \Omega$ if $\overline{G} \subset \Omega$ and G is a compact (that is, closed and bounded) subset of $\mathbb{R}^n$.

[2] By definition the modulus of the annulus $r \leq |z| \leq R$ is equal to $\frac{1}{2\pi} \ln \frac{R}{r}$.

has arbitrarily large modulus and surrounds $f_\epsilon(z), f_\epsilon(z')$. This is impossible if $f_\epsilon(z), f_\epsilon(z')$ lie in a bounded domain and their distance apart is bounded away from 0.

From the equicontinuity on compacta, it follows that the f_ϵ form a normal family. From the Carathéodory domain convergence of R_ϵ to U and of R'_ϵ to $\mathbb{D}$, it follows that every limit function f is a K-quasiconformal mapping of U onto $\mathbb{D}$ with $f(z_0) = 0$ and $f(z_1) > 0$. That $\mathbb{D} \supset f(U)$ follows from the fact that $f(z_0) = 0$.

To see that $\mathbb{D} = f(U)$, we pick $w_0 \in \mathbb{D}$. Let U be a subdomain of $\mathbb{D}$ with $0 \in U, w_0 \in U$, and $U_\epsilon \subset U$ for all sufficiently small $\epsilon > 0$. Consider now the restrictions of f_ϵ^{-1} to U; denote them by g_ϵ. Choose $\epsilon(n) \to 0+$ such that $f_\epsilon(n) \to f$ and $g_\epsilon(n) \to g$ uniformly on compacta. Now $U \supset g(U)$ because $g(0) = z_0$. It follows from $f_{\epsilon(n)}\left(g_{\epsilon(n)}(w_0)\right) = w_0$, using the uniform convergence of $f_\epsilon(n)$ near $g(w_0)$, that $f(g(w_0)) = w_0$. Hence $\mathbb{D} = f(U)$. Also f is one-to-one since the roles of U and $\mathbb{D}$ can be reversed.

The hexagonal packing lemma tells us that the simplicial mapping f_ϵ restricted to a fixed compact subset of U maps equilateral triangles to triangles of G'_ϵ that become arbitrarily close to equilateral as $\epsilon \to 0$. Therefore any limit function f of the f_ϵ will be 1-conformal and hence conformal. Since $f(z_0) = 0$ and $f(z_1) > 0$, we see that all limit functions are equal to the unique Riemann mapping with this normalization. □

2.4 The Riemann Map

While the Riemann mapping theorem assures us that any two simply connected domains with more than one boundary point can be mapped conformally to each other, it is not of much help when we are faced with the practical problem of finding the mapping function that transforms two given domains into each other.

This section endeavors to bridge the gulf between the theoretical approach given in previous sections and the more practical interest of the engineer, physicist, and applied mathematician, who are concerned with the actual construction of conformal maps.

In the present section we do not give the proof of the famous Riemann mapping theorem, but for every given simply connected region that is treated in the Riemann mapping theorem, we give a series of increasingly better approximations of the Riemann map in the Riemann mapping theorem. The idea is to approximate a domain by a polygon using a Whitney decomposition, then use the Schwarz-Christoffel formula to compute the map onto the polygonal approximation, and finally use the Carathéodory convergence theorem to obtain the Riemann map in the limit.

The proof of the Schwarz-Christoffel Theorem 2.4.5 *uses the existence of the desired conformal map*, so that it cannot be used in the proof of the Riemann mapping theorem.

2.4.1 The Schwarz-Christoffel Formula

Although the explicit conformal mapping between two given simply connected regions that is implied by the Riemann mapping theorem may be difficult to find in practice, there are cases of considerable importance where an explicit formula for conformal mapping can be given. One such is the case of a conformal mapping of $\mathbb{D}$ or a half plane onto the interior of a polygon. Since half planes and discs can easily be transformed into one another by linear fractional transformations, these problems are equivalent, and we discuss the mapping of the upper half plane $\mathbb{H}$ onto the interior of a polygon.

In theory, the Riemann mapping theorem is nice, but it is only an existence theorem. It does not specify how to find the conformal mapping. The next important breakthrough came with the works of Hermann Schwarz and Elwin Christoffel. In the 1860s, Schwarz and Christoffel independently developed a formula for a conformal mapping between the unit disc/upper half plane and a simple polygonal region in the complex plane. The formula is complicated and involves an integral in the complex plane. Furthermore, for most polygons, the integral can only be approximated numerically. Fortunately, for the special case of the square, the Schwarz-Christoffel formula can be reduced to an explicit analytical expression involving elliptic integrals[3] and elliptic functions [138].

Detailed discussion of Christoffel-Schwarz maps can be found in [138].

A (generalized) *polygon* P is defined by a collection of vertices $w_1, \ldots, w_n$ and real interior angles $\alpha_1\pi, \ldots, \alpha_n\pi$. It is convenient for indexing purposes to define $w_{n+1} = w_1$ and $w_0 = w_n$. The vertices, which lie in the extended complex plane $\mathbb{C} \cup \{\infty\}$, are given in counterclockwise order with respect to the interior of the polygon (i.e., locally the polygon is "to the left" as one traverses the edge from w_k to w_{k+1}).

---

[3] Modern mathematics defines an "*elliptic function*" as any function g that can be expressed in the form

$$g(x) = \int_c^x R\Big(t, \sqrt{p(t)}\Big)dt,$$

where R is a rational function of its two arguments, p is a polynomial of degree 3 or 4 with no repeated roots, and c is a constant. In general, integrals in this form cannot be expressed in terms of elementary functions. Incomplete elliptic integrals are functions of two arguments.

Integrals of the form $\int R(t, \sqrt{p(t)})dt$, where R is a rational function and p is a polynomial of degree 3 or 4 without multiple factors, are called *elliptic integrals*, because the first example occurs in the formula for the arc length of the ellipse. Leibniz considered the proper solution of an integration problem $\int f(x)dx$ to be a known function $g(x)$ such that $g'(x) = f(x)$. The functions then "known," and now called "elementary," were those composed from algebraic, circular, and exponential functions and their inverses. All efforts to express elliptic integrals in these terms failed, and as early as 1694 Jakob Bernoulli conjectured that the task was impossible. The conjecture was eventually confirmed by Liouville (1833), in the course of showing that a large class of integrals is nonelementary.

The *interior angle at vertex k* is defined as the angle swept from the outgoing side at w_k to the incoming side. If $|w_k| < \infty$, we have $\alpha_k \in (0, 2]$. If $\alpha_k = 2$, the sides incident on w_k are collinear, and w_k is the tip of a slit. The definition of the interior angle is applied on the Riemann sphere if $w_k = \infty$. In this case, $\alpha_k \in [-2, 0]$. Specifying α_k is redundant if w_k and its neighbors are finite, but otherwise α_k is needed to determine the polygon uniquely.

In addition to the preceding restrictions on the angles α_k, we require that the polygon makes a total turn of 2π. That is,

$$\sum_{k=1}^{n} (1 - \alpha_k) = 2$$

or, equivalently,

$$\sum_{k=1}^{n} \alpha_k = n - 2.$$

For now, we assume that P is bounded and without cusps or slits, so that $\alpha_j \in (0, 2)$ for each j.

The Schwarz-Christoffel integral formula provides a mapping function that transforms the upper half plane $\mathrm{Im} z > 0$ into a polygonal domain, that is, into a domain whose frontier consists of a finite number of line segments. Such transformations are of great use in many problems in applied mathematics.

We derive the Schwarz-Christoffel integral formula using the Poisson integral formula for the upper half plane:

$$u(x, y) = \frac{1}{\pi} \int_{-\infty}^{+\infty} u(\xi) \frac{y}{(x - \xi)^2 + y^2} \, d\xi. \tag{2.1}$$

Here $u(\xi)$ denotes the prescribed function giving the boundary values of the harmonic function on the real axis; it is supposed that $u(\xi)$ is such as to ensure convergence of the integral (2.1) and has only a finite number of (finite) discontinuities of the first kind along this axis. This result is called *Poisson's integral formula*; it gives the solution of Dirichlet's problem for the upper half plane.

Example 2.4.1 Suppose that $a_1 < a_2 < \cdots < a_n$, $a_j \in \mathbb{R}$, $a_0 = -\infty$ and $v(\xi) = v_j$ on (a_{j-1}, a_j), $a_{n+1} = \infty$. To find the potential in $\mathrm{Im}\, z > 0$, we can use Poisson's integral (2.1) directly:

$$\begin{aligned} V(x, y) &= \frac{1}{\pi} \int_{-\infty}^{+\infty} \frac{y \cdot d\xi}{(x - \xi)^2 + y^2} \\ &= \frac{v_0 y}{\pi} \int_{-\infty}^{a_1} \frac{d\xi}{(x - \xi)^2 + y^2} + \frac{v_1 y}{\pi} \int_{a_1}^{a_2} \frac{d\xi}{(x - \xi)^2 + y^2} + \cdots \end{aligned}$$

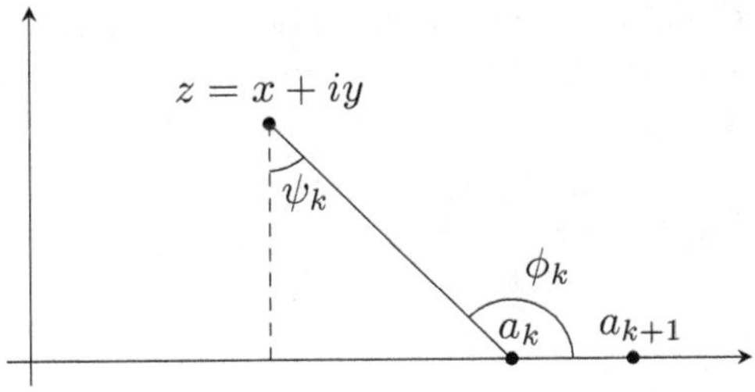

Fig. 2.5 The geometry of our argument

$$
\begin{aligned}
&+ \frac{v_n y}{\pi} \int_{a_n}^{+\infty} \frac{d\xi}{(x-\xi)^2 + y^2} \\
&= \frac{v_0}{\pi}\left(\arctan \frac{a_1 - x}{y} + \frac{\pi}{2}\right) + \frac{v_1}{\pi}\left(\arctan \frac{a_2 - x}{y} - \arctan \frac{a_1 - x}{y}\right) + \cdots \\
&+ \frac{v_n}{\pi}\left(\frac{\pi}{2} - \arctan \frac{a_n - x}{y}\right) \\
&= \frac{v_0 + v_n}{2} + \frac{v_0 - v_1}{\pi} \arctan \frac{a_1 - x}{y} + \frac{v_1 - v_2}{\pi} \arctan \frac{a_2 - x}{y} + \cdots \\
&+ \frac{v_{n-1} - v_n}{\pi} \arctan \frac{a_n - x}{y}.
\end{aligned}
$$

As shown by Fig. 2.5,

$$
\arctan \frac{a_k - x}{y} = \psi_k = \phi_k - \frac{\pi}{2},
$$

where $\phi_k = \arg(z - a_k)$ is the angle between the vector $z - a_k$ and the positive x-axis; accordingly,

$$
V(x, y) = v_n + \frac{v_0 - v_1}{\pi}\phi_1 + \cdots + \frac{v_{n-1} - v_n}{\pi}\phi_n \tag{2.2}
$$

$$
= v_n + \frac{v_0 - v_1}{\pi} \arg(z - a_1) + \cdots + \frac{v_{n-1} - v_n}{\pi} \arg(z - a_n). \tag{2.3}
$$

Theorem 2.4.2 *Let $f(z)$ be analytic in a simply connected domain $U \subset \mathbb{C}$. Then $F_0(z)$ given by*

$$
F_0(z) = \int_{z_0}^{z} f(\zeta)d\zeta \tag{2.4}
$$

is analytic in U and has a derivative given by

$$
F_0'(z) = f(z). \tag{2.5}
$$

Proof Let z be any fixed point in U, and let C be a circle with center at z lying, together with its interior, in U. Let $z + \eta$ be any point interior to C. Then z can be joined to $z + \eta$ by a straight line segment. Since $\int_{z_0}^{z} f(\zeta)d\zeta$ is independent of the path, we may take the integral $\int_{z_0}^{z+\eta} f(\zeta)d\zeta$ along the path from z_0 to z and then along the straight line from z to $z + \eta$. Utilizing (2.4), we obtain

$$F_0(z+\eta) - F_0(z) = \int_{z_0}^{z+\eta} f(\zeta)d\zeta - \int_{z_0}^{z} f(\zeta)d\zeta = \int_{z}^{z+\eta} f(\zeta)d\zeta.$$

Also

$$\frac{F_0(z+\eta) - F_0(z)}{\eta} - f(z) = \frac{1}{\eta}\int_{z}^{z+\eta} [f(\zeta) - f(z)]d\zeta.$$

Since $f(z)$ is continuous in U, for every $\epsilon > 0$, there exists a $\delta > 0$ such that

$$|f(\zeta) - f(z)| < \epsilon \quad \text{when } |\zeta - z| < \delta.$$

Hence we obtain

$$\left|\frac{F_0(\zeta+\eta) - F_0(z)}{\eta} - f(z)\right| < \epsilon$$

provided that $|\eta| < \delta$. Letting $\eta \to 0$, we see that $F_0'(z)$ exists and is equal to $f(z)$. Since z was any point in U, it follows that $F_0(z)$ is analytic at all points in the simply connected domain U. □

Theorem 2.4.3 *Let $f(z)$ be holomorphic in a simply connected domain U with $f(z) \neq 0$ for all z in U. Then there exists in U a single-valued and holomorphic function $\phi(z)$ such that $e^{\phi(z)} = f(z)$ for all z in U.*

Proof Let

$$\phi(z) = \int_{z_0}^{z} \frac{f'(\zeta)d\zeta}{f(\zeta)} + k, \tag{2.6}$$

where z_0 is any fixed point in U and k is a constant such that $e^k = f(z_0)$. Restricting the path of integration to lie in U, we see from Theorem 2.4.2 that $\phi(z)$ is holomorphic in U and

$$\phi'(z) = \frac{f'(z)}{f(z)}. \tag{2.7}$$

Let us consider the function

$$\psi(z) = e^{-\phi(z)} f(z).$$

Differentiating $\psi(z)$ and utilizing (2.7), we find that $\psi'(z) = 0$. Consequently,

$$e^{-\phi(z)} f(z) = M \quad (M \text{ constant}). \tag{2.8}$$

Setting $z = z_0$ in (2.8), noticing from (2.6) that $\phi(z_0) = k$, and recalling that $e^k = f(z_0)$, we find that

$$M = e^{-\phi(z_0)} f(z_0) = e^{-k} f(z_0) = \frac{f(z_0)}{f(z_0)} = 1.$$

Thus (2.8) may be written as $e^{\phi(z)} = f(z)$, and the theorem is established. □

Suppose $w = f(z)$ gives the desired conformal mapping of the upper half plane, onto the given polygonal domain Γ, so that $f'(z)$ cannot vanish in $\mathrm{Im}z > 0$. The existence of branches of the logarithm $\log f'$ is now obtained as a special case of Theorem 2.4.3. We find the imaginary part of a suitable regular branch of the logarithm of $f'(z)$ from the equation:

$$V(z) = \mathrm{Im}\left\{\ln f'(z)\right\}.$$

The function $V(z)$ is harmonic in $\mathrm{Im}\, z > 0$ and at each point of this half plane represents the corresponding angle of rotation in our mapping; it follows that, on every segment (a_k, a_{k+1}) of the x-axis (corresponding to a side $w_k w_{k+1}$ of the polygon), $V(z)$ assumes a constant value given by the angle, in the w-plane, between the vector

$$\overrightarrow{w_k w_{k+1}}$$

and the positive real axis. It is easily seen that the required function $V(z)$ is given by the solution (2.2) of the Dirichlet problem

$$V(z) = v_n + \frac{v_0 - v_1}{\pi} \arg(z - a_1) + \cdots + \frac{v_{n-1} - v_n}{\pi} \arg(z - a_n);$$

here, $v_0 = v_n$ is the value of V on the segments $(-\infty, a_1)$ and $(a_n, +\infty)$, and v_k is the value of V on the segment (a_k, a_{k+1}) $(k = 1, 2, \ldots, n-1)$. From the above it is clear that $v_k - v_{k-1}$ is equal to the difference of the angles at which the segments $[w_{k-1}, w_k]$ and $[w_k, w_{k+1}]$ are inclined to the real axis of the w-plane: We have $v_k - v_{k-1} = (1 - \alpha_k)\pi$, where $\alpha_k \pi$ is the interior angle of the polygon at w_k, so that

$$\begin{aligned} V(z) = \mathrm{Im}\left\{\ln f'(z)\right\} = c_2 + (\alpha_1 - 1)\arg(z - \alpha_1) + \ldots \\ + (\alpha_n - 1)\arg(z - a_n), \end{aligned}$$

where $c_2 = v_0 = v_n$ is some constant. As

$$\arg(z - a_k) = \operatorname{Im}\{\ln(z - a_k)\},$$

we can easily construct the corresponding analytic function

$$\ln f'(z) = (c_1 + ic_2) + (\alpha_1 - 1)\ln(z - a_1) + \cdots + (\alpha_n - 1)\ln(z - a_n)\ ;$$

then, taking exponentials and integrating, as before, we obtain the Schwarz-Christoffel formula for a half plane.

Theorem 2.4.4 *Let P be the interior of a polygon Γ having vertices $w_1, \ldots, w_n$ and interior angles $\alpha_1\pi, \ldots, \alpha_n\pi$ in counterclockwise order. Let f be any conformal map from the upper half plane $\mathbb{H}$ to P with $f(\infty) = w_n$. Then*

$$f(z) = A + C\int_{k=1}^{z} \prod_{k}^{n-1} (\zeta - a_k)^{\alpha_k - 1}\, d\zeta \tag{2.9}$$

for some complex constants A and C, where $w_k = f(a_k)$ for $k = 1, \ldots, n-1$.

An equally important version of the formula applies to the conformal map from the unit disc Δ.

Theorem 2.4.5 *Let P be the interior of a polygon Γ having vertices $w_1, \ldots, w_n$ and interior angles $\alpha_1\pi, \ldots, \alpha_n\pi$ in counterclockwise order. Let f be any conformal map from the unit disc Δ to P. Then the Schwarz-Christoffel formula for a disc is*

$$f(z) = A + C\int^{z} \prod_{k=1}^{n} \left(1 - \frac{\zeta}{z_k}\right)^{\alpha_k - 1} d\zeta \tag{2.10}$$

for some complex constants A and C, where $w_k = f(z_k)$ for $k = 1, \ldots, n$.

The Schwarz-Christoffel map is a formula for a one-to-one analytic function that maps the upper half of the complex plane onto the inside of a polygon. As such the Schwarz-Christoffel map can be used to translate problems set in polygonal domains to more manageable problem set in the upper half plane. Such applications of the Schwarz-Christoffel map are common in problems involving two-dimensional flows, diffusion, potentials, etc. (see [114, 369, 409]).

A statement of the Schwarz-Christoffel map can be found in most textbooks on elementary complex variables. One such statement is [138, p. 10].

Example 2.4.6 Find the integral representation of the univalent analytic function that conformally maps the upper half plane $\mathbb{H}$ (see Fig. 2.6a) onto the rectangular domain R (see Fig. 2.6b).

We may assume the vertices of R to be $-a, a, a+ib$, and $-a+ib$ $(a > 0, b > 0)$, as well as the existence of a conformal mapping from U onto R, such that the points on the x-axis: $-\frac{1}{k}, -1, 0, 1$, and $\frac{1}{k}(0 < k < 1, k$ indeterminate) correspond respectively to $-a + ib, -a, 0, a$, and $a + ib$.

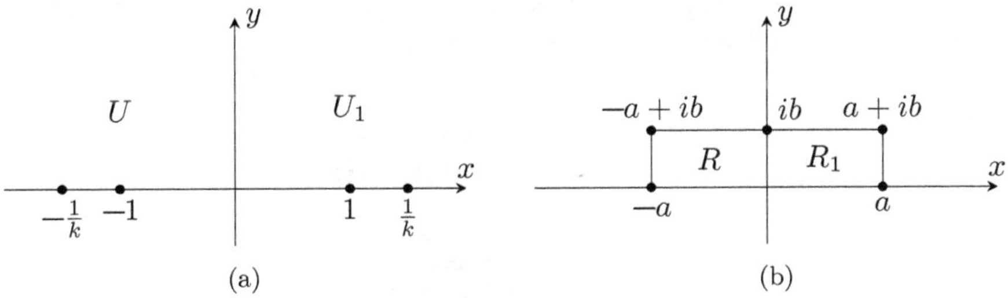

Fig. 2.6 The mapping of the upper half plane onto the rectangle

(1) Consider first the part of U in the first quadrant, U_1, and the part of R in the first quadrant, R_1. We know that the univalent analytic function $w = f_1(z)$ that conformally maps U_1 onto R_1 such that $0, 1, \infty$ correspond respectively to $0, a, ib$ exists and is unique; moreover, it maps the point $\frac{1}{k}$ on the positive x-axis into the point $a + ib$.

Since $w = f_1(z)$ maps the positive y-axis into the line segment $[0, ib]$, $f_1(z)$ can be analytically continued through the y-axis to U_2 (the other half of U). This process yields a function $f(z)$, such that $f(z) = f_1(z)$ inside U, and if z and z' are two points in U symmetric with respect to the y-axis, then $f(z)$ and $f(z')$ are symmetric with respect to the line segment $[0, ib]$. This function $w = f(z)$ is then the conformal mapping we seek.

(2) By formula (2.9), we have

$$f(z) = C_1 \int_0^z \frac{d\zeta}{\sqrt{(\zeta^2 - 1)(\zeta^2 - 1/k^2)}} + A_2$$
$$= C \int_0^z \frac{d\zeta}{\sqrt{(1 - \zeta^2)(1 - k^2\zeta^2)}}.$$

In order to determine C and k set $z = 1$, whereupon

$$a = C \int_0^1 \frac{dx}{\sqrt{(1 - x^2)(1 - k^2x^2)}}.$$

Letting $z = \frac{1}{k}$, we have

$$a + ib = C \left[\int_0^1 + \int_1^{1/k} \right] \frac{dx}{\sqrt{(1 - x^2)(1 - k^2x^2)}}$$
$$= a + iC \int_1^{1/k} \frac{dx}{\sqrt{(x^2 - 1)(1 - k^2x^2)}},$$

which then yields C and k.

If k $(0 < k < 1)$ is given, and the length and width of the rectangular domain R are chosen so that $C = 1$, i.e., if we choose

$$a = \int_0^1 \frac{dx}{\sqrt{(1 - x^2)(1 - k^2 x^2)}}$$

$$b = \int_1^{1/k} \frac{dx}{\sqrt{(x^2 - 1)(1 - k^2 x^2)}},$$

then the function

$$w = f(z) = \int_0^z \frac{d\zeta}{\sqrt{(1 - \zeta^2)(1 - k^2 \zeta^2)}}$$

maps the upper half plane $\mathbb{H}$ onto the rectangle R.

2.4.2 Carathéodory Convergence Theorem

A classical theorem of Carathéodory [104] plays a central role in geometric function theory. Let $U_1, U_2, \ldots$ be a sequence of simply connected domains in the complex plane $\mathbb{C}$, all containing the origin and none coinciding with $\mathbb{C}$. Let $w = f_n(z)$ be the conformal mapping of the unit disc $\mathbb{D} = \{z \in \mathbb{C} : |z| < 1\}$ onto U_n, normalized by the conditions $f_n(0) = 0$ and $f_n'(0) > 0$. Carathéodory's theorem connects the analytic behavior of the sequence of mapping functions f_n with the geometric behavior of their ranges U_n.

The theorem involves an unusual notion of convergence of the sequence of domains U_n. Two cases are distinguished. First suppose that the origin is an interior point of the intersection of the domains U_n, or equivalently that some disc $\{w \in \mathbb{C} : |w| < \rho\}$ lies in U_n for all n. Then the kernel of the sequence $\{U_n\}$ is defined to be the largest domain U containing the origin and having the property that each compact subset of U lies in all but a finite number of the domains U_n. A straightforward application of the Heine-Borel theorem shows that an arbitrary union of domains with this property must inherit the same property. This proves the existence of a largest domain U and shows that the kernel is well defined in this case. If the origin is not an interior point of the intersection, the kernel is defined as $U = \{0\}$. In either case, the sequence $\{U_n\}$ is said to *converge* to its kernel U if every subsequence of $\{U_n\}$ has the same kernel. This is indicated by the notation $U_n \to U$.

It is instructive to compare this situation with a more common definition of domain convergence. The sets

Fig. 2.7 The region D_n

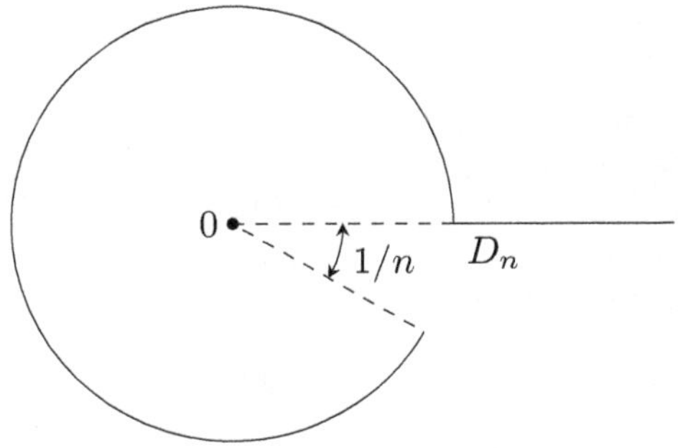

$$U_* = \liminf_{n\to\infty} U_n = \bigcup_{k=1}^{\infty}\bigcap_{n=k}^{\infty} U_n$$

and

$$U^* = \limsup_{n\to\infty} U_n = \bigcap_{k=1}^{\infty}\bigcup_{n=k}^{\infty} U_n$$

consist respectively of the points belonging to all but a finite number, and to infinitely many of the domains U_n. If $U_* = U^*$, the sequence $\{U_n\}$ is said to converge topologically to this common set.

If the sequence $\{U_n\}$ is increasing, so that $U_1 \subset U_2 \subset \cdots$, then clearly the kernel is $U = \bigcup_{n=1}^{\infty} U_n$ and $U_n \to U$. It is equally clear that $U_* = U^* = U$, so the two notions of convergence agree in this case.

On the other hand, the two notions need not agree even if the sequence $\{U_n\}$ is decreasing. Here is a typical example. Let U_n consist of the entire plane $\mathbb{C}$ minus the segment $1 \le x < \infty$ of the real axis and the arc $0 \le \theta \le 2\pi - 1/n$ of the unit circle. (See Fig. 2.7.) Then the kernel of $\{U_n\}$ is simply the unit disc $\mathbb{D}$, and $U_n \to \mathbb{D}$. However,

$$U_* = U^* = \mathbb{D} \cup \Delta \setminus \{z : \text{Im}z = 0, 1 \le \text{Re}z < \infty\},$$

the entire plane $\mathbb{C}$ minus the segment $\{z : \text{Im}z = 0, 1 \le \text{Re}z < \infty\}$ and the full unit circle. This is a disconnected set. Here Δ denotes the exterior of unit disc.

We are now prepared to state the basic theorem.

Theorem 2.4.7 (Carathéodory Convergence Theorem) *Let $\{U_n\}$ be a sequence of simply connected domains with $0 \in U_n \subsetneq \mathbb{C}$, $n = 1, 2, \ldots$. Let f_n map the* *unit disc* $\mathbb{D}$ *conformally onto* U_n *and normalized by the conditions* $f_n(0) = 0$ *and* $f_n'(0) > 0$.[4]

Let U *be the kernel of* $\{U_n\}$. *Then* $f_n \to f$ *uniformly on each compact subset of* $\mathbb{D}$ *if and only if* $U_n \to U \neq \mathbb{C}$. *In the case of convergence there are two possibilities. If* $U = \{0\}$, *then* $f = 0$. *If* $U \neq \{0\}$, *then* U *is a simply connected domain,* f *maps* $\mathbb{D}$ *conformally onto* U, *and the inverse functions* $f_n^{-1} \to f^{-1}$ *uniformly on each compact subset of* U.

Proof Suppose first that $f_n(z) \to f(z)$ uniformly on compact subsets of $\mathbb{D}$. Then f is analytic in $\mathbb{D}$, and by a well-known theorem f is either constant or univalent in $\mathbb{D}$. If f is constant, then $f(z) \equiv 0$ because $f_n(0) = 0$ for all n.

Case I: $f(z) \equiv 0$. Then we must show that $U = \{0\}$. Otherwise, some disc $\{w : |w| < \rho\}$ is contained in U_n for every n. The inverse functions φ_n are then defined in this disc and have the properties $\varphi_n(0) = 0$ and $|\varphi_n(w)| < 1$ there. It follows from the Schwarz lemma that $\left|\varphi_n'(0)\right| \leq 1/\rho$, or $\left|f_n'(0)\right| \geq \rho > 0$, which contradicts the assumption that $f_n(z) \to 0$ uniformly on compact sets. The same argument shows that every subsequence of $\{U_n\}$ has kernel $\{0\}$, so $U_n \to U = \{0\}$.

Case II: $f(z) \not\equiv 0$. Then f is univalent and maps $\mathbb{D}$ onto some domain Δ, with $f(0) = 0$ and normalized in such a way that $f'(0) > 0$. We must show that $\Delta = U$ and that $U_n \to U$.

We show first that $\Delta \subset U$. For this purpose, choose an arbitrary compact subset $E \subset \Delta$ and surround E by a rectifiable Jordan curve Γ in $\Delta \setminus E$. Let δ be the (positive) distance from E to Γ, and let $\gamma = f^{-1}(\Gamma)$. We shall now prove that $E \subset U_n$ for all n sufficiently large. Fix a point $w_0 \in E$ and observe that $|f(z) - w_0| \geq \delta$ for all $z \in \gamma$. By the uniform convergence, $|f_n(z) - f(z)| < \delta$ on γ for all $n \geq N$. Hence, by Rouché's theorem,

$$f_n'(z) - w_0 = [f(z) - w_0] + [f_n(z) - f(z)]$$

has the same number of zeros inside γ as does $[f(z) - w_0]$; namely, one zero. This shows that $w_0 \in U_n$ for all $n \geq N$, where N depends on E but not on w_0. In other words, $E \subset U_n$ for all $n \geq N$. In view of the definition of the kernel U, this proves that $\Delta \subset U$.

It follows, in particular, that the inverse functions $\varphi_n = f_n^{-1}$ are eventually defined (for all $n \geq N$) on E and are uniformly bounded there: $|\varphi_n(w)| \leq 1$. Now choose an expanding sequence of compact sets $E_m \subset \Delta$ and apply a diagonal argument to extract a subsequence $\left\{\varphi_{n_k}\right\}$ that converges uniformly on each compact subset of Δ to a function φ analytic in Δ with $\varphi(0) = 0$ and $\varphi'(0) \geq 0$.

[4] Generally, to normalize a function is to multiply it by some nonzero (real or complex) constant so that it satisfies some condition. Usually, one and only one constant will normalize any given function.

In fact,

$$0 < \frac{1}{f'(0)} = \lim_{n\to\infty} \frac{1}{f_n'(0)} = \lim_{n\to\infty} \varphi_n'(0) = \varphi'(0),$$

so φ is univalent in Δ.

The next step is to show that $\varphi = f^{-1}$. Fix $z_0 \in \mathbb{D}$ and let $w_0 = f(z_0)$. Choose $\varepsilon > 0$ so small that the circle C defined by $|z - z_0| = \varepsilon$ lies in $\mathbb{D}$, let $\Gamma = f(C)$, and let δ be the distance from w_0 to Γ. Then $|f(z) - w_0| \geq \delta$ on C, while $\left|f_{n_k}(z) - f(z)\right| < \delta$ on C for all $k \geq k_0$. Consequently, for each $k \geq k_0$ it follows from Rouchés theorem that $f_{n_k}(z_k) = w_0$ for some z_k inside C. Thus $|z_k - z_0| < \varepsilon$ and $z_k = \varphi_{n_k}(w_0)$. Therefore,

$$|\varphi(w_0) - z_0| \leq \left|\varphi(w_0) - \varphi_{n_k}(w_0)\right| + |z_k - z_0| < 2\varepsilon$$

for k sufficiently large. Letting $\varepsilon \to 0$, we conclude that $\varphi(w_0) = z_0$. Because z_0 was chosen arbitrarily in $\mathbb{D}$, this proves that $\varphi = f^{-1}$.

The preceding argument applies equally well to every subsequence of $\{\varphi_n\}$ and shows that some further subsequence converges to f^{-1} uniformly on compact subsets. It follows that $\varphi_n \to f^{-1}$ uniformly on each compact subset of Δ. In fact, the same argument shows that $\{\varphi_n\}$ converges uniformly on compact subsets of $\mathbb{D}$ to a univalent function ψ that satisfies $|\psi(w)| < 1$ there. This function ψ is an analytic continuation of f^{-1} from Δ to $\mathbb{D}$. However, f^{-1} already maps Δ conformally onto $\mathbb{D}$, so this is impossible unless $\Delta = \mathbb{D}$.

It remains to show that $U_n \to U$. But the entire argument can be repeated for any subsequence $\left\{U_{n_k}\right\}$ to conclude that f' maps $\mathbb{D}$ onto the kernel of $\left\{U_{n_k}\right\}$, which must therefore coincide with the kernel of $\{U_n\}$. Hence $U_n \to \mathbb{D}$. This completes the discussion of **Case II**.

Conversely, suppose $U_n \to U \neq \mathbb{C}$. Again there are two cases.

Case I: $U = \{0\}$. Then we claim that $f_n'(0) \to 0$. If not, there exist $\varepsilon > 0$ and a subsequence $\left\{f_{n_k}\right\}$ such that $f_{n_k}'(0) \geq \varepsilon$. In view of the Koebe one-quarter theorem, each domain U_{n_k} must then contain the disc $|w| < \varepsilon/4$, contradicting the assumption that each subsequence of $\{U_n\}$ has kernel $\{0\}$. Thus $f_n'(0) \to 0$. On the other hand, the growth theorem gives

$$|f_n(z)| \leq f_n'(0)\frac{|z|}{(1-|z|)^2}, \quad |z| < 1\,.$$

It follows that $f_n(z) \to 0$ uniformly on each compact subset of $\mathbb{D}$.

Case II: $U \neq \{0\}$, $U \neq \mathbb{C}$. Then we claim that $\left\{f_n'(0)\right\}$ is a bounded sequence. Indeed, if $f_{n_k}'(0) \to \infty$ for some subsequence, the Koebe one-quarter theorem would imply that $\left\{U_{n_k}\right\}$ has kernel $\mathbb{C}$. This contradiction shows that $\left\{f_n'(0)\right\}$ is bounded. We now infer from the growth theorem (as given above) that the functions f_n are uniformly bounded on each compact subset of $\mathbb{D}$ and therefore constitute a normal family. In order to conclude that $\{f_n\}$ converges uniformly on compact subsets of $\mathbb{D}$, it will now suffice to show that it converges pointwise.

Because the functions f_n form a normal family, however, two subsequences with different limits at some point $z_0 \in \mathbb{D}$ would have further subsequences $\{f_{n_k}\}$ and $\{f_{m_k}\}$ converging uniformly on compact sets to different functions f and $\widetilde{f}$, with $f(z_0) \neq \widetilde{f}(z_0)$. In view of what we have already proved, the corresponding sequences $\{U_{n_k}\}$ and $\{U_{m_k}\}$ would then have different kernels, the images of $\mathbb{D}$ under f and $\widetilde{f}$, respectively. But this contradicts the hypothesis that $U_n \to U$. Thus we have shown that $f_n \to f$ uniformly on each compact subset of $\mathbb{D}$. This completes the proof. □

2.4.3 Construction of the Riemann Map

The main idea is that any domain can be approximated by polygonal domains and for a polygonal domain there is a nice expression for a conformal map from the unit disc onto that domain that is given by the Schwarz-Christoffel formula. It is important to note that all constructive proofs give a series of increasingly better approximations to the Riemann map, but do not give the map itself.

Let U be any proper open subset of $\mathbb{C}$. We assume that U is bounded, simply connected and that $0 \in U$. Then there exist closed squares $\{S_j\}$, having pairwise disjoint interiors and sides parallel to the axes, such that:

(i) S_j has side length $\ell(S_j) = 2^{-n_j}$.
(ii) $U = \bigcup S_j$.
(iii) $\operatorname{diam}(S_j) \leq \operatorname{dist}(S_j, \partial U) < 4 \operatorname{diam}(S_j)$.

The squares $\{S_j\}$ are called *Whitney squares* (see [172, p. 21]). Here is one way to construct Whitney squares in the case $\operatorname{diam}(U) < \infty$: Let $2^{-N+1} \leq \operatorname{diam}(U) < 2^{-N+2}$ and partition the plane into squares having sides parallel to the axes and side length 2^{-N}. We call these squares "2^{-N} squares." Include in the family $\{S_j\}$ any 2^{-N} square $S \subset U$ that satisfies **(iii)**, and divide each of the remaining 2^{-N} squares into four squares of side 2^{-N-1}. Next include in $\{S_j\}$ any of these new 2^{-N-1} squares contained in U and satisfying **(iii)**, and continue (Fig. 2.8—the Whitney decomposition).

Set $U'_N := \{z \in U : \operatorname{dist}(z, \partial U) > \frac{1}{N}\}$, $N \in \mathbb{N}$. Let T_N be the union of a minimal collection of squares of side $\geq 2^{-N}$ that lie in U and such that the vertices of squares in T_N that belong to ∂T_N form a polygon Γ_N such that its interior $P_N \supset U'_N$. It is easy to see that one can arrange that $0 \in P_N \subset P_{N+1}$ for each $N \geq N_0$.

By Theorem 2.4.5 there is a conformal map from $\mathbb{D}$ onto P_N. Let f_N map the unit disc $\mathbb{D}$ conformally onto P_N.

Write $a_N = -f_N^{-1}(0)$. Let

$$\varphi_N(w) = \frac{w - a_N}{1 - w\overline{a}_N}.$$

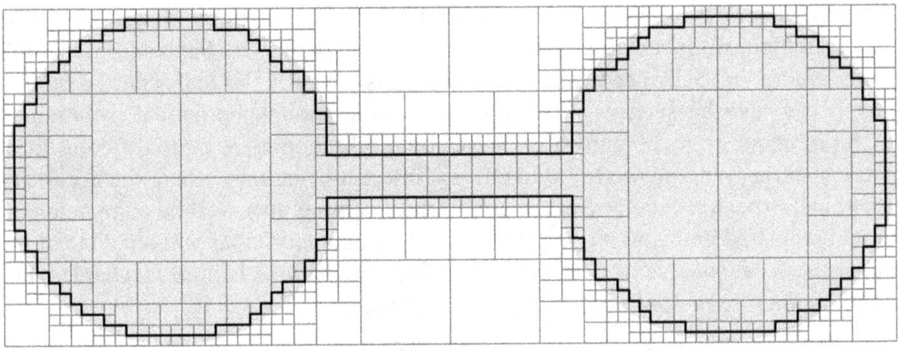

Fig. 2.8 Whitney squares U

Note that $\varphi_N(0) = -a_N = f_N^{-1}(0)$ and $\varphi_N'(0) = 1$.

Set $F_N = f_N \circ \varphi_N$. It is clear that $F_N(0) = f_N(\varphi_N(0)) = f_N(f_N^{-1}(0)) = 0$.

Put $\theta_N = -\arg f_N'(a_N)$; then

$$e^{i\theta_N} \cdot F_N'(0) = e^{i\theta_N} \cdot f_N'|_{a_N} \cdot \varphi'|_0 = |f_N'(a_N)| \cdot 1 > 0.$$

From the beginning, we can put $a_N = -f_N^{-1}(0)e^{-i\theta_N}$ and

$$\varphi_N(w) = e^{i\theta_N} \frac{w - a_N}{1 - w\overline{a}_N}.$$

Then $F_N(0) = 0$ and $F_N'(0) > 0$.

Since the sequence T_N is increasing, so that $T_N \subset T_{N+1}$, then clearly the kernel is $U = \bigcup_{N=1}^{\infty} T_N$ and $U \setminus T_N \to \emptyset$.

The proof now follows from Theorem 2.4.7.

Next consider the case $\operatorname{diam}(U) = \infty$. We assume that $0 \in U$. We construct inductively a sequence of regions

$$0 \in U_1 \subset U_2 \subset U_3 \subset \ldots \subset \cup_{N=1}^{\infty} U_N = U \subset \mathbb{C}$$

and conformal maps f_N from $\mathbb{D}$ onto U_N such that $f_N(0) = 0$ and $f_N'(0) > 0$.

Since $U \neq \mathbb{C}$ is a simply connected region, thus for all integer N the set $U \cap \{z \in \mathbb{C} : |z| < N\}$ has a simply connected component U_N, $\operatorname{diam}(U_N) < 2N$, such that $0 \in U_N$. It is clear that $U_N \subset U_{N+1}$ and $U = \cup_N U_N \subset \mathbb{C}$. Now from the first part of the proof and Theorem 2.4.7, it follows that there exists f_N a conformal map from $\mathbb{D}$ onto U_N, and therefore $f_N'(0) > 0$. Clearly $f_N(0) = 0$.

The proof now follows from Theorem 2.4.7.

Remark 2.4.8 For problems in which the Schwarz-Christoffel mapping is appropriate, it is very powerful. Many researchers have used Schwarz-Christoffel maps of simple geometries to get fully analytical solutions to applied problems. Banjai and Trefethen[39] invented a fast algorithm for computing the Schwarz-Christoffel maps. Numerical Schwarz-Christoffel mapping is typically fast and accurate enough that it can also be seen as providing an explicit, though numerical, solution to such problems. In this regard Schwarz-Christoffel mapping is quite different from conformal mapping methods based on, say, integral equations, where the necessary effort and expected accuracy of the conformal mapping may well be comparable to those associated with solving Laplace's equation on the original domain. One of the major classical sources of conformal mapping applications is fluid mechanics. For more on the role of conformal mapping in applications, see [231].

In [370, Chapter 8, Section 8.1, pp. 123–129], Marshall gives the geodesic zipper algorithm for computing a conformal map of a simply connected region whose boundary contains a prescribed (finite) collection of points, onto the upper half plane $\mathbb{H} = \{z \in \mathbb{C} : Re\, z > 0\}$. In [370, Chapter 8, Section 8.2] Marshall gave a somewhat constructive proof of the Riemann mapping theorem. The idea of the proof here is to construct, using the geodesic zipper algorithm, conformal maps from an increasing sequence of regions onto the disc, and then prove that the sequence converges to a conformal map of the union of these regions onto the disc.

The map from a simply connected domain U onto $\mathbb{D}$ is not unique; it can be composed with any Möbius transformation preserving the unit disc. The family of Möbius transformations is described by three real parameters: the real and imaginary parts of the point that is mapped to the origin and the angle of rotation. This means that, in general, we should be able to fix uniquely any three real parameters by the proper choice of a Möbius transformation.

In the standard formulation of the Riemann theorem, Ahlfors [12, Theorem 1, p. 230] normalizes the map by requiring that a fixed point z_0 is mapped to the origin and that the argument of the derivative at this point is zero. This choice corresponds exactly to fixing three real parameters, so it should not be a surprise that such a map is unique. We would like to point out that the argument with the number of parameters is just a rule of thumb, although a very good one, and each separate case requires a rigorous proof.

Chapter 3
The Ahlfors Map

One of the most recurrent and productive themes in complex analysis is the solution of extremal problems.

The Riemann map f_a associated to a point a in a simply connected domain $U \neq \mathbb{C}$ in the complex plane maps U one-to-one onto the unit disc $\mathbb{D}$ and is the solution to an extremal problem: Among all univalent holomorphic mappings of U into the unit disc, f_a is the map such that $f_a'(a)$ is real and as large as possible. The uniqueness follows from Schwarz's lemma.

What happens if we drop the condition that the competing functions are injective? That is, what happens if we maximize over the larger class of functions that are simply analytic and bounded by 1 in U? The answer is the same—that the solution to this new (relaxed?) extremal problem is again a conformal mapping of U onto $\mathbb{D}$ (unique if normalized as above).

This is not obvious—certainly Koebe's proof cannot be easily modified to eliminate the assumption of injectivity. That an apparently less restrictive extremal problem has the same solution as the original problem emerges as a special case of Ahlfors's theorem:

> The true equivalent of Schwarz's lemma for single-valued functions in a multiply-connected domain.

The quote is from Ahlfors's 1946 paper "Bounded analytic functions." He considers precisely the extremal problem above in the setting of multiply connected domains. Here is a partial statement of his result (we will give a more complete version later).

Theorem 3.1.1 (Ahlfors Mapping Theorem, First Version) *Let U be a domain of connectivity $n \geq 1$, none of whose boundary components reduces to a point. Fix a point $a \in U$ and consider the extremal problem*

$$\max \left| f'(a) \right| ,$$

P. V. Dovbush, S. G. Krantz, *The Geometric Theory of Complex Variables*,
https://doi.org/10.1007/978-3-031-77204-7_3

where f is analytic (and single-valued) in U with $|f| < 1$. The solution of this problem is an $n : 1$ map onto $\mathbb{D}$.

We see that the Riemann mapping theorem corresponds to $n = 1$, and, for that case, that Koebe's extremal problem can be relaxed to the family of functions that are analytic but need not be one-to-one. There is a slight catch here. Ahlfors needs a certain amount of boundary smoothness to form integrals over ∂U, and he uses some preliminary conformal mapping to assume that ∂U consists of analytic Jordan curves (a standard technique when working with finitely connected domains). This requires the Riemann mapping theorem! (The general methods in Bell's book [64] assume C^∞ boundaries.) To what extent these assumptions can be weakened or worked around we do not know. (And we do not really care—it cannot be an issue for our ultimate concerns, considering that we are going to take a discrete approximation to the boundary anyway!)

The problem of maximizing the derivative at a point is precisely the character of Schwarz's lemma, whence the opening quote. It is interesting that the same type of extremal problem was considered in generality for various classes of analytic functions in the famous paper by Ahlfors and Beurling on "Conformal invariants and function theoretic null sets." Indeed the "null set" aspect was already important for Ahlfors in the paper where the theorem above was proved (Painlevé's problem).

If U is a domain on the sphere whose complement relative to the sphere consists of exactly $m + 1$ (closed) components, each of which is nontrivial, then $m + 1$ applications of the Riemann mapping theorem produce a one-to-one holomorphic map of U onto a bounded domain whose boundary consists of $m + 1$ disjoint analytic simple closed curves. Since such a holomorphic map gives an isometry of the corresponding H^p spaces, we may assume initially that U is the second kind of domain. Specifically,

$$\partial U = \Gamma_0 \cup \cdots \Gamma_m,$$

where Γ_j is an analytic simple closed curve and $\Gamma_j \cap \Gamma_k = \emptyset$ if $j \neq k$. We may further suppose that Γ_0 is the boundary of the unbounded component of the complement of U.

(We shall assume $m \geq 1$ since the case $m = 0$ is just the case of the unit disc $\mathbb{D}$.)

The most intensively studied spaces of analytic functions are the Hardy spaces on the unit disc $\mathbb{D}$ in the complex plane. For each positive number p, the Hardy space $H^p(\mathbb{D})$ is the set of analytic functions f on the unit disc $\mathbb{D}$ such that

$$\sup_{0<r<1} \int_0^{2\pi} \left| f\left(re^{it}\right)\right|^p dt < \infty.$$

The Hardy space $H^\infty(\mathbb{D})$ is the set of bounded analytic functions on $\mathbb{D}$. A sample of the wealth of information that twentieth century mathematicians have discovered about these spaces can be found in the books of Hoffman [241], Duren [141], Koosis [313], and Garnett [170].

Let U be a connected, bounded open subset of the complex plane. To correspond to the case where U is the disc, it is clear that $H^\infty(U)$ should be defined to be the set of bounded analytic functions on U. For $0 < p < \infty$, the correct definition of $H^p(U)$ is not obvious. The space $H^p(\mathbb{D})$ is defined by integrations around circles. Correspondingly, we could take a set of curves in U tending to the boundary of U in an appropriate sense, and consider analytic functions f defined on U such that the set of integrals of $|f|^p$ over these curves is bounded. This procedure leads to an interesting theory, but a more fruitful and natural theory arises from another definition.

We say that a function g on a domain $U \subset \mathbb{C}$ has a harmonic majorant (resp. pluriharmonic majorant[1]) u if the harmonic (resp. pluriharmonic) function u satisfies $|g(z)| \leq u(z)$ for all U. It is a basic fact that a holomorphic function f on U lies in $H^p(U)$, $0 < p < \infty$, if and only if $|f|^p$ has a harmonic majorant (see [484], [328, Ch. 8]).

We define $(LH)^p(U)$ to be the class of all holomorphic functions $f : U \to \mathbb{C}$ such that $|f|^p \leq u$ for some pluriharmonic u. (Here $0 < p < \infty$.)

The definition of Hardy space where $|f|^p$ has a harmonic majorant is due to Lumer [366].

When $n = 1$, pluriharmonic is the same as harmonic, so that this definition coincides with the old one ([408, 446]) that involves harmonic majorants of $|f|^p$. But when $n > 1$, then $(LH)^p(U)$ is a proper subclass of what is usually called $H^p(U)$. (See, for example, [490].)

In the case of the disc it has been known that a function f analytic on $\mathbb{D}$ is in $H^p(\mathbb{D})$ if and only if there is a harmonic function u on $\mathbb{D}$ such that $|f(z)|^p \leqslant u(z)$ for every z in $\mathbb{D}$ (see [484], [328, Ch. 8]). This is in effect saying that $|f|^p$ is subharmonic.

Now it is clear how to define the Hardy spaces for more general regions U: For any positive number p, a function f analytic on U is said to be in $H^p(U)$ if and only if there is a harmonic function u on U such that $|f(z)|^p \leqslant u(z)$ for every z in U. For regions with smooth boundary, the Hardy space $H^p(U)$ will be the same set of functions as obtained from the definition alluded to in the previous paragraph (see [484, Theorem 1, p. 5] or [331, Theorem 8.3.6, p. 341]).

The function f is in $H^\infty(U)$ if it is both holomorphic on U and bounded there.

Clearly, $H^\infty(U) \subset H^p(U)$ for $0 < p < \infty$, and if $H^\infty(U)$ contains nonconstant functions, then so does $H^p(U)$.

Let U be a domain for which $H^\infty(U)$ consists of more than constants. Let $z \in U$ and put

$$\gamma = \sup\left\{|h'(z)| : h \in H^\infty(U), \|h\|_\infty \leqslant 1\right\}. \tag{3.1}$$

[1] The term pluriharmonic will always refer to real-valued functions. A pluriharmonic function is thus one whose domain is an open subset U of $\mathbb{C}^n$ and which is locally the real part of a holomorphic function.

A function $h \in H^\infty(U)$, $\|h\|_\infty = 1$, for which $h'(w) = \gamma$ will be called *extremal*. Clearly at least one extremal exists (apply a normal families argument).

Theorem 3.1.2 *There is precisely one extremal and it vanishes at w.*

Proof Let F be an extremal. We first show that $F(w) = 0$. To see this, let

$$g(z) = [F(z) - F(w)][1 - \overline{F(w)}F(z)]^{-1}, \quad z \in U.$$

A computation gives

$$g'(w) = \gamma \left(1 - |F(w)|^2\right)^{-1}.$$

Since $\|g\|_\infty \leqslant 1$, we must have $F(w) = 0$.

Next, if both F_1 and F_2 are extremals, then so is $\frac{1}{2}(F_1 + F_2)$. The proof will thus be complete if we show that any extremal F is an extreme point of the unit ball of H^∞. Suppose, then, that $g \in H^\infty(U)$ and $\|F \pm g\|_\infty \leqslant 1$. Consequently,

$$|F|^2 \pm 2\,\mathrm{Re}\, F\bar{g} + |g|^2 \leqslant 1$$

so that

$$|F|^2 + |g|^2 \leqslant 1.$$

Hence

$$|g|^2 \leqslant 1 - |F|^2 = (1 + |F|)(1 - |F|) \leqslant 2(1 - |F|).$$

Define $h = \frac{1}{2}(g^2)$; then $h \in H^\infty$ and

$$|F| + |h| \leqslant 1 \quad \text{on } U. \tag{3.2}$$

We shall show that (3.2) implies that h vanishes identically; hence, so does g, and F will be proved to be an extreme point of the unit ball of $H^\infty(U)$.

First note that $h(w) = 0$. For otherwise there is a choice of λ, $|\lambda| = 1$, which gives the function $F + \lambda F h$ a larger derivative at p than F, a contradiction since this function is bounded by 1 in U:

$$|F + \lambda F h| \leqslant |F| + |F||h| \leqslant |F| + |h| \leqslant 1.$$

If h is not identically zero, let r be the order of its zero at p, $r \geqslant 1$. Let ε be a small complex number and consider

$$F(z) + \varepsilon h(z)(z - p)^{-r+1} = G(z).$$

Then $|G'(w)| = |\gamma + \varepsilon h^{(r)}(w)/r!| > \gamma$, for an appropriate choice of the argument of ε. Furthermore, once out of a neighborhood of p, the quantity $|\varepsilon||z-p|^{1-r}$ is less than 1 so that $|G| \leqslant 1$ by the maximum principle. This contradiction establishes that $h \equiv 0$ and the theorem follows. □

The unique function specified by Theorem 3.1.2 will be called the *Ahlfors map* for p and U.

For an n-connected domain $U \subset \mathbb{C}$, and a point $p \in U$, the Ahlfors map f_p is a branched n-to-1 analytic function mapping U onto the unit disc, with $f_p(p) = 0$. Further, the Ahlfors map takes each component of the boundary of U one-to-one and onto the unit circle. Among all analytic functions mapping U into the unit disc, f_p is the unique function with maximum positive derivative at p.

There is a close relationship between the Ahlfors mapping function associated to an n-connected domain in the plane and the Bergman and Szegö kernels of the domain. The results show that the Ahlfors mapping plays a role in the multiply connected setting very similar to that of the Riemann mapping in the simply connected case (see [57]).

The Ahlfors map has a number of interesting properties, which we will see in the remainder of this section.

Definition 3.1.3 A subset S of the domain U is *dominating* if

$$\sup_{z\in S}|f(z)| = \sup_{z\in U}|f(z)| \quad \text{for all } f \in H^\infty(U).$$

Proposition 3.1.4 *There is a countable dominating set in U with no limit point in U.*

Proof Let $\{F_j\}$ be a sequence of compact subsets of U with $F_j \subset \operatorname{int}(F_{j+1})$ and $\cup_j F_j = U$. Let $G_n = F_n \backslash \operatorname{int}(F_{n-1})$ for $n = 2, 3, \ldots$ so that F_n is compact. Further

$$\max_{\mathfrak{I}_n}|f(z)| = \max_{\mathfrak{U}_n}|f(z)| \to \|f\|_\infty, \qquad \text{as } n \to \infty$$

for each $f \in H^\infty(U)$. The unit ball of $H^\infty(U)$ is equicontinuous on each G_n. Let δ_n be chosen so

$$|f(w) - f(z)| < \left(\frac{1}{2}\right)^n \quad \text{if} \quad |w-z| < \delta_n, \quad w, z \in G_n$$

and $\|f\|_\infty \leqslant 1$. Let G_n be a finite set in F_n such that each point of G_n is no more than $\frac{1}{2}\delta_n$ from some point of F_n. Then

$$\max_{G_n}|f(z)| \leqslant \max_{G_n}|f(z)| + \left(\frac{1}{2}\right)^n.$$

Set $G = \cup_{n=1}^\infty G_n$; then G is the desired set. □

We shall make use of dominating sets to prove the next result.

Theorem 3.1.5 *Let F be the Ahlfors function for U and $w \in U$. For each $h \in H^\infty$ we have*

$$\|F \cdot h\|_\infty = \|h\|_\infty. \tag{3.3}$$

Proof Let us define U' to consist of those points $a \in U$ for which there is some $h \in H^\infty(U)$ with $|h(a)| > 1$ yet $\|F \cdot h\| \leqslant 1$. Suppose $q \in U'$; let $F_1 = h \cdot F$ so that $\|F_1\| \leqslant 1$. However,

$$\left|F_1'(q)\right| = \left|F'(q)\right| |h(q)| > \left|F'(q)\right|,$$

a contradiction. Hence, $q \notin U'$. We shall show that U' is both open and closed in U. It will follow that U' is empty, and this will prove the theorem.

Clearly U' is open. To show U' is closed, we show that $\Omega \backslash U'$ is open. Let $q \in U \backslash U'$; then $\|F \cdot h\| \leqslant 1$ implies that $|h(q)| \leqslant 1$ if $h \in H^\infty(U)$. Let $\{z_j\}$ be a countable dominating sequence in U with no limit point in U; there is no loss in assuming no z_j is the point q. At this point we assume some familiarity with the maximal ideal space of a Banach algebra. Let X be the maximal ideal space of ℓ^∞; each function u that is bounded on a neighborhood of $\{z_j\}$ gives rise to an element $\hat{u}$ on X; in particular, this is true for functions in $H^\infty(U)$.

Our assumption that $q \notin U'$ implies that the linear functional from $F \cdot H^\infty$ into $\mathbb{C}$ given by

$$F \cdot h \mapsto h(q)$$

has norm 1. Hence, by the Hahn-Banach theorem, there is a measure λ on X of norm 1 with

$$\int \hat{F} \cdot \hat{h} d\lambda = h(q), \quad h \in H^\infty(U).$$

Thus,

$$1 = \int \hat{1}\hat{F} d\lambda \leqslant \int |\hat{F}\| d\lambda \,|\leqslant \|\lambda\| \leqslant 1$$

so that the measure $d\mu = \hat{F} d\lambda$ is nonnegative and has mass 1 and μ is supported on the set where $|\hat{F}| = 1$. For $\zeta \in U$ and ζ near q, let

$$s_\zeta(z) = (z - q)/(z - \zeta), \quad z \in U$$

and

$$u(\zeta) = \int \hat{s}_\zeta d\mu.$$

Clearly, u is continuous for ζ near q, and since $u(q) = 1$, we know $u(\zeta) \neq 0$ if ζ is near q. Let ζ be a point near q at which $u(\zeta) \neq 0$. Let $h \in H^\infty(U)$ and put

$$g(z) = \frac{h(z) - h(\zeta)}{z - \zeta}(z - q).$$

Then

$$0 = \int \hat{g} d\mu \quad \text{since } g(q) = 0.$$

Hence

$$h(\zeta)u(\zeta) = \int \hat{h}\hat{s}_\zeta d\mu,$$

which implies

$$|h(\zeta)| \leqslant C \sup\{|\hat{h}(m)| : m \in A\}, \quad A = \text{support}\, \mu,$$

where C is a constant independent of h. Now replace h by h^n, extract nth roots, and then let $n \to \infty$. We find that

$$|h(\zeta)| \leqslant \|\hat{h}\|_A = \sup\{|\hat{h}(m)| : m \in A\}.$$

Whence, if $\|hF\| \leqslant 1$, then $|\hat{h}| \leqslant 1$ on A and so $|h(\zeta)| \leqslant 1$. This shows $\zeta \notin U'$ and completes the proof that $\Omega \setminus U'$ is open. □

Definition 3.1.6 A point $x \in \partial U$ is *essential* if there is some $f \in H^\infty(U)$ and some $\varepsilon > 0$ such that f does not extend to be holomorphic in the disc $\{z : |z - x| < \varepsilon\}$.

There is a nice connection between the Ahlfors function and essential boundary points.

Theorem 3.1.7 *Let F be the Ahlfors function for U and $z \in U$. Then a point $x \in \partial U$ is essential if and only if*

$$\limsup\{|F(z)| : z \in U, z \to x\} = 1. \tag{3.4}$$

Proof Suppose first that x is essential but (3.4) fails, say $\limsup\{|F(w)| : w \to x\} = 1 - \delta, \delta > 0$. By Theorem [48, Theorem 2, p. 348], there is a function $h \in H^\infty(U)$ with

$$\limsup\{|h(w)| : w \to x\} = 1$$

$$\limsup\{|h(w)| : w \to y\} < 1, \quad y \in \partial U, \quad y \neq x.$$

Consider Fh. At x we have

$$\limsup\{|F(w)h(w)| : w \to x\} \leqslant 1 - \delta,$$

while at $y \neq x$, $y \in \partial U$, we have

$$\limsup\{|F(w)h(w)| : w \to y\} < 1$$

since $|F| \leqslant 1$ in U. Hence $\|Fh\|_\infty < 1 = \|h\|_\infty$, a contradiction to Theorem 3.1.5.

Conversely, suppose x is removable. Then there is a domain U^* containing U and x, and all functions in $H^\infty(U)$ extend to be in $H^\infty(U^*)$ without increasing their norm. In particular this holds for F, so that $1 = \|F\|_{H^\infty(U^*)}$. If (3.4) held, then $|F|$ would have an interior maximum on U^* and hence would be constant, a contradiction. □

The general notion of null set (no relation to the empty set) is this:

Definition 3.1.8 A compact subset K of $\mathbb{C}$ is a (Painlevé) *null set* if the only functions that are holomorphic and bounded on $\mathbb{S}^2 \setminus K$ are the constants.

Recall that by $\mathbb{S}^2$ we denote the Riemann sphere. We also use $\widehat{\mathbb{C}}$ for this purpose.

A few remarks concerning the general notion of null set are in order; the interested reader should consult the paper of Ahlfors and Beurling [9] for further information. Ahlfors [11] seems to have been the first to notice that the compact sets of analytic capacity zero are just the Painlevé null sets, i.e., sets whose complement in $\mathbb{S}^2$ supports no nonconstant bounded analytic function.

The Riemann mapping theorem shows that if a compact set K contains a closed connected subset with more than one point, then K is not a null set. Hence a null set is totally disconnected. To see this, note that if K is not totally disconnected, it must contain an arc γ. Then the Riemann map of $\mathbb{S}^2 \setminus \gamma$ onto the open unit disc is a nonconstant function bounded and analytic on $\mathbb{S}^2 \setminus K$.

The exact characterization of null sets is not known, but if K is a subset of a line, then such a characterization is easy: If K is a compact subset of the real line $\mathbb{R}$, then K is a *null set* if and only K has zero length.

The final result on the Ahlfors function in this generality is the following, which describes a property of its range. If K is a compact subset of an analytic curve γ in $\mathbb{C}$, then K is a null set for $H^\infty(U)$ if and only if K has arclength zero.

This result extends to the case when γ is smooth or even rectifiable, but the proof is considerably more difficult.

Definition 3.1.9 A domain U is *maximal* if each point in $\xi \in \partial U$ is essential.

Theorem 3.1.10 *Let U be maximal, let F be the Ahlfors function for U and $p \in U$, and let E consist of those points in $\mathbb{D}$ that are not in the range of F. Then $E \cap \{|z| \leqslant r\}$ is a null set for all $r < 1$.*

Proof Suppose the conclusion fails for some $r < 1$. Let $D = \mathbb{D} \setminus E_r$ and let ϕ be the Ahlfors function for $\mathbb{D}$ and 0. Then $\phi'(0) > 1$ for otherwise $\phi'(0) = 1$, and by uniqueness we would have $\phi(z) = z$, contradicting Theorem 3.1.7. But then

$$\begin{aligned}(\phi \circ F)'(p) &= \phi'(F(p))F'(p)\\ &= \phi'(0)F'(p)\\ &> F'(p)\end{aligned}$$

a contradiction. This establishes the theorem. □

Example 3.1.11 There is one interesting case in which the Ahlfors function can be given explicitly, although as far as we are concerned the demonstration that we have the correct function will be only half complete. Let K be a compact set in the real line $\mathbb{R}$ of finite, positive Lebesgue measure, and define a function h by

$$h(z) = \frac{1}{2}\int_K \frac{dt}{z-t}, \qquad z \notin K.$$

Then h maps $U = \mathbb{S}^2 \backslash K$ into the horizontal strip $|\operatorname{Im} z| < \pi/2$. Hence,

$$F(z) = \frac{e^{h(z)}-1}{e^{h(z)}+1}$$

maps U into $\mathbb{D}$ and $F(\infty) = 0$. Further,

$$F'(\infty) = \frac{1}{4}L \quad \text{where } L = \int_K dt.$$

Thus, $\frac{1}{4}L \leqslant \gamma(U)$. Pommerenke (1960) has shown that, in fact, $\frac{1}{4}L = \gamma(U)$, and so F must be the Ahlfors function for U and the point ∞. When U has a nice boundary, the Ahlfors function is particularly well behaved.

Theorem 3.1.12 *Let U be bounded by $m+1$ disjoint analytic simple closed curves $\Gamma_0, \ldots, \Gamma_m$ and let F be the Ahlfors function for U and $p \in U$. Then*

F maps U onto $\mathbb{D}$ exactly $m+1$ times ;

F extends analytically over each Γ_j and maps each Γ_j homeomorphically on $\partial\mathbb{D}$; and

F' is not zero on any Γ_j.

Proof Let $\Gamma = \partial U$ and let $A(U)$ denote those functions that are continuous on $U \cup \Gamma$ and holomorphic on U. Set

$$\alpha = \sup\left\{\left|f'(p)\right| : f \in A, \|f\| \leqslant 1\right\}.$$

There is a measure μ on ∂U of total variation α with

$$\int f d\mu = f'(p), \quad f \in A(U).$$

Let $g(z; p)$ be the Green's function on U with pole at p, let h be its harmonic conjugate, and put $Q = g + ih$. Recall from Propositions 19.3.6 and 19.3.7 that

$$iQ'(z)dz = d\omega_p(z), \quad z \in \Gamma.$$

We also know that

$$Q'(z) = -(z-p)^{-1} + w(z),$$

where w is holomorphic in a neighborhood of $\Omega \cup \partial U$. Thus,

$$\frac{i}{2\pi}\int_\Gamma f(z)Q'(z)dz = f(p), \quad f \in A(U).$$

Consequently,

$$0 = \int_\Gamma f(z)(z-p)\left\{\frac{i}{2\pi}\frac{Q'(z)}{z-p}dz - d\mu(z)\right\}, \quad f \in A(U);$$

it follows that the measure in the curly brackets is absolutely continuous with respect to dz, and so the same is true for $d\mu$. Moreover, because Q' does not vanish on Γ, we may write

$$d\mu(z) = r(z)\frac{i}{2\pi}Q'(z)dz, \quad z \in \partial U,$$

where r is in $L^1(\Gamma, ds)$. Hence,

$$0 = \int_\Gamma f(z)(z-p)\left\{\frac{1}{z-p} - r(z)\right\}\frac{i}{2\pi}Q'(z)dz, \quad f \in A(U).$$

We now conclude that

$$(1-(z-p)r(z))Q'(z) = h(z), \quad z \in \overline{U},$$

where $h \in H^1(U)$. Equivalently,

$$r(z) = \frac{1}{z-p} + \sum_1^m c_j\left(z-t_j\right)^{-1} + g(z), \quad z \in \overline{U},$$

where $g \in H^1(U)$, $c_1, \ldots, c_m$ are scalars and $t_1, \ldots, t_m$ are the critical points of Q, that is, the zeros of Q'. Thus if F is the Ahlfors function for p, then

$$\begin{aligned}\gamma = F'(p) &= \int_{\Gamma} F(z)r(z)\frac{i}{2\pi}Q'(z)dz \\ &\leqslant \int_{\Gamma} |F(z)||r(z)|\frac{i}{2\pi}Q'(z)dz \\ &\leqslant \|F\|_{\infty}\alpha = \alpha \leqslant \gamma.\end{aligned}$$

Hence,

$$\begin{aligned}F(z)r(z) &\geqslant 0 \text{ a. e. } |dz| \text{ on } \Gamma \\ |F(z)| &= 1 \text{ a. e. where } r \neq 0.\end{aligned}$$

However, each point $z \in \Gamma$ has a neighborhood $\mathcal{Q}$ such that $\mathcal{Q} \cap U$ is a conformal to the open unit disc $\mathbb{D}$ and r is in $H^1(U \cap \Omega)$. Thus we conclude that both F and r may be analytically continued over each point of Γ and certainly $|F| = 1$ on all of Γ. This means that the meromorphic function $r(z)F(z)$ is real and positive on Γ and so has as many poles as zeros in U. However, it has precisely $m + 1$ poles and so F has at most $m + 1$ zeros in U. On the other hand, the following argument shows $\arg F$ is increasing (locally) on Γ. Put $u = \log |F|$ near a point $x \in \Gamma$. Then $^*u = \arg F$, and the Cauchy-Riemann equations imply

$$0 \leqslant \frac{\partial u}{\partial n} = \frac{\partial^* u}{\partial \tau} \quad \text{on } \Gamma,$$

which is what we claimed. Since $\arg F$ must increase on each Γ_j by an integer multiple of 2π, we see F must have at least $m + 1$ zeros in U. Thus, F has exactly $m + 1$ zeros in U and $\arg F$ increases by exactly 2π on each Γ_j. This completes the proof. □

There are many ways in which the Ahlfors map can be thought of as the "Riemann mapping function for multiply connected domains." Indeed, the Ahlfors map associated to a point a in a multiply connected domain U is the unique holomorphic function f mapping U into the unit disc that makes $f'(a) > 0$ and as large as possible. Furthermore, the Ahlfors map also has mapping properties analogous to the Riemann map; it maps U onto the unit disc and maps each boundary curve of U one-to-one onto the unit circle. From the point of view of kernel functions, the Ahlfors map also takes on the role of a "Riemann map in multiply connected domains" (see [55] for more details).

Since conformal maps are bicontinuous bijections, conformally equivalent domains are also topologically equivalent. This means that there is no way to map, say, a simply connected domain onto a doubly connected domain.

All of the above facts can be found in [153].

Chapter 4
A Riemann Mapping Theorem for Two-Connected Domains in the Plane

The reader will, no doubt, wonder why we have not discussed mapping of multiply connected regions. For simply connected regions with more than one boundary point, the Riemann Mapping Theorem shows that $\mathbb{D}$ provides a canonical domain. For doubly connected regions (e.g., open annuli), there is an infinite one-parameter family of canonical domains. For regions of connectivity n, $n \geq 3$, there are $3n - 6$ parameters according to which two domains of connectivity n must agree in order for them to be mapped into one another. The mapping may be constructed as the limit of a convergent sequence of maps by a version of Koebe's construction of Theorem 2.1.2. Explicit consideration of multiply connected regions may be found in the cited book of Nehari [398, Chapter VII], and in Golusin [179, Chapter V and VI], where the existence and uniqueness of conformal mapping functions onto canonical domains (e.g., "Parallel slit domains") for regions of connectivity ≥ 2 are proved. Constructive methods are discussed in Gaier's [163] book already cited. A survey by Gaier appears in [164]. Infinitely connected regions are also discussed there.

In the previous section, we have shown that all nontrivial simply connected domains are conformally equivalent to the unit disc, and hence they all are conformally equivalent to each other. For multiply connected domains, this is no longer true. As an example of what may happen in regions of connectivity ≥ 2, we prove the following simple result.

Theorem 4.1.1 *Let $A(r, R) = \{z : r < |z| < R\}$ be an annulus with the smaller radius r and the larger radius R. In the case $0 < r_i < R_i < \infty$, there is a conformal map from $A_1 = A(r_1, R_2)$ onto $A_2 = A(r_2, R_2)$ if and only if $R_1/r_1 = R_2/r_2$. For degenerate annuli (where $r_i = 0$ and/or $R_i = \infty$), the situation is more complicated. The annulus $A(0, \infty)$ is not conformally equivalent to any other annulus and all annuli $A(0, R)$ and $A(r, \infty)$ with $r > 0$ and $R < \infty$ are conformally equivalent to each other and not equivalent to any other type of annuli.*

P. V. Dovbush, S. G. Krantz, *The Geometric Theory of Complex Variables*,
https://doi.org/10.1007/978-3-031-77204-7_4

Proposition 4.1.2 *Let A_1 and A_2 be two annuli as before. If there is a univalent map $f : A_1 \to A_2$, then $R_2/r_2 \geq R_1/r_1$.*

Proof As before, we can assume without loss of generality that $r_1 = r_2 = 1$ and that the inner circle is mapped to the inner circle, so that the outer circle is mapped to the outer circle. Since function f is analytic in an annulus, it can be written as the Laurent series

$$f(z) = \sum_{n \in \mathbb{Z}} a_n z^n.$$

We denote by $A(r)$ the area of a domain bounded by a Jordan curve $f\left(re^{i\theta}\right)$ where θ goes from 0 to 2π. By Green's formula for the area, we have

$$\begin{aligned} A(r) &= \frac{1}{2i} \int \bar{f}(z) df(z) = \frac{1}{2i} \int_{|z|=r} \bar{f}(z) f'(z) dz \\ &= \frac{1}{2i} \int_0^{2\pi} \left(\sum \bar{a}_n r^n e^{-i\theta n} \right) \left(\sum n a_n r^{n-1} e^{i\theta(n-1)} \right) rie^{i\theta} d\theta \\ &= \pi \sum_{n \in \mathbb{Z}} n \left| a_n \right|^2 r^{2n}. \end{aligned}$$

The last identity holds since $\int e^{in\theta} = 0$ unless $n = 0$. Passing to the limit as $r \to 1$, we have

$$\pi = \pi \sum n \left| a_n \right|^2.$$

Using this identity, we can write

$$A(r) - \pi r^2 = \pi r^2 \sum_{n \in \mathbb{Z}} n \left| a_n \right|^2 \left(r^{2n-2} - 1 \right) \geq 0,$$

where the last inequality holds termwise (indeed, n and $r^{2n-2} - 1$ have the same sign). Passing to the limit as $r \to R_1$, we obtain that $R_2 \geq R_1$.

For the proof of Theorem 4.1.1, we just use the proposition for f and f^{-1}.

Each doubly connected domain is conformally equivalent to an annulus, and the ratio of its radii completely determines the equivalence class. This is an important example of a conformal invariant: a quantity that does not change under conformal transformation. The standard conformal invariant of a doubly connected domain Ω that describes the equivalence class is the conformal modulus $M(\Omega)$ that is defined as

$$\frac{1}{2\pi} \log \frac{R}{r},$$

where r and R are the smaller and larger radii of an annulus that is conformally equivalent to Ω. This quantity is well defined and does not depend on a particular choice of an annulus.

Theorem 4.1.1 shows that not all doubly connected domains are conformally equivalent. This means that we cannot use the same uniformizing domain for all doubly connected domains.

It was noted in [56] that any two-connected domain Ω in the plane such that neither boundary component is a point can be mapped conformally via a mapping Φ to a representative domain of the form

$$\mathcal{A}_r := \{z \in \mathbb{C} : |z + 1/z| < 2r\},$$

where r is a constant greater than one. See [260] for a nice proof of this fact that also proves a much more general theorem.

See Fig. 4.1 for a picture of $\mathcal{A}_r$ with $r = 1.05$. The dotted circle in the figure is the unit circle. Notice that $1/z$ is an automorphism of $\mathcal{A}_r$ that fixes the unit circle.

We will show that the conformal map $\Phi : \Omega \to \mathcal{A}_r$ can be expressed simply in terms of an Ahlfors map of Ω. In fact, we shall show how to find a point a in Ω so that Φ is given by $cf_a + \sqrt{c^2 f_a^2 - 1}$ where f_a is the Ahlfors mapping associated to a, and c is a complex constant that we will determine. We will explain along the way how $cf_a + \sqrt{c^2 f_a^2 - 1}$ can be understood to be a single-valued holomorphic function on Ω. The Ahlfors map f_a is the solution to the extremal problem: Among all holomorphic mappings of Ω into the unit disc, f_a is the map such that $f_a'(a)$ is real and as large as possible. See [55, p. 49] or our earlier discussion for the basic properties of the Ahlfors map, including the property that $f_a(a) = 0$, which follows from the extremal condition.

Fig. 4.1 The domain $\mathcal{A}_r$ with $r = 1.05$

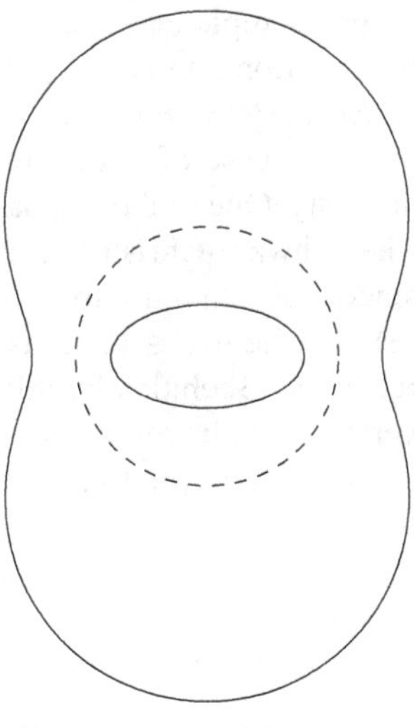

In this section, we show that the conformal map $\Phi : \Omega \to \mathcal{A}_r$ can be expressed simply in terms of an Ahlfors map of Ω.

Theorem 4.1.3 *Suppose that Ω is a two-connected domain in the plane such that neither boundary component is a point. Then there exists a point a in Ω and a complex constant c such that a one-to-one conformal mapping Φ of Ω onto the representative domain $\mathcal{A}_r$, where $r = |c|$, can be expressed in terms of the Ahlfors map f_a associated to the point via $\Phi(z) = cf_a + \sqrt{c^2 f_a^2 - 1}$.*

We shall reduce the proof of this mapping theorem to two key lemmas.

Lemma 4.1.4 *Suppose that Ω is a two-connected domain in the plane such that neither boundary component is a point. Then the Ahlfors map f_a associated to any point a in Ω is such that it has two distinct and simple branch points on the median of Ω.*

Lemma 4.1.4 was proved by McCullough and Mair in [368] and independently and by other means in [497].

Let a_ρ denote the annulus $\left\{z \in \mathbb{C} : \frac{1}{\rho} < |z| < \rho\right\}$ when $\rho > 1$, and let

$$J(z) := \frac{1}{2}\left(z + \frac{1}{z}\right)$$

denote the *Joukowsky map*.

Note that $\frac{1}{r}J(z)$ is a proper holomorphic map of $\mathcal{A}_r$ onto the unit disc that is a 2-to-1 branched covering map.

Lemma 4.1.5 *The Ahlfors map associated to the point i in $\mathcal{A}_r$ is given by $\frac{1}{r}J(z)$.*

Lemma 4.1.5 was proved in [131].

For completeness, and because it will not take much space, we will give in the next section alternative proofs of the lemmas that are simpler than the originals.

Rectangles and Elliptic Functions

In the case of a rectangle we have four vertices, three of them could be chosen arbitrary. One of the standard choices is to select them to be 0, $x_0 \in (0, 1)$, 1, and ∞. The Schwarz-Christoffel formula could be written in this setup, but it is not optimal since it does not use the existing symmetries. Instead, we choose them to be ± 1 and $\pm 1/k$, where $k \in (0, 1)$ is a parameter that will eventually define the shape of the rectangle. Technically, this corresponds to a pentagon where $\alpha_5 = 1$, so the fifth vertex could be removed. All angles are $\pi/2$ and $\beta_i = 1/2$. In this case, our integral can be written (up to a constant factor) as

$$F(z) = \int_0^z \frac{d\zeta}{\sqrt{\left(1 - \zeta^2\right)\left(1/k^2 - \zeta^2\right)}},$$

where arguments are chosen in such a way that the integrand is positive on $(-1, 1)$. This function can be written as

$$\begin{aligned} F(z) &= k \int_0^z \frac{d\zeta}{\sqrt{\left(1-\zeta^2\right)\left(1-k^2\zeta^2\right)}} \\ &= k \int_0^{\sin^{-1} z} \frac{d\theta}{\sqrt{1-k^2\sin^2\theta}} \\ &= kF\left(\arcsin(z); k^2\right), \end{aligned}$$

where $F\left(\arcsin(z); k^2\right)$ is the incomplete elliptic integral of the first kind.

Thus if

$$u = \int_0^{\varphi} \frac{d\theta}{\left(1-m\sin^2\theta\right)^{1/2}},$$

the angle φ is called the amplitude

$$\varphi = \operatorname{am} u,$$

and we define

$$\operatorname{sn}(u) = \sin\varphi, \qquad \operatorname{cn}(u) = \cos\varphi,$$

$$\operatorname{dn} u = \left(1-m\sin^2\varphi\right)^{1/2} = \Delta(\varphi)$$

(see [4, Chapter 16, p. 569] or [397, p. 280]).

Thus we have

$$\operatorname{sn}\left(k^{-1}F(z) \mid k^2\right) = z,$$

where $\operatorname{sn}(u \mid k^2)$ is the *Jacobi elliptic sine* with parameter k^2.

By the Schwarz-Christoffel formula, $F(z)$ maps $\mathbb{H}$ onto a rectangle with the vertices $F(\pm 1)$ and $F(\pm 1/k)$. First, we observe that this map could be continuously extended to the real line and that it maps $[-1, 1]$ onto $[-K, K]$, where

$$K = K(k) = \int_0^1 \frac{d\zeta}{\sqrt{\left(1-\zeta^2\right)\left(1-k^2\zeta^2\right)}}$$

is the complete elliptic integral of the first kind. To compute the images of two other points, we note that for $1 < \zeta < 1/k$ we have

$$F(z) = \int_0^1 \frac{d\zeta}{\sqrt{(1-\zeta^2)(1-k^2\zeta^2)}} + i \int_1^z \frac{d\zeta}{\sqrt{(\zeta^2-1)(1-k^2\zeta^2)}}$$

so $F(1/k) = K + iK'$, where

$$K' = K'(k) = \int_1^{1/k} \frac{d\zeta}{\sqrt{(\zeta^2-1)(1-k^2\zeta^2)}}$$

is the complementary complete elliptic integral of the first kind. It is a standard fact that $K'(k) = K(k')$ where $k' = \sqrt{1-k^2}$ is the complementary parameter. In other words, $F(z)$ maps $\mathbb{H}$ onto a rectangle with vertices $\pm K$ and $\pm K + iK'$. The ratio of the horizontal side to the vertical one is

$$\lambda(k) = \frac{2K(k)}{K'(k)}.$$

Several authors have found explicit biholomorphic maps between $a_\rho = \left\{z \in \mathbb{C} : \frac{1}{\rho} < |z| < \rho\right\}$ and $\mathcal{A}_r = \{z \in \mathbb{C} : |z + 1/z| < 2r\}$ and have determined the relationship between ρ and r (see [124, 131, 259, 260]).

Since we shall need such a map, and since we can construct one easily here using the method from [131], we shall include the construction for completeness.

The biholomorphic mapping $\Psi : a_\rho \to \mathcal{A}_r$ will be an explicit map involving the Jacobi *sn* function. There is a standard conformal map of the annulus $\{z : 1/\rho < |z| < 1\}$ onto the unit disc minus a slit from $-L$ to $+L$ along the real axis. (See Nehari [397, p. 293–295] for a construction where the relationship between L and ρ is given explicitly.) The mapping is given up to a constant factor by

$$\operatorname{sn}\left(K + \frac{2iK}{\pi} \log \rho z; \rho^{-4}\right),$$

where K is a constant that can be determined explicitly. The map takes the unit circle to the unit circle and the circle of radius $1/\rho$ to the slit from $-L$ to L. By composing with an inversion of the annulus, we get a conformal map that maps the annulus $\{z : 1/\rho < |z| < 1\}$ one-to-one onto the unit disc minus a slit from $-L$ to L and takes the unit circle to the slit from $-L$ to L and the circle of radius $1/\rho$ to the unit circle. The Joukowski map $J(z)$ is a biholomorphic mapping of $\{z \in \mathcal{A}_r : |z| > 1\}$ onto the disc of radius r minus a slit from -1 to $+1$. It takes the outer boundary of $\mathcal{A}_r$ to the circle of radius r and the unit circle two-to-one onto the slit. Thus, by choosing r properly, scaling, and then composing, we can write down a biholomorphic mapping from the annulus $\{z \in \mathbb{C} : 1/\rho < |z| < 1\}$ onto $\{z \in \mathcal{A}_r : |z| > 1\}$ that maps the unit circle to the unit circle. We may now extend this map by reflection to a biholomorphic map from a_ρ onto $\mathcal{A}_r$. Finally, we may compose with the automorphism $1/z$ and a rotation if necessary to obtain a conformal mapping $\Psi : a_\rho \to \mathcal{A}_r$ such that the outer boundary of a_ρ maps to the outer boundary of $\mathcal{A}_r$ and such that $\Psi(i) = i$. Note that Ψ is uniquely determined by these conditions. We shall need to know that $\Psi(-i) = -i$. Since $-1/\Psi(-1/z)$ satisfies the same properties, namely that it is a conformal map that takes the outer boundary to the outer boundary and fixes i, it must be that $\Psi(z) \equiv -1/\Psi(-1/z)$. From this, we deduce that $\Psi(-i) \equiv -1/\Psi(-i)$, and hence that $\Psi(-i)$ is a square root of -1. Because Ψ is one-to-one, it cannot be that $\Psi(-i)$ is i, and so $\Psi(-i) = -i$. To summarize, there is a biholomorphic mapping $\Psi : a_\rho \to \mathcal{A}_r$ such that

$$\Psi(i) = i \quad \text{and} \quad \Psi(-i) = (-i) \tag{4.1}$$

and which takes the unit circle to the unit circle, the outer boundary to the outer boundary, and the inner boundary to the inner boundary.

It is well known that any two-connected domain Ω in the plane such that neither boundary component is a point can be mapped biholomorphically via a map F_ρ to a unique annulus of the form a_ρ. Let C_1 denote the unit circle. We shall call ρ^2 the modulus of Ω, and we shall call the set $F_\rho^{-1}(C_1)$ the *median* of the two-connected domain. The median is the unique connected one-dimensional set that is left invariant under every automorphism of the two-connected domain. The biholomorphic map Ψ of a_ρ onto $\mathcal{A}_r$ takes the unit circle to itself. Hence, the median of $\mathcal{A}_r$ is also the unit circle.

***Proof (Proof of Theorem** 4.1.3)* Suppose that Ω is a two-connected domain in the plane such that neither boundary component is a point. We now describe the procedure to determine the mapping Φ, assuming the truth of the lemmas. Pick any point P in Ω. Lemma 4.1.4 guarantees the existence of a point a on the median of Ω such that $f_P'(a) = 0$. [Note that the Ahlfors map is eminently numerically computable if the boundary of Ω is sufficiently smooth (see [52, 272, 274, 512]), and the zeroes of a function with only two simple zeroes can easily be located numerically as well.] Now the Ahlfors map f_a associated to the point a is such that there are two distinct points p_1 and p_2 on the median of Ω such that $f_a'(p_i) = 0$ for $i = 1, 2$. The points p_i are also different from a because $f_a'(a) \neq 0$ (since $f_a'(a)$ is, in fact, the extremal value by definition of the Ahlfors map). We now claim that there exist an $r > 1$ and a biholomorphic mapping from Ω to $\mathcal{A}_r$ such that $\Phi(a) = i$. We may also arrange for the outer boundary of Ω to get mapped to the outer boundary of $\mathcal{A}_r$. To see this, we use the biholomorphic mapping F_ρ of Ω onto an annulus a_ρ. By composing with $1/z$, if necessary, we may map the outer boundary of Ω to the outer boundary of a_ρ. Since a is on the median, $F_\rho(a)$ is on the unit circle, and we may compose with a rotation so that $F_\rho(a) = i$. Now we may compose with the biholomorphic map Ψ (of a_ρ onto $\mathcal{A}_r$) constructed above, noting that i is fixed under Ψ and that outer boundaries are preserved, to obtain Φ. (At the moment, we can only say that the number r exists. We will explain how to find it momentarily.) Lemma 4.1.5 states that $\frac{1}{r}J(z)$ is the Ahlfors map of $\mathcal{A}_r$ associated to the point i. The extremal property of Ahlfors maps makes it easy to see that there is a unimodular complex number λ such that

$$\frac{1}{r}J(\Phi(z)) = \lambda f_a(z). \tag{4.2}$$

Since the branch points of $J(z)$ occur at ± 1, this last identity reveals that $\Phi\,(p_i) = \pm 1$ for $i = 1, 2$. We may assume that $\Phi\,(p_1) = -1$ and $\Phi\,(p_2) = 1$. Since $J(-1) = -1$ and $J(1) = 1$, Eq. (4.2) now gives $-\frac{1}{r} = \lambda f_a\,(p_1)$ and $\frac{1}{r} = \lambda f_a\,(p_2)$. Either one of these two equations determines both λ and r. We can now solve Eq. (4.2) locally for Φ to obtain

$$\Phi(z) = J^{-1}\left(cf_a(z)\right),$$

where $c = r\lambda$ and, of course, $J^{-1}(w) = w + \sqrt{w^2 - 1}$. This formula extends by analytic continuation from any point in Ω that is not a branch point of f_a. The apparent singularities at the branch points are removable because the construction aligned the two algebraic singularities of J^{-1} to occur at the two images of the branch points under cf_a. We shall explain how to compute Φ from its boundary values in Sect. 4.2. The modulus ρ^2 of Ω can now be obtained from r via any of the formulas relating ρ and r proved in [124, 131, 259, 260]. □

4.1 Proofs of the Lemmas

To prove Lemmas 4.1.4 and 4.1.5, we will make use of the simple fact that if $G : \Omega_1 \to \Omega_2$ is a biholomorphism between bounded domains, then the Ahlfors maps transform according to the following formula. Let $f_j(z; w)$ denote the Ahlfors map f_w associated to the point w in Ω_j. The extremal property of the Ahlfors maps can be exploited to show that

$$f_1(z; w) = \lambda f_2(G(z); G(w)), \tag{4.3}$$

where λ is a unimodular constant determined by the condition that the derivative of $f_1(z; w)$ in the z variable at $z = w$ must be real and positive.

***Proof of Lemma** 4.1.4* The biholomorphic map F_ρ from Ω to a_ρ together with Eq. (4.3) reveals that it is sufficient to prove Lemma 4.1.4 for an annulus a_ρ with base point $a > 0$. The function $h(z) = e^{iz}$ maps the horizontal strip $H = \{z : -\log\rho < \operatorname{Im} z < \log\rho\}$ onto a_ρ, takes the real line onto C_1, and is 2π-periodic. Consider $g(z) = f_a(h(z))$. Notice that g is 2π-periodic, maps H onto the unit disc, and maps the boundary of H onto the unit circle. It can be extended via a sequence of reflections to a function on $\mathbb{C}$. Because of these reflections, g has a second period of $(4\log\rho)i$, making g an elliptic function. Because f_a is two-to-one, g is a second order elliptic function.

Any second order elliptic function ϕ satisfies the equation $\phi(w - z) = \phi(z)$, where w is the sum of two noncongruent zeroes of ϕ (see [514, p. 474], [402, p. 74], or [236]). From [498], we know that if a is real, f_a has zeroes at a and $-1/a$. We will prove this fact below (Lemma 4.1.6) to make this discussion self-contained. It now follows that g will have zeroes at $-i \log a$ and $i \log a + \pi$. Thus, we can choose $w = \pi$. The mapping $z \mapsto \pi - z$ is symmetric with respect to $\pi/2$ and has $\pi/2$ as a fixed point, while the point $3\pi/2$ gets mapped to the congruent point $-\pi/2$. These two points will be branch points of g, and i and $-i$ will be the branch points of f_a, lying on the median of a_ρ.

***Proof of Lemma** 4.1.5* The biholomorphism Ψ from a_ρ onto $\mathcal{A}_r$ that we constructed in Sect. 10.2 takes i to i and $-i$ to $-i$ (see (4.2)). Let f_i denote the Ahlfors map of a_ρ with base point i. From [498], we know that f_i will have i and $-i$ as its zeroes (see Lemma 4.1.6). Equation (4.3) tells us that the Ahlfors map of $\mathcal{A}_r$ with base point i will have i and $-i$ as its zeroes. Because this Ahlfors map has the same zeroes as $\frac{1}{r}J(z)$, and both functions map the boundary of $\mathcal{A}_r$ to the unit circle, they are the same up to multiplication by a unimodular constant. Since $J'(i) = 1 > 0$, this constant must be 1. Lemma 4.1.5 is proved.

We shall now flesh out the proofs of the lemmas by giving a shorter and simpler proof than the one given in [498] of the following result.

Lemma 4.1.6 *The Ahlfors map f_a associated to a point a in a_ρ has simple zeroes at a and $-1/\bar{a}$. Thus f_a has simple zeroes at a and $-1/a$ if a is real, and f_i has simple zeroes at i and $-i$.*

Proof The Ahlfors map f_a always has a simple zero at a. Since the Ahlfors map is two-to-one (counting multiplicities), f_a must have exactly one other simple zero distinct from a. The Ahlfors map is given via

$$f_a(z) = S(z, a)/L(z, a), \tag{4.4}$$

where $S(z, a)$ is the Szegö kernel for a_ρ and $L(z, a)$ is the Garabedian kernel. For the basic properties of the Szegö and Garabedian kernels, see [55, p. 49]. We note here that $S(z, a) = \overline{S(a, z)}$ and that $L(z, a) \neq 0$ if $a \in \Omega$ and $z \in \overline{\Omega}$ with $z \neq a$. Also, $L(z, a)$ has a single simple pole at a in the z variable, and $L(z, a) = -L(a, z)$. It follows from Eq. (4.4) that the other zero of f_a is the one and only zero of $S(z, a)$ in the z variable. Let $Z(a)$ denote this zero. It is proved in [56] that $Z(a)$ is a proper antiholomorphic self-correspondence of the domain. In the two-connected case that we are in now, it is rather easy to see that $Z(a)$ is an antiholomorphic function of a that maps a_ρ one-to-one onto itself. To see this, let $S'(z, a)$ denote the derivative of $S(z, a)$ in the z variable and notice that the residue theorem yields

$$Z(a) = \frac{1}{2\pi i} \int_{z \in b\Omega} \frac{z S'(z, a)}{S(z, a)} dz.$$

(See [55] for proofs that $S(z, a)$ is smooth up to the boundary and nonvanishing on the boundary in the z variable when a is held fixed in a_ρ.) It can be read off from this last formula that $Z(a)$ is antiholomorphic in a. Now hold z fixed in a_ρ

and let a tend to a boundary point b. Equation (4.4) shows that $f_a(z)$ tends to $S(z,b)/L(z,b)$. But $S(z,b) = -\frac{1}{i}L(z,b)T(b)$, where $T(b)$ is the complex unit tangent vector at b (see [55, p. 107]), and so $f_a(z)$ tends to $iT(b)$, a unimodular constant. This shows that the zeroes of $f_a(z)$ must tend to the boundary as a tends to the boundary. Consequently, $Z(a)$ is a proper antiholomorphic self-mapping of the annulus. The only such maps are one-to-one and onto (see [386]). We shall say that $Z(a)$ is an antiholomorphic automorphism of a_ρ. Note that $Z(a)$ must satisfy $Z(Z(a)) = a$. The only such antiholomorphic automorphisms are $Z(a) = \lambda\overline{a}$, where λ is a unimodular constant, and $Z(a) = 1/\overline{a}$, and $Z(a) = -1/\overline{a}$. Since f_a does not have a double zero at a, $Z(a)$ cannot have a fixed point. Only $Z(a) = -1/\overline{a}$ is without fixed points in a_ρ, so f_a has zeroes at a and $-1/\overline{a}$. The proof is complete.

4.2 How to Compute Φ and the Median of Ω

Suppose that Ω is a bounded, two-connected domain bounded by two nonintersecting C^∞ Jordan curves. The Kerzman-Stein-Trummer method [272, 274, 275] can be used to compute the boundary values of Ahlfors mappings associated to Ω. (See [52] and [55, Chapter 26] for a description of how to do this in the multiply connected setting.) Thus we may pick any point P in Ω and compute f_P. We shall now explain how to compute the two branch points of f_P on the median of Ω. Let f denote the Ahlfors mapping associated to the point P and let P_1 and P_2 denote the two simple zeroes of f' on the median of Ω. The residue theorem yields that

$$P_1^n + P_2^n = \frac{1}{2\pi i}\int_{b\Omega} \frac{f''(z)z^n}{f'(z)}dz.$$

The two points can be determined from the values of $P_1^n + P_2^n$ for n equal to one and two. Indeed the coefficients A and B of the polynomial $z^2 - Az + B = (z-P_1)(z-P_2)$ are given as $A = P_1 + P_2$ and $B = \frac{1}{2}\left[(P_1+P_2)^2 - \left(P_1^2 + P_2^2\right)\right]$, and so P_1 and P_2 can be found via the quadratic formula.

Next, let a denote one of the two points P_1 or P_2. We may now compute the branch points p_1 and p_2 of f_a as we did above for f_P. We shall now show how to compute the function Φ using only the boundary values of the Ahlfors map f_a.

We may calculate $c = r\lambda = -1/f_a(p_1)$ by the procedure described in Sect. 4. There are two holomorphic branches of $J^{-1}(w) = w + \sqrt{w^2-1}$ near the boundary of the disc of radius r corresponding to the two choices of the square root. One branch j_o takes the circle of radius r one-to-one onto the outer boundary of $\mathcal{A}_r$, and the other branch j_i takes the circle of radius r one-to-one onto the inner boundary of $\mathcal{A}_r$. The mapping Φ that maps Ω one-to-one onto $\mathcal{A}_r$ can be determined from its boundary values on Ω. The boundary values of Φ on the outer boundary of Ω can be taken to be $j_o(cf_a(z))$ and the boundary values on the inner boundary are then $j_i(cf_a(z))$. The extension of Φ to Ω can be gotten from the boundary values of Φ via the Cauchy integral formula.

We now turn to a method to determine the median of Ω. Equation (4.2) shows that $J(\Phi(z)) = cf(z)$, where f is shorthand for f_a and $c = \lambda r$. This shows that f maps the median of Ω to the line segment $\mathcal{L}$ from $-1/c$ to $1/c$. Hence the median of Ω is given by $f^{-1}(\mathcal{L})$. We already know the points p_1, p_2, and a on the median. To compute other points w in this set such that $f(w) = \tau$ for some τ in $\mathcal{L}$ where τ is not one of the endpoints of $\mathcal{L}$, we may again use the residue theorem to obtain

$$w_1^n + w_2^n = \frac{1}{2\pi i} \int_{z \in b\Omega} \frac{f'(z) z^n}{f(z) - \tau} dz,$$

where w_1 and w_2 are the two points in $f^{-1}(\tau)$. Finally, we may use the Newton identity for $n = 2$ and the quadratic formula as we did above to determine p_1 and p_2 to get w_1 and w_2. Note that p_1 and p_2 correspond to the inverse images of the two endpoints of $\mathcal{L}$, and so we may generate the rest of the median by letting τ move between the endpoints.

Remark 4.1.7 A short combinatorial proof of *Newton's celebrated identities*

$$\sum_{r=0}^{k-1} (-1)^r \left(\sum_{1 < i_1 < \cdots < i_r < n} x_{i_1} \cdots x_{i_r} \right) \left(\sum_{i=1}^{n} x_j^{k-r} \right) + (-1)^k \left(\sum_{1 < i_1 < \cdots < i_k < n} x_{i_1} \cdots x_{i_k} \right) k = 0,$$

where n and k are positive integers and $x_1, \ldots, x_n$ are commuting indeterminates given by Zeilberger, D. in [539].

Chapter 5
Riemann Multiply Connected Domains

5.1 The Mappings on the Canonical Regions

Theorem 4.1.3 shows that domains $\mathcal{A}_r := \{z \in \mathbb{C} : |z + 1/z| < 2r\}$, where r is a constant greater than one, are the natural "standard" doubly connected domains. For the domains of higher connectivity, there is no natural unique choice of uniformizing domains. Instead, there are several somewhat standard families of canonical domains.

In this section, we discuss canonical domains and the corresponding uniformizing maps. Multiply connected domains are not the main subject of this book, so we are not going to study them extensively.

From the Riemann mapping theorem, it follows that all simply connected domains with more than one boundary point are conformally equivalent. For multiply connected regions of the same connectivity this is no longer true. If we pass now to doubly connected domains G, we have a new situation: Two such domains are conformally equivalent if and only if their so-called conformal modules are the same. This means in particular that every doubly connected domain G is conformally equivalent to an annulus with the same conformal module (see Theorem 4.1.1). (This annulus can degenerate.) One can find a detailed discussion of the corresponding questions in [340].

Therefore we must try to find a system of canonical regions with the property that each multiply connected region is conformally equivalent to one and only one canonical region. The choice of canonical regions is to a certain extent arbitrary, and there are several types with equally simple properties.

We will restrict ourselves here to the general case of conformal mappings of multiply connected domains of finite connectivity, including the case of connectivity greater than 2. Oded Schramm has found a way to treat domains of infinite connectivity.

P. V. Dovbush, S. G. Krantz, *The Geometric Theory of Complex Variables*,
https://doi.org/10.1007/978-3-031-77204-7_5

The main methods for proving the existence of canonical conformal mappings is the extremal problems combined with normal family arguments (the most elegant method, e.g., in the case of the parallel slit mappings).

Parallel slit domains. These are domains that are the complex sphere $\mathbb{C}$ without a finite union of intervals that are parallel to each other.

Definition 5.1.1 We say that f maps the domain $U \subset \mathbb{C}$ *conformally onto* $V \subset \mathbb{C}$ if:

(i) f is meromorphic in U.
(ii) f is injective (one-to-one), that is $z, z' \in U, z \neq z' \Rightarrow f(z) \neq f\left(z'\right)$.
(iii) $f(U) = V$.

We say that f maps U *conformally into* V if (iii) is replaced by $f(U) \subset V$. A univalent function is the same as a conformal map.

Note that we define conformal maps only for (connected) open sets. Here are some elementary properties (see, e.g., [15]):

(a) The inverse is also a conformal map.
(b) A conformal map is a homeomorphism, i.e., a continuous, injective map with continuous inverse.
(c) Conformal maps are locally univalent, i.e., the derivative does not vanish and there are only simple poles.
(d) Angles between curves, suitably oriented, are preserved by conformal maps.
(e) The conformal image of a (measurable) subset A has the area

$$\text{area } f(A) = \iint_A \left|f'(z)\right|^2 dxdy.$$

In particular, if $f(z) = a_0 + a_1 z + a_2 z^2 + \ldots$ is a conformal map of $\mathbb{D}$ into $\mathbb{C}$, then

$$\text{area } f(\mathbb{D}) = \pi \sum_{n=1}^{\infty} n \left|a_n\right|^2.$$

In this chapter, we always assume that U is a finitely connected domain in $\mathbb{C}$ such that its boundary components are analytic curves. This assumption is not restrictive since any domain could be mapped conformally onto a domain with analytic boundary.

To illustrate the basic approach, we shall treat one particular case: determining the conformal mapping of a multiply connected domain G onto a parallel slit domain (see Fig. 5.1). The latter consists of a subset of the extended complex plane, the complementary components of which are rectilinear slits making an angle θ with respect to the x-axis, or points (cf. Fig. 5.1).

Such a segment degenerates to a point only in the case that the corresponding boundary component of G as the preimage is a single point and therefore removable

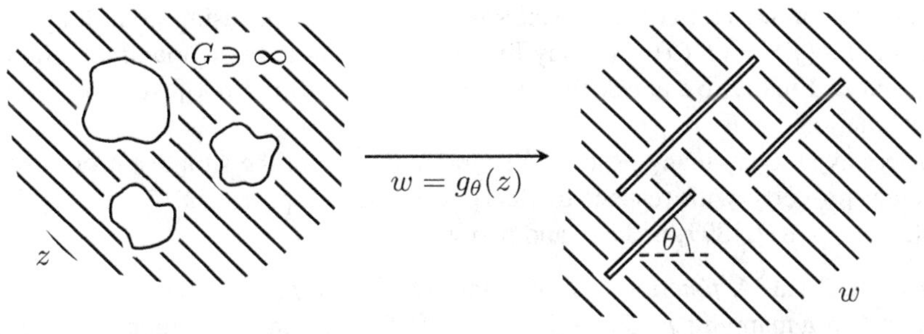

Fig. 5.1 Mapping the bounded, multiply connected domain onto a parallel slit domain

for the mapping function. The middle point and the length of the segments are uniquely determined by G. This is typical for canonical conformal mappings: One can prescribe only the "shape" (here the shape is a segment) of the images of the boundary components of G.

To take a simple example, note that, for $0 < \theta < 2\pi$, the function

$$g_\theta(z) = z + \frac{e^{2i\theta}}{z}$$

is a univalent mapping of $\{|z| > 1\}$ onto the extended complex plane with a rectilinear slit from $-2e^{i\theta}$ to $2e^{i\theta}$. Here, and in what follows, we use the term "univalent" to mean one-to-one analytic, except for a possible (simple) pole that may be at infinity.

We prove the following representative and classical example of the so-called parallel slit mapping. The result is that for every multiply connected domain G and θ, there is a univalent function mapping G onto a parallel slit domain where all slits form an angle θ with the real axis. The precise statement is given by the following theorem.

Theorem 5.1.2 *Every multiply connected domain G in $\mathbb{C}$ can be mapped by a univalent function onto a parallel slit domain, with slits of prescribed angle θ. Moreover, if ζ is a given point of G that is mapped to ∞, then the expansion of the mapping about $z = \zeta$ has the form*

$$w = \frac{1}{z-\zeta} + a_1(z-\zeta) + \dots \quad \textit{or} \quad w = z + \frac{a_1}{z} + \dots \tag{5.1}$$

according to whether ζ is finite or ∞.

The idea of the proof of the Riemann theorem is as follows. Consider the family $\mathcal{F}$ of holomorphic and univalent functions f in G bounded by one in absolute value, that is, those that map G into the unit disc $\mathbb{D}$. Then fix a point $a \in G$ and look for a function f that maximizes the dilation coefficient $|f'(a)|$ at the point a. Restricting ourselves to a sequentially compact subset $\mathcal{F}_1$ of $\mathcal{F}$ and using continuity of the functional $J(f) = |f'(a)|$, we may find a function f_0 with the maximal dilation at the point a. Finally, we check that f_0 maps G onto $\mathbb{D}$ and not just into $\mathbb{D}$ like the other functions in $\mathcal{F}$.

There exist today many proofs of Theorem 5.1.2, but the simplest proof uses the extremal property with respect to $\operatorname{Re} a_1$ in the development in (5.1); cf. [179, p. 179], [214, p. 82], [397, p. 346], and many other books [341].

Definition 5.1.3 A *functional* J of a family $\mathcal{F} = \{f\}$ of functions defined in a domain G is a mapping $J : \mathcal{F} \to \mathbb{C}$, that is, $J(f)$ is a complex number. A functional J is continuous if, given any sequence of functions $f_n \in \mathcal{F}$ that converges uniformly to a function $f_0 \in \mathcal{F}$ on any compact set $K \subset G$, we have

$$\lim_{n\to\infty} J(f_n) = J(f_0) \ .$$

Example 5.1.4 Let $\mathcal{O}(G)$ be the family of all functions f holomorphic in G and let a be an arbitrary point in G. Consider the p-th coefficient of the Taylor series at a:

$$c_p(f) = \frac{f^{(p)}(a)}{p!}.$$

This is a functional on the family $\mathcal{O}(G)$. Let us show that it is continuous. If $f_n \to f_0$ uniformly on every compact set $K \subset G$, then we may let K be the circle $\gamma = \{|z-a| = r\} \subset G$. Then, given any $\varepsilon > 0$, we may find N so that $|f_n(z) - f_0(z)| < \varepsilon$ for all $n > N$ and all $z \in \gamma$. The Cauchy formula for c_p gives that

$$c_p = \frac{1}{2\pi i} \int_\gamma \frac{f(z)}{(z-a)^{n+1}} dz$$

implies that

$$\left|c_p(f_n) - c_p(f_0)\right| \le \frac{\varepsilon}{r^n}$$

for all $n > N$, which in turn implies the continuity of the functional $c_p(f)$.

Definition 5.1.5 A compact family of functions $\mathcal{F}$ is *sequentially compact* if the limit of any sequence f_n that converges uniformly on every compact subset $K \subset G$ belongs to the family $\mathcal{F}$.

Theorem 5.1.6 *Any functional J that is continuous on a sequentially compact family $\mathcal{F}$ is bounded and attains its least upper bound. That is, there exists a function $f_0 \in \mathcal{F}$ so that we have*

$$|J(f_0)| = \sup_{f\in\mathcal{F}} |J(f)|.$$

Proof We let $A = \sup_{f \in \mathcal{F}} |J(f)|$—this is a number that might be equal to infinity. By definition of the supremum, there exists a sequence $f_n \in \mathcal{F}$ so that $|J(f_n)| \to A$. Since $\mathcal{F}$ is a sequentially compact family, there exists a subsequence f_{n_k} that converges to a function $f_0 \in \mathcal{F}$. Continuity of the functional J implies that

$$|J(f_0)| = \lim_{k \to \infty} \left| J\left(f_{n_k}\right) \right| = A\,.$$

This means that first $A < \infty$ and second, $|J(f_0)| \geq |J(f)|$ for all $f \in \mathcal{F}$.

We will consider below families of univalent functions in a domain G. The following theorem is useful to establish sequential compactness of such families.

Convergence of sequences of holomorphic functions $\left(f_j\right)$ on some domain Ω is always measured with respect to the Euclidean metric and is always assumed to be locally uniform. The most important results in this context are the theorems of Weierstrass and Hurwitz.

Theorem 5.1.7 (Weierstrass's Theorem) *Suppose f_j tends to f, locally uniformly on $U \subset \mathbb{C}$. Then f is holomorphic on U, and the sequence $\left(f_j'\right)$ tends to f', again locally uniformly on U.*

Theorem 5.1.8 (Hurwitz Theorem) *Let a sequence of functions f_j holomorphic in a domain $U \subset \mathbb{C}$ converge uniformly on any compact subset K of U to a function $f \neq const$. Then, if $f(z_0) = 0$, given any disc $\mathbb{D}(0, r) = \{|z - z_0| < r\}$, there exists N so that all functions f_j vanish at some point in $\mathbb{D}(0, r)$ when $n > N$.*

Proof The Weierstrass theorem implies that f is holomorphic in U. The uniqueness theorem implies that there exists a punctured disc $\{0 < |z - z_0| \leq \rho\} \subset U$ where $f \neq 0$ (we may assume that $\rho < r$). We write $\gamma = \{|z - z_0| = \rho\}$ and $\mu = \min_{z \in \gamma} |f(z)|$ and observe that $\mu > 0$. However, f_j converges uniformly to f on γ, and hence, there exists N so that

$$\left| f_j(z) - f(z) \right| < \mu$$

for all $z \in \gamma$ and all $j > N$. The Rouché theorem implies that for such j the function $f_j = f + \left(f_j - f\right)$ has as many zeros (with multiplicities) as f inside γ, that is, f_j has at least one zero inside $\mathbb{D}(0, \rho)$.

Corollary 5.1.9 *If a sequence of holomorphic and univalent functions f_j in a domain U converges uniformly on every compact subset K of U, then the limit function f is either a constant or is univalent.*

Proof Assume that $f(z_1) = f(z_2)$ but $z_1 \neq z_2$, $z_1, z_2 \in U$ and $f \neq$ const. Consider a sequence of functions $g_j(z) = f_j(z) - f_j(z_2)$ and a disc $\mathbb{D}(z_1, r) = \{|z - z_1| < r\} \subset U$ with $r < |z_1 - z_2|$. Since $z_2 \notin \mathbb{D}(z_1, r)$ and f_j are univalent, we have $g_j \neq 0$ in $\mathbb{D}(z_1, r)$. The limit function $g(z)$ vanishes at the point z_1. Hence, according to the Hurwitz theorem, all functions $g_j(z)$ starting with some $n \geq N$ vanish in $\mathbb{D}(z_1, r)$. The obtained contradiction proved the corollary.

The class $\mathcal{S}$ consists of all functions

$$f(z) = z + a_2 z^2 + a_3 z^3 + \dots,$$

holomorphic and univalent in $\mathbb{D}$. Many formulas take their nicest form with this normalization $f(0) = 0$, $f'(0) = 1$. These are the normalized "schlicht" functions in $\mathbb{D}$.

From the Koebe distortion theorem for $f \in \mathcal{S}$ follows (see, for example, Pommerenke [422, Theorem 1.3 p. 9])

$$|f(z)| \le \frac{|z|}{(1 - |z|)^2}, \quad z \in \mathbb{D}.$$

Hence $\mathcal{S}$ is locally bounded and thus a normal family. Moreover, if $\mathcal{S} \ni f_n \to f$ uniformly on compact subsets of $\mathbb{D}$, then f is holomorphic in U and either univalent or constant.

Let f_n be Schlicht and converge uniformly to f on compact sets. As $f_n'(0) \to f'(0)$, we have $f'(0) = 1$ so that f is univalent in U. Clearly $f(0) = 0$, and $f \in \mathcal{S}$, i.e., $\mathcal{S}$ is compact.

There is a complementary class of univalent functions Σ that admit the representation

$$g(\zeta) = \zeta + b_0 + b_1 \zeta^{-1} + \dots \quad (|\zeta| > 1).$$

If $f \in \mathcal{S}$, then

$$g(\zeta) = 1/f\left(\zeta^{-1}\right) = \zeta - a_2 + \left(a_2^2 - a_3\right)\zeta^{-1} + \dots \quad (|\zeta| > 1) \tag{5.2}$$

belongs to Σ and omits 0. Conversely, if $g \in \Sigma$ and $g(\zeta) \ne 0$ for $\zeta \in \mathbb{D}^* := \{z \in \mathbb{C} \mid |z| > 1\} \cup \{\infty\}$, then

$$f(z) = 1/g\left(z^{-1}\right) = z - b_0 z^2 + \left(b_0^2 - b_1\right) z^3 + \dots \quad (|z| < 1)$$

belongs to $\mathcal{S}$.

It is not difficult to show (see, e.g., [142, p. 29]) that

$$\text{area}\left(\mathbb{C}\backslash g\left(\mathbb{D}^*\right)\right) = \pi\left(1 - \sum_{n=1}^{\infty} n\,|b_n|^2\right) \quad \text{for} \quad g \in \Sigma.$$

This implies the *Area theorem*

$$|b_1|^2 \le \sum_{n=1}^{\infty} n\,|b_n|^2 \le 1 \quad \text{for} \quad g \in \Sigma; \tag{5.3}$$

equality holds if and only if $g(\zeta) \equiv \zeta + b_0 + b_1\zeta^{-1}$ with $|b_1| = 1$.

If $g \in \Sigma$ and $g(\zeta) \neq w$ for $\zeta \in \mathbb{D}^*$, then

$$h(\zeta) \equiv \sqrt{g\left(\zeta^2\right) - w} = \zeta + \frac{1}{2}\left(b_0 - w\right)\zeta^{-1} + \dots$$

is an odd function in Σ. Hence it follows from (5.3) that

$$|w - b_0| \le 2 \quad \text{for} \quad g \in \Sigma, \quad w \notin g\left(\mathbb{D}^*\right)$$

and thus from (5.2) that (see Bieberbach [83])

$$|a_2| \le 2 \quad \text{for} \quad f \in S\,.$$

Equality holds for the Koebe function

$$f_0(z) = \frac{z}{(1-z)^2} = \sum_{n=1}^{\infty} n z^n\,.$$

The Bieberbach conjecture that

$$|a_n| \le n \quad \text{for} \quad f \in S, \quad n = 2, 3, \dots$$

was proved by de Branges [93].

The class Σ is also normal and compact where it arises naturally in an application of conformal mappings.

The extremal method of Theorem 5.1.6 has also found significant application in problems of conformally mapping a multiply connected domain onto one of several canonical domains.

Given a multiply connected domain G in $\mathbb{C}$, consider the family

$$\begin{aligned} \mathcal{S}_\infty(G) = \{ & f(z) \text{ univalent in } G : f(z) \text{ has the Laurent expansion} \\ & f(z) = z + \frac{a_1}{z} + \frac{a_2}{z^2} + \dots \text{ in a neighbourhood of } z = \infty \Big\}. \end{aligned}$$

We intend to demonstrate that there is an extremal function in $\mathcal{S}_\infty(G)$ that attains the extremum

$$\max_{f \in \mathcal{S}_\infty(G)} \operatorname{Re}\left(e^{-2i\theta} a_1\right)$$

and that this function is the desired conformal mapping.

This line of reasoning dates back to de Possel [424]; cf., e.g., also Grötzsch [210] and Schiff [458].

First the following growth estimate is required.

Lemma 5.1.10 *If*

$$f(z) = z + a_0 + \frac{a_1}{z} + \dots$$

is univalent in $|z| > R$*, then*

$$|f(z) - a_0| \leq 2|z|.$$

Proof Given $|z_0| > R$, define

$$w = F(z) = \frac{1}{z_0} f(z_0 z) - \frac{a_0}{z_0} = z + \frac{a_1}{z_0^2 z} + \dots,$$

which is univalent in $|z| > 1$. Then consider the function

$$G(z) = \frac{1}{F\left(\frac{1}{z}\right) - c} = z + cz^2 + \dots$$

for any c not in the range of F. Clearly, G belongs to the Schlicht class $\mathcal{S}$, and as a consequence, its second Taylor coefficient satisfies $|c| \leq 2$. Therefore the entire boundary of $F(\{|z| > 1\})$ is contained in $|w| \leq 2$. Specifically, $|F(1)| \leq 2$, which gives the result

$$|f(z) - a_0| \leq 2|z|, \quad |z| > R.$$

□

Our main result also depends upon the following observation.

Lemma 5.1.11 *In the family*

$$\mathcal{S}_R = \left\{ f(z) \text{ univalent in } |z| > R : f(z) = z + \frac{a_1}{z} + \dots \right\},$$

given $\theta \in \mathbb{R}$*, the function* $f_\theta(z) = z + \frac{R^2 e^{2i\theta}}{2}$*, which maps* $|z| > R$ *onto the plane with a rectilinear slit at angle* θ*, uniquely solves the extremal problem*

$$\max_{f \in \mathcal{S}_R} \operatorname{Re}\left(e^{-2i\theta} a_1\right),$$

giving the maximum value R^2.

Proof Normalizing the functions of $\mathcal{S}_R$ by

$$F(z) = \frac{1}{R} f(Rz) = z + \frac{a_1}{R^2 z} + \dots$$

yields the functions $\{F(z)\}$ that are univalent in $|z| > 1$. At this stage we wish to invoke the *Area Theorem* (5.3) that asserts that any function $g(\zeta) = \zeta + \frac{\alpha_1}{\zeta} + \dots$ that is univalent in $|\zeta| > 1$ satisfies the condition

$$\sum_{n=1}^{\infty} n\,|\alpha_n|^2 \leq 1.$$

In particular, $|\alpha_1| \leq 1$ with equality holding if and only if $\alpha_2 = \alpha_3 = \cdots = 0$, i.e., for the function $g(\zeta) = \zeta + \frac{e^{i\phi}}{\zeta}$, discussed above. Applying this result to the present setting gives $|a_1| \leq R^2$, with equality holding if and only if $f(z) = z + \frac{R^2 e^{i\phi}}{z}$. Hence

$$\operatorname{Re}\left(e^{-2i\theta} a_1\right) \leq R^2$$

with equality holding if and only if $f(z) = z + \frac{R^2 e^{2i\theta}}{z}$. □

Proof of Theorem 5.1.2 Consider first the case $\zeta = \infty$. Then all of ∂G is contained in $|z| < R$ for R sufficiently large. Thus every f in $\mathcal{S}_\infty(G)$ is univalent in $|z| > R$ and has the expansion $f(z) = z + \frac{a_1}{z} + \dots$ there. In view of Lemma 5.1.10, $\mathcal{S}_\infty(G)$ is locally bounded in $R < |z| < \infty$, and if $|z| = R' > R$, then $|f(z)| \leq 2R'$. The latter implies that $|f(z)| \leq 2R'$ for all $z \in G \cap \left\{|z| < R'\right\}$ due to the univalence of $f \in \mathcal{S}_\infty(G)$. To see this, observe that if $|w| > 2R'$, then it follows from Rouché's theorem that the equations $f(z) = w$ and $f(z) = \infty$ have equally many roots in $|z| > R'$, i.e., exactly one. Since $f(z)$ is univalent in G, the equation $f(z) = w\ \left(|w| > 2R'\right)$ will have no other roots in G, and thus none in $G \cap \left\{|z| < R'\right\}$, so that $|f(z)| \leq 2R'$ there. We conclude that $\mathcal{S}_\infty(G)$ is normal in $G \setminus \{\infty\}$. Moreover, if $f_j \to f$ normally in $G \setminus \{\infty\}$, and $f_j(z) = z + \sum_{n=1}^{\infty} \frac{a_n^{(j)}}{z^n}$, $j = 1, 2, 3, \dots$ $(R < |z| < \infty)$, then $a_n^{(j)} \to a_n$ and f is analytic univalent, with $f(z) = z + \sum_{n=1}^{\infty} \frac{a_n}{z^n}$ $(R < |z| < \infty)$, that is, $\mathcal{S}_\infty(G)$ is compact in $G \setminus \{\infty\}$. Note also that the family of analytic univalent functions $f(z)$ defined on $G \setminus \{\infty\}$ with the expansion $f(z) = z + \frac{a_1}{z} + \dots$ in a neighborhood of ∞ coincides with $\mathcal{S}_\infty(G)$.

Next observe that the functional

$$\phi_\theta(f) = \operatorname{Re}\left(e^{-2i\theta} a_1\right), \quad f \in \mathcal{S}_\infty(G)$$

is continuous. An appeal to Theorem 5.1.6 yields the existence of $F(z) = z + \frac{A_1}{z} + \dots \in \mathcal{S}_\infty(G)$ such that

$$\max_{f \in \mathcal{S}_\infty(G)} \operatorname{Re}\left(e^{-2i\theta} a_1\right) = \operatorname{Re}\left(e^{-2i\theta} A_1\right).$$

It remains to show that the extremal function $F(z)$ is the desired conformal mapping. Suppose that the domain $G^* = F(G)$ has at least one boundary component β that is a continuum but not a rectilinear slit of inclination θ. Let G_1^* be the unbounded component of the complement of β. Then $G^* \subseteq G_1^*$ since G^* is connected, unbounded, and lies in the complement of β. Moreover, G_1^* is simply connected since its complement with respect to $\mathbb{C}$ is connected. Now let $w' = g(w)$ be a univalent mapping of G_1^* onto the w'-plane with a rectilinear slit of inclination θ, having the expansion

$$g(w) = w + \frac{a_\theta}{w} + \dots$$

in a neighborhood of $w = \infty$ (cf. Remarks 2.2.1 subsequent to the Riemann Mapping Theorem). Then the composition

$$F^*(z) = g(F(z))$$

maps the domain G univalently and has the expansion

$$F^*(z) = z + \frac{A_1 + a_\theta}{z} + \dots$$

near $z = \infty$, so that $F^* \in \mathcal{S}_\infty(G)$. Therefore

$$\begin{aligned} \operatorname{Re}\left(e^{-2i\theta}\left(A_1 + a_\theta\right)\right) &= \operatorname{Re}\left(e^{-2i\theta} A_1\right) + \operatorname{Re}\left(e^{-2i\theta} a_\theta\right) \\ &\leq \operatorname{Re}\left(e^{-2i\theta} A_1\right), \end{aligned}$$

giving $\operatorname{Re}\left(e^{-2i\theta} a_\theta\right) \leq 0$.

On the other hand, in view of Remarks 2.2.1, a domain $|W| > R'$ can be mapped univalently onto G_1^* by

$$h(W) = W + b_0 + \frac{b_1}{W} + \dots, \qquad |W| > R'.$$

Then the function

$$w' = g(h(W)) - b_0 = W + \frac{b_1 + a_\theta}{W} + \dots$$

maps $|W| > R'$ univalently onto the w'-plane with rectilinear slit at angle θ, whereas the mapping $h(W) - b_0$ maps $|W| > R'$ onto some other domain. In accordance with Lemma 5.1.11

$$\operatorname{Re}\left(e^{-2i\theta} b_1\right) < \operatorname{Re}\left(e^{-2i\theta} b_1 + e^{-2i\theta} a_\theta\right)$$

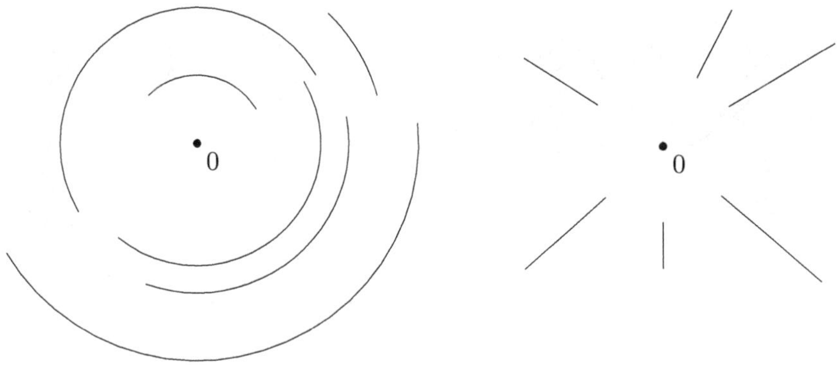

(a) a circular slit domain (b) a radial slit domain

Fig. 5.2 The canonical regions

implying Re $\left(e^{-2i\theta}a_\theta\right) > 0$. This contradicts the result of the preceding paragraph. That proves the first part of the theorem.

In the case of finite ζ, an application of the mapping $\frac{1}{z-\zeta} = z'$ transforms G onto a domain G' that contains the point $z' = \infty$, reducing the problem to the previous case. The proof of the theorem is herewith complete.

For the conformal mappings of a multiply connected domain onto other canonical regions, refer to [179, Chapter 5], [142, Chapter 10].

Circular and Radial Slit Domains These are two similar classes of slit domains consisting of the complex sphere without some slits. In the first case, we remove several arcs that lie on concentric circles centered at the origin. In the second case, we remove intervals that lie on rays emanating from the origin (on Fig. 5.2a design a circular slit domain, yet on Fig. 5.2b design a radial slit domain).

In both cases, we can normalize a map in such a way that two given points z_1 and z_2 from Ω are mapped to the origin and infinity. Let us consider a family of functions f that are univalent in Ω, $f(z_1) = 0$, and there is a simple pole of residue 1 at z_2. The function that maximizes $\left|f'(z_1)\right|$ maps the domain onto a circular slit domain, and the function that minimizes $\left|f'(z_1)\right|$ maps the domain onto a radial slit domain. We will not prove these results here. Interested readers can find the proofs and further discussion of uniformization of multiply connected domains in Nehari [398, Chapter VII].

The most natural generalization of the Riemann mapping theorem to the case of multiply connected domains is the famous:

Theorem 5.1.12 (Koebe's Kreisnormierung's Theorem) *For every domain G of finite connectivity, there exists a Schlicht conformal mapping onto a domain whose boundary components are circles. The mapping is unique up to a following Möbius transformation.*

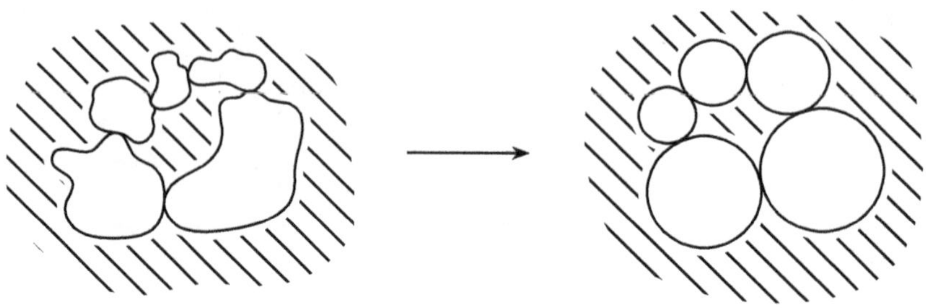

Fig. 5.3 A multiply connected domain is equivalent to a domain whose boundary components are circles

Koebe gave many proofs of this result. A proof with Koebe's method of continuity can be found in [179], [214, p. 114]; cf. many further references in [163, p. 182]; there are also proofs with iteration processes; another idea of a proof appears in [120, p.169] (Fig. 5.3).

Although Koebe's Kreisnormierung is so distinguished, an extremal property was found only in 1962 by Schiffer and Hawley. This paper was something of a sensation at that time. With this extremal property in [232, p. 488] also the existence was proved; in [135] the convergence of Koebe's iteration process for doubly connected domains is treated.

For more on this subject see Kühnau, R. [342].

In 1908, P. Koebe [302] posed the following conjecture, known as Koebe's Kreisnormierung's problem: *Any plane domain is conformally homeomorphic to a circle domain in* $\mathbb{C}$. When the domain is simply connected, this is the content of the Riemann mapping theorem. The conjecture was proved for finitely connected domains and certain symmetric domains by Koebe himself ([308, 309]); for domains with various conditions on the "limit boundary components" by R. Denneberg [132], H. Grötzsch [212], L. Sario [456], H. Meschowski ([377, 378]), K.L. Strebel ([491, 492]), L. Bers [78], A. Haas [224], and others; and for domains quasiconformally homeomorphic to a circle domain by R.J. Sibner [470, 471].

Z-X. He and O. Schramm [227] proved the Koebe uniformization conjecture for countably connected domains: *Any domain in the Riemann sphere* $\widehat{\mathbb{C}}$ *with countably many boundary components is conformally homeomorphic to a circle domain; i.e., to a domain in* $\widehat{\mathbb{C}}$ *whose boundary components are all circles and points.*

O. Schramm later [462] gave a conceptually different and simpler proof based on his transboundary extremal length, which also applies to certain classes of domains with uncountably many boundary components.

Chapter 6
Quasiconformal Mappings

6.1 Historical Development of Quasiconformal Mappings

Ahlfors's book [16] provides a classical introduction to quasiconformal maps and Teichmüller spaces. Quasiconformal maps appeared in the late 1920s and early 1930s in the papers of Grötzsch [209, 211] and Lavrentieff [351, 352] (although they were present implicitly already in work of Gauss and Beltrami related to the introduction of isothermal coordinates (that is, conformal-Euclidean coordinates on a two-dimensional surface)). It was discovered [209, 352] that many (even rather subtle) facts of complex analysis (for example, Picard's theorem) hold not only for holomorphic functions (just as the fundamental theorem of algebra on the existence of a root of a complex polynomial is a very important and very special case of the index theorem for a map) but for a more general class of functions. This observation was intensively developed in the investigations in the 1930s of Ahlfors, who constructed a geometric analog [7] of Nevanlinna's value distribution theory for holomorphic functions.

In the late 1930s and early 1940s, quasiconformal maps were at the center of the Teichmüller program [499–502] of investigating the moduli problem for Riemann surfaces, going back to Abel and Riemann. On the same topological surface, for instance, on a compact surface of genus g, one can realize many different conformal structures. The moduli problem is to find out how many parameters (moduli) determine this diversity of conformal structures and how these parameters can be introduced (it is also desirable to have a complex structure on these parameters). In simplified form, this is the question of how many conformally distinct Riemann surfaces there are of a given genus g. Here is the answer: for $g > 1$ the variety of these surfaces is determined by $6g - 6$ real or $3g - 3$ complex parameters. Teichmüller introduced a distance (Teichmüller metric) between distinct Riemann surfaces of the same genus (that is, between conformal structures or points in the parameter space) as follows. If surfaces of the same genus g are not conformally equivalent, they are at least diffeomorphic. Every diffeomorphism of a compactum is

P. V. Dovbush, S. G. Krantz, *The Geometric Theory of Complex Variables*,
https://doi.org/10.1007/978-3-031-77204-7_6

a quasiconformal map. Let k be the coefficient of quasiconformality of an extremal quasiconformal map deviating least from conformality. Then we take $\log k$ for the distance is question. In this way (with certain refinements skipped over here) the Teichmüller space of Riemann surfaces arises.

It should be noted that the first problem of finding a map between conformally distinct objects that deviates least from conformality (in such questions this problem plays the role of the omnipresent Schwarz lemma in the classical theory of analytic functions) was considered already by Grötzsch [211], bears his name, and consists in the following.

Let two non-isomorphic rectangles in the plane be given. The problem is to find the map deviating least from conformality among the maps of these rectangles with given correspondence of the vertices. Grötzsch proved that, in any reasonable sense, the corresponding linear map turns out to be extremal.

The program of Teichmüller was given new life and realized to a large degree in the works of Ahlfors and Bers [22]. The Teichmüller space was also endowed with a complex structure.

The Teichmüller space has in recent times become an attribute of theoretical physicists studying quantum field theory and the so-called string theory.

Quasiconformal mappings are fascinating objects in mathematics. They are flexible enough to be ubiquitous, yet they harbor enough subtle analytic and geometric properties so as to be useful in a variety of contexts. For example, the reason why quasiconformal mappings are so useful in iteration theory is that these mappings allow for the kind of complicated behavior that arises and at the same time admit a way to connect analytic information to geometric. The person who first realized the power of quasiconformal mappings in dynamical systems was D. Sullivan (1985), and it is now a standard tool. Actually, it is the main new idea in the field.

There are three main definitions for quasiconformal mappings in Euclidean spaces: metric, geometric, and analytic. We begin with the metric definition, which is the easiest to state and which makes sense in arbitrary metric spaces. It describes the property that "infinitesimal balls are transformed to infinitesimal ellipsoids of bounded eccentricity."

Let $f : X \to Y$ be a homeomorphism between two metric spaces. For $x \in X$ and $r > 0$, let

$$L_f(x, r) = \sup\{|f(x) - f(y)| : |x - y| \leq r\}$$

and

$$l_f(x, r) = \inf\{|f(x) - f(y)| : |x - y| \geq r\}.$$

(Here and later we use the Polish notation $|a - b|$ for the distance in any metric space.) The ratio $H_f(x, r) = L_f(x, r)/l_f(x, r)$ measures the eccentricity of the image of the ball $B(x, r)$ under f. We say that f is *H-quasiconformal*, $H \geq 1$, if

$$\limsup_{r \to 0} H_f(x, r) \leq H \tag{6.1}$$

for every $x \in X$.

Homeomorphisms that are 1-quasiconformal between domains in $\mathbb{R}^2 \cong \mathbb{C}$ are precisely the (complex analytic) conformal or anti-conformal mappings, by a theorem of Menshov from 1937. Homeomorphisms that are 1-quasiconformal between domains in $\mathbb{R}^n$, $n \geq 3$, are precisely the Möbius transformations, or compositions of inversions on spheres in the one-point compactification $\mathbb{R}^n \cup \{\infty\}$; this by the generalized Liouville theorem proved by Gehring and Reshetnyak in the 1960s. On the other hand, every diffeomorphism $f : \mathbb{R} \to \mathbb{R}$ is 1-quasiconformal according to the metric definition, as is every homeomorphism between discrete spaces. Surely not all such mappings deserve to be called quasiconformal. We will later remedy this situation.

Many early definitions for quasiconformality used some conformally invariant quantity and declared quasiconformal mappings to be those homeomorphisms that changed that quantity by a bounded amount.

Here is one such geometric definition. Let $f : \Omega \to \Omega'$ be a homeomorphism between two domains in $\mathbb{R}^n$, $n \geq 2$. Then f is said to be K-*quasiconformal*, $K \geq 1$, if

$$K^{-1} \operatorname{mod}(f(\Gamma)) \leq \operatorname{mod}(\Gamma) \leq K \operatorname{mod}(f(\Gamma)) \tag{6.2}$$

for every curve family Γ in Ω. The conformal modulus $\operatorname{mod}(\Gamma)$ of a family Γ of curves in $\mathbb{R}^n$ is the infimum of the numbers $\int_{\mathbb{R}^n} \rho^n dx$ over all nonnegative Borel functions $\rho : \mathbb{R}^n \to [0, \infty]$ such that $\int_\gamma \rho ds \geq 1$ for every $\gamma \in \Gamma$. The definition of modulus is admittedly not easy to digest at one glance, but once mastered it is a powerful tool in geometric function theory. The geometric definition (6.2) is a global requirement that quickly yields many strong properties of quasiconformal mappings; for example, the inverse of a quasiconformal mapping is automatically quasiconformal, which is not at all obvious from the metric definition.

Also, in the 1930s, an analytic definition for quasiconformal mappings was considered by Lavrentiev in connection with elliptic systems of partial differential equations. According to this definition, a homeomorphism $f : \Omega \to \Omega'$ between domains in $\mathbb{R}^n$, $n \geq 2$, is said to be K-*quasiconformal* if the first distributional partial derivatives of f are locally in the Lebesgue space L^n and if the formal differential matrix $Df = \left(\partial_i f_j\right)$ satisfies

$$\sup_{h \in \mathbb{R}^n, |h| \leq 1} |Df(x)(h)|^n \leq K \det Df(x) \tag{6.3}$$

for almost every $x \in \Omega$. The use of distributional derivatives is essential in this context; important compactness properties of quasiconformal mappings are lost if smooth mappings only are considered.

It is a deep fact that a homeomorphism $f : \Omega \to \Omega'$ between domains in $\mathbb{R}^n$, $n \geq 2$, is quasiconformal according to each of the three definitions—metric, geometric, and analytic—if it is quasiconformal according to any one of them. The parameters H and K depend only on each other and on n. The equivalence of these three definitions is a result of work by Gehring, Väisälä, and many others, in the 1950s and early 1960s.

Investigations of moduli of Riemann surfaces led to the recognition that Beltrami equations provide easy access to classical uniformization theory. The theory of Beltrami equations with bounded measurable coefficients has been initiated by Morrey [383]; a self-contained presentation will be found in a paper by Ahlfors and Bers [8].

There is a powerful existence theorem, proved by Morrey in 1938, that lends a special flavor to the two-dimensional theory: Given a measurable complex-valued function μ in $\mathbb{R}^2$ such that $\|\mu\|_\infty < 1$, there exists a quasiconformal homeomorphism $f : \mathbb{R}^2 \to \mathbb{R}^2$ (unique when properly normalized) such that $\bar{\partial} f = \mu \partial f$ in the sense of distributions, where $\bar{\partial} = \frac{1}{2}\left(\partial_x + i\partial_y\right)$ and $\partial = \frac{1}{2}\left(\partial_x - i\partial_y\right)$ in complex notation. That is, one can measurably preassign the eccentricity and angle of the ellipses that are carried to a circle by the almost everywhere defined derivative of f. This fact has had tremendous impact on complex analysis and dynamics, Teichmüller theory, and low-dimensional topology.

The interplay of all three aspects of quasiconformality (metric, geometric, and analytic) is an important feature of the theory; one cannot rely on just one of them. From the technical point of view, the quasiconformal mapping theory even in dimension $n = 2$ is much more part of real than complex analysis. Instead of power series, integral representations, or algebraic techniques, the theory relies on singular integrals, geometric measure theory, and Sobolev spaces.

Finally, it can be seen that the above quantities and notions are invariant with respect to any scale changes that depend only on the point and not on the direction (physicists refer to these changes as *gauge transformations*, and geometers call them *conformal changes of the metric*).

Another way to obtain the Beltrami equation: Recall that a Riemannian metric in a domain in the (x, y) plane is defined by a quadratic differential form

$$ds^2 = E(x, y)dx^2 + 2F(x, y)dxdy + G(x, y)dy^2 \tag{6.4}$$

with

$$EG - F^2 > 0, \quad E > 0. \tag{6.5}$$

This is in honor of Riemann, who proposed studies of geometry based only on the first fundamental form at the inaugural address at Göttingen University on June 10, 1854, at the age of 27. (Riemann spoke not only about the two-dimensional case, but also about the general n-dimensional case.)

Setting $x + iy = z$, $x - iy = \bar{z}$, (6.4) can be rewritten as

$$ds^2 = \Lambda(z)^2 |dz + \mu(z) d\bar{z}|^2, \tag{6.6}$$

where Λ is a real-valued and μ a complex-valued function. Condition (6.5) becomes

$$\Lambda > 0, \quad |\mu| < 1. \tag{6.7}$$

A mapping $x + iy = z \mapsto w = u + iv$ is called *conformal* with respect to ds if it preserves orientation and, except at isolated points, takes angles measured by the metric (6.4) into equal Euclidean angles.

Using the standard notations

$$\frac{\partial}{\partial z} = \frac{1}{2}\left(\frac{\partial}{\partial x} - i\frac{\partial}{\partial y}\right), \quad \frac{\partial}{\partial \bar{z}} = \frac{1}{2}\left(\frac{\partial}{\partial x} + i\frac{\partial}{\partial y}\right),$$

the requirement that w be *conformal with respect to the metric* (6.8) may be written as

$$\partial w / \partial \bar{z} = \mu(z) \partial w / \partial z. \tag{6.8}$$

This is the classical *Beltrami equation*, which goes over into the Cauchy-Riemann equations for u and $v(\partial w/\partial \bar{z} = 0)$ if $\mu(z) \equiv 0$. Note that Λ does not enter into (6.8), as was to be expected. From now on we call solutions of (6.8) *μ-conformal functions*.

(*Technical Remark:* Solutions of (6.8) are assumed to be continuous and to have distribution derivatives that are, locally, square integrable functions; Eq. (6.8) is assumed to hold almost everywhere.)

A homeomorphism w is called *quasiconformal* if it is μ-conformal for some μ with

$$\|\mu\|_\infty = \text{ess sup } |\mu(z)| < 1. \tag{6.9}$$

If so, the number

$$K(w) = (1 + \|\mu\|_\infty) \,/\, (1 - \|\mu\|_\infty) \tag{6.10}$$

is called the *dilatation* of w. If w is a diffeomorphism, $K(w)$ is the supremum of the ratios of the major to the minor axes of infinitesimal ellipses into which w takes infinitesimal circles in the domain considered. A similar definition can always be used, since every quasiconformal homeomorphism is differentiable almost everywhere.

(We have stated the so-called *analytic definition of quasiconformality*. Concerning the equivalent geometric definition, see [16, 355].)

A homeomorphism $f : S \to S'$ is called *μ-conformal* if, in terms of local parameters $z = x + iy$ on S and $\zeta = \xi + i\eta$ on S', it satisfies the Beltrami equation

$$\zeta_{\bar{z}} = \mu(z)\zeta_z.$$

Here the partial derivatives $\zeta_z = (\zeta_x - i\zeta_y)/2$ and $\zeta_{\bar{z}} = (\zeta_x + i\zeta_y)/2$, taken in the sense of distribution theory, are required to be measurable, locally square integrable functions. That a Beltrami equation can be solved without requiring any smoothness of μ was first recognized by Morrey [383]. This fact is very important in the applications.

An important and frequently used tool of the theory of quasiconformal maps in the real two-dimensional case is the theorem that one plane domain can be mapped onto another by solutions of the *Beltrami equation*

$$f_{\bar{z}} = \mu(z) f_z,$$

where $|\mu|(z) \leqslant q < 1$.

For $\mu(z) \equiv 0$ this Beltrami equation (system) reduces to the Cauchy-Riemann equation (system) $f_{\bar{z}} = 0$, and the theorem itself reduces to the classical Riemann theorem on conformal maps of plane domains.

We note that the problem of a conformal map of a two-dimensional surface (for instance, the graph of a function) onto a plane domain [and hence the problem of conformal-Euclidean (or isothermal) coordinates on the surface] can be immediately reduced to precisely the Beltrami equation.

Indeed, infinitesimal circles at the points (to be more exact, in the tangent planes) of a curved surface become under orthogonal projection of it to the plane a family of infinitesimal ellipses uniformly distributed over a plane domain. We now want to use an auxiliary map $w = f(z)$ to transform the domain obtained under the projection in such a way that the tangent map $dw = f_z dz + f_{\bar{z}} d\bar{z}$ transforms the corresponding ellipse to a circle at any point. However, this just means that the desired map of the plane domain satisfies the Beltrami equation in which the function $\mu(z) = f_{\bar{z}}/f_z$ corresponds to a given distribution.

We also note that, if the *Beltrami coefficient* $\mu_\zeta(z)$ in the Beltrami equation depends holomorphically on a complex parameter ζ, then the appropriately normalized solution of the equation also depends holomorphically on this parameter [16]. (The development of and the versions of the proof of this important existence theorem can be traced to [8, 121, 211, 383, 501, 515].) In particular, it becomes especially clear that Lavrentieff treated the notion of quasiconformality in a very broad sense, as subject to a certain fairly general system of equations or relations, and he referred to maps that are quasiconformal in the present sense as maps with *bounded distortion* [350].

One can readily see that the Beltrami coefficient is related to the coefficient of quasiconformality of a map at a point by the simple relation

$$k_f(z) = \frac{1 + |\mu|(z)}{1 - |\mu|(z)}.$$

Quasiconformal maps turned out also to be closely related to complex dynamics [380]. Let E be a set in the complex plane and let f_ζ be a family of injective maps of E into the same plane such that f_ζ depends holomorphically on the parameter ζ, $|\zeta| < \delta < 1$ (holomorphic evolution), where f_0 is the identity map. (Holomorphic dependence on the parameter ζ is understood to mean that the complex variable $f_\zeta(z)$ is a holomorphic function of the parameter ζ for any fixed $z \in E$.) It turns out that any such evolution of a set $E \subset \mathbb{C}$ can be extended to a holomorphic motion of the entire plane, and one can assume that the domain of the parameter is the unit disc. At any instant ζ the map f_ζ satisfies some Beltrami equation in which the Beltrami coefficient depends holomorphically on the parameter ζ. Moreover, $\left|\mu_\zeta\right|(z) < 1$, and $\mu_0(z) \equiv 0$ by assumption. By the Schwarz lemma we obtain $\left|\mu_\zeta\right|(z) < |\zeta|$ for any $z \in \mathbb{C}$ and any $|\zeta| < 1$. Thus, any holomorphic evolution is quasiconformal, and the coefficient of quasiconformality of the map f_ζ is bounded above by the quantity $(1+|\zeta|)/(1-|\zeta|)$. This enables one to use properties of quasiconformal maps when studying holomorphic dynamics. (See [380] and the literature therein.)

All of the foregoing, except for the very definition of quasiconformal map, is related to the real two-dimensional or complex one-dimensional case. For more on this topic and quasiconformal maps in the multidimensional case, see [540].

There are several reasons that quasiconformal (q.c.) mappings have come to play a very active part in the theory of analytic functions of a single complex variable:

1. The most superficial reason is that q.c. mappings are a natural generalization of conformal mappings. If this were their only claim to fame, then they would soon have been forgotten.
2. It was noticed at an early stage that many theorems on conformal mappings use only the quasiconformality. It is therefore of some interest to determine when conformality is essential and when it is not.
3. q.c. mappings are less rigid than conformal mappings and are therefore much easier to use as a tool. This was typical of the utilitarian phase of the theory. For instance, it was used to prove theorems about the conformal type of simply connected Riemann surfaces (now mostly forgotten).
4. q.c. mappings play an important role in the study of certain elliptic partial differential equations.
5. Extremal problems in q.c. mappings lead to analytic functions connected with regions or Riemann surfaces. This was a deep and unexpected discovery due to Teichmüller.
6. The problem of moduli was solved with the help of q.c. mappings. They also throw light on Fuchsian and Kleinian groups.
7. Conformal mappings degenerate when generalized to several variables, but q.c. mappings do not.

6.2 The Planar Theory

Recently there have been considerable advances made in the analytic theory of quasiregular mappings, and it is our aim here to relate these advances, together with many of the classical aspects of the theory, and to show the utility of these results and the ideas that go into their proofs in other areas of nonlinear analysis and PDEs. Thus our approach here is analytical. We therefore do not develop the theory of moduli of curve families, nor do we use this approach to establish any important results. The reader interested in developments along these lines is warmly referred to the recent monograph of Rickman [433] and the well-known notes of Väisälä [513]. Other aspects appear in the tribute to Gehring [140].

The theory of mappings was born in the plane, and so we begin with a brief overview of the planar theory, pointing out some of the highlights and the analogies, where they exist, with classical function theory. Precise definitions can be found in the body of the text, along with proofs of many of the mentioned results. We hope that providing a reasonable discussion of this classical setting will motivate and clarify a number of the definitions, aims and objectives, and, importantly, a number of the proofs given later in this book.

Unfortunately, the geometric definition of quasiconformal mapping, while aesthetically pleasing, is difficult to work with. These days the following analytic definition of quasiconformality is more common.

Definition 6.2.1 A homeomorphism $f : \Omega \to \Omega'$ is called *K-quasiconformal* if f lies in the Sobolev class $W^{1,2}_{loc}(\Omega)$ of functions whose first derivatives are square integrable, and if its directional derivatives satisfy

$$\max_{\alpha} |\partial_\alpha f(z)| \leq K \min_{\alpha} |\partial_\alpha f(z)| \tag{6.11}$$

for almost every $z \in \Omega$.

In particular, a mapping f is 1-quasiconformal if and only if it is conformal, that is, a univalent holomorphic mapping.

It is not difficult to see the relationship between quasiconformality and bounded distortion for diffeomorphisms. It was in the smooth setting that planar quasiconformal mappings were first studied around 1928 by Grötzsch [209]. The term "quasiconformal" was coined by Ahlfors in 1935 [7, 19, 37] when this class of mappings proved an integral tool in his geometric development of Nevanlinna theory, based on the "length-area" method. They were also successfully used by Drasin in his solution to the inverse Nevanlinna problem [137]. Teichmüller found a fundamental connection between quasiconformal mappings and quadratic differentials in his studies of extremal mappings between Riemann surfaces [500] or [503] around 1939. Developments of the length-area method led to the definition of quasiconformal mappings in terms of the distortion of the modulus of curve families by Pfluger [411, 412] and then were systematically studied in their own right by Ahlfors in 1953 [14].

The class of quasiconformal diffeomorphisms is not closed under uniform limits. Thus the generalization to Sobolev spaces is necessary if one is to solve various extremal problems. After making this generalization, we find that the limit of a bounded sequence of quasiconformal mappings is either quasiconformal or constant. Thus the Sobolev setting provides more flexibility and has a greater range of applications.

The equivalence of the geometric definition and the analytic definition was shown by Gehring and Lehto in 1959 [174]. The connection between quasiconformal mappings, Teichmüller theory, and quadratic differentials has been intensively investigated by Ahlfors and Bers, by Reich and Strebel, and by Lehto and others; see [168, 338, 353] and the references therein.

There is another route to the theory of planar quasiconformal mappings. Directly from the analytic definition of a quasiconformal mapping, we see that there is a measurable function μ defined in U such that

$$\bar{\partial} f(z) = \mu(z) \partial f(z), \tag{6.12}$$

where $\bar{\partial} = \frac{1}{2}\left(\partial_x + i\partial_y\right)$ and $\partial = \frac{1}{2}\left(\partial_x - i\partial_y\right)$. Indeed,

$$\|\mu\|_\infty = \frac{K-1}{K+1} < \infty. \tag{6.13}$$

Equation (6.12) is called the *complex Beltrami equation*. Notice that when $\mu = 0$, or equivalently $K = 1$, we obtain the usual Cauchy-Riemann system. The Beltrami equation has a long history. Gauss first studied the equation, with smooth μ, in the 1820s while investigating the problem of existence of isothermal coordinates on a given surface. In fact we shall see later a strong connection between the Beltrami coefficient μ and Riemannian structures on domains Ω or more generally Riemannian manifolds in higher dimensions. The complex Beltrami equation was studied intensively by Morrey in the late 1930s, and he established the existence of homeomorphic solutions for measurable μ. It took another 20 years before Bers recognized that homeomorphic solutions are quasiconformal mappings. The function $\mu_f = \partial f / \bar{\partial} f$ is called the *Beltrami coefficient* of f or the *complex dilatation* of f.

Suppose that f is an orientation-preserving linear mapping of $\mathbb{C}$. Then a little geometric calculation will show that $f(z) = az + b\bar{z}$, $|a| > |b|$, $\mu_f(z) = b/a$ and f maps the unit circle to an ellipse; and the ratio of the major and minor axes of this ellipse is

$$K = \frac{|a| + |b|}{|a| - |b|} = \frac{1 + |\mu_f|}{1 - |\mu_f|}.$$

In this way we view the Beltrami coefficient of a quasiconformal $f : U \to \mathbb{C}$ as defining an ellipse field on U via the affine approximation to f at each point z_0,

$$f(z) = z_0 + \partial f(z_0)(z - z_0) + \bar{\partial} f(z_0)\left(\overline{z - z_0}\right) + O\left(|z - z_0|^2\right).$$

There is a strong interaction between linear and nonlinear elliptic systems in the plane and quasiconformal mappings. The most general first-order linear (over the real numbers) elliptic system takes the form

$$\bar{\partial} f = \mu_1 \partial f + \mu_2 \overline{\partial f},$$

where μ_1 and μ_2 are complex-valued measurable functions such that

$$|\mu_1(z)| + |\mu_2(z)| \leq \frac{K-1}{K+1} < 1 \quad \text{almost everywhere.}$$

The complex Beltrami equation is simply that equation that is linear over the complex numbers.

These sets of equations are particular cases of the genuine nonlinear first-order system

$$\bar{\partial} f = H(z, \partial f), \tag{6.14}$$

where $H : U \times \mathbb{C} \to \mathbb{C}$ is assumed to be Lipschitz in the second variable,

$$|H(z, \zeta) - H(z, \xi)| \leq \frac{K-1}{K+1} |\zeta - \xi| .$$

The principal feature of (6.14) is that the difference of two solutions need not solve the same equation, but it is still K-quasiregular (which in the plane means that it is the composition of a quasiconformal homeomorphism and an analytic map). Thus quasiconformal mappings become the central tool in establishing the a priori estimates needed for the existence and uniqueness for these nonlinear systems of PDEs.

In studying solutions to the Beltrami equation, an operator, analogous to the Hilbert transform, was introduced. This operator is now known as the Beurling-Ahlfors transform [9, 256, 515]. It is defined as a singular integral of Calderón-Zygmund type,

$$S\omega(z) = \frac{1}{2\pi i} \iint_{C} \frac{\omega(\zeta) d\zeta \wedge d\bar{\zeta}}{(\zeta - z)^2}, \tag{6.15}$$

for all functions $\omega \in L^p(\mathbb{C})$, $1 < p < \infty$. More precisely, the integral is understood by means of the Cauchy principal value.

The Cauchy principal value for contour integrals is defined in the following manner. A contour integral of a complex-valued function with a pole c on the contour strictly does not exist. However, it may have the Cauchy principal value if this concept is extended to this case. For this purpose, let L be a closed or open regular curve (i.e., it has continuous curvature at every point). Enclose the point c by a small circle of radius ε with center c. Let L_ε denote the part of the contour outside this circle. If a complex-valued function $f(z)$, $z = x + iy$, is integrable along L_ε, however small the positive number ε, then the limit

$$\lim_{\varepsilon\to 0}\int_{L_\varepsilon} f(z)dz\,,$$

if it exists, is called the *Cauchy principal value* and is denoted by

$$p.v.\int_L f(z)dz.$$

We call this operator S. Note that the operator $S : L^p(\mathbb{C}) \to L^p(\mathbb{C})$ is bounded for all $1 < p < \infty$ and is an isometry in $L^2(\mathbb{C})$. We denote the p-norms of the operator S by $\|S\|_p$, so $\|S\|_2 = 1$. The characteristic property of this operator, and the property that makes it very important to complex analysis, is that it intertwines the ∂ and $\bar\partial$ derivatives:

$$S\circ\frac{\partial}{\partial\bar z} = \frac{\partial}{\partial z}. \tag{6.16}$$

In fact it was Bojarski [92] who first gave the elegant analytic proof of the existence and uniqueness of the Beltrami equation using this operator. He developed the L^p theory of planar quasiconformal mappings, the generalizations of which form a large part of this book. It is fair to say he, along with Gehring in higher dimensions, was the pioneer of the L^p theory of quasiconformal mappings.

One of the major recent advances in the higher-dimensional theory of quasiconformal mappings was based on finding an operator analogous to S and developing its theory [136, 255, 256]. Closely related to the operator S is the complex Riesz potential

$$T\omega(z) = \frac{1}{2\pi i}\iint_{\mathbb{C}} \frac{\omega(\zeta)d\zeta\wedge d\bar\zeta}{\zeta - z}\,. \tag{6.17}$$

Green's formula gives $\bar\partial T\omega = \omega$ and $\partial T\omega = S\omega$ for smooth ω with compact support. The density of smooth functions in L^p and the L^p boundedness of these operators imply that all the above formulas hold whenever $\omega \in L^p(\mathbb{C})$. Now a solution f to the complex Beltrami equation with compactly supported μ can be found in the form $f(z) = z + T\omega(z)$, where

$$\omega = \mu + \mu S\omega \quad \text{almost everywhere}\,. \tag{6.18}$$

The integral equation at (6.18) is solved by the Neumann iteration procedure in $L^p(\mathbb{C})$ if

$$\|\mu\|_x\|S\|_p < 1\,. \tag{6.19}$$

In fact it is the invertibility of the Beltrami operator $I - \mu S : L^p(\mathbb{C}) \to L^p(\mathbb{C})$ that determines the L^p theory of solutions to the Beltrami equation, as was first observed by Bojarski [92]. The above representation of the solution f to the Beltrami equation shows that $f \in W^{1,p}_{\text{loc}}$ for all $p > 2$ such that $\|\mu\|_x \|S\|_p < 1$. Since $\|S\|_2 = 1$ and $\|\mu\|_x < 1$, this implies a higher degree of regularity than the initial assumption that f lies in $W^{1,2}_{loc}(\mathbb{C})$. It is not too difficult to go from the representation formula to the existence theorem. The existence theorem for quasiconformal mappings, more recently called the "measurable Riemann mapping theorem," is one of the most fundamental results in the theory and has come to play a central role in modern complex analysis, Teichmüller theory, and complex dynamics.

Theorem 6.2.2 *Let μ be a measurable function defined in a domain $U \subset \mathbb{C}$ and suppose $\|\mu\|_x < 1$. Then there is a quasiconformal mapping $g : U \to \mathbb{C}$ whose complex distortion is equal to μ almost everywhere. Moreover, every $W^{1,2}_{loc}(U, \mathbb{C})$ solution f to the Beltrami equation is of the form*

$$f(z) = F(g(z)),$$

where $F : g(U) \to \mathbb{C}$ is a holomorphic function.

In the book by Kari Astala, Tadeusz Iwaniec, and Gaven Martin's [34, Chapter 5, p.161] has been given a complete proof, in as elementary fashion as possible, of the classical existence theory for quasiconformal mappings, or what has come to be known as the measurable Riemann mapping theorem, originally due to Morrey in 1938 [383]. This theorem is a cornerstone in the various interactions among PDEs, complex analysis and geometry, the basic themes of the monograph [34].

Roughly, it states that any bounded measurable conformal structure—or any measurable ellipse field—is equivalent to the usual (round) structure via a quasiconformal change of coordinates. The result underpins the relationship between quasiconformal mappings and Teichmüller theory, holomorphic dynamics, Kleinian groups, as well as the interactions between quasiconformal mappings and nonlinear elliptic PDEs in two dimensions.

The significant generalizations of the existence and uniqueness theorem (Theorem 6.2.2), following from the work of David [128], are given in [257].

6.3 The Problem and Definition of Grötzsch

The notion of a quasiconformal mapping, but not the name, was introduced by H. Grötzsch in 1928. If Q is a square and R is a rectangle, not a square, there is no conformal mapping of Q onto R that maps vertices on vertices. Instead, Grötzsch asks for the most nearly conformal mapping of this kind. This calls for a measure of approximate conformality, and in supplying such a measure Grötzsch took the first step toward the creation of a theory of q.c. mappings.

All the work of Grötzsch was late to gain recognition, and this particular idea was regarded as a curiosity and allowed to remain dormant for several years. It reappears in 1935 in the work of Lavrentiev, but from the point of view of partial differential equations. In 1936 Ahlfors included a reference to the q.c. case in his theory of covering surfaces. From then on the notion became generally known, and in 1937 Teichmüller began to prove important theorems by use of q.c. mappings, and later theorems about q.c. mappings.

We return to the definition of Grötzsch. Let $w = f(z)$ ($z = x + iy$, $w = u + iv$) be a C^1 homeomorphism from one region to another. At a point z_0, it induces a linear mapping of the differentials

$$\begin{aligned} du &= u_x dx + u_y dy \\ dv &= v_x dx + v_y dy, \end{aligned} \tag{6.20}$$

which we can also write in the complex form

$$dw = f_z dz + f_{\bar{z}} d\bar{z} \tag{6.21}$$

with

$$f_z = 1/2\left(f_x - i f_y\right), \quad f_{\bar{z}} = 1/2\left(f_x + i f_y\right). \tag{6.22}$$

Geometrically, (6.20) represents an affine transformation from the (dx, dy) to the (du, dv) plane. It maps circles about the origin into similar ellipses. We wish to compute the ratio between the axes as well as their direction. In classical notation, one writes

$$du^2 + dv^2 = E dx^2 + 2F dx dy + G dy^2 \tag{6.23}$$

with

$$E = u_x^2 + v_x^2, \quad F = u_x u_y + v_x v_y, \quad G = u_y^2 + v_y^2.$$

The eigenvalues are determined from

$$\begin{vmatrix} E - \lambda & F \\ F & G - \lambda \end{vmatrix} = 0 \tag{6.24}$$

and are

$$\lambda_1, \lambda_2 = \frac{E + G \pm \left[(E - G)^2 + 4F^2\right]^{1/2}}{2}. \tag{6.25}$$

The ratio $a : b$ of the axes is

$$\left(\frac{\lambda_1}{\lambda_2}\right)^{1/2} = \frac{E + G + \left[(E - G)^2 + 4F^2\right]^{1/2}}{2\left(EG - F^2\right)^{1/2}}. \tag{6.26}$$

The complex notation is much more convenient. Let us first note that

$$\begin{aligned} f_z &= 1/2\left(u_x + v_y\right) + i/2\left(v_x - u_y\right) \\ f_{\bar z} &= 1/2\left(u_x - v_y\right) + i/2\left(v_x + u_y\right). \end{aligned} \tag{6.27}$$

This gives

$$|f_z|^2 - |f_{\bar z}|^2 = u_x v_y - u_y v_x = J, \tag{6.28}$$

which is the Jacobian. The Jacobian is positive for sense-preserving and negative for sense-reversing mappings. For the moment we shall consider only the sense-preserving case. Then $|f_{\bar z}| < |f_z|$. It now follows immediately from (6.21) that

$$(|f_z| - |f_{\bar z}|)\,|dz| \le |dw| \le (|f_z| + |f_{\bar z}|)\,|dz|, \tag{6.29}$$

where both limits can be attained. We conclude that the ratio of the major to the minor axis is

$$D_f = \frac{|f_z| + |f_{\bar z}|}{|f_z| - |f_{\bar z}|} \ge 1. \tag{6.30}$$

This is called the *dilation* at the point z. It is often more convenient to consider

$$d_f = \frac{|f^{-1}|}{|f_z|} < 1 \tag{6.31}$$

related to D_f by

$$D_f = \frac{1 + d_f}{1 - d_f}, \quad d_f = \frac{D_f - 1}{D_f + 1}. \tag{6.32}$$

The mapping is conformal at z if and only if $D_f = 1$, $d_f = 0$. The maximum is attained when the ratio

$$\frac{f_{\bar z} d\bar z}{f_z dz}$$

is positive, and the minimum when it is negative. We introduce now the *complex dilation*

$$\mu_f = \frac{f_{\bar{z}}}{f_z} \tag{6.33}$$

with $|\mu_f| = d_f$. The maximum corresponds to the direction

$$\arg dz = \alpha = 1/2 \arg \mu, \tag{6.34}$$

and the minimum to the direction $a \pm \pi/2$. In the dw-plane the direction of the major axis is

$$\arg dw = \beta = 1/2 \arg \nu, \tag{6.35}$$

where we have set

$$\nu_f = \frac{f_{\bar{z}}}{\overline{f_{\bar{z}}}} = \left(\frac{f_z}{|f_z|}\right)^2 \mu_f. \tag{6.36}$$

The quantity ν_f may be called the *second complex dilation*. We will illustrate this idea by the following self-explanatory figure:

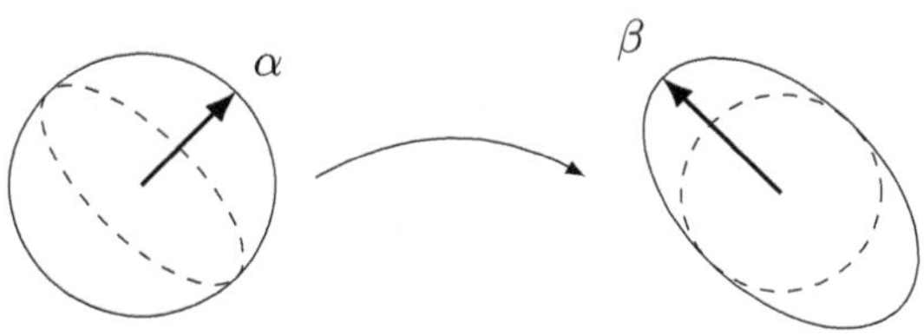

Observe that $\beta - \alpha = \arg f_z$.

Definition 6.3.1 The mapping f is said to be *quasiconformal* if D_f is bounded. It is K-quasiconformal if $D_f \leq K$.

The condition $D_f \leq K$ is equivalent to $d_f \leq k = (K - 1)/(K + 1)$. A 1-quasiconformal mapping is conformal. Let it be said at once that the restriction to C^1-mappings is most unnatural. One of our immediate aims is to dispense with this restriction. For the moment, however, we prefer to push this difficulty aside.

6.4 Solution of Grötzsch's Problem

We pass to Grötzsch's problem and give it a precise meaning by saying that f is most nearly conformal if $\sup D_f$ is as small as possible. Let R, R' be two rectangles with sides a, b and a', b'. We may assume that $a : b \leq a' : b'$ (otherwise, interchange a and b). The mapping f is supposed to take a-sides into a-sides and b-sides into b-sides.

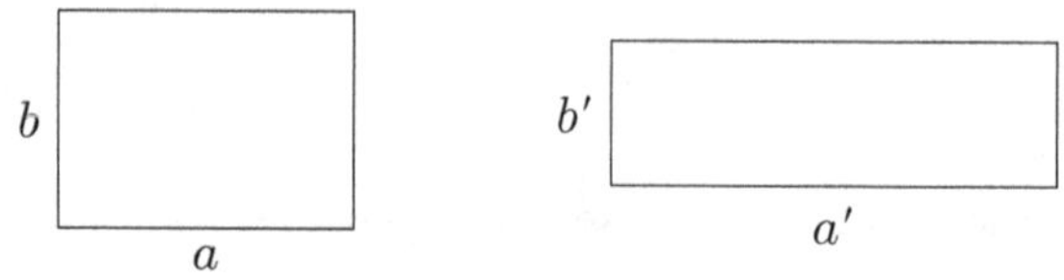

The computation is as follows:

$$
\begin{aligned}
a' &\leq \int_0^a |df(x+iy)| \leq \int_0^a (|f_z| + |f_{\bar{z}}|)\, dx \\
a'b &\leq \int_0^a \int_0^b (|f_z| + |\, f_{\bar{z}})\, dxdy \\
a'^2 b^2 &\leq \int_0^a \int_0^b \frac{|f_z| + |f_{\bar{z}}|}{|f_z| - |f_{\bar{z}}|} dxdy \int_0^a \int_0^b \left(|f_z|^2 - |f_{\bar{z}}|^2\right) dxdy \\
&= a'b' \int_0^a \int_0^b D_f dxdy
\end{aligned}
$$

or

$$
\frac{a'}{b'} : \frac{a}{b} \leq \frac{1}{ab} \iint_R D_f dxdy \tag{6.37}
$$

and in particular

$$
\frac{a'}{b'} : \frac{a}{b} \leq \sup D_f .
$$

The minimum is attained for the affine mapping that is given by

$$
f(z) = 1/2 \left(\frac{a'}{a} + \frac{b'}{b} \right) z + 1/2 \left(\frac{a'}{a} - \frac{b'}{b} \right) \bar{z}.
$$

Theorem 6.4.1 *The affine mapping has the least maximal and the least average dilation. The ratios $m = a/b$ and $m' = a'/b'$ are called the modules of R and R' (taken with an orientation). We have proved that there exists a K-q.c. mapping of R on R' if and only if*

$$
\frac{1}{K} \leq \frac{m'}{m} \leq K. \tag{6.38}
$$

6.5 Composed Mappings

We shall determine the complex derivatives and complex dilations of a composed mapping $g \circ f$. There is the usual trouble with the notation that is most easily resolved by introducing an intermediate variable $\zeta = f(z)$.

The usual rules are applicable, and we find that

$$
\begin{aligned}
(g \circ f)_z &= (g_\zeta \circ f)\, f_z + \left(g_{\overline{\zeta}} \circ f\right) \overline{f}_z \\
(g \circ f)_{\overline{z}} &= (g_\zeta \circ f)\, f_{\overline{z}} + \left(g_{\overline{\zeta}} \circ f\right) \overline{f}_{\overline{z}} .
\end{aligned}
\tag{6.39}
$$

When solved, these equations give

$$
\begin{aligned}
g_\zeta \circ f &= \frac{1}{J}\left[(g \circ f)_z \overline{f}_{\overline{z}} - (g \circ f)_{\overline{z}} \overline{f}_z\right] \\
g_{\overline{\zeta}} \circ f &= \frac{1}{J}\left[(g \circ f)_{\overline{z}} f_z - (g \circ f)_z f_{\overline{z}}\right],
\end{aligned}
\tag{6.40}
$$

where $J = |f_z|^2 - |f_{\overline{z}}|^2$.

For $g = f^{-1}$ the formulas become

$$
\begin{aligned}
\left(f^{-1}\right)_\zeta \circ f &= \overline{f}_{\overline{z}}/J, \\
\left(f^{-1}\right)_{\overline{\zeta}} \circ f &= -f_{\overline{z}}/J.
\end{aligned}
\tag{6.41}
$$

One derives, for instance,

$$
\mu_{f^{-1}} = -\nu_f \circ f^{-1} \tag{6.42}
$$

and, on passing to the absolute values,

$$
d_{f^{-1}} = d_f \circ f^{-1}. \tag{6.43}
$$

In other words, inverse mappings have the same dilations at corresponding points.

From (6.40) we obtain

$$
\mu_g \circ f = \frac{f_z}{\overline{f_{\overline{z}}}} \frac{\mu_{g \circ f} - \mu_f}{1 - \overline{\mu}_f \mu_{g \circ f}}. \tag{6.44}
$$

If g is conformal, then $\mu_g = 0$, and we find

$$
\mu_{g \circ f} = \mu_f. \tag{6.45}
$$

If f is conformal, $\mu_f = 0$ and

$$\mu_g \circ f = \left(\frac{f'}{|f'|}\right)^2 \mu_{g \circ f}, \tag{6.46}$$

which can also be written as

$$\nu_g \circ f = \nu_{g \circ f}\,. \tag{6.47}$$

In any case, the dilation is invariant with respect to all conformal transformations.

If we set $g \circ f = h$, we find from (6.44) that

$$\mu_{h \circ f^{-1}} \circ f = \frac{f_z}{\overline{f}_{\overline{z}}} \frac{\mu_h - \mu_f}{1 - \overline{\mu}_f \mu_h}. \tag{6.48}$$

For the dilation

$$d_{h \circ f^{-1}} \circ f = \left| \frac{\mu_h - \mu_f}{1 - \mu_f \overline{\mu}_h} \right|, \tag{6.49}$$

we see that

$$\log D_{h \circ f^{-1}} \circ f = \left[\mu_h, \mu_f\right] \tag{6.50}$$

is the non-Euclidean distance (with respect to the metric $ds = \frac{2|dw|}{1-|w|^2}$ in G).

We can obviously use $\sup \left[\mu_h, \mu_f\right]$ as a *distance between the mappings* f and h (the *Teichmüller distance*). It is a metric provided one identifies mappings that differ by a conformal transformation.

The composition of a K_1-q.c. and a K_2-q.c. mapping is $K_1 K_2$-q.c.

6.6 Extremal Length

Let Γ be a family of curves in the plane. Each $\gamma \in \Gamma$ shall be a countable union of open arcs, closed arcs, or closed curves, and every closed subarc shall be rectifiable. We shall introduce a geometric quantity $\lambda(\Gamma)$, called the *extremal length* of Γ, which is a sort of average minimal length. Its importance for our topic lies in the fact that it is invariant under conformal mappings and quasi-invariant under q.c. mappings (the latter means that it is multiplied by a bounded factor).

A function ρ, defined in the whole plane, will be called *allowable* if it satisfies the following conditions:

1. $\rho \geq 0$ and measurable.
2. $A(\rho) = \iint \rho^2 dxdy \neq 0, \infty$ (the integral is over the whole plane).

For such a ρ, set

$$L_\gamma(\rho) = \int_\gamma \rho|dz|$$

if ρ is measurable on γ, $L_\gamma(\rho) = \infty$ otherwise.[1] We introduce

$$L(\rho) = \inf_{\gamma \in \Gamma} L_\gamma(\rho)$$

and

Definition 6.6.1 we have

$$\lambda(\Gamma) = \sup_\rho \frac{L(\rho)^2}{A(\rho)}$$

for all allowable ρ.

We shall say that $\Gamma_1 < \Gamma_2$ if every γ_2 contains a γ_1 (the γ_2 are fewer and longer).

Remark 6.6.2 Observe that $\Gamma_1 \subset \Gamma_2$ implies $\Gamma_2 < \Gamma_1$!

Theorem 6.6.3 *If $\Gamma_1 < \Gamma_2$, then $\lambda(\Gamma_1) \leq \lambda(\Gamma_2)$.*

Proof If $\gamma_1 \subseteq \gamma_2$, then

$$L_{\gamma_1}(\rho) \leq L_{\gamma_2}(\rho)$$
$$\inf L_{\gamma_1}(\rho) \leq \inf L_{\gamma_2}(\rho),$$

and it follows at once that $\lambda(\Gamma_1) \leq \lambda(\Gamma_2)$.

Example 6.6.4 Let Γ be the set of all arcs in a closed rectangle R that joins a pair of opposite sides.

For any ρ,

$$\int_0^a \rho(x+iy)dx \geq L(\rho)$$
$$\int\int_R \rho dxdy \geq bL(\rho)$$
$$b^2L(\rho)^2 \leq ab\int\int_R \rho^2 dxdy \leq abA(\rho)$$
$$\frac{L(\rho)^2}{A(\rho)} \leq \frac{a}{b}.$$

This proves $\lambda(\Gamma) \leq a/b$.

[1] As a function of arclength.

On the other hand, take $\rho = 1$ in R, $\rho = 0$ outside. Then $L(\rho) = a$, $A(\rho) = ab$, hence $\lambda(\Gamma) \geq a/b$. We have proved

$$\lambda(\Gamma) = \frac{a}{b}.$$

Example 6.6.5 The collection Γ is the set of all arcs in an annulus $r_1 \leq |z| \leq r_2$ that join the boundary circles.

Computation shows that:

$$\int_{r_1}^{r_2} \rho dr \geq L(\rho)\,, \qquad \iint \rho dr d\theta \geq 2\pi L(\rho)\,,$$

$$4\pi^2 L(\rho)^2 \leq 2\pi \log \frac{r_2}{r_1} \iint \rho^2 r dr d\theta\,,$$

$$\frac{L(\rho)^2}{A(\rho)} \leq \frac{1}{2\pi} \log \frac{r_2}{r_1}\,.$$

Equality holds for $\rho = 1/r$.

Example 6.6.6 *The module of an annulus.*

Let U be a doubly connected region in the finite plane with C_1 the bounded, C_2 the unbounded component of the complement. We say the closed curve γ in U *separates* C_1 and C_2 if γ has nonzero winding number about the points of C_1. Let Γ be the family of closed curves in U that separate C_1 and C_2. The module $M(U)$ equals $\lambda(\Gamma)^{-1}$. Consider, for example, the annulus $A = \{r_1 \leq |z| \leq r_2\}$. We have

$$L(\rho) \leq \int_0^{2\pi} \rho\left(re^{i\theta}\right) r d\theta\,,$$

$$\frac{L(\rho)}{r} \leq \int_0^{2\pi} \rho d\theta\,,$$

$$L(\rho) \log\left(\frac{r_2}{r_1}\right) \leq \iint \rho dr d\theta\,,$$

$$L(\rho)^2 \log^2\left(\frac{r_2}{r_1}\right) \leq 2\pi \log\left(\frac{r_2}{r_1}\right) \iint \rho^2 r dr d\theta\,,$$

$$\frac{L(\rho)^2}{A(\rho)} \leq \frac{2\pi}{\log(r_2/r_1)}\,.$$

Once again $\rho = 1/2\pi r$ gives equality. Indeed, for any $\gamma \in \Gamma$, we have

$$1 \leq |n(\gamma, 0)| = \frac{1}{2\pi} \left|\int_\gamma \frac{dz}{z}\right| \leq \frac{1}{2\pi} \int_\gamma \frac{|dz|}{|z|} = L_\rho(\gamma),$$

so $L(\rho) = 1$ and $A(\rho) = \frac{1}{2\pi} \log(r_2/r_1)$. We conclude that $M(G) = \frac{1}{2\pi} \log(r_2/r_1)$.

Suppose that all $\gamma \in \Gamma$ are contained in a region U and let ϕ be a K-quasiconformal mapping of U on U'. Let Γ' be the image set of Γ.

Theorem 6.6.7 *We have that* $K^{-1}\lambda(\Gamma) \leq \lambda\left(\Gamma'\right) \leq K\lambda(\Gamma)$.

Proof For a given $\rho(z)$, define $\rho'(\zeta) = 0$ outside U' and

$$\rho'(\zeta) = \frac{\rho}{|\phi_z| - |\phi_{\overline{z}}|} \circ \phi^{-1}$$

in U'. Then

$$\int_{\gamma'} \rho' |d\zeta| \geq \int_{\gamma} \rho |dz|$$

$$\iint \rho'^2 d\xi d\eta = \iint_{\Omega} \rho^2 \frac{|\phi_z| + |\phi_{\overline{z}}|}{|\phi_z| - |\phi_{\overline{z}}|} dxdy \leq KA(\rho).$$

This proves $\lambda' \geq K^{-1}\lambda$, and the other inequality follows by considering the inverse. □

Corollary 6.6.8 $\lambda(\Gamma)$ *is a conformal invariant.*

Let X be a topological space.

A path in X is a continuous map $\gamma : [0, 1] \to X$. It is important to note that a path in X is not just a set of points but is a function.

For example, the two paths $[0, 1] \to \mathbb{R}$ given by $t \mapsto t$ and $t \mapsto t^2$ are distinct paths in $\mathbb{R}$ joining 0 to 1.

A composition of paths may be defined when the endpoint of one agrees with the initial point of the other.

Given two paths α and β in X, we define their composition $\alpha + \beta$ to be the path in X such that

$$(\alpha + \beta)(s) = \begin{cases} \alpha(2s) & \text{if } s \in [0, 1/2], \\ \beta(2s - 1) & \text{if } s \in [1/2, 1]. \end{cases}$$

The reader should be warned that addition of paths is not commutative: If $\beta + \alpha$ is defined, then $\alpha + \beta$ need not be defined. And even if both $\beta + \alpha$ and $\alpha + \beta$ are defined, they are in general unequal.

There are two important composition principles.

Let Γ_1 and Γ_2 be families of curves in the plane. Then:

(i) $\Gamma_1 + \Gamma_2 := \{\gamma_1 + \gamma_2 \mid \gamma_1 \in \Gamma_1, \gamma_2 \in \Gamma_2\}$[2]
(ii) $\Gamma_1 \cup \Gamma_2 := \{\gamma \mid \gamma \in \Gamma_1 \text{ or } \gamma \in \Gamma_2\}$.

[2] $\gamma_1 + \gamma_2$ means "γ_1 followed by γ_2."

Theorem 6.6.9

(a) $\lambda\left(\Gamma_1+\Gamma_2\right) \geq \lambda\left(\Gamma_1\right)+\lambda\left(\Gamma_2\right)$.
(b) $\lambda\left(\Gamma_1 \cup \Gamma_2\right)^{-1} \geq \lambda\left(\Gamma_1\right)^{-1}+\lambda\left(\Gamma_2\right)^{-1}$

if Γ_1, Γ_2 *lie in disjoint measurable sets.*

Proof **Proof of (a)**. We may assume that $0<\lambda\left(\Gamma_1\right), \lambda\left(\Gamma_2\right)<\infty$ for otherwise the inequality is trivial.

We may normalize so that

$$
\begin{aligned}
L_1\left(\rho_1\right) & =A\left(\rho_1\right) \\
L_2\left(\rho_2\right) & =A\left(\rho_2\right) .
\end{aligned}
$$

Choose $\rho=\max \left(\rho_1, \rho_2\right)$. Then

$$
\begin{aligned}
& L(\rho) \geq L_1\left(\rho_1\right)+L_2\left(\rho_2\right)=A\left(\rho_1\right)+A\left(\rho_2\right) \\
& A(\rho) \leq A\left(\rho_1\right)+A\left(\rho_2\right)
\end{aligned}
$$

$\lambda=\sup \frac{L(\rho)^2}{A(\rho)} \geq A\left(\rho_1\right)+A\left(\rho_2\right)=\frac{L_1\left(\rho_1\right)^2}{A\left(\rho_1\right)}+\frac{L_2\left(\rho_2\right)^2}{A\left(\rho_2\right)}$. It follows that $\lambda \geq \lambda_1+\lambda_2$.

Proof of (b). If $\lambda=\lambda\left(\Gamma_1 \cup \Gamma_2\right)=0$, there is nothing to prove. Consider an admissible ρ with $L(\rho)>0$ and set $\rho_1=\rho$ on E_1, $\rho_2=\rho$ on E_2, 0 outside (where E_1 and E_2 are complementary measurable sets with $\Gamma_1 \subseteq E_1$, $\Gamma_2 \subseteq E_2$).

Then $L_1\left(\rho_1\right) \geq L(\rho)$, $L_2\left(\rho_2\right) \geq L(\rho)$, and $A(\rho)=A\left(\rho_1\right)+A\left(\rho_2\right)$. Thus

$$
\frac{A(\rho)}{L(\rho)^2} \geq \frac{A\left(\rho_1\right)}{L_1\left(\rho_1\right)^2}+\frac{A\left(\rho_2\right)}{L_2\left(\rho_2\right)^2}
$$

and hence

$$
\lambda^{-1} \geq \lambda_1^{-1}+\lambda_2^{-1} .
$$

□

6.7 Extremal Length on Riemann Surfaces

Let X be a Riemann surface. A Borel metric ρ on X is locally of the form $\rho(z)|dz|$ where $\rho \geq 0$ is a Borel measurable function. If γ is a rectifiable path on X, then its ρ-length is defined by

$$
\ell_\rho(\gamma)=\int_\gamma \rho(z)|dz| .
$$

Similarly the ρ-area of X is given by

$$\text{area}_\rho(X) = \int_X \rho(z)^2 |dz|^2.$$

Now let Γ be a collection of paths on X. Setting

$$\ell_\rho(\Gamma) = \inf_\Gamma \ell_\rho(\gamma),$$

we define the *extremal length* of Γ by

$$\lambda(\Gamma) = \sup_\rho \frac{\ell_\rho(\Gamma)^2}{\text{area}_\rho(X)}. \tag{6.51}$$

The supremum is taken over all Borel metrics of finite area. Clearly $\lambda(\Gamma)$ is a conformal invariant of the pair (X, Γ).

A metric is *extremal* if it realizes the supremum in (6.51).

6.8 Beurling's Criterion for an Extremal Metric

Theorem 6.8.1 (See [17, §4.7]) *Suppose the measure ρ^2 on X lies in the closed convex hull of the measures*

$$\{\rho \mid \gamma : \gamma \in \Gamma \text{ and } \ell_\rho(\gamma) = \ell_\rho(\Gamma)\}.$$

Then ρ is extremal for Γ.

Proof Consider any other Borel metric α. We may assume both α and ρ are normalized to give X area 1.

Now for any $\gamma \in \Gamma$, we have

$$\ell_\alpha(\Gamma) \le \int_\gamma \alpha = \int_\gamma (\alpha/\rho)\rho = \langle \alpha/\rho, \rho \mid \gamma \rangle,$$

where the last expression is the pairing between functions and measures. Since the probability measure ρ^2 is a convex combination of the probability measures $(\rho \mid \gamma)/\ell_\rho(\Gamma)$, we have

$$\frac{\ell_\alpha(\Gamma)}{\ell_\rho(\Gamma)} \le \left\langle \alpha/\rho, \rho^2 \right\rangle = \int_X \alpha\rho \le \left(\int \alpha^2 \int \rho^2 \right)^{1/2} = 1$$

by Cauchy-Schwarz. Thus $\ell_\alpha(\Gamma) \le \ell_\rho(\Gamma)$, and therefore ρ maximizes the ratio (6.51) of length-squared to area. □

6.8.1 *Extremal Length and Quasiconformal Mappings*

Theorem 6.8.2 (Cf. [355, §IV.3.3]) *If Γ and Γ' are related by a K-quasiconformal mapping, then*

$$\frac{1}{K}\lambda\left(\Gamma'\right) \leq \lambda(\Gamma) \leq K\lambda\left(\Gamma'\right).$$

Proof Let $f : X \to X'$ be a quasiconformal mapping sending Γ to Γ'. For each conformal metric ρ on the domain, we get a metric $\rho' = f_*(\rho)$ on X' with the same lengths and areas as ρ; in particular, $\text{area}_{\rho'}\left(X'\right) = \text{area}_\rho(X)$ and $\ell_{\rho'}\left(\Gamma'\right) = \ell_\rho(\Gamma)$.

However ρ' is generally not conformal. To make it conformal, we expand its infinitesimal unit balls (which are ellipses of oblateness at most K) to round balls. This does not decrease the ρ'-length of Γ', and it increases the ρ'-area of X by at most K. Thus $\lambda\left(\Gamma'\right) \geq \lambda(\Gamma)/K$, and the theorem follows by symmetry.

Corollary 6.8.3 *If $f : X \to Y$ is a K-quasiconformal map, then* $\text{mod} f(Q) \leq K \text{ mod } Q$ *for every quadrilateral Q on X.* □

In fact the converse is true: A map that distorts the modulus of every quadrilateral by at most K is K-quasiconformal. The converse is clear for linear maps, so it follows easily for smooth quasiconformal maps. The general case depends on the a.e. differentiability of quasiconformal maps. Using this fact, we have:

Theorem 6.8.4 *Any homeomorphism f that is a uniform limit of K-quasiconformal mappings is itself K-quasiconformal.*

Chapter 7
Manifolds

A formal prerequisite for an effective reading of this book, especially the chapter on Teichmüller spaces, is familiarity with the most basic facts of set theory, general topology, complex analysis of one and several variables, Sobolev spaces, and linear algebra. The purpose of this chapter is not to establish these results but to clarify terminology and notation, and to give the reader a survey of the material that will be assumed as known in the sequel. In addition, some of the literature is noted out where adequate information and further references can be found.

7.1 Principal Spaces of Functions

Let $\mathbb{R}$ and $\mathbb{C}$ denote the fields of real and complex numbers, respectively, with their usual topologies, and let K denote either of these fields.

Let $\Omega \subseteq \mathbb{C}$ be an open set, and let $u : \Omega \to \mathbb{C}$ be a C^1 function. Then we use the abbreviation

$$\overline{\partial} u = \frac{\partial u}{\partial \overline{z}} = \frac{1}{2}\left(\frac{\partial u}{\partial x} + i\frac{\partial u}{\partial y}\right),$$

where x, y are the canonical real coordinates on $\mathbb{C}$ and $z = x + iy$. The function $\overline{\partial} u$ is called the *Cauchy-Riemann derivative* of u. By the *Cauchy-Riemann criterion*, u is holomorphic if and only if $\overline{\partial} u = 0$.

Turning to $\mathbb{C}^n = \mathbb{R}^{2n}$ with coordinates $z_j = x_j + \sqrt{-1} y_j$, one introduces the partial differential operators

$$\frac{\partial}{\partial z_j} = \frac{1}{2}\left(\frac{\partial}{\partial x_j} + \frac{1}{i}\frac{\partial}{\partial y_j}\right), \quad \frac{\partial}{\partial \bar{z}_j} = \frac{1}{2}\left(\frac{\partial}{\partial x_j} - \frac{1}{i}\frac{\partial}{\partial y_j}\right),$$

where $\bar{z}_j := x_j - iy_j$.

P. V. Dovbush, S. G. Krantz, *The Geometric Theory of Complex Variables*,
https://doi.org/10.1007/978-3-031-77204-7_7

These operators $\partial/\partial z_j$ and $\partial/\partial\bar{z}_j$ will also be denoted by ∂_j and $\bar{\partial}_j$, respectively. The following rules are easily verified:

$$\overline{\partial f/\partial z_j} = \partial\bar{f}/\partial\bar{z}_j, \quad \overline{\partial f/\partial\bar{z}_j} = \partial\bar{f}/\partial z_j.$$

A multiorder is an n-tuple $k = (k_1, \ldots, k_n)$ of nonnegative integers. For every multiorder k, we write $z^k := z_1^{k_1} \ldots z_n^{k_n}$ if $z \in \mathbb{C}^n$, $k! := k_1! \ldots k_n!$, $\partial^k := \partial_1^{k_1} \circ \ldots \circ \partial_n^{k_0}$, and $\overline{\partial^n} := \bar{\partial}_1^{k_1} \circ \ldots \circ \bar{\partial}_n^{k_n}$. If $\Omega \subseteq \mathbb{C}^n$ is open and $f \in C^0(\Omega)$, then we define

$$\partial f := \sum_{j=1}^{n} \frac{\partial f}{\partial z_4} dz_j \quad \text{and} \quad \bar{\partial} f := \sum_{j=1}^{n} \frac{\partial f}{\partial\bar{z}_j} d\bar{z}_q \text{ in } \Omega.$$

Then the differential df of f can be expressed as $df = \partial f + \overline{\partial} f$ in Ω.

(Remark. In general, ∂f and $\bar{\partial} f$ are differential forms whose coefficients are distributions. However, in this book, we need $\partial f(\bar{\partial} f)$ only if $\partial f(\bar{\partial} f)$ is continuous.)

Let Ω be open in $\mathbb{C}^n$. The definition of $\overline{\partial}$ shows that a function $h \in \mathcal{C}^1(\Omega)$ is holomorphic if and only if $\overline{\partial} h = 0$. Thus $\overline{\partial} h = 0$ is the Cauchy-Riemann equation.

The class of all holomorphic functions with domain Ω will be denoted by $\mathcal{O}(\Omega)$.

If Ω is an open subset of K^n, we shall be concerned with the following function spaces on Ω: **(a)** $K = \mathbb{C}$:

(1) $\mathcal{O}(\Omega)$ will denote the complex-valued holomorphic functions on Ω, i.e., if $(z_1, \ldots, z_n)$ are coordinates in $\mathbb{C}^n$, then $f \in \mathcal{O}(\Omega)$ if and only if near each point $z^0 \in \Omega$, f can be represented by a convergent power series of the form

$$f(z) = f(z_1, \ldots, z_n) = \sum_{\alpha_1, \ldots, \alpha_n = 0}^{\infty} a_{\alpha_1, \ldots, \alpha_n} \left(z_1 - z_1^0\right)^{\alpha_1} \cdots \left(z_n - z_n^0\right)^{\alpha_n}.$$

(See, e.g., Gunning and Rossi [218, Chap. I], Hörmander [242, Chap. II], or Krantz [331] for the elementary properties of holomorphic functions on an open set in $\mathbb{C}^n$.)

$\mathcal{O}(\Omega)$ is a Frechét space, see, for example, [509, Chapter 10, Example 2, p. 89].

(b) $K = \mathbb{R}$:

(2) For any nonnegative integer m, let $\mathcal{E}^m(\Omega)$ denote the vector space consisting of all functions ϕ that, together with all their partial derivatives $\partial^\alpha \phi$ of orders $|\alpha| \leq m$, are continuous on Ω. We abbreviate $\mathcal{E}^0(\Omega) \equiv \mathcal{E}(\Omega)$. Let $\mathcal{E}^\infty(\Omega) = \bigcap_{m=0}^{\infty} \mathcal{E}^m(\Omega)$.

$\mathcal{E}^m(\Omega)$ is a Frechét space, see, for example, [509, Chapter 10, Example 1, p. 85].

The subspaces $\mathcal{E}_0(\Omega)$ and $\mathcal{E}_0^\infty(\Omega)$ consist of all those functions in $\mathcal{E}(\Omega)$ and $\mathcal{E}^\infty(\Omega)$, respectively, that have compact support in Ω.

(3) $\mathcal{E}^\infty(\Omega)$ will denote the real-valued infinitely differentiable functions on Ω, which we shall simply call $\mathcal{C}^\infty$ functions on Ω, i.e., $f \in \mathcal{E}(\Omega)$ if and only if f is a real-valued function such that partial derivatives of all orders exist and are continuous at all points of Ω ($\mathcal{E}^\infty(\Omega)$ is often denoted by $\mathcal{C}^\infty(\Omega)$).

We have that $\mathcal{A}(\Omega)$ will denote the real-valued, real-analytic functions on Ω; i.e., $\mathcal{A}(\Omega) \subset \mathcal{E}^\infty(\Omega)$, and $f \in \mathcal{A}(\Omega)$ if and only if the Taylor expansion of f converges to f in a neighborhood of any point of Ω.

It is a standard remark that the $\mathcal{E}^\infty$ functions on an open subset of a Euclidean space do not form a Banach space. They form a Frechét space (denumerably many norms instead of one). Note that the implicit function theorem is not true in Frechét space.

7.2 Definition of a Fréchet Space

A seminorm on a vector space F is a real-valued function $\|\cdot\| : F \to \mathbb{R}$ such that:

(i) $\|f\| \geqslant 0$ for all vectors f.
(ii) $\|f + g\| \leqslant \|f\| + \|g\|$ for all vectors f and g.
(iii) $\|cf\| = |c| \cdot \|f\|$ for all scalars c and vectors f.

A collection of seminorms $\{\|\cdot\|_n : n \in N\}$ defines a unique topology such that a sequence or net $f_j \to f$ if and only if $\|f_j - f\|_n \to 0$ for all $n \in N$.

A locally convex topological vector space is a vector space with a topology that arises from some collection of seminorms. The topology is *Hausdorff* if and only if $f = 0$ when all $\|f\|_n = 0$. The topology is *metrizable* if and only if it may be defined by a countable collection of seminorms $\{\|\cdot\|_n\}$. In this case we may always use sequences instead of nets. A sequence f_j is *Cauchy* if $\|f_j - f_k\|_n \to 0$ as j and $k \to \infty$ for all n. The space F is (sequentially) complete if every Cauchy sequence converges to an element of the space.

Definition 7.2.1 A *Fréchet space* is a complete Hausdorff metrizable locally convex topological vector space.

Example 7.2.2 Every Banach space is a Fréchet space. The collection of norms contains only one element.

Example 7.2.3 Let $\mathcal{C}(\mathbb{R})$ denote the vector space of all continuous functions on the real line. Put

$$\|f\|_n = \sup\{|f(x)| : -n \leqslant x \leqslant n\}.$$

Then $\mathcal{C}(\mathbb{R})$ is a Fréchet space.

Example 7.2.4 Let $\mathcal{HC}$ denote the vector space of entire holomorphic functions. Put

$$\|f\|_n = \sup\{|f(z)| : |z| \leqslant n\}.$$

Then $\mathcal{HC}$ is a Fréchet space.

Example 7.2.5 Let $\mathcal{C}^\infty[a, b]$ be the vector space of smooth functions on $a \leqslant x \leqslant b$. Put

$$\|f\|_n = \sum_{j=0}^{n} \sup_{x\in[a,b]} \left|D^j f(x)\right|.$$

Then $\mathcal{C}^\infty[a, b]$ is a Fréchet space.

Example 7.2.6 More generally, let X be a compact manifold and V a vector bundle over X. Let $\mathcal{C}^\infty(X, V)$ be the vector space of smooth sections of the bundle over X. Choose Riemannian metrics and connections on the bundles TX and V and let $D^j f$ denote the j th covariant derivative of a section f of V. Put

$$\|f\|_n = \sum_{j=0}^{n} \sup_{x\in X} \left|D^j f(x)\right|.$$

Then $\mathcal{C}^\infty(X, V)$ is a Fréchet space.

Properties of Fréchet Spaces A seminorm is a norm if $f = 0$ whenever $\|f\| = 0$. Some Fréchet spaces admit continuous norms and others admit none. If a Fréchet space admits one norm, then all the seminorms in a collection defining the topology may be taken to be norms (by adding the one norm to them all).

Example 7.2.7 A Banach space, $\mathcal{C}^\infty[a, b]$, $\mathcal{C}^\infty(X, V)$ for X compact, and $\mathcal{H}$ all admit norms, while $\mathcal{C}(\mathbb{R})$ does not.

A closed subspace of a Fréchet space is also a Fréchet space, as is a quotient of a Fréchet space by a closed subspace. The direct sum of two Fréchet spaces is a Fréchet space. A space G is a (topological) direct summand of F if there exists a third space H such that F is isomorphic to the direct sum $G \oplus H$. If a Fréchet space admits a norm, then so does any closed subspace, but a quotient by a closed subspace may not.

The dual of a Banach space is again a Banach space. However, the dual of a Fréchet space that is not itself a Banach space is never a Fréchet space. Hence in general the space of linear maps of one Fréchet space to another will not form a third Fréchet space. For this reason we shall always avoid taking the space of linear maps. This causes some differences from the Banach space theory as it is usually presented.

The Hahn-Banach theorem holds for Fréchet spaces. Thus if F is a Fréchet space and f is a nonzero vector in F, then we can find a continuous linear functional $l : F \to \mathbb{R}$ (or $\mathbb{C}$) such that $l(f) = 1$. Hence if $l(f) = l(g)$ for all continuous linear functionals, then $f = g$. This is a very useful tool for reducing theorems to the real-valued case. Also the open mapping theorem holds. Thus if F and G are Fréchet spaces and if $L : F \to G$ is continuous, linear, and invertible, then $L^{-1} : G \to F$ is also continuous.

Definition of the Directional Derivative.

Let F and G be Fréchet spaces, U an open subset of F, and $P : U \subseteq F \to G$ a continuous, possibly nonlinear map.

Definition 7.2.8 The derivative of P at the point $f \in U$ in the direction $h \in F$ is defined by

$$DP(f)h = \lim_{t \to 0} \frac{P(f + th) - P(f)}{t}.$$

We say that P is differentiable at f in the direction h if the limit exists. We say that P is *continuously differentiable* (or $\mathcal{C}^1$) on U if the limit exists for all $f \in U$ and all $h \in F$ and if $DP : (U \subseteq F) \times F \to G$ is continuous (jointly as a function on a subset of the product).

Note well that this definition of a $\mathcal{C}^1$ map does not agree with the usual definition for a Banach space. The derivative will of course be the same map, but our continuity requirement is weaker.

Similar definitions apply to the higher derivatives. The nth derivative

$$D^n P(f)\{h_1, h_2, \dots, h_n\}$$

will be regarded as a map

$$D^n P : (U \subseteq F) \times F \times \cdots \times F \to G.$$

We say that P is of class C^n if $\mathbb{D}^n P$ exists and is continuous (jointly as a function on the product space).

Example 7.2.9 Let $f : (a, b) \subseteq R \to F$ be a parametrized curve. The simple vector space R contains the distinguished vector $\mathbf{1}$ that spans the space, and $f'(t) = Df(t)\mathbf{1}$. Thus our two notions of differentiability coincide.

Example 7.2.10 Let L be a continuous linear map. Then L is $\mathcal{C}^1$ and $DL(f)h = Lh$.

Proof By linearity $L(f + th) = L(f) + tL(h)$. Then by the definition

$$DL(f)h = \lim_{t \to 0}[L(f + th) - L(f)]/t = L(h).$$

□

Example 7.2.11 Let $P : U \subseteq \mathbb{R}^2 \to \mathbb{R}^2$ and write

$$\begin{pmatrix} z \\ w \end{pmatrix} = P \begin{pmatrix} x \\ y \end{pmatrix}.$$

Then

$$DP\begin{pmatrix} x \\ y \end{pmatrix}\begin{pmatrix} u \\ v \end{pmatrix} = \begin{pmatrix} \partial z/\partial x & \partial z/\partial y \\ \partial w/\partial x & \partial w/\partial y \end{pmatrix}\begin{pmatrix} u \\ v \end{pmatrix},$$

and the directional derivative is given by the matrix of partial derivatives.

Example 7.2.12 Let U be a relatively open subset of $[a, b] \times \mathbb{R}$ and let $p(x, y)$ be a smooth function on U. Let $\widetilde{U}$ be the open subset of the Fréchet space $\mathcal{C}^\infty([a, b])$ consisting of all functions whose graphs $y = f(x)$ lie in U. Define a nonlinear map $P : \widetilde{U} \subseteq \mathcal{C}^\infty[a, b] \to \mathcal{C}^\infty[a, b]$ by setting

$$P(f)(x) = p(x, f(x))$$

Then the map P is $\mathcal{C}^1$ (and indeed $\mathcal{C}^\infty$) and

$$[DP(f)h](x) = D_y p(x, f(x))h(x).$$

We call such a map P a *vector bundle operator*. Choose local coordinates x on X and Y in the fiber of V and z in the fiber of W. Then p is given locally by $z = p(x, y)$. A section f of the bundle V is given locally by $y = f(x)$. The operator P is given locally by

$$[P(f)](x) = p(x, f(x)).$$

Except for a fancier setting, the situation is the same as before. The derivative is given locally by

$$[DP(f)h](x) = D_y p(x, f(x))h(x),$$

where $D_y p$ is interpreted as the matrix of partial derivatives with respect to the y variable and is applied to the vector h. It is easy to see that P is $\mathcal{C}^0$. For a proof that P is $\mathcal{C}^1$ and for a proof that P is $\mathcal{C}^\infty$, see, e.g., [510].

7.3 Distributions and Weak Derivatives

Sobolev spaces are vector spaces whose elements are functions defined on domains in n-dimensional Euclidean space $\mathbb{R}^n$ and whose partial derivatives satisfy certain integrability conditions. In order to develop and elucidate the properties of these spaces and mappings between them, we require some of the machinery of general topology and real and functional analysis. We assume that readers are familiar with the concept of a vector space over the real or complex scalar field, and with the related notions of dimension, subspace, linear transformation, and convex set. We also expect that the reader will have some familiarity with the concept of topology on a set, at least to the extent of understanding the concepts of an open set and continuity of a function.

We require in subsequent chapters some of the basic concepts and techniques of the Schwartz theory of distributions [464], and we present here a brief description of those aspects of the theory that are relevant for our purposes. Of special importance is the notion of weak or distributional derivative of an integrable function.

One of the standard definitions of Sobolev spaces is phrased in terms of such derivatives.

(*Test Functions*) Let Ω be a domain in $\mathbb{R}^n$. A sequence $\{\phi_j\}$ of functions belonging to $C_0^\infty(\Omega)$ is said to *converge* in the sense of the space $\mathcal{D}(\Omega)$ to the function $\phi \in C_0^\infty(\Omega)$ provided the following conditions are satisfied:

(i) There exists $K \Subset \Omega$ such that $\operatorname{supp}\left(\phi_j - \phi\right) \subset K$ for every j.
(ii) $\lim_{j\to\infty} D^\alpha \phi_j(x) = D^\alpha \phi(x)$ uniformly on K for each multi-index α.

There is a locally convex topology on the vector space $C_0^\infty(\Omega)$ with respect to which a linear functional T is continuous if and only if $T\left(\phi_j\right) \to T(\phi)$ in $\mathbb{C}$ whenever $\phi_j \to \phi$ in the sense of the space $\mathcal{D}(\Omega)$. Equipped with this topology, $C_0^\infty(\Omega)$ becomes a topological vector space called $\mathcal{D}(\Omega)$ whose elements are called *test functions*. The space $\mathcal{D}(\Omega)$ is not a normable space. (We ignore the question of uniqueness of the topology asserted above. It uniquely determines the dual of $\mathcal{D}(\Omega)$ that is sufficient for our purposes.)

(*Schwartz Distributions*) The dual space $\mathcal{D}'(\Omega)$ of $\mathcal{D}(\Omega)$ is called the *space of (Schwartz) distributions* on Ω. The space $\mathcal{D}'(\Omega)$ is given the weak-star topology as the dual of $\mathcal{D}(\Omega)$ and is a locally convex topological vector space with that topology.

We summarize the vector space and convergence operations in $\mathcal{D}'(\Omega)$ as follows: If S, T, T_j belong to $\mathcal{D}'(\Omega)$ and $c \in \mathbb{C}$, then

$$
\begin{aligned}
(S+T)(\phi) &= S(\phi) + T(\phi), \quad && \phi \in \mathcal{D}(\Omega),\\
(cT)(\phi) &= cT(\phi), && \phi \in \mathcal{D}(\Omega),
\end{aligned}
$$

$T_j \to T$ in $\mathcal{D}'(\Omega)$ if and only if $T_j(\phi) \to T(\phi)$ in $\mathbb{C}$ for every $\phi \in \mathcal{D}(\Omega)$.

(*Locally Integrable Functions*) A function u defined almost everywhere on Ω is said to be *locally integrable* on Ω provided that $u \in L^1(U)$ for every open $U \Subset \Omega$. In this case we write $u \in L^1_{loc}(\Omega)$. Corresponding to every $u \in L^1_{loc}(\Omega)$ there is a distribution $T_u \in \mathcal{D}'(\Omega)$ defined by

$$
T_u(\phi) = \int_\Omega u(x)\phi(x)dx, \quad \phi \in \mathcal{D}(\Omega). \tag{7.1}
$$

Evidently T_u thus defined is a linear functional on $\mathcal{D}(\Omega)$. To see that it is continuous, suppose that $\phi_j \to \phi$ in $\mathcal{D}(\Omega)$. Then there exists $K \Subset \Omega$ such that $\operatorname{supp}\left(\phi_j - \phi\right) \subset K$ for all j. Thus

$$\left|T_u\left(\phi_j\right) - T_u(\phi)\right| \leq \sup_{x \in K} \left|\phi_j(x) - \phi(x)\right| \int_K |u(x)| dx.$$

The right side of the above inequality tends to zero as $j \to \infty$ since $\phi_j \to \phi$ uniformly on K.

Not every distribution $T \in \mathcal{D}'(\Omega)$ is of the form T_u defined by (7.1) for some $u \in L^1_{\text{loc}}(\Omega)$.

Indeed, if $0 \in \Omega$, there can be no locally integrable function δ on Ω such that for every $\phi \in \mathcal{D}(\Omega)$

$$\int_\Omega \delta(x)\phi(x)dx = \phi(0).$$

However, the linear functional δ defined on $\mathcal{D}(\Omega)$ by

$$\delta(\phi) = \phi(0) \tag{7.2}$$

is easily seen to be continuous and hence a distribution on Ω. It is called a *Dirac distribution*.

(*Derivatives of Distributions*) Let $u \in \mathcal{C}^1(\Omega)$ and $\phi \in \mathcal{D}(\Omega)$. Since ϕ vanishes outside some compact subset of Ω, we obtain by integration by parts in the variable x_j

$$\int_\Omega \left(\frac{\partial}{\partial x_j} u(x)\right) \phi(x) dx = - \int_\Omega u(x) \left(\frac{\partial}{\partial x_j} \phi(x)\right) dx .$$

Similarly, if $u \in C^{|\alpha|}(\Omega)$, then integration by parts $|\alpha|$ times leads to

$$\int_\Omega \left(D^\alpha u(x)\right) \phi(x) dx = (-1)^{|\alpha|} \int_\Omega u(x) D^\alpha \phi(x) dx.$$

By α we have denoted a multi-index $\alpha = (\alpha_1, \ldots, \alpha_n)$. By $|\alpha|$, we have denoted $\alpha_1 + \cdots + \alpha_n$.

This motivates the following definition of the derivative $D^\alpha T$ of a distribution $T \in \mathcal{D}'(\Omega)$:

$$\left(D^\alpha T\right)(\phi) = (-1)^{|\alpha|} T \left(D^\alpha \phi\right). \tag{7.3}$$

Since $D^\alpha \phi \in \mathcal{D}(\Omega)$ whenever $\phi \in \mathcal{D}(\Omega)$, $D^\alpha T$ is a functional on $\mathcal{D}(\Omega)$, and it is clearly linear. We show that it is continuous, and hence a distribution on Ω. To this end, suppose that $\phi_j \to \phi$ in $\mathcal{D}(\Omega)$. Then

$$\operatorname{supp}\left(D^\alpha \left(\phi_j - \phi\right)\right) \subset \operatorname{supp}\left(\phi_j - \phi\right) \subset K$$

for some $K \Subset \Omega$. Moreover,

$$D^{\beta}\left(D^{\alpha}\left(\phi_j-\phi\right)\right)=D^{\beta+\alpha}\left(\phi_j-\phi\right)$$

converges to zero uniformly on K as $j \to \infty$ for each multi-index β. Hence $D^{\alpha}\phi_j \to D^{\alpha}\phi$ in $\mathcal{D}(\Omega)$. Since $T \in \mathcal{D}'(\Omega)$, it follows that

$$D^{\alpha}T\left(\phi_j\right)=(-1)^{|\alpha|}T\left(D^{\alpha}\phi_j\right)\to(-1)^{|\alpha|}T\left(D^{\alpha}\phi\right)=D^{\alpha}T(\phi)$$

in $\mathbb{C}$. Thus $D^{\alpha}T \in \mathcal{D}'(\Omega)$.

We have shown that every distribution in $\mathcal{D}'(\Omega)$ possesses derivatives of all orders in $\mathcal{D}'(\Omega)$ in the sense of definition (7.3). Furthermore, the mapping D^{α} from $\mathcal{D}'(\Omega)$ into $\mathcal{D}'(\Omega)$ is continuous; if $T_j \to T$ in $\mathcal{D}'(\Omega)$ and $\phi \in \mathcal{D}(\Omega)$, then

$$D^{\alpha}T_j(\phi)=(-1)^{|\alpha|}T_j\left(D^{\alpha}\phi\right)\to(-1)^{|\alpha|}T\left(D^{\alpha}\phi\right)=D^{\alpha}T(\phi)\,.$$

Example 7.3.1 If $0 \in \Omega$ and $\delta \in \mathcal{D}'(\Omega)$ is the Dirac distribution defined by (7.2), then $D^{\alpha}\delta$ is given by

$$D^{\alpha}\delta(\phi)=(-1)^{|\alpha|}D^{\alpha}\phi(0).$$

Example 7.3.2 Here is the most famous example in the theory: If $\Omega = \mathbb{R}$ (i.e., $n = 1$) and $H \in L^1_{\text{loc}}(\mathbb{R})$ is the Heaviside function defined by

$$H(x)=\begin{cases}1 & \text{if } x \geq 0\\ 0 & \text{if } x<0,\end{cases}$$

then the derivative $(T_H)'$ of the corresponding distribution T_H is δ. To see this, let $\phi \in \mathcal{D}(\mathbb{R})$ have support in the interval $[-a, a]$. Then

$$\left(T_H\right)'(\phi)=-T_H\left(\phi'\right)=-\int_0^a \phi'(x)dx=\phi(0)=\delta(\phi).$$

(*Weak Derivatives*) We now define the concept of a function being the weak derivative of another function. Let $u \in L^1_{loc}(\Omega)$. There may or may not exist a function $v_\alpha \in L^1_{loc}(\Omega)$ such that $T_{v_\alpha} = D^{\alpha}T_u$ in $\mathcal{D}'(\Omega)$. If such a v_α exists, it is unique up to sets of measure zero and is called the *weak or distributional partial derivative* of u and is denoted by $D^{\alpha}u$. Thus $D^{\alpha}u = v_\alpha$ in the weak (or distributional) sense provided $v_\alpha \in L^1_{\text{loc}}(\Omega)$ satisfies

$$\int_{\Omega}u(x)D^{\alpha}\phi(x)dx=(-1)^{|\alpha|}\int_{\Omega}v_\alpha(x)\phi(x)dx$$

for every $\phi \in \mathcal{D}(\Omega)$.

If u is sufficiently smooth to have a continuous partial derivative $D^\alpha u$ in the usual (classical) sense, then $D^\alpha u$ is also a weak partial derivative of u. Of course $D^\alpha u$ may exist in the weak sense without existing in the classical sense. Certain functions having weak derivatives (those in Sobolev spaces) can be suitably approximated by smooth functions.

Let us note in conclusion that distributions in Ω can be multiplied by smooth functions. If $T \in \mathcal{D}'(\Omega)$ and $\omega \in C^\infty(\Omega)$, then the *product* $\omega T \in \mathcal{D}'(\Omega)$ is defined by

$$(\omega T)(\phi) = T(\omega\phi), \quad \phi \in \mathcal{D}(\Omega).$$

If $T = T_u$ for some $u \in L^1_{loc}(\Omega)$, then $\omega T = T_{\omega u}$. The Leibniz rule is easily checked to hold for $D^\alpha(\omega T)$.

7.4 Definitions of Sobolev Spaces

Let Ω be a domain in $\mathbb{R}^n$. We denote by $L^2(\Omega)$ the class of all measurable functions u, defined on Ω, for which

$$\int_\Omega |u(x)|^2 dx < \infty. \tag{7.4}$$

We identify in $L^2(\Omega)$ functions that are equal almost everywhere on Ω. The elements of $L^2(\Omega)$ are thus actually equivalence classes of measurable functions satisfying (7.4), two functions being equivalent if they are equal a.e. in Ω. For convenience, however, we ignore this distinction and write $u \in L^2(\Omega)$ if u satisfies (7.4), and $u = 0$ in $L^p(\Omega)$ if $u(x) = 0$ a.e. in Ω. It is clear that if $u \in L^2(\Omega)$ and $c \in \mathbb{C}$, then $cu \in L^2(\Omega)$. Moreover, if $u, v \in L^2(\Omega)$, then since

$$|u(x) + v(x)|^2 \le (|u(x)| + |v(x)|)^2 \le 4\left(|u(x)|^2 + |v(x)|^2\right),$$

then $u + v \in L^2(\Omega)$, so $L^2(\Omega)$ is a vector space.

The functional $\|\cdot\|_2$ defined by

$$\|u\|_2 = \left\{\int_\Omega |u(x)|^2 dx\right\}^{1/2}$$

is a norm on $L^2(\Omega)$.

Sobolev spaces are defined over an arbitrary domain $\Omega \subset \mathbb{R}^n$ and are vector subspaces of various Lebesgue spaces $L^p(\Omega)$.

(*The Sobolev Norms*) We define a functional $\|\cdot\|_S$, where S is a positive integer, as follows:

$$\|u\|_S = \left(\sum_{0 \le |\alpha| \le S} \|D^\alpha u\|_2^2 \right)^{1/2}, \tag{7.5}$$

for any function u for which the right side makes sense, $\|\cdot\|_2$ being, of course, the norm in $L^2(\Omega)$. By α we have denoted an n-tuple $\alpha = (\alpha_1, \ldots, \alpha_n)$ of positive integers; we shall denote by $|\alpha|$ the integer $\alpha_1 + \cdots + \alpha_n$.

Evidently, (7.5) defines a norm on any vector space of functions on which the right side takes finite values provided functions are identified in the space if they are equal almost everywhere in Ω.

The norm (7.5) turns $H^S(\Omega)$ into a Hilbert space (see, for example, [509, Chapter 31, Proposition 31.1, Corollary p. 322]), hence a Fréchet space.

We let $H^S(\Omega) = H^S(\Omega, \mathbb{R})$ denote the linear space of all (real valued) functions on Ω whose derivatives up to order S exist in the sense of distributions and are square integrable functions. We say that these functions belong to the *Sobolev class* H^S.

The reader is referred to [5, 339, 376, 509] for more complete treatments of the spaces $\mathcal{D}(\Omega)$, $\mathcal{D}'(\Omega)$ and Sobolev spaces introduced above, as well as useful generalizations of these spaces.

7.5 Definition of Manifolds

We begin with the most elementary aspects and primitive motivations.

We say that a *topological structure* or, simply, a *topology* is defined on a set X if there is given a class of subsets of X that contains the union of any collection in the class and the intersection of any finite collection in the class. A set endowed with a topological structure is called a *topological space*, its elements—points, and the sets of the given class—are *open sets*.

A subset of a topological space is *closed* if its complement is open. The class of closed sets contains the intersection of any collection of sets from the class, and it contains the union of any finite collection of sets from the class. Moreover, given any class of subsets of a set X with these properties, there exists a unique topology on X such that the given class is the class of all closed sets.

A *neighborhood* of a point of a topological space is any open set containing the given point. A neighborhood of a subset of a topological space is any open set containing the given subset.

The topological spaces that are studied in differential geometry are those that look, locally, like Euclidean space.

We shall begin this section with some basic definitions in which we shall use the following standard notations. A more detailed discussion of the basic theory of manifolds can be found in Griffiths-Harris [205], Narasimhan [394], and Wells [523].

These particular classes of functions will be used to define the particular classes of manifolds that we shall be interested in.

A topological *n-manifold* is a Hausdorff topological space with a countable basis that is locally homeomorphic to an open subset of $\mathbb{R}^n$. The integer n is called the *topological dimension* of the manifold. Suppose that $\mathcal{S}$ is one of the three K-valued families of functions defined on the open subsets of K^n described above, where we let $\mathcal{S}(\Omega)$ denote the functions of $\mathcal{S}$ defined on Ω, an open set in K^n. (That is, $\mathcal{S}(\Omega)$ is either $\mathcal{E}(\Omega)$, $\mathcal{A}(\Omega)$, or $\mathcal{O}(\Omega)$. We shall only consider these three examples in this chapter.)

Definition 7.5.1 An $\mathcal{S}$-structure, $\mathcal{S}_M$, on a K-manifold M is a family of K-valued continuous functions defined on the open sets of M such that:

(a) For every $p \in M$, there exists an open neighborhood U of p and a homeomorphism $h : U \to U'$, where U' is open in K^n, such that for any open set $V \subset U$

$$f : V \longrightarrow K \in \mathcal{S}_M \text{ if and only if } f \circ h^{-1} \in \mathcal{S}(h(V)).$$

(b) If $f : U \to K$, where $U = \cup_i U_i$ and U_i is open in M, then $f \in \mathcal{S}_M$ if and only if $f|_{U_i} \in \mathcal{S}_M$ for each i.

It follows clearly from **(a)** that if $K = \mathbb{R}$, the dimension, k, of the topological manifold is equal to n, and if $K = \mathbb{C}$, then $k = 2n$. In either case n will be called the K-dimension of M, denoted by $\dim_K M = n$ (which we shall call the *real dimension* and *complex-dimension*, respectively). A manifold with an $\mathcal{S}$-structure is called an $\mathcal{S}$-manifold, denoted by $(M, \mathcal{S}_M)$, and the elements of $\mathcal{S}_M$ are called $\mathcal{S}$ functions on M. An open subset $U \subset M$ and a homeomorphism $h : U \to U' \subset K^n$ as in **(a)** above are called an $\mathcal{S}$-coordinate system.

For our three classes of functions, we have defined:

(a) $\mathcal{S} = \mathcal{E}$: A differentiable (or $\mathcal{C}^\infty$) manifold, and the functions in $\mathcal{E}_M$ are called $\mathcal{C}^\infty$ *functions* on open subsets of M.
(b) $\mathcal{S} = \mathcal{A}$: A real-analytic manifold, and the functions in $\mathcal{A}_M$ are called *real-analytic functions* on open subsets of M.
(c) $\mathcal{S} = \mathcal{O}$: A complex analytic (or simply complex) manifold, and the functions in $\mathcal{O}_M$ are called *holomorphic* (or *complex analytic*) *functions* on open subsets of M.

We shall refer to $\mathcal{E}_M$, $\mathcal{A}_M$, and $\mathcal{O}_M$ as differentiable, real-analytic, and complex structures, respectively.

Remark 7.5.2 **(d)** Abstract manifolds.

The usual definition of a manifold generalizes directly to Fréchet space calculus.

Definition 7.5.3 A Fréchet manifold is a Hausdorff topological space with an atlas of coordinate charts taking their values in Fréchet spaces, such that the coordinate transition functions are all smooth maps between Fréchet spaces.

Example 7.5.4 Let X be a compact finite-dimensional manifold. A *bundle* over X is another finite-dimensional manifold B with a smooth projection map $\pi : B \to X$ whose derivative is everywhere surjective. A *section* of the bundle is a smooth map $f : X \to B$ such that $\pi f = 1$ is the identity. The space of all sections of the bundle $\mathcal{C}^\infty(X, B)$ is a Fréchet manifold (at least if it is not empty). Associated to each section f is a vector bundle over X called the *vertical tangent bundle* to B at f, which we denote by $T_v B(f)$. Its fiber at a point $x \in X$ consists of all the tangent vectors of B at $f(x)$ that lie in the null space of the derivative of π.

It is easy to construct a diffeomorphism from a neighborhood of the zero section of the vector bundle $T_v B(f)$ to a neighborhood of the image of f in B that takes fibers into fibers over the same point. This provides a one-to-one correspondence between sections near zero in the Fréchet space $\mathcal{C}^\infty(X, T_v B(f))$ and sections near f in the manifold $\mathcal{C}^\infty(X, B)$, and these maps serve as our coordinate charts. The coordinate transition functions are clearly vector bundle maps in the sense of Example 7.2.12. Hence, $\mathcal{C}^\infty(X, B)$ is a Fréchet manifold.

Example 7.5.5 Let X and Y be manifolds with X compact. The space of all smooth maps of X into Y forms a Fréchet manifold $\mathfrak{M}(X, Y)$. This is a special case of the preceding example, as a map of X into Y is a section of the product bundle $X \times Y$ over X.

In the same way one can define a Banach manifold.

It is a manifold modeled on Banach spaces. A Banach manifold is a topological space in which each point has a neighborhood homeomorphic to an open set in a Banach space. Banach manifolds are one possibility for extending manifolds to infinite dimensions.

A $\mathcal{C}^\infty$ *Banach manifold* M is a paracompact topological Hausdorff space, together with a maximal collection of open subsets $(U_i)_{i\in I}$, $\bigcup_{i\in I} U_i = M$, and of homeomorphisms $\varphi_i : U_i \to E$, where E is a Banach space, such that $\varphi_i \circ \varphi_j^{-1}$ is $\mathcal{C}^\infty$ whenever these are defined (i.e., whenever $U_i \cap U_j \neq \emptyset$).

A further generalization is to Fréchet manifolds, replacing Banach spaces by Fréchet spaces. On the other hand, a Hilbert manifold is a special case of a Banach manifold in which the manifold is locally modeled on Hilbert space.

If E is a Hilbert space, then M is called a $\mathcal{C}^\infty$ Hilbert manifold.

Similarly, manifolds can be defined, which are modeled after any topological vector space E (say, e.g., a Fréchet space), and which are of any differentiability class by stipulating that $\varphi_i \circ \varphi_j^{-1}$ should be of the corresponding class. See more on this subject in [3, 538].

Definition 7.5.6

(a) An $\mathcal{S}$-*morphism* $F : (M, \mathcal{S}_M) \to (N, \mathcal{S}_N)$ is a continuous map, $F : M \to N$, such that

$$f \in \mathcal{S}_N \text{ implies } f \circ F \in \mathcal{S}_M.$$

(b) An $\mathcal{S}$-*isomorphism* is an $\mathcal{S}$-morphism $F : (M, \mathcal{S}_M) \to (N, \mathcal{S}_N)$ such that $F : M \to N$ is a homeomorphism, and

$$F^{-1} : (N, \mathcal{S}_N) \to (M, \mathcal{S}_M) \text{ is an } \mathcal{S}\text{-morphism.}$$

It follows from the above definitions that if, on an $\mathcal{S}$-manifold $(M, \mathcal{S}_M)$, we have two coordinate systems $h_1 : U_1 \to K^n$ and $h_2 : U_2 \to K^n$ such that $U_1 \cap U_2 \neq \varnothing$, then

$$h_2 \circ h_1^{-1} : h_1 (U_1 \cap U_2) \to h_2 (U_1 \cap U_2)$$
$$\text{is an } \mathcal{S}\text{-isomorphism on open subsets of } \left(K^n, \mathcal{S}_{K^n}\right). \tag{7.6}$$

Conversely, if we have an open covering $\{U_\alpha\}_{\alpha \in A}$ of M, a topological manifold, and a family of homeomorphisms $\left\{h_\alpha : U_\alpha \to U'_\alpha \subset K^n\right\}_{\alpha \in A}$ satisfying (7.6), then this defines an $\mathcal{S}$-structure on M by setting $\mathcal{S}_M = \{f : U \to K\}$ such that U is open in M and $f \circ h_\alpha^{-1} \in \mathcal{S}(h_\alpha (U \cap U_\alpha))$ for all $\alpha \in A$. That is, the functions in $\mathcal{S}_M$ are pullbacks of functions in $\mathcal{S}$ by the homeomorphisms $\{h_\alpha\}_{\alpha \in A}$. The collection $\{(U_\alpha, h_\alpha)\}_{\alpha \in A}$ is called an *atlas* for $(M, \mathcal{S}_M)$.

In our three classes of functions, the concepts of an $\mathcal{S}$-morphism and $\mathcal{S}$-isomorphism have special names:

(a) $\mathcal{S} = \mathcal{E}$: differentiable mapping and diffeomorphism of M to N.
(b) $\mathcal{S} = \mathcal{A}$: real-analytic mapping and real-analytic isomorphism (or bianalytic mapping) of M to N.
(c) $\mathcal{S} = \mathcal{O}$: holomorphic mapping and biholomorphism (biholomorphic mapping) of M to N.

It follows immediately from the definition above that a differentiable mapping

$$f : M \longrightarrow N,$$

where M and N are differentiable manifolds, is a continuous mapping of the underlying topological space that has the property that, in local coordinate systems on M and N, f, can be represented as a matrix of $\mathcal{C}^\infty$ functions. This could also be taken as the definition of a differentiable mapping. A similar remark holds for the other two categories.

Let N be an arbitrary subset of an $\mathcal{S}$-manifold M; then an $\mathcal{S}$-*function* on N is defined to be the restriction to N of an $\mathcal{S}$ function defined in some open set containing N, and $\mathcal{S}_M|_N$ consists of all the functions defined on relatively open subsets of N that are restrictions of $\mathcal{S}$ functions on the open subsets of M.

Definition 7.5.7 Let N be a closed subset of an $\mathcal{S}$-manifold M; then N is called an $\mathcal{S}$-*submanifold* of M if for each point $x_0 \in N$, there is a coordinate system $h : U \to U' \subset K^n$, where $x_0 \in U$, with the property that $U \cap N$ is mapped onto

$U' \cap K^k$, where $0 \leq k \leq n$. Here $K^k \subset K^n$ is the standard embedding of the linear subspace K^k into K^n, and k is called the K-dimension of N, and $n-k$ is called the K-codimension of N.

It is easy to see that an $\mathcal{S}$-submanifold of an $\mathcal{S}$-manifold M is itself an $\mathcal{S}$-manifold with the $\mathcal{S}$-structure given by $\mathcal{S}_M|_N$. Since the implicit function theorem is valid in each of our three categories (see [336]), it is easy to verify that the above definition of submanifold coincides with the more common one that an $\mathcal{S}$-submanifold (of k dimensions) is a closed subset of an $\mathcal{S}$-manifold M that is locally the common set of zeros of $(n-k)$ $\mathcal{S}$ functions whose Jacobian matrix has maximal rank.

It is clear that an n-dimensional complex structure on a manifold induces a $2n$-dimensional real-analytic structure, which, likewise, induces a $2n$-dimensional differentiable structure on the manifold. One of the questions we shall be concerned with is how many different (i.e., non-isomorphic) complex analytic structures induce the same differentiable structure on a given manifold. The analogous question of how many different differentiable structures exist on a given topological manifold is an important problem in differential topology.

What we have actually defined is a category wherein the objects are $\mathcal{S}$-manifolds and the morphisms are $\mathcal{S}$-morphisms. We leave to the reader the proof that this actually is a category, since it follows directly from the definitions. In the course of what follows, then we shall use three categories—the differentiable ($\mathcal{S} = \mathcal{E}$), the real-analytic ($\mathcal{S} = \mathcal{A}$), and the holomorphic ($\mathcal{S} = \mathcal{O}$) categories—and the above remark states that each is a subcategory of the former.

We are especially interested in:

Definition 7.5.8 Let X be a connected, second countable, Hausdorff space. A *complex analytic chart* on X is a pair (U, z) where U is open in X and $z : U \to z(U) \subset \mathbb{C}^n$ is a homeomorphism onto an open subset of $\mathbb{C}^n$.

A *system of charts/atlas* is given by a collection of pairs $\{(U_j, \varphi_j)\}_{j\in I}$ where:

(1) $X = \bigcup U_j$.
(2) The maps $\varphi_j \circ \varphi_k^{-1} : \varphi_k\left(U_j \cap U_k\right) \to \varphi_j\left(U_j \cap U_k\right)$ are biholomorphic for all j, k such that $U_j \cap U_k \neq \emptyset$. Two atlases $\{(U_j, \varphi_j)\}_{j\in I}$ and $\{(W_\lambda, \psi_\lambda)\}_{\lambda\in\Lambda}$ are equivalent if the maps $\psi_\lambda \circ \varphi_j^{-1} : \varphi_j\left(U_j \cap W_\lambda\right) \to \psi_\lambda\left(U_i \cap W_\lambda\right)$ are biholomorphic for all j, k with $U_j \cap W_\lambda \neq \emptyset$.

A manifold M is said to be *orientable* if the covering can be chosen so that $\det D\left(\varphi_i \circ \varphi_j^{-1}\right)\left(\varphi_j(x)\right)$ is always positive. If M is orientable, then, subject to this property, it has two possible maximal coverings. A choice of one is called an *orientation*.

A manifold is called *connected*, *compact*, *simply connected*, etc., if the underlying topological manifold has this property. Clearly, any open subset of a complex manifold is in a natural way a complex manifold itself.

Definition 7.5.9 Let $\phi_1 : U_1 \to V_1$ and $\phi_2 : U_2 \to V_2$ be two complex charts on X. We say that ϕ_1 and ϕ_2 are *compatible* if either $U_1 \cap U_2 = \emptyset$, or

$$\phi_2 \circ \phi_1^{-1} : \phi_1 (U_1 \cap U_2) \to \phi_2 (U_1 \cap U_2)$$

is holomorphic.

Note that the definition is symmetric: If $\phi_2 \circ \phi_1^{-1}$ is holomorphic on $\phi_1 (U_1 \cap U_2)$, then $\phi_1 \circ \phi_2^{-1}$ will be holomorhic on $\phi_2 (U_1 \cap U_2)$. The function $T = \phi_2 \circ \phi_1^{-1}$ is called the *transition function* between the two charts; it is a bijection in any case.

That is, if $z_1, \dots, z_n$ are local coordinates in one such neighborhood and if $w_1, \dots, w_n$ are local coordinates in another neighborhood, then where they are both defined $w_k = u_k (z_1, \dots, z_n)$ where each u_k is a complex analytic function of the z's.

We now want to give some examples of various types of manifolds.

Example 7.5.10 (Euclidean Space) K^n, $(\mathbb{R}^n, \mathbb{C}^n)$. For every $p \in K^n, U = K^n$ and $h =$ identity. Then $\mathbb{R}^n$ becomes a real-analytic (hence differentiable) manifold and $\mathbb{C}^n$ is a complex analytic manifold.

Example 7.5.11 Let $\mathbb{S}^2$ denote the unit 2-sphere in $\mathbb{R}^3$, i.e.,

$$\mathbb{S}^2 = \left\{(x, y, w) \in \mathbb{R}^3 \mid x^2 + y^2 + w^2 = 1\right\}.$$

Note that the sphere is Hausdorff and connected. Consider the $w = 0$ plane as a copy of the complex plane $\mathbb{C}$, with $(x, y, 0)$ being identified with $z = x + iy$. Let $\phi_1 : \mathbb{S}^2 \setminus \{(0, 0, 1)\} \to \mathbb{C}$ be defined by projection from $(0, 0, 1)$. Specifically,

$$\phi_1(x, y, w) = \frac{x}{1 - w} + i\frac{y}{1 - w}.$$

The inverse to ϕ_1 is

$$\phi_1^{-1}(z) = \left(\frac{2\,\mathrm{Re}(z)}{|z|^2 + 1}, \frac{2\,\mathrm{Im}(z)}{|z|^2 + 1}, \frac{|z|^2 - 1}{|z|^2 + 1}\right).$$

Define $\phi_2 : \mathbb{S}^2 \setminus \{(0, 0, -1)\} \to \mathbb{C}$ by projection from $(0, 0, -1)$ followed by a complex conjugation:

$$\phi_2(x, y, w) = \frac{x}{1 + w} - i\frac{y}{1 + w}.$$

The inverse to ϕ_2 is

$$\phi_2^{-1}(z) = \left(\frac{2\,\mathrm{Re}(z)}{|z|^2 + 1}, \frac{-2\,\mathrm{Im}(z)}{|z|^2 + 1}, \frac{1 - |z|^2}{|z|^2 + 1}\right).$$

The common domain is $\mathbb{S}^2 \setminus \{(0, 0, \pm 1)\}$ and is mapped by both ϕ_1 and ϕ_2 bijectively onto $\mathbb{C}^* = \mathbb{C} \setminus \{0\}$. The composition is $\phi_2 \circ \phi_1^{-1}(z) = 1/z$, which is holomorphic. Thus the two charts are compatible.

Stereographic projection. There is a geometric identification between the unit sphere $\mathbb{S}^2 \subset \mathbb{R}^3$ and the Riemann sphere $\widehat{\mathbb{C}}$ in which the north pole N becomes ∞ and the rest of the sphere is projected linearly to $\mathbb{C} = \mathbb{R}^2 \times \{0\}$.

Theorem 7.5.12 *Stereographic projection is a conformal map that sends circles to circles.*

Proof Consider two vectors at a point $p \neq N$ on $\mathbb{S}^2$. Construct a pair of circles tangent to these vectors at p and passing through N. Then these circles meet in the same angles at p and N. On the other hand, each circle is the intersection of the sphere with a plane. These planes meet $\mathbb{C}$ in the same angle they meet a plane tangent to the sphere at the north pole N. Thus stereographic projection preserves angles. □

Theorem 7.5.13 *The Euclidean metric on $\mathbb{S}^2$ is transported, by stereographic projection, to the metric $2|dz|/\left(1 + |z|^2\right)$ on $\widehat{\mathbb{C}}$.*

Proof Reducing dimensions by one, we obtain the stereographic projection $p : \mathbb{S}^1 \setminus N \to \mathbb{R}$. If $\theta \in [-\pi, \pi]$ is the angle on the sphere normalized so $\theta(N) = \pi$, then we find $x = p(\theta) = \tan(\theta/2)$. Thus $dx = (1/2)\sec^2(\theta/2)d\theta = (1/2)\left(1 + x^2\right)d\theta$, which gives $d\theta = 2dx/\left(1 + x^2\right)$. The case of $\mathbb{S}^2$ follows by conformality and rotation invariance. □

Two atlases $\left\{\left(U_j, \varphi_j\right)\right\}_{j \in I}$ and $\{(W_\lambda, \psi_\lambda)\}_{\lambda \in \Lambda}$ are *equivalent* if the maps $\psi_\lambda \circ \varphi_j^{-1} : \varphi_j\left(U_j \cap W_\lambda\right) \to \psi_\lambda\left(U_i \cap W_\lambda\right)$ are biholomorphic for all j, k with $U_j \cap W_\lambda \neq \emptyset$ (Fig. 7.1).

A *complex structure* on X is an equivalence class of atlases, and a *complex manifold* M is a space X as above, together with a complex structure.

It is convenient to provide oneself from the very outset with as many coordinate systems on M as possible. A complex chart $\varphi : U \to \mathbb{C}^n$ on a complex manifold

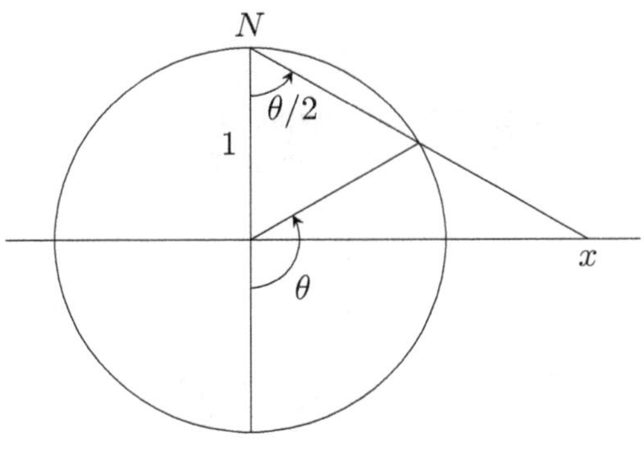

Fig. 7.1 Stereographic projection on $\mathbb{S}^1$

M is said to be *analytic* if it can be added to $\{(U_j, \varphi_j)\}_{j \in I}$, say, without destroying the analyticity of this atlas. This means that the transition functions between the coordinates of φ and those of any chart of $\{(U_j, \varphi_j)\}_{j \in I}$ are analytic. Any atlas made up of complex analytic charts on M is analytic. Moreover, the atlas consisting of all complex analytic charts on M is a *maximal analytic atlas*: The adjunction of any new chart makes it non-analytic. This maximal analytic atlas is called a *complex analytic structure* on M. Further on, unless stated to the contrary, a local coordinate system on M will always mean a complex coordinate system corresponding to a complex analytic chart on M.

Remark 7.5.14 Complex manifolds are defined in a way that reflects their local structure, which is the same as that of a ball in $\mathbb{C}^n$. In particular, the use of coordinates locally reduces the study of manifolds to the theory of analytic functions of n variables. For Riemann surfaces, functions of one variable normally suffice.

Example 7.5.15 Define projective space $\mathbb{CP}^n = \mathbb{C}^{n+1}/\sim$, where $(w_0, \ldots, w_m) \sim (\lambda w_0, \ldots, \lambda w_j)$ for all $\lambda \in \mathbb{C}^*$. Write points of projective space as $(w_0 : \cdots : w_j)$ (homogeneous coordinates).

Now let us discuss the quotient topology.

Charts: Let $U_j = \{(w_0 : \cdots : w_j) \in \mathbb{CP}^n : w_j \neq 0\}$. Define $\varphi_j : U_j \cong \mathbb{C}^n$ by

$$(w_0 : \ldots : w_j) \overset{\varphi_j}{\to} \left(\frac{w_0}{w_j}, \ldots, \frac{w_{j-1}}{w_j}, \frac{w_{j+1}}{w_j}, \ldots, \frac{w_j}{w_j}\right).$$

The U_j forms an open cover. For $j < k$,

$$\varphi_j (U_j \cap U_k) = \{(z_1, \ldots, z_n) : z_k \neq 0\},$$

where $z_i = \frac{w_i}{w_j}$.

Transition functions:

$$\varphi_k \varphi_j^{-1} (z_1, \ldots, z_n) = \left(\frac{z_1}{z_k}, \ldots, \underbrace{\frac{1}{z_k}}_{j}, \ldots, \frac{z_{k-1}}{z_k}, \frac{z_{k+1}}{z_k}, \ldots, \frac{z_n}{z_k} \right).$$

This is clearly biholomorphic on $U_k \cap U_j$.

Since $\mathbb{CP}^n$ is a continuous image of

$$\left\{ (w_0, \ldots, w_n) \in \mathbb{C}^{n+1} : \sum_0^n |w_i|^2 = 1 \right\},$$

i.e., $S^{2n+1} \subset \mathbb{R}^{2n+2}$, we see that $\mathbb{CP}^n$ is a compact, complex manifold.

7.6 Vector Bundles

The study of vector bundles on manifolds has been motivated primarily by the desire to linearize nonlinear problems in geometry, and their use has had a profound effect on various modern fields of mathematics. In this section we want to introduce the concept of a vector bundle and give various examples. We shall also discuss some of the now classical results in differential topology (the classification of vector bundles, for instance) that form a motivation for some of our constructions later in the context of holomorphic vector bundles.

We shall use the same notation as in Sect. 7.1. In particular $\mathcal{S}$ will denote one of the three structures on manifolds ($\mathcal{E}$, $\mathcal{A}$, $\mathcal{O}$) studied there, and $K = \mathbb{R}$ or $\mathbb{C}$.

Definition 7.6.1 A continuous map $\pi : E \to X$ of one Hausdorff space, E, onto another, X, is called a *K-vector bundle* of rank r if the following conditions are satisfied:

(a) $E_p := \pi^{-1}(p)$, for $p \in X$, is a K-vector space of dimension r (E_p is called the *fiber* over p).
(b) For every $p \in X$ there is a neighborhood U of p and a homeomorphism.

$$h : \pi^{-1}(U) \longrightarrow U \times K^r \text{ such that } h\left(E_p\right) \subset \{p\} \times K^r,$$

and h^p, defined by the composition

$$h^p : E_p \xrightarrow{h} \{p\} \times K^r \xrightarrow{\text{proj.}} K^r,$$

is a K-vector space isomorphism (the pair (U, h) is called a *local trivialization*).

For a K-vector bundle $\pi : E \to X$, E is called the *total space* and X is called the *base space*, and we often say that E is a vector bundle over X. Notice that for two local trivializations (U_α, h_α) and $\left(U_\beta, h_\beta\right)$ the map

$$h_\alpha \circ h_\beta^{-1} : \left(U_\alpha \cap U_\beta\right) \times K^r \longrightarrow \left(U_\alpha \cap U_\beta\right) \times K^r$$

induces a map

$$g_{\alpha\beta} : U_\alpha \cap U_\beta \longrightarrow GL(r, K), \tag{7.7}$$

where

$$g_{\alpha\beta}(p) = h_\alpha^p \circ \left(h_\beta^p\right)^{-1} : K^r \longrightarrow K^r.$$

The functions $g_{\alpha\beta}$ are called the *transition functions* of the K-vector bundle $\pi : E \to X$ (with respect to the two local trivializations above).

The transition functions $g_{\alpha\beta}$ satisfy the following compatibility conditions:

$$g_{\alpha\beta} \circ g_{\beta\gamma} \circ g_{\gamma\alpha} = I_r \quad \text{on } U_\alpha \cap U_\beta \cap U_\gamma, \tag{7.8}$$

and

$$g_{\alpha\alpha} = I_r \quad \text{on } U_\alpha, \tag{7.9}$$

where the product is a matrix product and I_r is the identity matrix of rank r. This follows immediately from the definition of the transition functions.

Definition 7.6.2 A K-vector bundle of rank $r, \pi : E \to X$, is said to be an $\mathcal{S}$-*bundle* if E and X are $\mathcal{S}$-manifolds, π is an $\mathcal{S}$-morphism, and the local trivializations are $\mathcal{S}$-isomorphisms.

Note that the fact that the local trivializations are $\mathcal{S}$-isomorphisms is equivalent to the fact that the transition functions are $\mathcal{S}$-morphisms. In particular, then we have differentiable vector bundles, real-analytic vector bundles, and holomorphic vector bundles (K must equal $\mathbb{C}$).

Remark 7.6.3 Suppose that on a $\mathcal{S}$-manifold we are given an open covering $\mathfrak{A} = \{U_\alpha\}$ and that to each ordered non-empty intersection $U_\alpha \cap U_\beta$ we have assigned an $\mathcal{S}$ function

$$g_{\alpha\beta} : U_\alpha \cap U_\beta \longrightarrow GL(r, K)$$

satisfying the compatibility conditions (7.8) and (7.9). Then one can construct a vector bundle $E \xrightarrow{\pi} X$ having these transition functions. An outline of the construction is as follows: Let

$$\widetilde{E} = \bigcup_\alpha U_\alpha \times K^r \quad \text{(disjoint union)}$$

equipped with the natural product topology and $\mathcal{S}$-structure. Define an equivalence relation in $\widetilde{E}$ by setting

$$(x, v) \sim (y, w), \text{ for } (x, v) \in U_\beta \times K^r, (y, w) \in U_\alpha \times K^r$$

if and only if[1]

$$y = x \quad \text{and} \quad w = g_{\alpha\beta}(x)v\,.$$

[1] Note that the transition function $g_{\alpha\beta}(p)$ is a linear mapping from the U_β trivialization to the U_α trivialization. The order is significant.

The fact that this is a well-defined equivalence relation is a consequence of the compatibility conditions (7.8) and (7.9). Let $E = \widetilde{E}/\sim$ (the set of equivalence classes), equipped with the quotient topology, and let $\pi : E \to X$ be the mapping that sends a representative (x, v) of a point $p \in E$ into the first coordinate. One then shows that an E so constructed carries an $\mathcal{S}$-structure and is an $\mathcal{S}$-vector bundle. In the examples discussed below we shall see more details of such a construction.

Example 7.6.4 (Trivial Bundle) Let M be an $\mathcal{S}$-manifold. Then

$$\pi : M \times K^n \longrightarrow M,$$

where π is the natural projection, is an $\mathcal{S}$-bundle called a *trivial bundle*.

Example 7.6.5 (Tangent Bundle) Let M be a differentiable manifold. Then we want to construct a vector bundle over M whose fiber at each point is the linearization of the manifold M, to be called the *tangent bundle* to M. Let $p \in M$. Then we let

$$\mathcal{E}_{M,p} := \varinjlim_{\substack{p \in U \subset M \\ \text{open}}} \mathcal{E}_M(U)$$

be the algebra (over $\mathbb{R}$) of germs of differentiable functions at the point $p \in M$, where the inductive limit[2] is taken with respect to the partial ordering on open neighborhoods of p given by inclusion. Expressed differently, we can say that if f and g are defined and $\mathcal{C}^\infty$ near p and they coincide on some neighborhood of p, then they are equivalent. The set of equivalence classes is easily seen to form an algebra over $\mathbb{R}$ and is the same as the inductive limit algebra discussed above; an equivalence class (element of $\mathcal{E}_{M,p}$) is called a *germ* of a $\mathcal{C}^\infty$ function at p. A *derivation* of the algebra $\mathcal{E}_{M,p}$ is a vector space homomorphism $D : \mathcal{E}_{M,p} \to \mathbb{R}$ with the property that $D(fg) = D(f) \cdot g(p) + f(p) \cdot D(g)$, where $g(p)$ and $f(p)$ denote evaluation of a germ at a point p (which clearly makes sense). The *tangent space* to M at p is the vector space of all derivations of the algebra $\mathcal{E}_{M,p}$, which we denote by $T_p(M)$. Since M is a differentiable manifold, we can find a diffeomorphism h defined in a neighborhood U of p where

$$h : U \longrightarrow U' \subset_{\text{open}} \mathbb{R}^n$$

and where, letting $h^* f(x) = f \circ h(x)$, h has the property that, for $V \subset U'$,

$$h^* : \mathcal{E}_{\mathbb{R}^n}(V) \longrightarrow \mathcal{E}_M\left(h^{-1}(V)\right)$$

[2] We denote by $\varinjlim$ the inductive (or direct) limit and by $\varprojlim$ the projective (or inverse) limit of a partially ordered system.

is an algebra isomorphism. It follows that h^* induces an algebra isomorphism on germs, i.e., (using the same notation),

$$h^* : \mathcal{E}_{\mathbb{R}^n, h(p)} \xrightarrow{\cong} \mathcal{E}_{M,p},$$

and hence induces an isomorphism on derivations:

$$h_* : T_p(M) \overset{\cong}{\rightrightarrows} T_{h(p)}\left(\mathbb{R}^n\right).$$

It is easy to verify that:

(a) $\partial/\partial x_j$ are derivations of $\mathcal{E}_{\mathbb{R}^n, h(p)}$, $j = 1, \ldots, n$, and that
(b) $\{\partial/\partial x_1, \ldots, \partial/\partial x_n\}$ is a basis for $T_{h(p)}\left(\mathbb{R}^n\right)$

and thus that $T_p(M)$ is an n-dimensional vector space over $\mathbb{R}$, for each point $p \in M$. (The derivations are, of course, simply the classical directional derivatives evaluated at the point $h(p)$.) Suppose that $f : M \to N$ is a differentiable mapping of differentiable manifolds. Then there is a natural map

$$df_p : T_p(M) \longrightarrow T_{f(p)}(N)$$

defined by the diagram below.

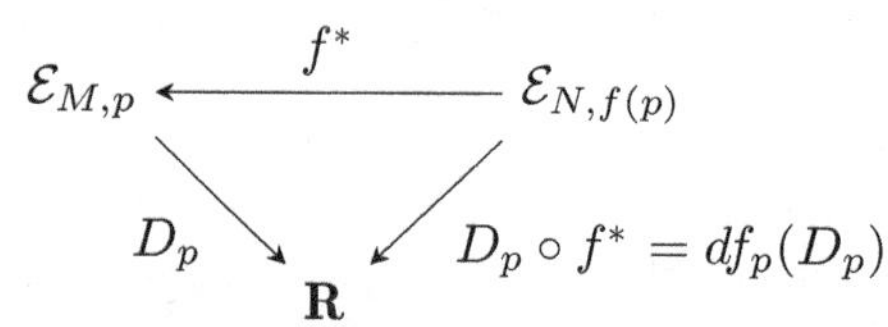

Let $D_p \in T_p(M)$. The mapping df_p is a linear mapping and can be expressed as a matrix of first derivatives with respect to local coordinates. The coefficients of such a matrix representation will be $\mathcal{C}^\infty$ functions of the local coordinates. Classically, the mapping df_p (the derivative mapping, differential mapping, or tangent mapping) is called the *Jacobian* of the differentiable map f. The tangent map represents a first-order linear approximation (at p) to the differentiable map f.

We are now in a position to construct the tangent bundle to M. Let

$$T(M) = \bigcup_{p \in M} T_p(M) \quad \text{(disjoint union)}$$

and define

$$\pi : T(M) \longrightarrow M$$

by

$$\pi(v) = p \quad \text{if } v \in T_p(M).$$

We can now make $T(M)$ into a vector bundle. Let $\{(U_\alpha, h_\alpha)\}$ be an atlas for M, and let $T(U_\alpha) = \pi^{-1}(U_\alpha)$ and

$$\psi_\alpha : T(U_\alpha) \longrightarrow U_\alpha \times \mathbb{R}^n$$

be defined as follows: Suppose that $v \in T_p(M) \subset T(U_\alpha)$. Then $dh_{\alpha,p}(v) \in T_{h\alpha(p)}(\mathbb{R}^n)$. Thus

$$dh_{\alpha,p}(v) = \sum_{j=1}^{n} \xi_j(p) \frac{\partial}{\partial x_j}\bigg|_{h_\alpha(p)},$$

where $\xi_j \in \mathcal{E}_M(U_\alpha)$ (the fact that the coefficients are C^∞ follows easily from the proof that $\{\partial/\partial x_1, \ldots, \partial/\partial x_n\}$ is a basis for the tangent vectors at a point in $\mathbb{R}^n$). Now let

$$\psi_\alpha(v) = (p, \xi_1(p), \ldots, \xi_n(p)) \in U_\alpha \times \mathbb{R}^n.$$

It is easy to verify that ψ_α is bijective and fiber preserving and moreover that

$$\psi_\alpha^p : T_p(M) \xrightarrow{\psi_\alpha} \{p\} \times \mathbb{R}^n \xrightarrow{\text{proj.}} \mathbb{R}^n$$

is an $\mathbb{R}$-linear isomorphism. We can define transition functions

$$g_{\alpha\beta} : U_\beta \cap U_\alpha \longrightarrow GL(n, \mathbb{R})$$

by setting

$$g_{\alpha\beta}(p) = \psi_\alpha^p \circ \left(\psi_\beta^p\right)^{-1} : \mathbb{R}^n \longrightarrow \mathbb{R}^n.$$

Moreover, it is easy to check that the coefficients of the matrices $\{g_{\alpha\beta}\}$ are C^∞ functions in $U_\alpha \cap U_\beta$, since $g_{\alpha\beta}$ is a matrix representation for the composition $dh_\alpha \circ dh_\beta^{-1}$ with respect to the basis $\{\partial/\partial x_1, \ldots, \partial/\partial x_n\}$ at $T_{h_\beta(p)}(\mathbb{R}^n)$ and $T_{h_\alpha(p)}(\mathbb{R}^n)$, and that the tangent maps are differentiable functions of local coordinates. Thus the $\{(U_\alpha, \psi_\alpha)\}$ become the desired trivializations. We have only to put the right topology on $T(M)$ so that $T(M)$ becomes a differentiable manifold. We simply require that $U \subset T(M)$ be open if and only if $\psi_\alpha(U \cap T(U_\alpha))$ is open in $U_\alpha \times \mathbb{R}^n$ for every α. This is well defined since

$$\psi_\alpha \circ \psi_\beta^{-1} : (U_\alpha \cap U_\beta) \times \mathbb{R}^n \longrightarrow (U_\alpha \cap U_\beta) \times \mathbb{R}^n$$

is a diffeomorphism for any α and β such that $U_\alpha \cap U_\beta \neq \varnothing$ (since $\psi_\alpha \circ \psi_\beta^{-1} = \mathrm{id} \times g_{\alpha\beta}$, where id is the identity mapping). Because the transition functions are diffeomorphisms, this defines a differentiable structure on $T(M)$ so that the projection π and the local trivializations ψ_α are differentiable maps.

Example 7.6.6 (Tangent Bundle to a Complex Manifold) Let $X = (X, \mathcal{O}_x)$ be a complex manifold of complex dimension n, let

$$\mathcal{O}_{X,x} := \varinjlim_{\substack{x\in U\subset X\\ \text{open}}} \mathcal{O}(U)$$

be the $\mathbb{C}$-algebra of germs of holomorphic functions at $x \in X$, and let $T_x(X)$ be the derivations of this $\mathbb{C}$-algebra (defined exactly as in Example 7.6.5). Then $T_x(X)$ is the holomorphic (or complex) tangent space to X at x. In local coordinates, we see that $T_x(X) \cong T_x(\mathbb{C}^n)$ (abusing notation) and that the complex partial derivatives $\{\partial/\partial z_1, \dots, \partial/\partial z_n\}$ form a basis over $\mathbb{C}$ for the vector space $T_x(\mathbb{C}^n)$. In the same manner as in Example 7.6.5, we can make the union of these tangent spaces into a holomorphic vector bundle over X, i.e., $T(X) \to X$, where the fibers are all isomorphic to $\mathbb{C}^n$.

Let M be a manifold.

Definition 7.6.7 The manifold M is called *orientable* if TM is an orientable vector bundle.

An orientation of M means an orientation of TM; an oriented manifold is a pair (M, ω) where ω is an orientation of M. We define $-\omega$ to be the orientation of M such that $(-\omega)_x = -\omega_x$ (these are orientations of M_x) for all $x \in M$. If M is connected and orientable, then it has exactly two orientations, ω and $-\omega$. Every simply connected manifold is orientable.

An alternative definition of "orientable manifold" is:

Definition 7.6.8 The manifold M is *orientable* if it has an atlas whose coordinate changes have positive Jacobian determinants at all points. A maximal atlas of this kind is an *oriented differential structure*.

In other words we say that M is orientable if there exists a local system of coordinates $(U_\alpha, \varphi_\alpha)_{\alpha\in A}$ such that, for any $\alpha, \beta \in A$, the determinant of the Jacobian of $\varphi_\alpha^{-1} \circ \varphi_\beta$ is positive in every point where $\varphi_\alpha^{-1} \circ \varphi_\beta$ is defined.

By considering natural vector bundle charts, it is easy to see that the two definitions are equivalent.

Let (M, ω) and (N, θ) be oriented manifolds. A diffeomorphism $f : M \to N$ is called *orientation preserving* if $Tf : (TM, \omega) \to (TN, \theta)$ preserves orientation; in this case, we write $f(\omega) = \theta$. On the other hand, f is called *orientation reversing* if Tf reverses orientation. Notice that when M is connected, f must have one of these properties; to determine which one, it suffices to see whether a single $T_x f$ preserves orientation.

Now let M be a connected orientable manifold and let $g : M \to M$ be a diffeomorphism. Let $\omega, -\omega$ be the two orientations of M. Then either $g(\omega) = \omega$ and $g(-\omega) = -\omega$, or else $g(\omega) = -\omega$ and $g(-\omega) = \omega$. In other words g either preserves both orientations or reverses both orientations. We call g orientation preserving or orientation reversing, accordingly, independent of any choice of orientation for M. If Φ is an oriented differential structure for M, then f preserves orientation if $f^*\Phi = \Phi$, that is, if the derivative of f at any point, expressed by charts in Φ, has positive Jacobian.

Example 7.6.9 Let $M \subset \mathbb{R}^{m+1}$ be defined by $f(x) = c$, with $df(a) \neq 0$ if $f(a) = c$. By the inverse function theorem, M is a manifold, and moreover, if $\partial f/\partial x_i \neq 0$, then $x_1, \ldots, x_{i-1}, x_{i+1}, \ldots, x_{m+1}$ are local coordinates. Consider, on such a coordinate patch,

$$\omega = (-1)^i \frac{1}{\partial f/\partial x_i} dx_1 \wedge \ldots \wedge dx_{i-1} \wedge dx_{i+1} \cdots \wedge dx_{m+1} \,. \tag{7.10}$$

This is nonvanishing.

Now M is defined by $f(x) = c$ so that, on M,

$$\sum_j \frac{\partial f}{\partial x_j} dx_j = 0$$

and, if $\partial f/\partial x_j \neq 0$, then

$$dx_j = -\frac{1}{\partial f/\partial x_j} (\partial f/\partial x_i dx_i + \ldots) \,.$$

Substituting into (7.10), we get

$$\omega = (-1)^j \frac{1}{\partial f/\partial x_j} dx_1 \wedge \ldots \wedge dx_{j-1} \wedge dx_{j+1} \cdots \wedge dx_{m+1} \,.$$

The formula (7.10) therefore defines for all coordinate charts a nonvanishing m-form, so M is orientable.

The obvious example is the sphere S^m with

$$\omega = (-1)^j \frac{1}{x_j} dx_1 \wedge \cdots \wedge dx_{j-1} \wedge dx_{j+1} \cdots \wedge dx_{m+1} \,.$$

There are two obvious necessary conditions on the topological structure of the space of a complex manifold:

1. It is even dimensional.
2. It is orientable.

The first is obvious. We shall prove the second by showing that complex structure induces a natural orientation on the space. To do this it suffices to show that there is a real differential form of degree $2n$ defined on the whole space up to a positive factor.

If z_k are local coordinates, then we set $z_k = x_k + iy_k$. Then xs and ys are real local coordinates and $dz_k = dx_k + idy_k$. We define two linear operators on $\mathbb{C}^n$:

$$\frac{\partial}{\partial z_k} = \frac{1}{2}\left(\frac{\partial}{\partial x_k} - i\frac{\partial}{\partial y_k}\right)$$

$$\frac{\partial}{\partial \overline{z}_k} = \frac{1}{2}\left(\frac{\partial}{\partial x_k} + i\frac{\partial}{\partial y_k}\right). \tag{7.11}$$

The functional determinant $\det\left(\frac{\partial w_k}{\partial z_l}\right)$ will be nonvanishing

$$\left(\text{since } \det\left(\frac{\partial w_k}{\partial z_l}\right)\det\left(\frac{\partial z_k}{\partial w_l}\right) = 1\right)$$

and continuous, therefore always of one sign.

Let U be a domain in $\mathbb{C}^n$ and $\phi_1, \phi_2 : U \to \mathbb{C}^n$ be a C^1 mapping. In the case of a negative (positive) Jacobian, the mappings ϕ_1, ϕ_2 will be called *oppositely* (*similarly*) oriented.

Let $\overline{z}_k = x_k - iy_k$, so that $d\overline{z}_k = dx_k - idy_k$. Then $dz_k \wedge d\overline{z}_k = -2idx_k \wedge dy_k$. Hence

$$dx_1 \wedge \cdots \wedge dx_n \wedge dy_1 \wedge \cdots \wedge dy_n =$$
$$= \left(\frac{i}{2}\right)^n dz_1 \wedge \cdots \wedge dz_n \wedge d\overline{z}_1 \wedge \cdots \wedge d\overline{z}_n.$$

It follows that the right side is a real form. We show that a change in local coordinates changes this form by a positive factor. If w_k is another set of local coordinates, then

$$dw_1 \wedge \cdots \wedge dw_n = J dz_1 \wedge \cdots \wedge dz_n,$$

where $J = \det\left(\frac{\partial w_k}{\partial z_l}\right)$. Then

$$d\overline{w}_1 \wedge \cdots \wedge d\overline{w}_n = \overline{J} d\overline{z}_1 \wedge \cdots \wedge d\overline{z}_n.$$

Therefore

$$dw_1 \wedge \cdots \wedge dw_n \wedge d\overline{w}_1 \wedge \cdots \wedge d\overline{w}_n =$$
$$= J\overline{J} dz_1 \wedge \cdots \wedge dz_n \wedge d\overline{z}_1 \wedge \cdots \wedge d\overline{z}_n$$

so $(i/2)^n dz_1 \wedge \cdots \wedge dz_n \wedge d\overline{z_1} \wedge \cdots \wedge d\overline{z}_n$ differs from the corresponding expression in the z's by a positive factor $|J|^2$.

Now we can give a direct proof that a Riemann surface is an oriented manifold. For complex charts φ_α and φ_β with transition function $h = \varphi_\alpha(\varphi_\beta^{-1}(z))$, h can be considered as a map from an open set of $\mathbb{R}^2$ to $\mathbb{R}^2$ whose Jacobian at a point z is just the real linear map given by multiplication by the complex number $h'(z)$. However, the real determinant of multiplication by a complex number α equals $|\alpha|^2$, so the Jacobian of h has positive determinant. Consequently, the Riemann surface is an oriented manifold.

Remark 7.6.10 The same technique used to construct the tangent bundles in the above examples can be used to construct other vector bundles. For instance, suppose that we have $\pi : E \to X$, where X is an $\mathcal{S}$-manifold and π is a surjective map, so that:

(a) Each E_p is a K-vector space.
(b) For each $p \in X$, there is a neighborhood U of p and a bijective map:

$$h : \pi^{-1}(U) \longrightarrow U \times K^r \text{ such that } h\left(E_p\right) \subset \{p\} \times K^r.$$

(c) $h_p : E_p \to \{p\} \times K^r \overset{\text{proj.}}{\longrightarrow} K^r$ is a K-vector space isomorphism.

If for every (U_α, h_α), $\left(U_\beta, h_\beta\right)$ as in (b) we have $h_\alpha \circ h_\beta^{-1}$ is an $\mathcal{S}$-isomorphism, then we can make E into an $\mathcal{S}$-bundle over X by giving it the topology that makes h_α a homeomorphism for every α.

7.7 Almost Complex Manifolds and the $\overline{\partial}$-Operator

In this section we want to introduce certain first-order differential operators that act on differential forms on a complex manifold and that intrinsically reflect the complex structure. The most natural context in which to discuss these operators is from the viewpoint of almost complex manifolds, a generalization of a complex manifold that has the first-order structure of a complex manifold (i.e., at the tangent space level). We shall first discuss the concept of a $\mathbb{C}$-linear structure on an $\mathbb{R}$-linear vector space and will apply the (linear algebra) results obtained to the real tangent bundle of a differentiable manifold.

Let V be a real vector space and suppose that J is an $\mathbb{R}$-linear isomorphism $J : V \overset{\sim}{\longrightarrow} V$ such that $J^2 = -I$ (where I = identity). Then J is called a *complex structure* on V. Suppose that V and a complex structure J are given. Then we can equip V with the structure of a complex vector space in the following manner:

$$(\alpha + i\beta)v := \alpha v + \beta J v, \quad \alpha, \beta \in \mathbb{R}, \quad i = \sqrt{-1}.$$

Thus scalar multiplication on V by complex numbers is defined, and it is easy to check that V becomes a complex vector space. Conversely, if V is a complex vector space, then it can also be considered as a vector space over $\mathbb{R}$, and the operation of multiplication by i is an $\mathbb{R}$-linear endomorphism of V onto itself, which we can call J, and is a complex structure. Moreover, if $\{v_1, \ldots, v_n\}$ is a basis for V over $\mathbb{C}$, then $\{v_1, \ldots, v_n, Jv_1, \ldots, Jv_n\}$ will be a basis for V over $\mathbb{R}$.

Example 7.7.1 Let $\mathbb{C}^n$ be the usual Euclidean space of n-tuples of complex numbers, $\{z_1, \ldots, z_n\}$, and let $z_j = x_j + iy_j, j = 1, \ldots, n$, be the real and imaginary parts. Then $\mathbb{C}^n$ can be identified with $\mathbb{R}^{2n} = \{x_1, y_1, \ldots, x_n, y_n\}$, $x_j, y_j \in \mathbb{R}$. Scalar multiplication by i in $\mathbb{C}^n$ induces a mapping $J : \mathbb{R}^{2n} \longrightarrow \mathbb{R}^{2n}$ given by

$$J(x_1, y_1, \ldots, x_n, y_n) = (-y_1, x_1, \ldots, -y_n, x_n)$$

and, moreover, $J^2 = -1$. This is the standard complex structure on $\mathbb{R}^{2n}$. The coset space $GL(2n, \mathbb{R})/GL(n, \mathbb{C})$ determines all complex structures on $\mathbb{R}^{2n}$ by the mapping $[A] \longrightarrow A^{-1}JA$, where $[A]$ is the equivalence class of $A \in GL(2n, \mathbb{R})$.

Example 7.7.2 Let X be a complex manifold and let $T_x(X)$ be the (complex) tangent space to X at x. Let X_0 be the underlying differentiable manifold of X (i.e., X induces a differentiable structure on the underlying topological manifold of X) and let $T_x(X_0)$ be the (real) tangent space to X_0 at x. Then we claim that $T_x(X_0)$ is canonically isomorphic with the underlying real vector space of $T_x(X)$ and that, in particular, $T_x(X)$ induces a complex structure J_x on the real tangent space $T_x(X_0)$. To see this, we let (h, U) be a holomorphic coordinate system near x. Then $h : U \longrightarrow U' \subset \mathbb{C}^n$, and hence, by taking real and imaginary parts of the vector-valued function h, we obtain

$$\widetilde{h} : U \longrightarrow \mathbb{R}^{2n}$$

given by

$$\widetilde{h}(x) = (\operatorname{Re} h_1(x), \operatorname{Im} h_1(x), \ldots, \operatorname{Re} h_n(x), \operatorname{Im} h_n(x)),$$

which is a real-analytic (and, in particular, differentiable) coordinate system for X_0 near x. Then it suffices to consider the claim above for the vector spaces $T_0(\mathbb{C}^n)$ and $T_0\left(\mathbb{R}^{2n}\right)$ at $0 \in \mathbb{C}^n$, where $\mathbb{R}^{2n}$ has the standard complex structure. Let $\{\partial/\partial z_1, \ldots, \partial/\partial z_n\}$ be a basis for $T_0(\mathbb{C}^n)$ and let $\{\partial/\partial x_1, \partial/\partial y_1, \ldots, \partial/\partial x_n, \partial/\partial y_n\}$ be a basis for $T_0\left(\mathbb{R}^{2n}\right)$. Then we have the diagram

$$
\begin{array}{ccc}
T_0\left(\mathbb{C}^n\right) & \cong_{\mathbb{C}} & \mathbb{C}^n \\
\alpha \;\; \wr\!\|_{\mathbb{R}} & & \wr\!\|_{\mathbb{R}} \\
T_0\left(\mathbb{R}^{2n}\right) & \cong_{\mathbb{R}} & \mathbb{R}^{2n},
\end{array}
$$

where α is the $\mathbb{R}$-linear isomorphism between $T_0\left(\mathbb{R}^{2n}\right)$ and $T_0\left(\mathbb{C}^n\right)$ induced by the other maps, and thus the complex structure of $T_0\left(\mathbb{C}^n\right)$ induces a complex structure on $T_0\left(\mathbb{R}^{2n}\right)$, just as in Example 7.7.1. We claim that the complex structure J_x induced on $T_x\left(X_0\right)$ in this manner is independent of the choice of local holomorphic coordinates. To check that this is the case, consider a biholomorphism f defined on a neighborhood N of the origin in $\mathbb{C}^n$, $f : N \longrightarrow N$, where $f(0) = 0$. Then, letting $\zeta = f(z)$ and expressing things in terms of real and imaginary coordinates, we have the corresponding diffeomorphism expressed in real coordinates:

$$
\begin{aligned}
\xi &= u(x, y)\,, \\
\eta &= v(x, y)\,,
\end{aligned}
\tag{7.12}
$$

where $\xi, \eta, x, y \in \mathbb{R}^n$ and $\xi + i\eta = \zeta \in \mathbb{C}^n, x + iy = z \in \mathbb{C}^n$. The map $f(z)$ corresponds to a holomorphic change of coordinates on the complex manifold X; the pair of mappings u, v corresponds to the change of coordinates for the underlying differentiable manifold. The Jacobian matrix (differential) of these mappings corresponds to the transition functions for the corresponding trivializations for $T(X)$ and $T\left(X_0\right)$, respectively. Let J denote the standard complex structure in $\mathbb{C}^n$; we shall show that J commutes with the Jacobian of the real mapping. The real Jacobian of (7.7.1) has the form of an $n \times n$ matrix of 2×2 blocks,

$$
M = \begin{bmatrix} \frac{\partial u_\alpha}{\partial x_\beta} & \frac{\partial u_\alpha}{\partial y_\beta} \\ \frac{\partial v_\alpha}{\partial x_\beta} & \frac{\partial v_\alpha}{\partial y_\beta} \end{bmatrix}, \quad \alpha, \beta = 1, \ldots, n,
$$

which, by the Cauchy-Riemann equations (since f is a holomorphic mapping), is the same as

$$
\begin{bmatrix} \frac{\partial v_\alpha}{\partial y_\beta} & \frac{\partial u_\alpha}{\partial y_\beta} \\ -\frac{\partial u_\alpha}{\partial y_\beta} & \frac{\partial v_\alpha}{\partial y_\beta} \end{bmatrix}, \quad \alpha, \beta = 1, \ldots, n\,.
$$

Thus the Jacobian is an $n \times n$ matrix consisting of 2×2 blocks of the form

$$
\begin{bmatrix} a & b \\ -b & a \end{bmatrix}.
$$

Moreover, J can be expressed in matrix form as an $n \times n$ matrix of 2×2 blocks with matrices of the form

$$\begin{bmatrix} 0 & 1 \\ -1 & 0 \end{bmatrix}$$

along the diagonal and zero elsewhere. It is now easy to check that $MJ = JM$. It follows then that J induces the same complex structure on $T_x(X_0)$ for each choice of local holomorphic coordinates at x.

Let V be a real vector space with a complex structure J, and consider $V \otimes_{\mathbb{R}} \mathbb{C}$, the complexification of V. The $\mathbb{R}$-linear mapping J extends to a $\mathbb{C}$-linear mapping on $V \otimes_{\mathbb{R}} \mathbb{C}$ by setting $J(v \otimes \alpha) = J(v) \otimes \alpha$ for $v \in V$, $\alpha \in \mathbb{C}$. Moreover, the extension still has the property that $J^2 = -I$, and it follows that J has two eigenvalues $\{i, -i\}$. Let $V^{1,0}$ be the eigenspace corresponding to the eigenvalue i and let $V^{0,1}$ be the eigenspace corresponding to $-i$. Then we have

$$V \otimes_{\mathbb{R}} \mathbb{C} = V^{1,0} \oplus V^{0,1}.$$

Moreover, conjugation on $V \otimes_{\mathbb{R}} \mathbb{C}$ is defined by $\overline{v \otimes \alpha} = v \otimes \overline{\alpha}$ for $v \in V$ and $\alpha \in \mathbb{C}$. Thus $V^{1,0} \cong_{\mathbb{R}} V^{0,1}$ (conjugation is a conjugate-linear mapping). It is easy to see that the complex vector space obtained from V by means of the complex structure J, denoted by V_J, is $\mathbb{C}$-linearly isomorphic to $V^{1,0}$, and we shall identify V_J with $V^{1,0}$ from now on.

We now want to consider the exterior algebras of these complex vector spaces. Namely, denote $V \otimes_{\mathbb{R}} \mathbb{C}$ by V_c and consider the exterior algebras

$$\wedge V_c, \quad \wedge V^{1,0}, \quad \text{and} \quad \wedge V^{0,1}.$$

Then we have natural injections (and we let $\wedge^{p,q} V$ be the subspace of $\wedge V_c$ generated by elements of the form $u \wedge w$, where $u \in \wedge^p V^{1,0}$ and $w \in \wedge^q V^{0,1}$).

$$\begin{array}{ccc} \wedge V^{1,0} & \searrow & \\ & & \wedge V_c \\ \wedge V^{0,1} & \nearrow & \end{array}$$

Thus we have the direct sum (letting $n = \dim_c V^{1,0}$)

$$\wedge V_c = \sum_{r=0}^{2n} \sum_{p+q=r} \wedge^{p,q} V.$$

We now want to carry out the above algebraic construction on the tangent bundle to a manifold. First, we have a definition.

Definition 7.7.3 Let X be a differentiable manifold of dimension $2n$. Suppose that J is a differentiable vector bundle isomorphism

$$J : T(X) \longrightarrow T(X)$$

such that $J_x : T_x(X) \longrightarrow T_x(X)$ is a complex structure for $T_x(X)$; i.e., $J^2 = -I$, where I is the identity vector bundle isomorphism acting on $T(X)$. Then J is called an *almost complex structure* for the differentiable manifold X. If X is equipped with an almost complex structure J, then (X, J) is called an *almost complex manifold.*

We see that a differentiable manifold having an almost complex structure is equivalent to prescribing a $\mathbb{C}$-vector bundle structure on the $\mathbb{R}$-linear tangent bundle.

Proposition 7.7.4 *A complex manifold X induces an almost complex structure on its underlying differentiable manifold.*

Proof As we saw in Example 7.7.2, for each point $x \in X$ there is a complex structure induced on $T_x(X_0)$, where X_0 is the underlying differentiable manifold. What remains to check is that the mapping

$$J_x : T_x(X_0) \longrightarrow T_x(X_0), \quad x \in X_0$$

is, in fact, a $\mathcal{C}^\infty$ mapping with respect to the parameter x. To see that J is a $\mathcal{C}^\infty$ vector bundle mapping, choose local holomorphic coordinates (h, U) and obtain a trivialization for $T(X_0)$ over U, i.e.,

$$T(X_0)|_U \cong h(U) \times \mathbb{R}^{2n},$$

where we let $z_j = x_j + iy_j$ be the coordinates in $h(U)$ and $(\xi_1, \eta_1, \ldots, \xi_n, \eta_n)$ be the coordinates in $\mathbb{R}^{2n}$. Then the mapping $J|_U$ is defined (with respect to this trivialization) by

$$id \times J : h(U) \times \mathbb{R}^{2n} \longrightarrow h(U) \times \mathbb{R}^{2n},$$

where

$$J(\xi_1, \eta_1, \ldots, \xi_n, \eta_n) = (-\eta_1, \xi_1, \ldots, -\eta_n, \xi_n)$$

as before, That is, in this trivialization, J is a constant mapping, and hence $\mathcal{C}^\infty$. Since differentiability is a local property, it follows that J is a differentiable bundle mapping. □

Remark 7.7.5 There are various examples of almost complex structures that do not arise from complex structures. The 2-sphere $\mathbb{S}^2$ carries a complex structure $[\cong \mathbb{P}_1(\mathbb{C})]$, and the 6-sphere $\mathbb{S}^6$ carries an almost complex structure induced on it by the unit Cayley numbers in $\mathbb{S}^7$ (see Steenrod [482]). However, this almost complex structure does not come from a complex structure (it is not integrable; see the discussion below). Moreover, it is unknown whether $\mathbb{S}^6$ carries any complex structure. A theorem of Borel and Serre [90] asserts that only $\mathbb{S}^2$ and $\mathbb{S}^6$ admit almost complex structures among the even-dimensional real spheres. For more information about almost complex structures on manifolds, consult, e.g., Kobayashi and Nomizu [299] or Helgason [229].

7.8 Tensor Fields

Definition 7.8.1 Let M be a manifold and $\tau_M : TM \to M$ its tangent bundle. We call $T^r_s(M) = T^r_s(TM)$ the *vector bundle of tensors* contravariant of order r and covariant of order s, or simply of type (r, s). We identify $T^1_0(M)$ with TM and call $T^0_1(M)$ the *cotangent bundle* of M also denoted by $\tau^*_M : T^*M \to M$. The zero section of $T^r_s(M)$ is identified with M.

Recall that a section of a vector bundle assigns to each base point b a vector in the fiber over b and the addition and scalar multiplication of sections take place within each fiber. In the case of $T^r_s(M)$ these vectors are called *tensors*. The $\mathcal{C}^\infty$ sections of $\pi : E \to B$ were denoted $\Gamma^\infty(\pi)$, or $\Gamma^\infty(E)$. Recall that $\mathcal{F}(M)$ denotes the set of mappings from M into $\mathbb{R}$ that are of class $\mathcal{C}^\infty$ (the standard local manifold structure being used on $\mathbb{R}$) together with its structure as a ring; namely, $f + g, cf, fg$ for $f, g \in \mathcal{F}(M), c \in \mathbb{R}$ are defined by

$$(f+g)(x) = f(x)+g(x), \quad (cf)(x) = c(f(x)), \quad \text{and} \quad (fg)(x) = f(x)g(x).$$

Finally, recall that a vector field on M is an element of $\mathfrak{X}(M) = \Gamma^\infty(TM)$.

Definition 7.8.2 A *tensor field* of type (r, s) on a manifold M is a $\mathcal{C}^\infty$ section of $T^r_s(M)$. We denote by $\mathcal{T}^r_s(M)$ the set $\Gamma^\infty\left(T^r_s(M)\right)$ together with its (infinite-dimensional) real vector space structure. *A* covector field or differential one-form is an element of $\mathfrak{X}^*(M) = \mathcal{T}^0_1(M)$.

If $f \in \mathcal{F}(M)$ and $t \in \mathcal{T}^r_s(M)$, let $ft : M \to T^r_s(M)$ be defined by $m \mapsto f(m)t(m)$. If $X_i \in \mathfrak{X}(M)$, $i = 1, \ldots, s$, $\alpha^j \in \mathfrak{X}^*(M)$, $j = 1, \ldots, r$, and $t' \in \mathcal{T}^{r'}_{s'}(M)$ define

$$t\left(\alpha^1, \ldots, \alpha^r, X_1, \ldots, X_s\right) : M \to \mathbb{R} \quad \text{by} \quad m \mapsto t(m)\left(\alpha_1(m), \ldots, X_s(m)\right)$$

and

$$t \otimes t' : M \to T^{r+r'}_{s+s'}(M) \quad \text{by} \quad m \mapsto t(m) \otimes t'(m).$$

Proposition 7.8.3 *With f, t, X_i, α^j, and t' as in Definition 7.8.2,*

$$ft \in \mathcal{T}^r_s(M), \quad t\left(\alpha^1, \ldots, X_s\right) \in \mathcal{F}(M), \quad \textit{and} \quad t \otimes t' \in \mathcal{T}^{r+r'}_{s+s'}.$$

Proof The differentiability is evident in each case from the product rule in local representation.

For the tangent bundle TM, a natural chart is obtained by taking $T\varphi$, where φ is an admissible chart of M. This in turn induces a chart $(T\varphi)_*$ on $T^r_s M$. We shall call these the *natural charts* of $T^r_s M$. □

Chapter 8
Riemann Surfaces

Among a number of wonderful ideas we owe to Riemann, the idea of Riemann surface is, without doubt, the most beautiful, everlasting, intensively developing, and unifying. It fertilizes a number of other ideas, penetrating the whole body of mathematics and, in turn, many branches of physics. This idea originated from the germ of algebraic functions or elliptic integrals (differentials) and their local inverses: elliptic functions.

Riemann surfaces are of fundamental importance in many areas of mathematics and theoretical physics. This chapter is a survey of the basic notions and main results of the theory of Riemann surfaces. A detailed exposition, and proofs, can be found in Ahlfors-Sario [23], Forster [158], Springer [481], and Weyl [525].

Riemann surfaces were first introduced by Riemann in an attempt to understand multivalued functions of a complex variable. Such multivalued functions occur because the analytic continuation of a given holomorphic function element along different paths leads in general to different branches of that function. It was the idea of Riemann to replace the domain of the function with a many-sheeted covering of the complex plane. If the covering is constructed so that it has as many points lying over any given point in the plane as there are function elements at that point, then on this "covering surface" the analytic function becomes single valued.

The modern era in the study of Riemann surfaces, and, indeed, of all manifolds, opened in 1913 with the publication of Hermann Weyl's book *The Concept of a Riemann Surface*. His definition is purely geometric and also independent of any embedding in Euclidean 3-space.

Now, forgetting the fact that these surfaces are "spread out" over the complex plane (or the Riemann sphere), we get the notion of an abstract Riemann surface, and these may be considered as the natural domain of definition of analytic functions in one complex variable.

P. V. Dovbush, S. G. Krantz, *The Geometric Theory of Complex Variables*,
https://doi.org/10.1007/978-3-031-77204-7_8

8.1 The Evolution of the Concept of a Riemann Surface

Riemann surfaces were first introduced by Riemann in an attempt to understand multivalued functions of complex variables. The equation $X^2 - Y^2 = 0$ does not define X as a single-valued function of Y. However, if X and Y are chosen to have real values, a choice of sign of $\sqrt{Y}$ may be consistently made if Y is not permitted to assume the value zero. At $X = Y = 0$, the differential of $P(X, Y) = X^2 - Y^2$ vanishes, that is, $(0, 0)$ is a singularity of the equation $P(X, Y) = 0$. Avoiding the singularities is not sufficient to permit such a choice if X and Y are permitted to have complex values (Figs. 8.1 and 8.2).

Using a construction that may be found in any elementary complex variables text, Riemann resolved the problem by constructing sheets above the complex plane. On each sheet, a choice may be made. The sheets are then glued together to form a natural domain S_1 of definition for the function $\sqrt{X^2}$. Notice that the modification is made on the domain of the function not on the range. Riemann initially thought of his sheets as lying over the complex plane in $\mathbb{R}^3$. Much of the subsequent development of the notion of Riemann surface was done to remove the artificiality and arbitrary character of Riemann's embedding in $\mathbb{R}^3$. As Weyl noted, S_1 is a natural parameter space for the variety S. In this sense Riemann had a uniformization theorem, but it was not the first. The first uniformization theorem was proved by Puiseux in 1850.

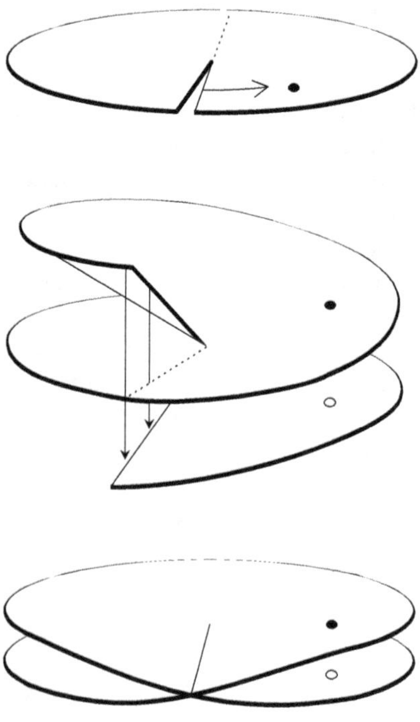

Fig. 8.1 Construction of a Riemann surface

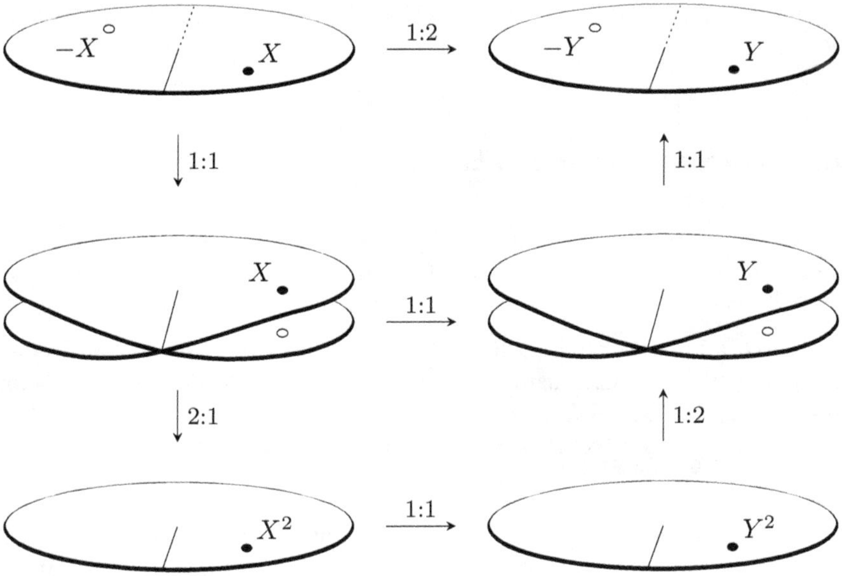

Fig. 8.2 Riemann surface $Y = \sqrt{X^2}$

The method discovered by Weierstrass for studying multivalued functions is quite different from Riemann's. It is probably the first step in the development of the modem notions of abstract manifold and sheaf. Weierstrass's original paper on the subject seems to have been written in 1842, although it did not appear in print until 1894. The basic idea is to construct a Riemann surface, in this context called an analytic configuration, by starting with a power series $f(z)$ centered at some z_0 and having positive radius of convergence. One then considers all meromorphic continuations of f in $\widehat{\mathbb{C}}$. Each continuation of f to a point z is given by a chain of discs in which the continuation is possible. The set of points to which f may be continued is subject to an equivalence relation, and a natural topology may be given to the equivalence classes (see, e.g., Ahlfors [18] or Hille [236] for details). Singular points admitting uniformizations by Puiseux series are usually added to give complete analytic configurations, which are Riemann surfaces in the sense of Weyl (see below). The analytic configurations are Riemann surfaces which lie in the sheaf of germs of holomorphic or meromorphic functions on $\widehat{\mathbb{C}}$. Overlapping discs define the topology, a concept that underlies the modern notion of manifold.

In his 1907 paper on uniformization, Poincaré removed the reliance of Weierstrass's construction on the initial power series $f(z)$. He proved the uniformization theorem for collections of overlapping discs on which there exists some meromorphic function in the sense of Weierstrass. The geometric entity became paramount; the role of the function was secondary.

The modem era in the study of Riemann surfaces, and, indeed, of all manifolds, opened in 1913 with the publication of Hermann Weyl's book *The Concept of a Riemann Surface* [525]. Here, for the first time, no defining function is assumed. The definition is purely geometric and also independent of any embedding in Euclidean 3-space.

8.1.1 Riemann Surfaces: The Classical Approach

Weyl's definition of *Riemann surfaces*, in modern language and without unnecessary hypotheses, is the following: A Riemann surface is a one-dimensional complex manifold. That is,

Definition 8.1.1 A Riemann surface M is a real two-dimensional (one complex dimensional) $\mathcal{C}^\infty$ oriented manifold, together with a collection of local homeomorphisms $\varphi_i : U_i \to \mathbb{C}$ in the given orientation with $\cup U_i = M$ such that $\varphi_i \circ \varphi_j^{-1} : U_{ij} \to \mathbb{R}^2$ are holomorphic maps, where defined, i.e., if $\emptyset \neq U_{ij} := \varphi_j \left(U_i \cap U_j\right) \subset \mathbb{R}^2$.

Here, $\mathbb{R}^2$ is to be considered as $\mathbb{C}$. The collection $c := \{(U_i, \varphi_i) \mid i \in I\}$ is called a *complex structure* for M. If M has a fixed complex structure, we write (M, c) to denote M with its given complex structure.

Note that we do not a priori assume a Riemann surface S to be a second countable space. It is however a theorem of Radó that every Riemann surface is second countable (for a proof, see [248, Section 1.3]). Moreover every Riemann surface is automatically orientable (see, for instance, [206, Page 18]).

Definition 8.1.2 A mapping $f : M \to N$, where (M, c), $\left(N, c'\right)$, are Riemann surfaces is called *holomorphic* iff for all charts $\left(V_j, \psi_j\right) \in c'$, $(U_i, \varphi_i) \in c$, the mapping $\psi_j \circ f \circ \varphi_i^{-1}$ is holomorphic (where defined). A similar definition holds for $\mathcal{C}^\infty$ maps between $\mathcal{C}^\infty$ manifolds. Note that the definition does not depend on the choice of the charts.

Notice now that the Euclidian plane $\mathbb{C}$ is a Riemann surface, where $\mathbb{C}$ has the usual topology, and a function is analytic if it is analytic in the usual sense.

A Riemann surface is a connected complex analytic manifold of dimension 1. Furnished with this definition, one can immediately extend local properties and objects from $\mathbb{C}$ to any Riemann surface; in particular the concepts of a holomorphic or a meromorphic function or form on a Riemann surface and holomorphic and biholomorphic mappings (isomorphisms) between such surfaces now acquire meaning.

A map $f : S \to S'$ from one Riemann surface to a second is holomorphic if for each point P of S, there are coordinate neighborhoods (U, z) of P and $\left(U', z'\right)$ of $f(P)$ such that $z' \circ f \circ z^{-1} : z(U) \to z\left(U'\right)$ is holomorphic. An isomorphism of Riemann surfaces is a bijective holomorphic map whose inverse is also holomorphic.

The sheaf of germs of holomorphic functions over the surface is called the *complex structure* of the Riemann surface.

Example 8.1.3 Let T be the torus $\mathbb{R}^2/\mathbb{Z}^2$. We shall see that there are infinitely many non-isomorphic complex structures on T.

8.2 Examples of Riemann Surfaces

Example 8.2.1 The simplest examples of Riemann surfaces are any domain (connected open subset) $U \subset \mathbb{C}$ in the complex plane, the complex plane $\mathbb{C}$ itself, and the extended complex plane (or Riemann sphere) $\widehat{\mathbb{C}} = \mathbb{CP} = \mathbb{C} \cup \{\infty\} = \mathbb{S}^2$. The complex structures on U and $\mathbb{C}$ are defined by single coordinate charts $(U, \text{ id})$ and $(\mathbb{C}, \text{ id})$.

The holomorphic and meromorphic functions on U with this structure of a Riemann surface are just the usual holomorphic and meromorphic functions.

Example 8.2.2 Suppose that S is a Riemann surface and $S' \subset S$ is a domain, i.e., a connected open subset. Then S' has a natural complex structure which makes it a Riemann surface. Namely, one takes as its atlas all those complex charts $\varphi : U \to V$ on S, where $U \subset S'$. In particular, every domain $S' \subset \mathbb{C}$ is a Riemann surface.

Hence $\mathbb{C}$, $\mathbb{H} = \{z \in \mathbb{C} : \operatorname{Im} z > 0\}$, $\mathbb{D} = \{z \in \mathbb{C} : |z| < 1\}$, and $\mathbb{D}^* = \{z \in \mathbb{C} : 0 < |z| < 1\}$ are Riemann surfaces.

Example 8.2.3 (Graphs of Holomorphic Functions) Let $V \subset \mathbb{C}$ be a connected open subset of the complex plane, and let g be a holomorphic function defined on all of V. Consider the graph X of g, as a subset of $\mathbb{C}^2$:

$$X = \{(z, g(z)) \mid z \in V\}.$$

Give X the subspace topology, and let $\pi : X \to V$ be the first projection; note that π is a homeomorphism, whose inverse simply sends the point $z \in V$ to the ordered pair $(z, g(z))$. Thus π is a complex chart on X, whose domain covers all of X. Hence we have a complex atlas on X, composed of a single chart; this gives X the structure of a Riemann surface.

This example can be immediately generalized to any finite collection of holomorphic functions $g_1, \dots, g_n$ on V; simply take X to be the graph in $\mathbb{C}^{n+1}$:

$$X = \{(z, g_1(z), \dots, g_n(z)) \mid z \in V\}.$$

Example 8.2.4 (Smooth Affine Plane Curves) This is a further generalization of the graph construction introduced above. We would like to consider a locus $X \subset \mathbb{C}^2$ which is locally a graph, but perhaps not globally. The most natural way to do this is to define a locus X by requiring a complex polynomial of two variables $f(z, w)$ to vanish. Morally speaking, this should cut the complex dimension down by 1, and we have a chance of producing a Riemann surface this way.

A Riemann surface is defined as a non-singular surface in $\mathbb{C}^2$ given by an equation of the form

$$f(z, w) = 0, \tag{8.1}$$

where $f(z, w)$ is an analytic function of z and w (for instance, a polynomial in z and w).

Writing $f = u + iv$, we see that (8.1) is equivalent to the two real equations $u = 0$ and $v = 0$, so that it defines a two-dimensional surface in $\mathbb{R}^4$ $(\cong \mathbb{C}^2)$.

By analogy with the real case, we define the complex gradient, $\text{grad}_{\mathbb{C}} f$, of the function f, by

$$\text{grad}_{\mathbb{C}} f = \left(\frac{\partial f}{\partial w}, \frac{\partial f}{\partial z} \right), \tag{8.2}$$

and we say that a point (w_0, z_0) on the curve (8.1) is non-singular if $\text{grad}_{\mathbb{C}} f\big|_{(w_0, z_0)} \neq 0$. The following theorem (whose proof we omit) is the complex analog of the implicit function theorem (for the proof see [328, Theorem 1.4.11, p. 54]).

Theorem 8.2.5 (The Complex Implicit Function Theorem) *Let $f(w, z)$ be a complex analytic function of the variables w, z, and let (w_0, z_0) be a point at which $\text{grad}_{\mathbb{C}} f \neq 0$. If, for instance, $\partial f / \partial w \neq 0$ at the point, then in a sufficiently small neighborhood (in $\mathbb{C}^2$) of the point (w_0, z_0), the equation $f(w, z) = 0$ has a unique solution $w = w(z)$ (i.e., $f(w(z), z) \equiv 0$ in the neighborhood); moreover this (unique) solution is analytic, i.e., $\partial w / \partial \overline{z} \equiv 0$ in the neighborhood (where $\partial / \partial \overline{z}$ is defined in (7.11)).*

Hence the condition for non-singularity, which makes the surface a one-dimensional complex manifold (i.e., complex curve), is as follows:

$$\text{grad}_{\mathbb{C}} f \neq 0.$$

As an (important) example, consider the case that $f(w, z)$ is a polynomial in the variables w, z. We call the totality of all solutions in the form $w = w(z)$, of the equation $f(w, z) = 0$, a multiple valued algebraic function and the surface (i.e., complex curve) $f(w, z) = 0$ the graph or Riemann surface of this multiple valued function.

An important special case is that of a hyper-elliptic curve, defined as the Riemann surface given by an equation of the form

$$f(w, z) = w^2 - P_n(z) = 0, \tag{8.3}$$

where $P_n(z)$ is a polynomial in one variable of degree n which is not a perfect square. Then the polynomial $f(w, z) = w^2 - P_n(z)$ is irreducible.

If we solve Eq. (8.1) for w, it may happen that we obtain a multivalued function; for instance:

(i) If $f(z,w) = w^2 - P_n(z)$, where $P_n(z)$ is a polynomial without multiple roots, we obtain the two-valued function $w = \sqrt{P_n(z)}$ (a "hyperelliptic" Riemann surface), where the square root sign is to be interpreted ambiguously.
(ii) If $f(z,w) = e^w - z$, we obtain $w = \ln z = \ln|z| + i \arg z + 2\pi i n$.

(The geometric meaning of multivaluedness of $w(z)$ is that (some of) the surfaces $z = \text{const.}$ meet the surface $f(z,w) = 0$ in more than one point.)

Lemma 8.2.6 *The surface (8.3) is free of singularities if and only if the polynomial $P_n(z)$ has no multiple roots.*

Proof Calculating the gradient of the function f in (8.3), we obtain

$$\mathrm{grad}_{\mathbb{C}} f = \left(\frac{\partial f}{\partial w}, \frac{\partial f}{\partial z}\right) = \left(2w, -\frac{dP_n(z)}{dz}\right).$$

Thus a point (w_0, z_0) is a singular point of the curve if and only if

$$2w_0 = 0, \qquad \left.\frac{dP_n(z)}{dz}\right|_{z_0} = 0, \quad w_0^2 - P_n(z_0) = 0,$$

i.e., if and only if $w_0 = 0$, and z_0 is a root of both the polynomial $P_n(z)$ and its derivative. The lemma then follows from the well-known equivalence of this condition on z_0, with the condition that z_0 must be a multiple root of $P_n(z)$. □

The complex implicit function theorem allows us to introduce the local coordinate z on the surface (8.3), at (i.e., in neighborhoods of) those points where $\partial f/\partial w = 2w \neq 0$, i.e., where $P_n(z) \neq 0$. At those points z for which $P_n(z) = 0$, we may take w as a local coordinate, provided that we do not also have $\partial f/\partial z = -dP_n(z)/dz = 0$, i.e., provided that z is not a multiple root of $P_n(z)$.

8.3 Mappings of Riemann Surfaces

All mappings of Riemann surfaces will be assumed holomorphic. The geometric classification is reflected in maps between Riemann surfaces, as detailed in Liouville's theorem and the Little Picard theorem: Maps from hyperbolic to parabolic to elliptic are easy, but maps from elliptic to parabolic or parabolic to hyperbolic are very constrained (indeed, generally constant!).

There are inclusions of the disc in the plane and then in the sphere:

$$\mathbb{D} \subset \mathbb{C} \subset \widehat{\mathbb{C}},$$

but any holomorphic map from the sphere to the plane is constant, any holomorphic map from the plane into the unit disc is constant (Liouville's theorem), and in fact any holomorphic map from the plane into the plane minus two points is constant (Little Picard theorem)!

These statements are clarified by considering the type of a Riemann sphere $\widehat{\mathbb{C}}$ with a number of punctures. With no punctures, it is the Riemann sphere, which is elliptic. With one puncture, which can be placed at infinity, it is the complex plane, which is parabolic. With two punctures, it is the punctured plane or alternatively annulus or cylinder, which is parabolic. With three or more punctures, it is hyperbolic—compare a pair of points. One can map from one puncture to two, via the exponential map (which is entire and has an essential singularity at infinity, so not defined at infinity, and misses zero and infinity), but all maps from zero punctures to one or more or one or two punctures to three or more are constant.

Every non-compact Riemann surface admits nonconstant holomorphic functions (with values in $\mathbb{C}$). In fact, every non-compact Riemann surface is a *Stein manifold.*

In contrast, on a compact Riemann surface S, every holomorphic function with values in $\mathbb{C}$ is constant due to the maximum principle.

It is not trivial, but possible, to prove that every Riemann surface in the sense of Weyl carries a pair of nonconstant linearly independent meromorphic functions and may be embedded in $\mathbb{R}^3$. It is worth comparing this statement with the fact that any orientable 2-manifold can be embedded in $\mathbb{R}^3$. It then follows that, up to holomorphic equivalence, the notions of Riemann surfaces developed by Riemann (suitably generalized), Weierstrass, and Weyl are equivalent.

When are two Riemann surfaces to be considered the same? The answer is of course the natural one.

Definition 8.3.1 An *isomorphism* (or biholomorphism) between Riemann surfaces is a holomorphic map $f : S_1 \to S_2$ which is bijective and whose inverse $f^{-1} : S_1 \to S_2$ is holomorphic. A self-isomorphism $f : S_1 \to S_1$ is called an *automorphism* of S_1. If there exists an isomorphism between S_1 and S_2, we say that S_1 and S_2 are *isomorphic* (or biholomorphic).

Lemma 8.3.2 *The Riemann sphere $\mathbb{S}^2$ and the projective line $\mathbb{CP}^1$ are isomorphic.*

Proof The function from $\mathbb{CP}^1$ to the Riemann sphere sending $[z : w]$ to

$$\left(2\,\mathrm{Re}(z\overline{w}), 2\,\mathrm{Im}(z\overline{w}), |z|^2 - |w|^2\right) / \left(|z|^2 + |w|^2\right) \in \mathbb{S}^2$$

is an isomorphism onto $\widehat{\mathbb{C}}$. □

8.4 Construction of Riemann Surfaces

Riemann surfaces can be described in many completely different ways. Interrelations between these descriptions make for an essential part of the theory. The basic examples of Riemann surfaces we are going to discuss now are exactly these foundation stones on which the entire theory is based.

Uniformization theory exploits the fact that Riemann surfaces can be constructed in three seemingly different ways.

(i) Let $w = a_0 + a_1 (z - z_0) + a_2 (z - z_0)^2 + \ldots$ be a convergent nonconstant power series. The set of (equivalence classes of) all convergent Puiseux series from which w can be obtained by analytic continuation is, in a natural way, a Riemann surface.

(ii) Let $\widetilde{S}$ be a Riemann surface and G a properly discontinuous group of automorphisms (conformal self-mappings) of $\widetilde{S}$. The quotient $\widetilde{S}/G$ is, in a natural way, a Riemann surface.

(iii) Let S be an oriented surface, with a Riemannian metric. The surface S is, in a natural way, a Riemann surface.

In fact it is known that every Riemann surface arises in one of these three ways. We now discuss these three methods in some detail.

8.4.1 Branch Points

The introduction of power series helped to make mathematicians conscious of the function concept by drawing attention to the generality of the expression $a_0 + a_1 x + a_2 x^2 + \cdots$. However, not every function $f(x)$ is expressible as a power series $a_0 + a_1 x + a_2 x^2 + \cdots$. This is obvious in the case of functions that tend to infinity as $x \to 0$, since the power series has value a_0 when $x \to 0$. For other functions, such as $f(x) = x^{1/2}$, the behavior at 0 disallows a power series expansion for a more subtle reason. These functions have branching behavior at 0; they are many-valued, and hence they are not functions in the strict sense. The function $x^{1/2}$, for example, is two-valued because each number has two square roots, one the negative of the other.

Such behavior is not reflected in a power series $a_0 + a_1 x + a_2 x^2 + \cdots$, which can be assigned only one value for each value of x. All fractional powers of x are many-valued—$x^{1/3}$ is three-valued, $x^{1/4}$ is four-valued, and so on—and many-valued behavior is typical of algebraic functions in general. We say that y is an algebraic function of x if x and y satisfy a polynomial equation $p(x, y) = 0$. It follows from the impossibility of solving most polynomial equations by radicals that algebraic functions are not generally expressible by radicals, that is, by finite expressions built from $+, -, \times, \div$, and fractional powers.

Nevertheless, it was the remarkable discovery of Newton (1671) that any algebraic function y can be expressed as a fractional power series in x:

$$y = a_0 + a_1 x^{r_1} + a_2 x^{r_2} + a_3 x^{r_3} + \cdots ,$$

where $r_1, r_2, r_3, \ldots$ are rational numbers. Furthermore, the series can be rewritten in the form

$$\begin{aligned} a_0 + b_1 x^{s_1} &\left(c_{00} + c_{01}x + c_{02}x^2 + \cdots\right) \\ + b_2 x^{s_2} &\left(c_{10} + c_{11}x + c_{12}x^2 + \cdots\right) \\ \vdots & \\ + b_n x^{s_n} &\left(c_{n0} + c_{n1}x + c_{n2}x^2 + \cdots\right), \end{aligned}$$

that is, as a finite sum of ordinary power series with fractional powers of x as multipliers. This means that, in a neighborhood of $x = 0$, the behavior of y is like that of a finite sum of fractional powers.

For example, if $y^2(1+x)^2 = x$, we have

$$\begin{aligned} y &= \frac{x^{1/2}}{1+x} \\ &= x^{1/2}\left(1 - x + x^2 - x^3 + \cdots\right), \end{aligned}$$

and near the origin, y has behavior similar to $x^{1/2}$; in particular there are two values of y for each x. Newton's contribution was an ingenious algorithm for obtaining the successive powers of x. The fractional powers themselves were not properly understood until the variables x and y were taken to be complex. This was done in the nineteenth century, and on this basis a more rigorous derivation of Newton's series was given by Puiseux (1850). For this reason, the fractional power series expansions of algebraic functions are now called *Puiseux expansions*.

The key to the topological form of a complex curve $p(x, y) = 0$ lies in its branch points, the points α where the Newton-Puiseux expansion of y begins with a fractional power of $(x - \alpha)$. The nature of branch points was first described by Riemann (1851) as part of a revolutionary new geometric theory of complex functions. Riemann's idea, one of the most illuminating in the history of mathematics, was to represent a relation $p(x, y) = 0$ between complex x and complex y by covering a plane (or sphere) representing the x variable by a surface representing the y variable, the point or points of the y surface over a given point $x = \alpha$ being those values of y that satisfy $p(\alpha, y) = 0$.

If the equation $p(\alpha, y) = 0$ is of degree n in y, there will in general be n distinct y values for a given α, consequently n "sheets" of the y surface lying over the x-plane in the neighborhood of $x = \alpha$. At finitely many exceptional values of x, sheets merge due to coincidence of roots, and the Newton-Puiseux theory says that the surface at such a point y behaves like a fractional power of x at 0. Our main problem, therefore, is to understand the behavior of the Riemann surface for $y = x^{m/n}$ in the neighborhood of 0. The idea can be grasped sufficiently well from seeing the special

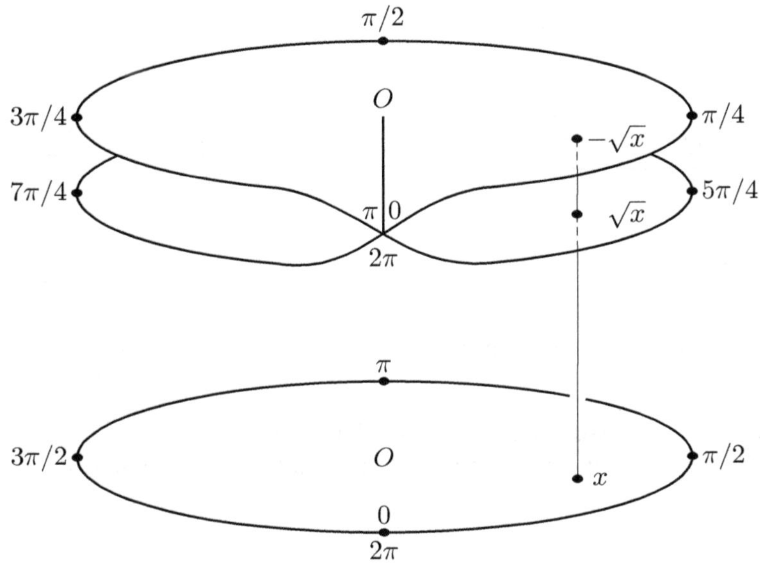

Fig. 8.3 Branch point for the square root

case $y = x^{1/2}$. If we consider the unit disc in the y-plane and try to deform it so that the points $y = \pm\sqrt{x}$ lie above the point x in the unit disc of x-plane, then the result is something like Fig. 8.3.

The angles θ on the disc boundaries are the arguments of the corresponding points $e^{i\theta}$. If

$$x = e^{i\theta} = e^{i(\theta+2\pi)},$$

then

$$y = e^{i\theta/2}, \quad e^{i(\theta/2+\pi)}$$

giving the values shown.

It should be noted that the awkward appearance of the branch point, in particular, the line of self-intersection, is a consequence of representing the relation $y^2 = x$ in fewer dimensions than the 4 it really requires. If we similarly attempt to represent the relation $y^2 = x$ between real x and y by lying the y-axis along the x-axis so that $y = \pm\sqrt{x}$ are on top of x, then the result is an awkward folded "branch point" at 0 (Fig. 8.4). This is a consequence of trying to represent the relation in one dimension. In reality, as the second part of the figure shows, when viewed as a curve in the plane the relation is just as smooth at 0 as anywhere else.

Example 8.4.1 Let $\widehat{\mathbb{C}} = \mathbb{C} \cup \{\infty\}$ be the Riemann sphere. Puiseux series are a generalization of power series that allow for negative and fractional exponents of

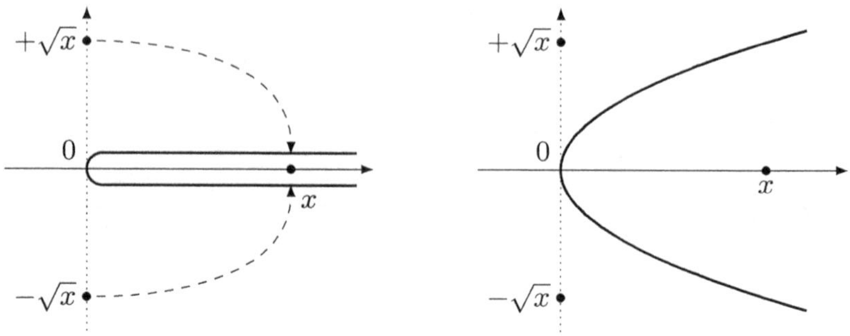

Fig. 8.4 A one-dimensional branch point

the indeterminate. Puiseux series were first introduced by Isaac Newton in 1676 [1] and rediscovered by Victor Puiseux in 1850 [426]. A formal *Puiseux series* with center $z_0 \in \widehat{\mathbb{C}}$ is a series of the form

$$\sum_{n=m}^{\infty} a_n (z - z_0)^{n/\nu} \text{ if } z_0 \neq \infty, \qquad \sum_{n=m}^{\infty} a_n z^{-n/\nu} \text{ if } z_0 = \infty. \tag{8.4}$$

Here the a_n are complex numbers, $a_m \neq 0$, ν (the *ramification index*) is a positive integer, and it is assumed that one cannot, by omitting terms with $a_n = 0$, rewrite the series with a smaller value of ν.

The value w_0 of the series (at z_0) is defined as a_0 if $m = 0$, as 0 if $m > 0$, and as ∞ if $m < 0$.

The series is called *convergent* if the power series $a_0 + a_1 t + a_2 t^2 + \ldots$ has a positive radius of convergence and nonconstant if either $m \neq 0$ or $m = 0$, and there is a $\mu > m$ with $a_\mu \neq 0$.

The definition of a Puiseux series includes that the denominators (here ν) of the exponents must be bounded. So a Puiseux series becomes a Laurent series in a νth root of the indeterminate. For example, (8.4) is a Laurent series in $(z - z_0)^{1/\nu}$ if $z_0 \neq \infty$ and a Laurent series in $z^{1/\nu}$ if $z_0 = \infty$. Because a complex number has ν νth roots, a *convergent* Puiseux series typically defines ν functions in a neighborhood of z_0.

If $z_1 \in \mathbb{C}$ is distinct from but close to the center z_0 of a convergent Puiseux series, the latter induces, at z_1, ν distinct germs of holomorphic functions, obtained by choosing, near z_1, a single-valued branch of $(z - z_0)^{1/\nu}$ and then summing the resulting series.

Two Puiseux series with index ν and center z_0 and with coefficients a_n and $\widehat{a}_n$, respectively, are called equivalent if $\widehat{a}_n = w^n a_n$ with $w^\nu = 1$. A function element θ is an equivalence class of convergent nonconstant Puiseux series. A *regular function element*, i.e., one with $\nu = 1$ and $m \geq 0$, is simply a germ of a nonconstant holomorphic function at z_0.

The set $\mathfrak{W}$ of all function elements is topologized by calling a subset open if, whenever it contains a function element θ, it also contains all (regular) function elements induced by θ at points sufficiently close to the center of θ. A neighborhood of θ can be parametrized by a complex number t (*distinguished local parameter*) as follows. Let θ be defined by a series (8.4). Assign to $t = 0$ the function element $\theta_0 = \theta$ and to $t \neq 0$, $|t|$ small, the regular function element θ_t, the germ at z_1 of the holomorphic function $w = \varphi(z)$, where $z_1 = z + t_1^\nu$ (if $z_0 \neq \infty$) or $z_1 = t_1^{-\nu}$ (if $z_0 = \infty$). The function $w = \varphi(z)$ admits, for small values of $|z - z_1|$, the expansion

$$\varphi(z) = \sum_{n=m}^{\infty} a_n \left(t_1^\nu + (z - z_0)\right)^{n/\nu} = \sum_{j=0}^{\infty} b_j (z - z_0)^j .$$

Here

$$b_j = t^{j\nu} \sum_{n=m}^{\infty} \binom{n/\nu}{j} a_n t^n \qquad (\text{ if } z_0 \neq \infty)$$

or

$$\varphi(z) = \sum_{n=m}^{\infty} a_n \left(t^{-\nu} + (z - z_1)\right)^{-n/\nu} = \sum_{j=0}^{\infty} b_j (z - z_1)^j ,$$

where

$$b_j = t^{j\nu} \sum_{n=m}^{\infty} \binom{-n/\nu}{j} a_n t^n \qquad (\text{ if } z_0 = \infty) .$$

Every component S of $\mathfrak{W}$ is made into a Riemann surface by calling a function f, defined near $\theta \in S$, holomorphic if it is so when considered as a function of a distinguished local parameter belonging to θ. Note that every component S comes equipped with two meromorphic functions—the value and the center of the function elements belonging to S.

8.5 Coverings of Riemann Surfaces

Let S be a Riemann surface. As a topological surface, S has a "universal covering space." That is, there is a simply connected surface $\widetilde{S}$ and a "projection" $\pi : \widetilde{S} \to S$ that is a covering space in the topological sense of the word. This means by definition that every point $p \in S$ has a neighborhood U_p with $\pi^{-1}\left(U_p\right)$ equaling a disjoint union of open sets $\left\{V_j\right\}$ such that $\pi|_{V_j} : V_j \to S$ is a homeomorphism onto U_p. One says here that U_p is evenly covered.

The topological surface $\widetilde{S}$ can be given a Riemann surface structure by "pull-back," that is, by declaring $\pi|_{V_j}$ to be holomorphic. Thus $\widetilde{S}$ becomes a simply connected Riemann surface, and $\pi : \widetilde{S} \to S$ is a holomorphic mapping. Koebe-Poincaré's celebrated uniformization theorem says this:

Theorem 8.5.1 *Any simply connected Riemann surface S is conformally (thus biholomorphically) equivalent to one of the three following surfaces:*

(a) $\mathbb{C}$, *the complex plane*
(b) $\mathbb{D}$ *(this is conformally equivalent to the upper half plane* $\mathbb{H}$*)*
(c) $\mathbb{S}^2 \simeq \mathbb{CP}^1$

Theorem 8.5.1 asserts a more general result than the Riemann mapping theorem by identifying the simply connected Riemann surfaces. Now it is a basic fact of topology, more precisely covering space theory [189], that any Riemann surface Σ admits a universal cover $\widetilde{\Sigma}$, a simply connected Riemann surface, for which the deck transformations (the fundamental group) will act as a discrete group Γ of conformal transformations without fixed points. The discrete groups of conformal transformations of $\mathbb{C}$ and $\widehat{\mathbb{C}}$ are easily identified. Any homeomorphism of $\widehat{\mathbb{C}}$ has a fixed point. A conformal transformation of $\mathbb{C}$ is a similarity. A similarity without fixed points is a translation. It follows with a little work that $\mathbb{C}$ is the universal cover only for Tori $\approx \mathbb{S} \times \mathbb{S}$ and cylinders $\mathbb{C}\backslash\{0\} \approx \mathbb{S} \times \mathbb{R}$. The sphere can only cover the sphere. Hence every other Riemann surface therefore admits the disc as its universal cover.

Theorem 8.5.1 asserts that $\widetilde{S}$ can only be one of the three things (up to biholomorphic equivalence): the plane, the unit disc, or the Riemann sphere. And, as we shall see, the plane and the sphere arise only in a few special cases. So the uniformization theorem says that all but a finite number (topologically) of Riemann surfaces have simply connected covering space biholomorphic to the disc.

It is a standard and easily checked fact from topological covering space theory (for which see, e.g., [167, 189, 479]) that the projection $\pi : \widetilde{S} \to S$ can be interpreted as a quotient by a group action. There is a group Γ, the group of covering transformations, acting as homeomorphisms on $\widetilde{S}$, such that S is the orbit space of the action of Γ. In the Riemann surface case, Γ consists of homeomorphisms that are holomorphic, i.e., elements of Aut $(\widetilde{S})$.

The possibilities $\widetilde{S} = \mathbb{C}$ and $\widetilde{S} = \mathbb{C} \cup \{\infty\}$ yield few Riemann surfaces. It is again part of general covering space theory that the (nonidentity) elements of Γ act without fixed points, that is, if $\sigma \in \Gamma$ and $\sigma(p) = p$ for some $p \in \widetilde{S}$, then σ is the identity map of $\widetilde{S}$ to $\widetilde{S}$. Now direct calculation shows that the linear fractional transformations that are automorphisms of $\mathbb{C} \cup \{\infty\}$ all have fixed points—this is just elementary algebra. Thus $\mathbb{C} \cup \{\infty\}$ has no covering space quotients except itself. In other words, in the language of the uniformization theorem, the sphere can cover only the sphere.

When $\widetilde{S} = \mathbb{C}$, quotients are possible. The automorphisms of $\mathbb{C}$, $\{z \mapsto az + b : a \neq 0, a, b \in \mathbb{C}\}$, do include maps without fixed points. Specifically, any map of the form $z \mapsto z + b$ has no fixed point; these are in fact the only ones, since, if $a \neq 1$, then $z = b/(1 - a)$ is a fixed point of $z \mapsto az + b$. A familiar analysis shows that the possible Riemann surface quotients that can thus arise are $\mathbb{C}\setminus\{0\}$, topologically a cylinder, and surfaces of genus 1, topologically tori. And these possibilities do indeed occur. But these are the only topological possibilities.

All other Riemann surfaces must be quotients of the disc $\mathbb{D}$. Clearly the possibilities for groups acting on $\mathbb{D}$ to yield Riemann surface covering quotients must be many and varied. But, as we shall see, much can be said about this situation in spite of its generality.

8.6 Möbius Transformations

Möbius transforms may be associated with complex matrices by way of the following correspondence:

$$z \to M(z) = \frac{az + b}{cz + d}$$

$$\updownarrow$$

$$\left\{ \begin{pmatrix} a & b \\ c & d \end{pmatrix} : a, b, c, d \in \mathbb{C}, ad - bc \neq 0 \right\} = GL(2, \mathbb{C}).$$

These mappings have many beautiful properties, and they find many and varied application throughout complex analysis. Despite their apparent simplicity, Möbius transformations lie at the heart of several exciting areas of modern mathematical research.

The long Chapter 3 of Needham's book [396] remains the most complete and the most geometrical treatment of these transformations that we have seen in any introductory text.

The reason Needham lavished such extravagant attention upon these deceptively simple transformations is that they are possessed of magical powers, manifesting themselves in multifaceted guises across mathematics and physics. Again, it was Poincaré who was the first to recognize this phenomenon. They are the famous Lorentz transformations (isometries) of Minkowski and Einstein's four-dimensional spacetime, and when the constants are all integers and $ad - bc = 1$, they form the modular group, describing the symmetries of the modular functions so important in modern number theory, and the list goes on ....

The restriction to $ad - bc = 1$ gives the special linear group $SL(2, \mathbb{C})$. Since we obtain the same Möbius transform with $\{-a, -b, -c, -d\}$ as with $\{a, b, c, d\}$, one can quotient to obtain the projective special linear group $\mathrm{PSL}(2, \mathbb{C}) = SL(2, \mathbb{C})/\{I, -I\}$. This is the group of automorphisms of the Riemann sphere $\widehat{\mathbb{C}}$. The automorphism groups for the three simply connected Riemann surfaces are summarized below:

$$\mathrm{Aut}\left(\widehat{\mathbb{C}}\right) \cong PSL_2(\mathbb{C})$$
$$\mathrm{Aut}(\mathbb{C}) = \{f(z) = az + b : a, b \in \mathbb{C}, a \neq 0\}$$
$$\mathrm{Aut}(\mathbb{D}) = \left\{f(z) = \frac{az+b}{\overline{b}z+\overline{a}} : a, b \in \mathbb{C}, |a|^2 + |b|^2 = 1\right\} \cong$$
$$\mathrm{Aut}(\mathbb{H}) = \left\{f(z) = \frac{az+b}{cz+d} : a, b, c, d \in \mathbb{R}, ad - bc = 1\right\} / \{I, -I\}.$$

Note that the last two are automatically isomorphic since, due to the uniformization theorem, $\mathbb{D}$ is conformally equivalent to $\mathbb{H} = \{z \in \mathbb{C} : \ \mathrm{Im} z > 0\}$.

Theorem 8.6.1 *We have* $\mathrm{Aut}(\widehat{\mathbb{C}}) = \mathrm{PSL}_2(\mathbb{C})$.

Proof Given $f \in \mathrm{Aut}(\widehat{\mathbb{C}})$, pick $g \in \mathrm{PSL}_2(\widehat{\mathbb{C}})$ so that $g(f(\infty)) = \infty$. Then $g \circ f \in \mathrm{Aut}(\mathbb{C}) \subset \mathrm{PSL}_2(\mathbb{C})$, so $f \in \mathrm{PSL}_2(\mathbb{C})$. □

Corollary 8.6.2 *The action of* $\mathrm{Aut}(\widehat{\mathbb{C}})$ *is uniquely triply transitive: Any distinct triple of points can be sent to* $(0, 1, \infty)$ *by a unique Möbius transformation.*

Corollary 8.6.3 *The group* $\mathrm{PSL}_2(\mathbb{C})$ *is isomorphic, as a complex manifold, to the space of distinct triples of points on* $\widehat{\mathbb{C}}$.

Circles and lines. A circle $C \subset \widehat{\mathbb{C}}$ is either a line through ∞ or an ordinary Euclidean circle in $\mathbb{C}$.

Theorem 8.6.4 *Möbius transformations send circles to circles, and any two circles are equivalent.*

Proof A circle $x^2 + y^2 + Ax + By + C = 0$ is also given by $r^2 + r(A\cos\theta + B\sin\theta) + C = 0$, and it is easy to transform the latter under $z \mapsto 1/z$, which replaces r by $1/r$ and θ by $-\theta$. □

A group G of topological self-mappings of a Hausdorff space X is called *properly discontinuous* if, for every compact set $K \subset X$, $g(K) \cap K = \emptyset$ for all but finitely many $g \in G$.

Theorem 8.6.5 *Let* U *be a domain in* $\widehat{\mathbb{C}}$, *and let* G *be a group of Möbius transformations which leaves* U *invariant and acts discontinuously in* U. *Then* U/G *is a Riemann surface.*

Proof We know that U/G is a topological space with the quotient topology and that the quotient map $\pi : U \to U/G$ is continuous. As U is connected and π is continuous, it follows that U/G is connected (in fact, arcwise connected). It is also clear that π is an open map for if $A \subset U$, then

$$\pi^{-1}(\pi A) = \bigcup_{g \in G} g(A);$$

thus if A (and therefore $g(A)$) is open, then so is $\pi(A)$.

We now show that U/G is Hausdorff. First, choose distinct points z_1 and z_2 in U, and choose a positive r so that the discs $\mathbb{D}(z_1, r)$, $\mathbb{D}(z_2, r)$ lie in U. If for every $j \in \mathbb{N}$,

$$\pi\left(\mathbb{D}(z_1, r/j)\right) \cap \pi\left(\mathbb{D}(z_2, r/j)\right) \neq \varnothing,$$

then there are some w_j in $\mathbb{D}(z_1, r/j)$ and some g_j in G with $g_j\left(w_j\right) \in \mathbb{D}(z_2, r/j)$. This implies that

$$g_j(K) \cap K \neq \varnothing,$$

where $K = \mathbb{D}(z_1, r) \cup \mathbb{D}(z_1, r)$ (which is compact), and it follows (from discontinuity) that the set $\{g_1, g_2, \ldots\}$ is finite. On a suitable subsequence, $g_j = g$, say, and

$$g\left(z_1\right) = \lim_j g_j\left(w_j\right) = z_2.$$

To prove that U/G is Hausdorff, consider two distinct points, say $\pi\left(z_1\right)$ and $\pi\left(z_2\right)$ in U/G. Thus z_1 and z_2 are in U but not equivalent under G.

It follows that for some n, the disjoint sets $\pi\left(\mathbb{D}(z_1, r/j)\right)$ and $\pi\left(\mathbb{D}(z_2, r/j)\right)$ separate $\pi\left(z_1\right)$ and $\pi\left(z_2\right)$, and these sets are open as π is an open map.

Now we construct an atlas for U/G. For each z in U, we select an open disc $\mathbb{D}(z, R)$ such that $\overline{\mathbb{D}(z, R)} \subset U$ with the properties

$$\begin{aligned} g\left(\mathbb{D}(z, R)\right) &= \mathbb{D}(z, R) && \text{if } g(z) = z \\ g\left(\mathbb{D}(z, R)\right) \cap \mathbb{D}(z, R) &= \varnothing && \text{if } g(z) \neq z; \end{aligned}$$

see [45, Theorems 5.3.11 and 5.3.12, p. 100].

Observe that $\mathbb{D}(z, R) \setminus \{z\}$ contains no fixed points of G. Indeed, if $h(\neq I)$ fixes a point in $\mathbb{D}(z, R)$, then (because of the definition of $\mathbb{D}(z, R)$) h fixes z. The inverse point of z with respect to $\mathbb{D}(z, R)$ is also fixed by h, so there are no fixed points of h in $\mathbb{D}(z, R) \setminus \{z\}$. Recall that if h fixes z, then h is elliptic.

For each w in U, let σ be a Möbius transformation which maps w to zero and $\mathbb{D}(w, Q)$ to the unit disc $\mathbb{D}$. The stabilizer of w in G is of order n, say, and is generated by some elliptic g where

$$\sigma g \sigma^{-1}(z) = z \exp(2\pi i/n), \quad z \in \mathbb{D}.$$

Now let $q(z) = z^j$: This maps $\mathbb{D}$ onto itself and has the property that for all k and for all z in $\mathbb{D}(w, Q)$, we have

$$q\sigma g^k(z) = \left[\sigma g^k \sigma^{-1}(\sigma z)\right]^j = [\sigma(z)\exp(2\pi i k/n)]^j = \sigma(z)^j. \tag{8.5}$$

Observe that this is independent of the integer k.

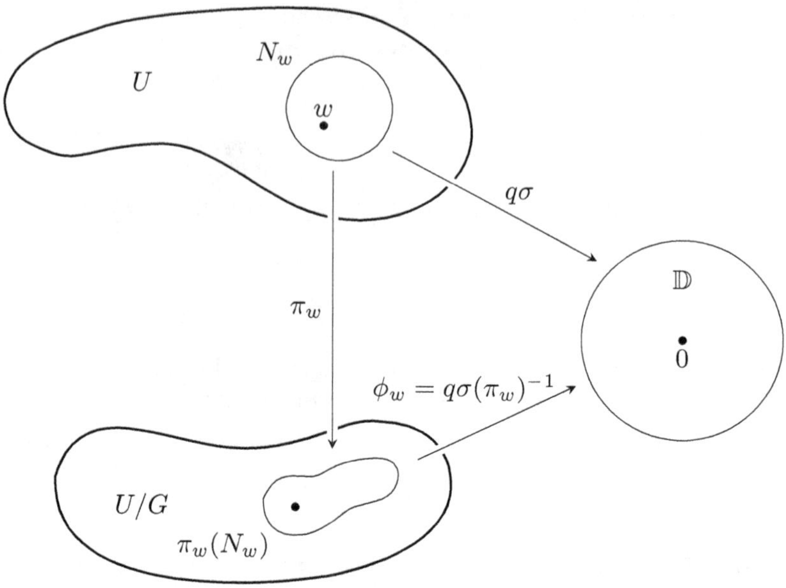

Fig. 8.5 Charts for the pairs

We shall take as charts for U/G the pairs

$$\left(q\sigma\,(\pi_w)^{-1}, \pi_w\,(\mathbb{D}(w, R))\right),$$

where π_w is the restriction of π to $\mathbb{D}(w, Q)$: see Fig. 8.5.

Each point in $\pi_w\,(\mathbb{D}(w, R))$ is mapped by $(\pi_w)^{-1}$ into n points $g^k(z)$, say, where $k = 0, 1, \ldots, n-1$ in $\mathbb{D}(w, R)$. According to (8.5), these map under $q\sigma$ to the same point in $\mathbb{D}$, and thus

$$\phi_w = q\sigma\,(\pi_w)^{-1}$$

is a bijection of $\pi_w\,(\mathbb{D}(w, Q))$ onto $\mathbb{D}$. As the maps q, σ, and π_w are both open and continuous, we see that each ϕ_w is a homeomorphism.

In order to check that the transition maps are analytic, we must first study the maps

$$(\pi_v)^{-1}\,\pi_u, \quad u \neq v. \tag{8.6}$$

Suppose that $\zeta_u \in N_u, \zeta_v \in N_v$, and

$$\pi\,(\zeta_u) = \pi\,(\zeta_v) = \zeta,$$

say: Then for some g in G, we have

$$\zeta_v = g(\zeta_u).$$

Suppose now that ζ_u, and hence ζ_v, are not elliptic fixed points. Then π_v is one-to-one in some neighborhood of ζ_v, and therefore there is a local inverse $(\pi_v)^{-1}$ mapping ζ to ζ_v. The two maps

$$\pi_v g, \quad \pi_u$$

agree with π, and hence with each other, on some neighborhood of ζ_u and take values in $\pi_v(N_v)$. Applying $(\pi_v)^{-1}$, we see that

$$g = (\pi_v)^{-1}\pi_u$$

near ζ_u. We deduce that the maps (8.6) are analytic near points which are not elliptic fixed points of G.

We now show that the transition maps

$$\phi_v(\phi_u)^{-1} \quad (u \neq v)$$

are analytic (where defined): Writing

$$\phi_v = q_v\sigma_v(\pi_v)^{-1}$$

and similarly for u, the situation is illustrated in Fig. 8.6.

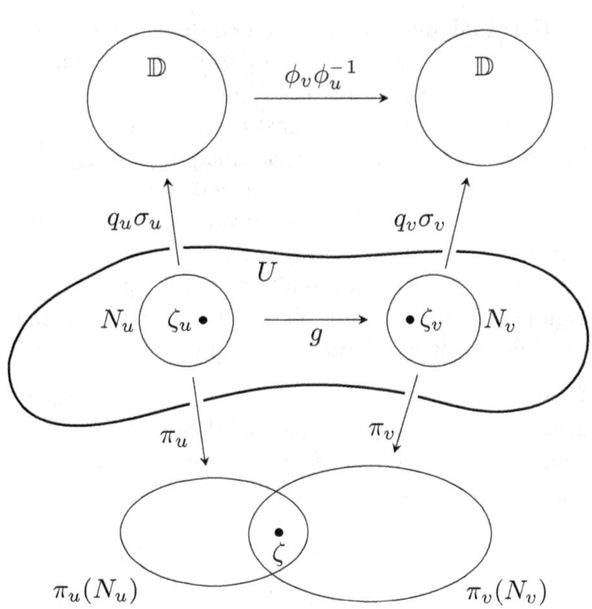

Fig. 8.6 Transition maps are analytic

At points corresponding to the non-fixed points of G, we can compute $\phi_v(\phi_u)^{-1}$ by choosing a single-valued branch of $(q_u)^{-1}$, and the map $\phi_v(\phi_u)^{-1}$ is a composition of analytic maps. At points corresponding to elliptic fixed points, the homeomorphism $\phi_v(\phi_u)^{-1}$ is analytic in a deleted neighborhood of the point in question (by the previous remark) and hence has a removable singularity at this point.

8.6.1 *Fuchsian Groups*

Definition 8.6.6 A group G of Möbius transformations is a *Fuchsian group* if and only if there is some G-invariant disc in which G acts discontinuously.

It turns out that every Riemann surface admits a representation as $\widetilde{S}/G$ where either $\widetilde{S} = \widehat{\mathbb{C}}$ or $\widetilde{S} = \mathbb{C}$ or $\widetilde{S} = \mathbb{H}$ and G acts freely. The group G is a group of Möbius transformations

$$z \to \frac{az+b}{cz+d} \quad (ad - bc = 1),$$

with real coefficients in the case $\widetilde{S} = \mathbb{H}$, and is determined by S, up to conjugation. Furthermore, $\widetilde{S} = \mathbb{H}$ except if $S = \widehat{\mathbb{C}}$ (then $\widetilde{S} = \widehat{\mathbb{C}}$, $G = 1$) or $S = \mathbb{C}$, or $\mathbb{C}\backslash\{0\}$, or S is homeomorphic to a torus (then $\widetilde{S} = \mathbb{C}$).

This is the main special case of the Poincaré-Klein limit circle theorem, conjectured by Klein and by Poincaré in 1882 but completely proved only in 1907, by Poincaré and Koebe.

The usual proof proceeds in two steps. First one constructs the universal covering surface $\widetilde{S}$ of S and the covering group G, so that $S = \widetilde{S}/G$. Then one transfers to $\widetilde{S}$ the complex structure of S. Now $\widetilde{S}$ becomes a Riemann surface, and the covering group G, which is by construction freely acting and properly discontinuous, becomes a group of conformal mappings. Since $\widetilde{S}$ is simply connected, it remains to show that every simply connected Riemann surface is conformally equivalent to $\widehat{\mathbb{C}}$, $\mathbb{C}$, or $\mathbb{H}$. Once this generalized Riemann mapping theorem is established, the remaining assertions follow easily.

The second, more difficult, step is the proof of the mapping theorem. Today an argument based on Perron's method of subharmonic functions and on an idea of M. Heins is standard, see the presentation in Ahlfors's book *Conformal Invariants*.

Example 8.6.7 Whenever $\widetilde{S} = \mathbb{H}$, G is a Fuchsian group that is a properly discontinuous group of non-Euclidean motions (real Möbius transformations). Of course, one can replace $\mathbb{H}$ by another half plane or by a disc, for instance, by the unit disc with the metric $ds = 2|dz|\left(1 - |z|^2\right)^{-1}$.

The group G is called of the *first kind* if it does not act properly discontinuously on any open subset of $\widehat{\mathbb{R}} = \mathbb{R} \cup \{\infty\}$ (or of $|z| = 1$, if $\mathbb{H}$ is replaced by the unit disc). This happens if and only if $S' = S \backslash$(ramification points) has no ideal boundary curves, that is, cannot be represented as $S' \subset S''$, S'' another Riemann surface, with the boundary of S' relative to S containing Jordan arcs.

If G is a Fuchsian group of the first kind, and $\varphi(z)$, $z \in U$, a meromorphic automorphic function for G, then φ is singular at every boundary point of $\mathbb{H}$. Thus the boundary of $\mathbb{H}$ is a natural boundary or "limit circle" for φ hence the name of the theorem.

The canonical Riemannian metric on S is obtained by transferring to $S = \widetilde{S}/G$ the canonical metric on $\widetilde{S}$, that is, either the metric on $\widehat{\mathbb{C}}$ obtained by stereographic projection from the standard metric on $\mathbb{S}^2$, or the Euclidean metric on $\mathbb{C}$, or the Poincaré metric $|dz|/\operatorname{Im} z$ on $\mathbb{H}$. Of course, the first fundamental form of a surface discussed in Sect. 6.1 of Chap. 6 is an important example of Riemannian metrics, but here we will give two examples of Riemannian metrics defined directly in a domain U. These are important for application.

Example 8.6.8 Let U be the unit disc $\mathbb{D}$, and let

$$ds^2 = 4\frac{(dx)^2 + (dy)^2}{\left\{1 - \left(x^2 + y^2\right)\right\}^2}. \tag{8.7}$$

We call this the *Poincaré metric*. If we use the complex coordinate $z = x + iy$, we can write the above metric as

$$ds^2 = \frac{4dzd\overline{z}}{\left(1 - |z|^2\right)^2}, \tag{8.8}$$

where $dz = dx + idy$, $d\overline{z} = dx - idy$.

Example 8.6.9 Let U be the upper half plane defined by $\mathbb{H}$. In terms of the complex coordinate $z = x + iy$, $\mathbb{H}$ is the set of complex numbers with positive imaginary part. On $\mathbb{H}$, consider a Riemannian metric

$$ds^2 = \frac{(dx)^2 + (dy)^2}{y^2} = \frac{dzd\overline{z}}{y^2}. \tag{8.9}$$

We call this also the Poincaré metric.

Remark 8.6.10 Let us explain that the above two examples are essentially the same, although they look different. As some readers may know from complex function theory the transformation

$$z = i\frac{1 - w}{1 + w} \tag{8.10}$$

gives a one-to-one correspondence between $\mathbb{D}$ and $\mathbb{H}$. In fact, since

$$2yi = z - \overline{z} = i\frac{1-w}{1+w} + i\frac{1-\overline{w}}{1+\overline{w}} = \frac{2(1-w\overline{w})}{(1+w)(1+\overline{w})}i, \tag{8.11}$$

we have $y > 0$ if and only if $|w|^2 < 1$. Also, solving (8.10) for w, we get the inverse map of (8.10):

$$w = \frac{i-z}{i+z}. \tag{8.12}$$

Thus we see that (8.10) and (8.12) give a one-to-one complex analytic correspondence between $\mathbb{D}$ and $\mathbb{H}$. From (8.12) we get

$$dw = \frac{-2idz}{(i+z)^2}, \quad d\overline{w} = \frac{2id\overline{z}}{(-i+\overline{z})^2}, \quad \frac{1}{\left(1-|w|^2\right)^2} = \frac{(i+z)^2(-i+\overline{z})^2}{(4y)^2}, \tag{8.13}$$

and by substituting these into (8.8), we get

$$\frac{4dwd\overline{w}}{\left(1-|w|^2\right)^2} = \frac{dzd\overline{z}}{y^2}. \tag{8.14}$$

This shows that under the correspondence between $\mathbb{D}$ and $\mathbb{H}$ given by (8.8) or (8.12), the Riemannian metric (8.8) on $\mathbb{D}$ corresponds to the Riemannian metric (8.9) on $\mathbb{H}$.

In general, given a domain Ω in the (u, v)-plane with a Riemannian metric

$$Edudu + 2Fdudv + Gdvdv \tag{8.15}$$

and a domain Ω' in the (x, y)-plane with a Riemannian metric

$$E'dxdx + 2F'dxdy + G'dydy\,, \tag{8.16}$$

if there is a one-to-one correspondence between Ω and Ω' by differentiable functions

$$u = u(x, y), \quad v = v(x, y), \tag{8.17}$$

which, substituted into (8.15), yields (8.16), then we call the correspondence an *isometry*. If (8.15) and (8.16) correspond to each other under an isometry, then they are regarded as essentially the same.

Example 8.6.11 A *Riemannian metric* on an oriented differentiable surface S is defined by a line element

$$ds^2 = E(dx)^2 + 2Fdxdy + G(dy)^2;$$

here (x, y) are any local coordinates defined in a coordinate patch, and the E, F, G are measurable functions of (x, y), which transform like the components of a symmetric covariant tensor, and the matrix

$$\begin{bmatrix} E & F \\ -F & G \end{bmatrix}$$

is positive definite a.e.

Local coordinates (x, y) are called *isothermal* if $E = G, F = 0$. It is known that for a given Riemannian metric there is always, locally, an isothermal coordinate system, but the proof of this fact is not easy. An isothermal coordinate system is essentially the same as a complex coordinate system (see, e.g., [298]).

An *orientable surface* is a complex manifold (of dimension 1). We suppose the surface to be C^∞ and define on it a positive definite Riemannian metric. Isothermal coordinates on a Riemannian manifold are local coordinates where the metric is conformal to the Euclidean metric. This means that, in isothermal coordinates, the Riemannian metric locally has the form

$$ds^2 = \exp^{2\beta}((dx)^2 + (dy)^2),$$

where β is a smooth function. Korn and Lichtenstein proved that isothermal coordinates exist around any point on a two-dimensional Riemannian manifold (see [317, 362]).

The n-dimensional differential manifold is a notion which is much more general than the notion of analytical (complex) manifold. At this point, the reader may pose the question:

> When is an (orientable) two-dimensional (Riemannian) differential manifold M a Riemann surface, that is, when one can introduce analytical coordinates (holomorphic atlas) on M?

Let (S, g) be a two-real dimensional orientable differential manifold with a metric g. In local coordinate $(x, y) : U \subset S \to \mathbb{R}^2$, one has

$$g = adx^2 + 2bdxdy + cdy^2, \quad a > 0, c > 0, ac - b^2 > 0\,. \tag{8.18}$$

Definition 8.6.12 Two metrics g and $\widetilde{g}$ are called *conformally equivalent* if they differ by a function on S

$$g \sim \widetilde{g} \Leftrightarrow g = f\widetilde{g}, \quad f : S \to \mathbb{R}_+\,. \tag{8.19}$$

The relation (8.19) defines the classes of conformally equivalent metrics.

Note that the angles between tangent vectors are the same for conformally equivalent metrics.

We show that there is one-to-one correspondence between the conformal equivalence classes of metrics on an orientable 2-manifold S and the complex structures on S. In terms of the complex variable[1] $z = x + iy$, one rewrites the metric as

$$g = Adz^2 + 2Bdzd\overline{z} + \overline{A}d\overline{z}^2, \quad A \in \mathbb{C}, B \in S, B > |A| \tag{8.20}$$

with

$$a = 2B + A + \overline{A}, \quad b = i(A - \overline{A}), \quad c = 2B - A - \overline{A}\,. \tag{8.21}$$

Definition 8.6.13 A coordinate $w : U \to \mathbb{C}$ is called *conformal* if the metric in this coordinate is of the form

$$g = e^{\phi} dw d\overline{w}, \tag{8.22}$$

i.e., it is conformally equivalent to the standard metric of $\mathbb{R}^2 = \mathbb{C}$:

$$dwd\overline{w} = du^2 + dv^2, \quad w = u + iv\,.$$

We start with the following theorem:

Theorem 8.6.14 *Every compact Riemann surface admits a conformal Riemannian metric.*

Proof Each point $P \in S$ possesses a local parameter $z(P) : U(P) \to \mathbb{D}(P, r) \subset \mathbb{C}$, where $\mathbb{D}(P, r)$ is a small open disc. Since S is compact, there exists a finite covering $\cup_{i=1}^{n} U(P_i) = S$. For each i, choose a smooth function $g_i : \mathbb{D}(P_i, r_i) \to S$ with

$$g_i > 0 \quad \text{on } \mathbb{D}(P_i, r_i), \quad g_i = 0 \quad \text{on } \mathbb{C}\backslash\mathbb{D}(P_i, r_i);$$

$g_i dz(P_i) d\overline{z}(P_i)$ is a conformal metric on $U(P_i)$. The sum of these metrics over $i = 1, \dots, n$ yields a conformal metric on S.

Let us show how one finds conformal coordinates. The metric (8.20) can be written as follows (we suppose $A \neq 0$):

$$g = s(dz + \mu d\overline{z})(d\overline{z} + \overline{\mu} dz), \quad s > 0, \tag{8.23}$$

where

$$\mu = \frac{\overline{A}}{2B}\left(1 + |\mu|^2\right), \quad s = \frac{2B}{1 + |\mu|^2}.$$

Here $|\mu|$ is a solution of the quadratic equation

[1] Note that the complex coordinate z is not compatible with the complex structure we will define on S with the help of g.

$$|\mu| + \frac{1}{|\mu|} = \frac{2B}{|A|},$$

which can be chosen $|\mu| < 1$

$$|\mu| = \frac{1}{|A|}\left(B - \sqrt{B^2 - |A|^2}\right), \tag{8.24}$$

Comparing (8.23) and (8.22), we get

$$dw = \lambda(dz + \mu d\overline{z})$$

or

$$dw = \lambda(d\overline{z} + \overline{\mu} dz).$$

In the first case the map $w(z, \overline{z})$ satisfies the equation

$$w_{\overline{z}} = \mu w_z \tag{8.25}$$

and preserves the orientation $w : U \subset \mathbb{C} \to V \subset \mathbb{C}$ since $|\mu| < 1$: For the map $z \to w$ written in terms of the real coordinates

$$z = x + iy, \quad w = u + iv,$$

one has

$$du \wedge dv = |w_z|^2 \left(1 - |\mu|^2\right) dx \wedge dy.$$

In the second case $w : U \to V$ inverses the orientation. □

Definition 8.6.15 The equation

$$w_{\overline{z}} = \mu w_z \tag{8.26}$$

is called the *Beltrami equation*, and $\mu(z, \overline{z})$ is called the *Beltrami coefficient*.

Let us postpone for a moment the discussion of the proof of existence of solutions to the Beltrami equation, and let us assume that this equation can be solved in a small neighborhood of any point of S.

Theorem 8.6.16 *Let S be a two-dimensional orientable manifold with a metric g and an oriented atlas $\left((x_\alpha, y_\alpha) : U_\alpha \to \mathbb{R}^2\right)_{\alpha \in A}$ on S. Let $(x, y) : U \subset S \to \mathbb{R}^2$ be one of these coordinate charts with a point $P \in U, z = x + iy, \mu(z, \overline{z})$—the Beltrami coefficient, and $w_\beta(z, \overline{z})$ be a solution to the Beltrami equation (8.26) in a* *neighborhood* $V_\beta \subset V = z(U)$ *with* $P \in U_\beta = z^{-1}(V_\beta)$. *Then the coordinate* w_β *is conformal, and the atlas* $\left(w_\beta : U_\beta \to \mathbb{C}\right)_{\beta\in B}$ *defines a complex structure on* S.

Proof To prove the holomorphicity of the transition function, let us consider two local parameters $w : U \to \mathbb{C}$, $\widetilde{w} : \widetilde{U} \to \mathbb{C}$ with a non-empty intersection $U \cap \widetilde{U} \neq \emptyset$. Both coordinates are conformal

$$g = e^{\phi} dw d\overline{w} = e^{\widetilde{\phi}} d\widetilde{w} d\overline{\widetilde{w}},$$

which happens in one of the two cases

$$\frac{\partial \widetilde{w}}{\partial \overline{w}} = 0 \text{ or } \frac{\partial \overline{\widetilde{w}}}{\partial w} = 0 \tag{8.27}$$

only. The transition function $\widetilde{w}(w)$ is holomorphic and not antiholomorphic since the map $w \to \widetilde{w}$ preserves orientation. □

Repeating the arguments of the proof of Theorem 8.6.16, one immediately observes that conformally equivalent metrics generate the same complex structure. Finally, we obtain the following:

Theorem 8.6.17 *Conformal equivalence classes of metrics on an orientable 2-manifold S are in one-to-one correspondence with the complex structures on S.*

Remark 8.6.18

1. The assumption of orientability is (obviously) not needed.

   This observation is due to Korn and Lichtenstein. The problem reduces to the theorem on (local) solution of (nonlinear) elliptic equations of second order with smooth coefficients.

   One can find isothermal coordinates near a given point of S, provided the three functions E, F, G are real-analytic (Gauss, [477, pp. 463-475]) or Hölder continuous (Korn [317], 1914, Lichtenstein [362], 1916) or even only measurable and satisfying a.e. an inequality of the form $(E+G)^2 / \left(EG - F^2\right) < \text{const}$ (Morrey [383], 1938). The geometric meaning of this inequality is that the eccentricity of the indicatrix of the metric is bounded.

   An oriented surface with a Riemannian metric is made into a Riemann surface by calling a function defined near a point P holomorphic if it is so as a function of $z = x + iy$, where (x, y) are isothermal coordinates defined near P.

   An elementary and rather straightforward proof of their theorem is given by S.S. Chern [110]. A similar proof was given by L. Bers in his mimeographed notes on Riemann surfaces, New York University, 1951–1952, p. 15–35.

   There exist local isothermal parameters x, y so that locally the metric can be written as

   $$ds^2 = \lambda^2 \left(dx^2 + dy^2\right), \quad \lambda > 0$$

or $ds^2 = \lambda^2 dz d\overline{z}$, where $z = x+iy$, the orientation of the manifold being defined by $dx \wedge dy = \frac{i}{2} dz \wedge d\overline{z}$. If w is another local coordinate, then we will have

$$ds^2 = \lambda^2 dz d\overline{z} = \mu^2 dw d\overline{w}$$

because ds^2 is globally defined. It follows that dw is a multiple of dz or $d\overline{z}$. If we assume that the complex coordinates z and w define the same orientation, then dw must be a multiple of dz. This means that w is a holomorphic function of z, and the surface becomes a complex manifold.

2. Clearly the kernel of the theorem and the major difficulty of the proof reside in the verification that there exist isothermal coordinates.
3. In general, the differential mapping $f : M \to M'$ of Riemann manifolds (M, g), (M', g') is conformal if

$$g'_{f(x)}(df \cdot X, df \cdot Y) = \lambda(x) g_x(X, Y), \quad \text{for } x \in M, X, Y \in TM_x, \lambda(x) > 0 .$$

The conformal property of f means that the mapping f preserves angles between tangent vectors (thus between intersecting curves.)

4. A holomorphic function on a compact domain in $\mathbb{C}$ (or locally holomorphic mapping of Riemann surface) is, obviously, conformal. This observation was a cornerstone of the "geometric theory of functions," that is, the theory of Riemann surfaces. At this point we should recall the central theorem of the theory of Riemann surfaces, which is

8.7 Covering Spaces and Automorphisms in General

While much of the material of the previous two sections was very specific, underlying the considerations there and illustrated by them were some general principles. In this section, we shall make these principles explicit. This is well worthwhile in spite of the tedium of the notation involved because these principles have wide applications.

Definition 8.7.1 Given two covering spaces $p_1 : Y_1 \to X$ and $p_2 : Y_2 \to X$ of the same space X, a *covering transformation* $F : (Y_1, p_1) \to (Y_2, p_2)$ is a continuous map $F : Y_1 \to Y_2$ such that $p_1 = p_2 \circ F$.

Proposition 8.7.2 *Suppose that $\widetilde{M}$ and M are complex manifolds with $\widetilde{M}$ simply connected, and further suppose that $\pi : \widetilde{M} \to M$ is a holomorphic mapping which is also a covering in the sense of topology so that, in particular, π is a local holomorphic diffeomorphism. Let Γ be the subgroup of $\mathrm{Aut}(\widetilde{M})$ consisting of the covering transformations of $\widetilde{M}$ over M. In this case these transformations are necessarily holomorphic. Then:*

(a) *For each $f \in Aut(M)$, there is an element $\widehat{f}$ of $Aut(\widetilde{M})$ such that $f \circ \pi \equiv \pi \circ \widehat{f}$. Any such $\widehat{f}$ maps each orbit of Γ bijectively onto another orbit of Γ. Moreover, given $p \in M$, $\widehat{p} \in \widetilde{M}$ with $\pi(\widehat{p}) = p$, and $\widehat{q} \in \widetilde{M}$ with $\pi(\widehat{q}) = f(p)$, there is exactly one such $\widehat{f}$ with $\widehat{f}(\widehat{p}) = \widehat{q}$.*

(b) *Conversely, if $F \in Aut(\widetilde{M})$ and if F maps each orbit of Γ bijectively onto another orbit of Γ, then there is exactly one $f \in Aut(M)$ such that $\pi \circ F \equiv f \circ \pi$.*

Proof If $f : M \to M$ is a holomorphic mapping, whether an automorphism or not, then there is exactly one continuous mapping $\widehat{f} : \widetilde{M} \to \widetilde{M}$ with $f \circ \pi = \pi \circ \widehat{f}$ and $\widehat{f}(\widehat{p}) = \widehat{q}$. This is just the standard lifting argument for covering spaces $\widetilde{M} \to M$ with $\widetilde{M}$ simply connected. Since π is a holomorphic local diffeomorphism, it follows that $\widehat{f}$ must be holomorphic. To see that such an $\widehat{f}$ is an automorphism of $\widetilde{M}$, let $\widehat{g}$ be the lift to $\widetilde{M}$ of $f^{-1} : M \to M$ with $\widehat{g}(\widehat{q}) = \widehat{p}$, i.e., $\pi \circ \widehat{g} = f^{-1} \circ \pi$. Now, since π is a local diffeomorphism and since $g \circ f$ is the identity, it follows that $\widehat{g} \circ \widehat{f}$ is the identity in a neighborhood of $\widehat{p}$. By uniqueness of analytic continuation, $\widehat{g} \circ \widehat{f}$ is the identity on all of $\widetilde{M}$. Similarly, $(\widehat{f} \circ \widehat{g})(\widehat{q}) = \widehat{f}(\widehat{p}) = \widehat{q}$ and $\widehat{f} \circ \widehat{g}$ is also the identity on $\widetilde{M}$. So $\widehat{f} \in \text{Aut}(\widetilde{M})$.

Now, for $x \in \widetilde{M}$ and $\gamma \in \Gamma$,

$$\pi \widehat{f}(\gamma(x)) = f\pi(\gamma(x)) = f(\pi(x)).$$

So $\widehat{f}(\gamma(x))$ is in the Γ-orbit of $\widehat{f}(x)$ since the orbits of Γ are exactly the sets of the form π^{-1} (points of $\{M\}$). Now, with $\widehat{g}$ as given, the same holds for $\widehat{g}$ since $\widehat{g}$ is a lift of f^{-1}: The map $\widehat{g}$ sends Γ-orbits to Γ-orbits. Since $\widehat{f}$ and $\widehat{g}$ are inverses, the maps by $\widehat{f}$ of Γ-orbits to Γ-orbits must each be bijective.

For the second statement, note that under the hypothesis that $F : \widetilde{M} \to \widetilde{M}$ maps Γ-orbits (i.e., sets of the form $\pi^{-1}(\{p\})$ to other sets of the same form), set-theoretic considerations show that F induces a map on M first, set $f(x) = \pi \circ F(\widehat{x})$ for any $\widehat{x} \in \pi^{-1}(\{x\})$; this f is well defined by hypothesis. The full hypothesis on F, including the bijectivity on inverse image sets, means that F^{-1} also pushes down to a well-defined map $g : M \to M$. It is easy to check that $f \circ g$ and $g \circ f$ are the identity on M, so that $f \in \text{Aut}(M)$. This "general nonsense" argument becomes clearer if one thinks of it in terms of the orbits of the covering transformation group Γ acting on $\widetilde{M}$. The hypothesis of the second statement says then that F maps each Γ-orbit bijectively onto another Γ-orbit, and so does F^{-1}. The latter condition is needed, since the orbits are in general infinite sets. As an instance, the map $z \mapsto 2z$ maps orbits of Γ for $\exp : \mathbb{C} \to \mathbb{C}\setminus\{0\}$ into themselves: If $z_1 - z_2 = 2\pi i k$, then $2z_1 - 2z_2 \in 2\pi i\mathbb{Z}$. But the induced map on $\mathbb{C}\setminus\{0\}$ is $z \mapsto z^2$, which has no inverse.

Now the set of all $F \in \text{Aut}(\widetilde{M})$ such that both F and F^{-1} map orbits to orbits is a subgroup of $\text{Aut}(\widetilde{M})$. Moreover, this subgroup is closed. One can see this from the definition of covering space and the group Γ. Suppose that the sequence $\{F_j\}$ converges uniformly on compact sets to some F_0. For fixed $\gamma_0 \in \Gamma$ and $p_0 \in \widetilde{M}$, the images $F_j(\gamma_0(p_0))$ by hypothesis belong to the same orbit of $F_j(p_0)$ so that $F_j(\gamma_0(p_0)) = \gamma_j(F_j(p_0))$ for some unique $\gamma_j \in \Gamma$. Now choose a neighborhood of $F_0(\gamma_0(p_0))$, say U, (sufficiently small) so that the conditions $\sigma_1(U) \cap \sigma_2(U) \neq \emptyset$ and $\sigma_1, \sigma_2 \in \Gamma$ imply $\sigma_1 = \sigma_2$. This is possible by the definition of covering space. Then, for j_0 so large that $j \geq j_0$ implies that $F_j(\gamma_0(p_0)) \in U$, it must be that γ_j is independent of j. For clearly $F_0(p_0) \in \gamma_{j_1}^{-1}(U) \cap \gamma_{j_2}^{-1}(U)$ for any $j_1, j_2 > j_0$. Thus $F_0(\gamma_0(p_0)) = \gamma_\infty(F_0(p_0))$, where γ_∞ is the eventually constant value of γ_j. In particular, F_0 maps orbits to orbits. (A similar argument applies to F^{-1} to show that F^{-1} also maps orbits to orbits.) □

Proposition 8.7.3 *With $\pi : \widetilde{M} \to M$ as in Proposition 8.7.2, let G be the subset of $Aut(\widetilde{M})$ consisting of those $F \in Aut(\widetilde{M})$ such that F maps each orbit of Γ bijectively to another orbit of Γ. The covering transformation group Γ of $\widetilde{M}$ over M is a closed, normal subgroup of G, and $Aut(M)$ is homeomorphically isomorphic to G/Γ.*

Proof Almost everything follows from arguments already presented. We check only the normality of Γ in G. Suppose that $F \in G$ and $\gamma \in \Gamma$. We need to see that $F \circ \gamma \circ F^{-1}$ is a covering transformation. Obviously $F \circ \gamma \circ F^{-1} \in \mathrm{Aut}(\widetilde{M})$, so we need only check that $\pi \circ \left(F \circ \gamma \circ F^{-1}\right) = \pi$. Now, for $p \in \widetilde{M}$, we see that $\gamma \circ F^{-1}(p)$ is an element of the set of Γ-orbits of $F^{-1}(p)$, and hence $F\left(\gamma\left(F^{-1}(p)\right)\right)$ is an element of the Γ-orbit of $F\left(F^{-1}(p)\right)$. This is so because F applied to a point of an orbit of $q \in \widetilde{M}$ is equal to the orbit of the point $F(q)$ by hypothesis. Note that

$$\pi\left(\Gamma\text{-orbit of } F\left(F^{-1}(p)\right)\right) = \pi(\Gamma\text{-orbit of } p) = \{\pi(p)\}.$$

The remainder of the proof of the proposition, which follows the patterns already established, is left to the reader. □

Corollary 8.7.4 *If $Aut(\widetilde{M})$ is a Lie group, then $Aut(M)$ is a Lie group.*

Proof Let G be the same as in Proposition 8.7.3. Since a closed subgroup of a Lie group is a Lie group, G is a Lie group. A quotient of a Lie group by a closed, normal subgroup is again a Lie group. Hence G/Γ is a Lie group. □

Proposition 8.7.5 *If $Aut(\widetilde{M})$ has compact isotropy subgroups, then so does $Aut(M)$.*

Proof Given $p \in M$, choose $\widehat{p} \in \widetilde{M}$ such that $\pi(\widehat{p}) = p$. By Proposition 8.7.2 (a), we see that, for each $f \in I_p$, there is exactly one $\widehat{f} \in I_{\widehat{p}}$ with $\pi \circ \widehat{f} = f \circ \pi$. And $\widehat{f}$ maps each orbit of Γ bijectively onto another orbit. Also, by 8.7.2 (b), the set of all $F \in I_{\widehat{p}}$ that map each Γ-orbit bijectively to a Γ-orbit is exactly the set of such $\widehat{f}$. Moreover, this set of F is a closed subgroup, say $G_{\widehat{p}}$ of $I_{\widehat{p}}$, by the proof of Proposition 8.7.3. $G_{\widehat{p}}$ is compact since $I_{\widehat{p}}$ is, by hypothesis. And with G as in Proposition 8.7.3, $G_{\widehat{p}} \subset G$ maps, under the continuous map $G \to G/\Gamma$, onto a subgroup corresponding to I_p under the isomorphism $\mathrm{Aut}(M) \cong G/\Gamma$. Indeed $G_{\widehat{p}}$ maps isomorphically onto the subgroup of G/Γ corresponding to I_p. In particular, since $G \to G/\Gamma$ is continuous and $G_{\widehat{p}}$ is compact, it follows that I_p is compact.

□

In the case when M has an $\text{Aut}(M)$-invariant metric, all the previous considerations become considerably simplified.

Theorem 8.7.6 *Let M be a compact, complex manifold. Suppose that $d : M \times M \to \mathbb{R}$ is a metric space structure on M, with the property that the metric space topology for d is the same as the manifold topology. If d is $Aut(M)$ invariant in the sense that $d(f(p), f(q)) = d(p, q)$ for all $f \in Aut(M)$ and all $p, q \in M$, then $Aut(M)$ is a compact Lie group.*

Proof If $\{f_j\}$ is a sequence in $\text{Aut}(M)$, then, by the invariance of d, the sequence is d-equicontinuous and indeed d-Lipschitz continuous with constant 1. By the Arzelá-Ascoli theorem and the compactness of M, there is a subsequence $\{f_{j_k}\}$ that converges uniformly on M to a continuous d-isometric map $f_0 : M \to M$. Passing again to a subsequence if necessary, we can suppose that the sequence $\left\{f_{j_k}^{-1}\right\}$, which is also d-equicontinuous, converges uniformly on M to a limit $g_0 : M \to M$, also d-isometric. It is now elementary to see that $f_0 \circ g_0$ and $g_0 \circ f_0$ are both the identity map of M to M. Also, d-uniform convergence implies uniform-on-compact-subsets convergence in local coordinates, since the d-topology is equivalent to the manifold topology. So f_0 is holomorphic and invertible, and $f_0 \in \text{Aut}(M)$.

Since $\text{Aut}(M)$ is compact, it is certainly locally compact. By the Bochner-Montgomery theorem (see [190, Theorem 1.3.11, p. 13]), $\text{Aut}(M)$ is a Lie group. □

We turn now to some general circumstances under which a complex manifold M has an $\text{Aut}(M)$-invariant metric.

Proposition 8.7.7 *If $\pi : \widetilde{M} \to M$ is a holomorphic covering space with $\widetilde{M}$ simply connected and if $\widetilde{H}$ is a Hermitian metric on $\widetilde{M}$ that is invariant under the subgroup Γ of $Aut(\widetilde{M})$ consisting of the covering transformations of $\pi : \widetilde{M} \to M$, then there is a Hermitian metric H on M such that π is a local isometry, and H is uniquely determined by this condition. If, in addition, $\widetilde{H}$ is $Aut(\widetilde{M})$-invariant, then H is $Aut(M)$-invariant.*

Proof The proof of the existence and uniqueness of H follows a by-now familiar pattern. To determine H at $p \in M$, choose $\widehat{p}$ with $\pi(\widehat{p}) = p$. Then define H at p by requiring $d\pi|_{\widehat{p}}$ to be isometric from $\widetilde{H}$ at $\widehat{p}$ to H at p. Since $d\pi|_{\widehat{p}}$ is a vector space isomorphism, this gives one and only one H at p. The metric H is well defined since any other element of $\pi^{-1}(p)$ has the form $\gamma(\widehat{p})$ for some $\gamma \in \Gamma$ and γ is an isometry of $\widetilde{H}$.

For the second statement, it suffices to apply Proposition 8.7.2. To wit, if $f \in \text{Aut}(M)$, then there is an $\widehat{f} \in \text{Aut}(\widetilde{M})$ with $f \circ \pi = \widehat{f} \circ \pi$. Since $\widehat{f}$ is $\widetilde{H}$-isometric and since π is locally $\widetilde{H}$-to- H isometric, we see that f must be H-isometric. □

Proposition 8.7.7 was more or less obvious. What is somewhat less obvious is that the completeness of $\widetilde{H}$ and H happens together.

An even more general result holds:

Proposition 8.7.8 *Let $\widetilde{M}$ and M be Riemannian manifolds equipped with metrics $\widetilde{H}$ and H, respectively. Suppose that $\pi : \widetilde{M} \to M$ is a covering and that π is* *locally isometric. Then the $\widetilde{M}$ metric $\widetilde{H}$ is complete if and only if the M metric H is complete.*

Remark 8.7.9 This is a standard result in Riemannian geometry.

Sketch of the Proof. Use the fact that completeness in the Cauchy sense is equivalent for a Riemannian manifold to infinite extendibility of geodesics (see [299]). Geodesics in M can be lifted to geodesics in $\widetilde{M}$ because continuous curves can always be lifted. It is also the case that the image under π of a geodesic in $\widetilde{M}$ is a geodesic in M. Thus infinite extendibility in $\widetilde{M}$ happens precisely when it does so in M.

The most obvious application of Proposition 8.7.8 is to quotients of $\mathbb{D}$ with its Poincaré metric.

Corollary 8.7.10 *If $\pi : \mathbb{D} \to M$ is a holomorphic covering space of a Riemann surface M, then the H-metric (in our previous terminology) on M is complete.*

The Poincaré metric on the disc $\mathbb{D}$ is of course complete. To check the corollary by hand, from the geodesic extendibility viewpoint, it is enough to deal with geodesics through 0. These are, up to parameter, straight lines: The form of the metric in polar coordinates shows immediately that these are minimal-length connections, hence geodesics. With Poincaré metric arclength parameter, they have the form in polar coordinates given[2] by

$$\sigma_{\theta_0}(t) = \left(\frac{e^t - 1}{e^t + 1}, \theta_0\right), \qquad \theta_0 \in [0, \pi).$$

In particular, the geodesics σ_{θ_0} are defined on $(-\infty, +\infty)$. So $\mathbb{D}$ is complete. This completeness is also clear from the compactness of closed metric balls (the equivalence of all these different notions of completeness is the famous Hopf-Rinow theorem, treated in all Riemannian geometry texts, e.g., [229, 299, 410]). Explicitly

$$\left\{z \in \mathbb{C} : \mathrm{dis}_{\widetilde{H}}(0, z) \leq r\right\} = \left\{z \in \mathbb{C} : |z| \leq \left(e^r - 1\right) / \left(e^r + 1\right)\right\},$$

[2] To check for arclength parameterization, calculate

$$\frac{dr}{dt} = \frac{d}{dt}\left(\frac{e^t - 1}{e^t + 1}\right) = \frac{2e^t}{(1 + e^t)^2};$$

hence, at the point $\left(\left(e^t - 1\right) / \left(e^t + 1\right), \theta_0\right)$, the length of the vector $\sigma'(t)$ is

$$\left(\frac{2e^t}{(1 + e^t)^2}\right) \Big/ \left(\frac{2}{1 - \left(\frac{e^t - 1}{e^t + 1}\right)^2}\right) = 1.$$

which is compact in $\mathbb{D}$. So the (quotient) H-metrics on Riemann surfaces covered by $\mathbb{D}$ are necessarily complete.

8.7.1 Compact Quotients of $\mathbb{D}$ and Their Automorphisms

In this section we shall investigate the automorphism groups of compact Riemann surfaces. We have already discussed the situation for the Riemann sphere and for the torus quotients of $\mathbb{C}$. In actuality, all other compact Riemann surfaces are quotients of $\mathbb{D}$ or, in other words, their universal covering space is biholomorphic to $\mathbb{D}$. We almost have this result already, but not quite. The "not quite" arises from the fact that we need to show that a torus can only arise as a quotient of $\mathbb{C}$; it cannot have $\mathbb{D}$ as a (holomorphic) universal cover. Before proving this statement, we quickly remind the reader of some basic results about the topology of compact surfaces (see, e.g., [371]).

According to the nineteenth century work of Jordan and Möbius, every compact orientable surface M (i.e., topological 2-manifold) is homeomorphic either to the 2-sphere S^2 or to the 2-sphere with g handles attached (here g is a finite, positive integer). The number g is uniquely determined; g is called the *genus* of the surface.[3] The homology groups of the surface M are:

- $H_0(M, \mathbb{Z}) \cong \mathbb{Z}$
- $H_1(M, \mathbb{Z}) \cong \mathbb{Z} \oplus \cdots \oplus \mathbb{Z}$ (2g times)
- $H_2(M, \mathbb{Z}) \cong \mathbb{Z}$
- $H_k(M, \mathbb{Z}) = 0$ if $k > 2$

In particular, the Euler characteristic of M is $2-2g$.[4] No two surfaces of different genus are homeomorphic or even of the same homotopy type. This assertion follows immediately from the homology results, e.g., because the first homology group is the abelianization of the first homotopy group (i.e., the fundamental group) and that group is clearly a homotopy type invariant. Or one can use directly the homotopy type invariance of homology. If $g = 0$, then the fundamental group of a surface M of genus g is the one-element group $\{0\}$. If $g \geq 1$, then the fundamental group has $2g$ generators $a_1, b_1, \ldots, a_g, b_g$ subject to the one relation

$$a_1 b_1 a_1^{-1} b_1^{-1} a_2 b_2 a_2^{-1} b_2^{-1} \cdots a_g b_g a_g^{-1} b_g^{-1} = 1.$$

The group $\pi_1(M)$ is noncommutative if $g \geq 2$.

We now turn to the result that shows the relationship between the topology of the surface and the complex structure of the universal cover.

[3] By convention, the sphere S^2, having no handles, is considered to have genus 0.

[4] Note that when $g = 0$ or M is the sphere, then $H_1(M, \mathbb{Z}) = 0$, so the formulae still apply.

Theorem 8.7.11 (Uniformization of Compact Riemann Surfaces: Poincaré and Koebe) *Let M be a compact Riemann surface of genus g and $\pi : \widetilde{M} \to M$ its (holomorphic) universal cover. Then:*

(1) If $g = 0$, then $\widetilde{M} = M$ and M is conformally equivalent to the Riemann sphere $\mathbb{C} \cup \{\infty\}$;
(2) If $g = 1$, then $\widetilde{M}$ is conformally equivalent to $\mathbb{C}$ and M is obtained from $\mathbb{C}$ by quotienting out by a group of the form $\Gamma = \{m\omega_1 + n\omega_2 : m, n \in \mathbb{Z}\}$, with ω_1 and ω_2 being $\mathbb{R}$-linearly independent complex numbers.
(3) If $g \geq 2$, then $\widetilde{M}$ is conformally equivalent to $\mathbb{D}$ and M is $\mathbb{D}/\Gamma$. Here Γ is a subgroup of Aut($\mathbb{D}$).

Proof We have already covered all the issues involved with this argument save one: We need to show that a surface of genus 1 cannot arise as a quotient of $\mathbb{D}$. This can be established by working with the geometry of $\mathbb{D}$ and its holomorphic quotients in terms of the group actions of covering transformations, but it is much quicker to use a result of classical differential geometry.

The torus, the topological surface of genus 1, has Euler characteristic equal to 0. The Gauss-Bonnet theorem thus yields that $\int K dA = 0$ for any Riemannian metric on the torus, where K is the Gauss curvature and dA denotes integration with respect to the Riemannian metric area element. Now the Gauss curvature of the Poincaré metric is $\equiv -1$, and hence so is the Gauss curvature of the H-metric on any covering space quotient of $\mathbb{D}$. Thus no topological torus admits such an H-metric. □

The sudden incursion of a result from pure Riemannian geometry may at first be startling. So we outline how the impossibility of covering a torus holomorphically by the unit disc $\mathbb{D}$ can be proved using only direct ideas about the quotienting possibilities that could arise. Refer to Fig. 8.7.

Specifically, if M is a Riemann surface that is topologically $S^1 \times S^1$, then associated with this situation we could find a surface $\widetilde{M}$ covering $S^1 \times S^1$ where $\widetilde{M}$ is topologically $\mathbb{R} \times S^1$. This is done simply by "unwinding" the first S^1 factor and leaving the second factor alone. Now if M were holomorphically covered by $\mathbb{D}$, then $\widetilde{M}$ could also be taken to be holomorphically covered by $\mathbb{D}$ and the covering $\widetilde{M} \to M$ to be holomorphic as well. Both coverings would be isometric for the "H-metrics" as usual. Now choose a sequence $\{p_j\}$ diverging to infinity in one end of the topological cylinder $\widetilde{M}$ and $\{q_j\}$ diverging to infinity in the other end. Choose a closed curve C in $\widetilde{M}$ going around the cylinder in the obvious sense of those words so that $\widetilde{M} \backslash C$ has two components, both unbounded. Let, for each j, γ_j be an H-metric geodesic from p_j to q_j with minimal length. Then γ_j intersects C at least once. Let ξ_j be one such point of intersection. Now parameterize γ_j by arclength with $\gamma_j(0) = \xi_j$. By compactness of C, there is an arclength parameter geodesic $\gamma_0 : (-\infty, +\infty) \to \widetilde{M}$ with $\gamma_0(0)$ a point of C which is minimizing between any two of its points. It is in a sense a minimal connection between the two ends of $\widetilde{M}$. [**Note:** To get γ_0, pass to a subsequence γ_{j_k} of the γ_js with ξ_j converging and with $\gamma_j'(0)$ converging. Then let γ_0 be the geodesic with $\gamma_0(0) = \lim \gamma_j(0)$

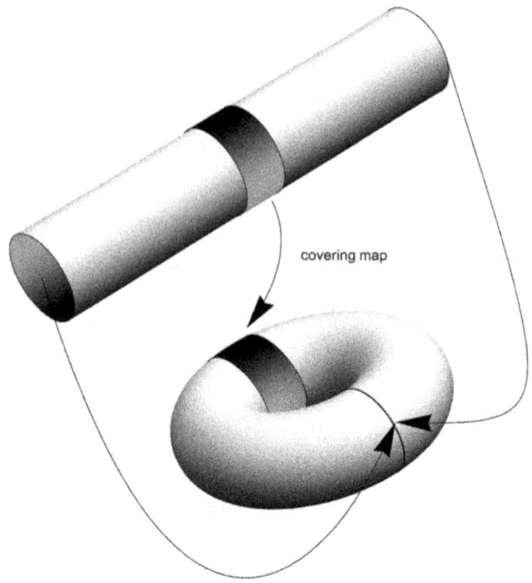

Fig. 8.7 The covering of the torus

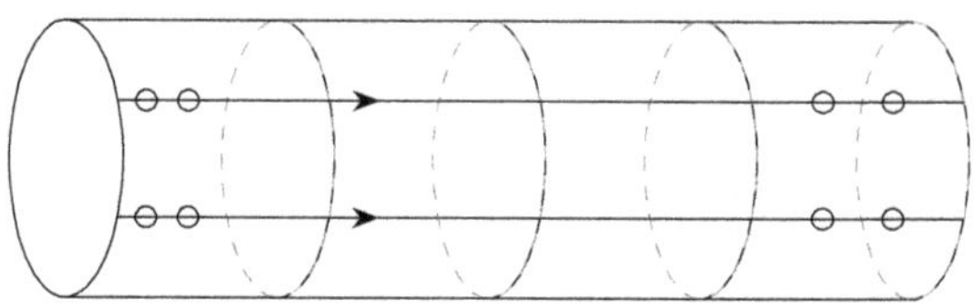

Fig. 8.8 Behavior of geodesics on $S^1 \times R$

and $\gamma_0'(0) = \lim \gamma_j'(0)$.] Then $\widetilde{M} \setminus \{\gamma_0(t) : t \in (-\infty, +\infty)\}$ is isometric to an open subset of $\mathbb{D}$ (in the Poincaré metric) bounded by two nonintersecting geodesics, say σ and τ. Then $\widetilde{M}$ is obtained by gluing the two geodesics together in the same way that a circular cylinder is obtained by gluing together two parallel lines that bound a strip in the plane. We can parameterize σ and τ so that $\sigma(t)$ is glued to $\tau(t)$ to give $\gamma_0(t)$. See Fig. 8.8.

Two such nonintersecting (Poincaré metric) geodesics in $\mathbb{D}$ can be related in only one of the two ways: The geodesics diverge from each other at both ends, or they converge at one end but diverge at the other. These cases correspond to the Euclidean circles perpendicular to the boundary, which are the Poincaré metric geodesics in $\mathbb{D}$ having distinct boundary endpoints or having one boundary endpoint in common. [Two boundary endpoints in common are not possible; the geodesics would then coincide.] In the convergence-at-one-end case, one gets an immediate contradiction of our situation: In this case, one obtains arbitrarily short curves from one point far out on the geodesic (near the common boundary point) to the corresponding point on the "other side" of $\widetilde{M}$ cut open. This corresponds on M to arbitrarily short curves representing the free homotopy class of "once around" the first factor S^1. Such an eventuality is impossible: Such a curve cannot be shorter than the injectivity radius of M.

The case of divergence of the pair of geodesics at both ends is also impossible, but the reason is slightly more subtle. Note that, by a simple compactness argument,

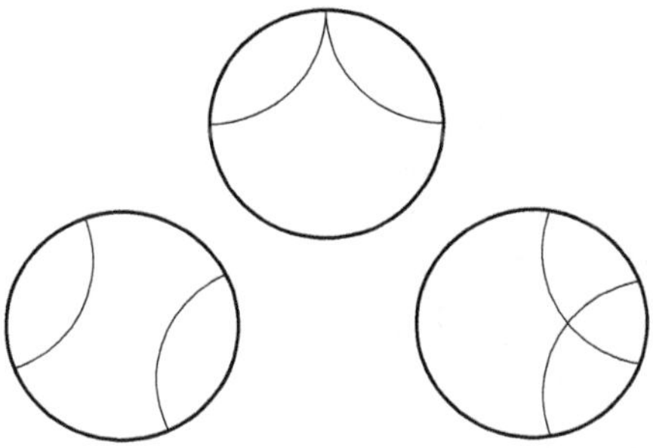

Fig. 8.9 The behavior of geodesics in $\mathbb{D}$

no covering spaces needed, there is a number L such that, for each $p \in M$, there is a curve starting and ending at p with length $\leq L$ which lies in the free homotopy class of "once around the first factor" of $S^1 \times S^1$. Now choose p very far out on the geodesic γ_0 in the direction of divergence of the two copies σ, τ of γ_0 bounding the region in $\mathbb{D}$ that glues together at the edge to give $\widetilde{M}$. Choose a closed curve of length $\leq L$ at the projection of p into M in the free-homotopy class of "once around" the first S^1 factor. The lift of this curve to $\widetilde{M}$ that starts at a point $\sigma(t)$ projecting to p must end at $\tau(t)$ (or vice versa depending on direction). But this lift has length $\leq L$. This is impossible if p is sufficiently far out on γ since σ and τ diverge from each other. [**Note:** One could use this argument twice to show convergence of σ and τ at both ends and detour around the argument we have used for the impossibility of the convergence-at-one-end case.] See Fig. 8.9.

This is impossible if p is sufficiently far out on γ since σ and τ diverge from each other. See Fig. 8.9.

These arguments are intricate to express in words. But if the reader will consider suitable pictorial representations, then the ideas will become clear.

The geometric analysis of possible groups Γ of holomorphic covering transformations acting on $\mathbb{D}$ can be pushed much further. It can be shown that, for such a Γ, every abelian subgroup is cyclic. This recovers the part of Theorem 8.7.11 ruling out the covering by $\mathbb{D}$ of a topological torus. For if M were a torus and $\pi : \mathbb{D} \to M$ a holomorphic covering, then the fundamental group $\pi_1(M)$ would be isomorphic to the group Γ of covering transformations. But π_1 of a torus is $\mathbb{Z} \oplus \mathbb{Z}$, abelian but not cyclic. The cyclic nature of abelian subgroups actually holds in much more general situations: It applies to subgroups of the fundamental group of a compact Riemannian manifold of any dimension having negative Riemannian sectional curvature (see Preissmann's theorem [425]), or, for a recent treatment, see [410].

It is not difficult to see by direct construction using hyperbolic geometry that, for each $g = 2, 3, 4, \ldots$, there is a (compact) covering space quotient of $\mathbb{D}$ with genus g. This amounts just to realizing in hyperbolic geometry the representation of the topological surface of genus g via a polygon of $4g$ sides with suitable identifications. However, this is far from showing that every Riemann surface of genus $g \geq 2$ is holomorphically a covering space quotient of $\mathbb{D}$. There are infinitely many complex structures on a compact surface of genus $g \geq 2$ (or indeed $g \geq 1$).

The direct geometric construction of some of them does not a priori guarantee that all are obtained by that means. To see that all are uniformized by $\mathbb{D}$ requires the uniformization theorem (Theorem 8.7.11) or at least some method beyond pure hyperbolic geometry.[5]

We turn now to the subject of the automorphism groups $\mathrm{Aut}(M)$ for compact Riemann surfaces M of genus ≥ 2.

Proposition 8.7.12 *If M is a compact Riemann surface of genus $g \geq 2$, then $\mathrm{Aut}(M)$ is a finite group.*

Remark 8.7.13 A compact Riemann surface can certainly have an infinite automorphism group; even a compact surface with an automorphism-invariant metric can have such a group. For instance, the quotient T of $\mathbb{C}$ by the integer lattice $\{m + ni : m, n \in \mathbb{Z}\}$, which is topologically a torus, has a positive-dimensional automorphism group induced by the translations $z \mapsto z + \alpha$, $\alpha \in \mathbb{C}$. Thus the condition that the genus be at least 2 in the proposition is essential.

Proof of the Proposition By Theorem 8.7.6 and the existence of the invariant H-metric on M (Proposition 8.7.7 with $\widetilde{M} = \mathbb{D}$), $\mathrm{Aut}(M)$ is a compact Lie group. Thus $\mathrm{Aut}(M)$ is a finite group if and only if the identity component of $\mathrm{Aut}(M)$ consists of the identity element alone (i.e., the group must be topologically discrete). By standard Lie group theory, this happens if and only if M contains no nontrivial one-parameter subgroup. For a proof by contradiction, suppose that $\{\varphi_t\}$ is a nontrivial one-parameter subgroup of $\mathrm{Aut}(M)$. Then $X(p) = d\varphi_t(p)/\,dt|_{t=0}$ for all $p \in M$ is a C^∞ vector field on M. [It is a part of the Bochner-Montgomery theorem ([89]) that the action of the Lie group $\mathrm{Aut}(M)$ on M is smooth, so that X is indeed defined and C^∞.] The vector field X on M lifts to a C^∞ vector field $\widehat{X}$ on $\mathbb{D}$ by setting $\widehat{X}(w)$ equal to the unique vector $\mathbf{v}$ at w such that $d\pi|_w(\mathbf{v}) = X(\pi(w))$, where $\pi : \mathbb{D} \to M$ is the universal covering of M. The vector field $\widehat{X}$ generates a one-parameter group $\{\psi_t\}$ of diffeomorphisms of $\mathbb{D}$; indeed, one obtains (complete) integral curves for $\widehat{X}$ by lifting the integral curves of X. Clearly the elements of the group $\{\psi_t\}$ are isometries of $\mathbb{D}$. In particular, the vector field $\widehat{X}$ can have at most one zero. For if $\widehat{X}(w_0) = 0$, then $\{\psi_t\}$ consists of a one-parameter group of Poincaré rotations around the fixed point w_0, and such a one-parameter group has generator that vanishes only at w_0—as one sees by writing it out explicitly (setting $w_0 = 0$ without loss of generality). On the other hand, if $\widehat{X}(w_0) = 0$, then $\widehat{X}(\gamma(w_0)) = 0$ for all $\gamma \in \Gamma$. Since Γ is infinite by the compactness of M, it must be that $\widehat{X}$, having at most one zero, has in fact no zeros at all. Thus X itself has no zeros. But a compact manifold with a nowhere vanishing vector field has Euler characteristic 0 (by the Poincaré-Hopf theorem (cf. [387])), while M has Euler characteristic $2 - 2g < 0$. That is a contradiction. □

[5] Proofs that use differential equations without appeal to uniformization are also known—see [72, 266].

8.7.2 The Automorphism Group of a Riemann Surface of Genus at Least 2

The proof in the previous section that the automorphism group $\mathrm{Aut}(M)$ of a compact Riemann surface M of genus $g \geq 2$ is necessarily a finite group involved very general considerations: The group $\mathrm{Aut}(M)$ was compact because the universal cover of M was $\mathbb{D}$ and hence M inherited an $\mathrm{Aut}(M)$-invariant metric from the $\mathrm{Aut}(\mathbb{D})$-invariant Poincaré metric of $\mathbb{D}$. The identity component of $\mathrm{Aut}(M)$ was the identity map alone because otherwise M would have had a zero-free vector field, contradicting the fact that its Euler characteristic $2 - 2g$ was nonzero. The specific structure of M played no role in these considerations, and the result was correspondingly general: finiteness of $\mathrm{Aut}(M)$ with no explicit estimate of the number of elements in $\mathrm{Aut}(M)$. It is quite surprising in this context that an explicit estimate is possible in the same generality.

Theorem 8.7.14 (Hurwitz) *If M is a compact Riemann surface of genus $g \geq 2$, then the order of the finite group $\mathrm{Aut}(M)$ does not exceed* $84(g-1)$.

This theorem seems mysterious at first sight. And the apparent mystery is deepened by the fact that the estimate is sharp. In fact there are Riemann surfaces with genus $g > 2$ for which the order of $\mathrm{Aut}(M)$ is precisely equal to $84(g-1)$ (see, for instance, [148, p. 325, VII.3.10]). The exact origin of the specific number 84 will become apparent from the proof that follows.

The proof of the Hurwitz theorem that we shall present is closely related to Hurwitz's original argument. The subject has tended meanwhile to become burdened with abstract Riemann surface theory. In our treatment we shall, like Hurwitz himself, stick to the basics. However, we shall use our metric viewpoint rather than Hurwitz's analogous pure function theory. The theorem will then turn out to be completely geometric in nature.

Before beginning the proof itself, we need to develop some general ideas about quotients by group actions. Specifically, we want to consider the orbit space of $\mathrm{Aut}(M)$ on M. An orbit of $\mathrm{Aut}(M)$ is as usual, by definition, for a given point p, the set $\{\gamma(p) : \gamma \in \mathrm{Aut}(M)\}$. The manifold M is a disjoint union of orbits, because $\mathrm{Aut}(M)$ is a group, and "being in the same orbit" is an equivalence relation, with each orbit an equivalence class.

Let $\mathcal{M}$ be the collection of orbits. Then $\mathcal{M}$ has a natural topology: the quotient topology. In other words, a set of orbits is said to be open if and only if the union of the orbits as sets is an open subset of M. For an arbitrary group action on a manifold, the topological orbit space can be complicated. But in our case, where $\mathrm{Aut}(M)$ is finite, $\mathcal{M}$ is a decent topological space; in particular, it is necessarily a Hausdorff space, as is easily checked. This is a special property of orbit spaces for finite groups acting on manifolds (or even just on Hausdorff topological spaces).[6]

Lemma 8.7.15 *If M is a compact, orientable Riemannian 2-manifold and if Γ is a finite group of orientation-preserving isometries of M, then the orbit space M/Γ is an orientable 2-manifold.*

Proof If two points $p, q \in M$ lie in the same Γ-orbit, then $I_p = \{\gamma \in \Gamma : \gamma(p) = p\}$ and $I_q = \{\gamma \in \Gamma : \gamma(q) = q\}$ are conjugate subgroups of Γ first, and if $\sigma(p) = q, \sigma \in \Gamma$, then $\sigma \circ \gamma \circ \sigma^{-1} \in I_q$ if and only if $\gamma \in I_p$. With this idea in mind, we divide the orbits of Γ (that is to say, the points of M/Γ) into two classes:

(a) Nonsingular orbits $\mathcal{O}$: For these, $I_p = \{\mathrm{id}_M\}$ if $p \in \mathcal{O}$.
(b) Singular orbits $\mathcal{O}$: For these, $I_p \neq \{\mathrm{id}_M\}$ if $p \in \mathcal{O}$.

Similarly, we call points of M itself singular or nonsingular according to whether they belong to singular or nonsingular orbits. These ideas are all well defined by our conjugacy observation made at first.

The next step in the proof of Lemma 8.7.15 is the following subsidiary result.

Lemma 8.7.16 *The singular orbits of M/Γ and the set of singular points of M are both finite sets.*

Proof By standard results from Riemannian geometry, there is an $\epsilon > 0$ such that, for any two points $p, q \in M$ with $\mathrm{dist}_M(p, q) < \epsilon$, there is a unique geodesic from p to q of length $\mathrm{dist}_M(p, q)$. Now suppose that $p_1, p_2, \ldots$ were an infinite sequence of distinct singular points of M. Since Γ is finite, we can suppose, by passing to a subsequence, that there is a fixed nonidentity element $\gamma_0 \in \Gamma$ such that $\gamma_0 \in I_{p_j}$ for all j. By the compactness of M, there are two points p_{j_1} and p_{j_2} with $\mathrm{dist}_M\left(p_{j_1}, p_{j_2}\right) < \epsilon$. Clearly γ fixes the (short) geodesic $c(t)$ from p_{j_1} to p_{j_2}. But the only nonidentity isometry that fixes every point of a geodesic segment in dimension 2 is locally given by reflection in that geodesic. Such an isometry is orientation reversing. Thus by contradiction the set of singular points of Γ in M is finite, and hence so is the set of singular orbits. □

Returning now to the proof of Lemma 8.7.15, we need to exhibit a Euclidean neighborhood of each point of $\mathcal{M}$, that is, a neighborhood that is homeomorphic to an open subset of $\mathbb{R}^2$. For a nonsingular orbit $\mathcal{O}$, this is easy: Choose a point $p \in \mathcal{O} \subset M$, and choose a Euclidean neighborhood U of p in M with every point of U nonsingular (this is possible by Lemma 8.7.16) and with the natural map $U \to \mathcal{M}$ injective. The latter choice is possible by the finiteness of Γ and the fact that p is nonsingular. For if $\{(p_j, q_j)\}$ were Γ-equivalent pairs of distinct points

[6] However, for a general finite group action on a manifold, the quotient need not be a manifold. For example, the orbit space of the action $\{x \mapsto x, x \mapsto -x\}$ on $\mathbb{R}$ is homeomorphic to $\{x \in \mathbb{R} : x \geq 0\}$. This is not a manifold. This difficulty does not arise in our situation.

with $\lim_j p_j = \lim_j q_j = p$, then there would be an element $\gamma \neq \mathrm{id}_M$ in I_p.[7] Thus U maps homeomorphically onto a neighborhood of $\mathcal{O}$, giving a Euclidean neighborhood as required.

The situation is more complicated when we want to find a Euclidean neighborhood of a singular orbit $\mathcal{O}$, but it is only a little more complicated. Let $p \in \mathcal{O}$. The group I_p, being a group of orientation-preserving isometries that all fix the point p, has a simple structure—both as a group and as a space acting on a neighborhood of p. Choose a small ball B around p, and let (r, θ) be a geodesic polar coordinate system centered at p on this ball. The I_p acts in this coordinate system as a finite group of Euclidean rotations. In particular, I_p is cyclic, say of order m_p.

Recall from Lemma 8.7.16 that if the ball B around p is small enough, then every point of the ball except p is nonsingular. Moreover, $B \to M/\Gamma$ is exactly m_p-to-1, except on the orbit $\mathcal{O}$ corresponding to p (which has only the pre-image p). The same limit argument as in the nonsingular case shows that, for B small enough, no quotienting occurs on B except via I_p. But it is then clear by Euclidean geometry that $\mathcal{O}$ has a Euclidean neighborhood, since a disc modulo a finite group of rotations around its center is locally Euclidean—it is a one-nappe circular cone with p as vertex.

The orientability is clear since Γ contains only orientation-preserving transformations, and the singular orbits do not separate the collection of orbits $\mathcal{M}$, even locally. □

Geometrically, $\mathcal{M}$ is a C^∞ manifold with C^∞ metric pushed down from M except for a finite number of cone-point singularities of the structure already described. Topologically, $\mathcal{M}$ remains a compact, orientable manifold as indicated, although it is likely to be a quite different manifold from M itself. Let g^* denote the genus of $\mathcal{M}$.

Now we want to describe a relationship between the genus g of M, the genus g^* of $\mathcal{M}$, and the action of Γ. To express this idea, we need some notation. Write $|\Gamma|$ for the order of Γ, let $\mathcal{O}_1, \ldots, \mathcal{O}_m$ be the singular orbits (finite in number by Lemma 8.7.16) of Γ, and let $k_1, \ldots, k_m$ be the orders of the corresponding isotropy subgroups $I_{p_1}, \ldots, I_{p_m}$ for $p_1 \in \mathcal{O}_1, p_2 \in \mathcal{O}_2, \ldots, p_m \in \mathcal{O}_m$. [The groups I_{p_j} are determined only up to conjugacy, but their orders are of course uniquely determined.]

Lemma 8.7.17 *In the notation just given,*

$$2 - 2g = |\Gamma| \cdot (2 - 2g^*) - |\Gamma| \cdot \sum_{j=1}^{m} \left(1 - \frac{1}{k_j}\right).$$

Lemma 8.7.17, which will turn out to be the crucial ingredient for the proof of Theorem 8.7.14, is a special case of the well-known Riemann-Hurwitz formula

---

[7] More precisely, if $q_j = \gamma_j(p_j), \gamma_j \neq id_M$, then by the finiteness of Γ, and passing to a subsequence if necessary, γ_j is independent of j and (by continuity) $\gamma_j \in I_p$.

for branched coverings. But an exact description of the Riemann-Hurwitz general setup would be almost as long as the proof in our case. So we give the direct proof instead. In fact, we shall give two proofs. The first one is a bit disingenuous, since we shall assume without proof the existence of a triangulation of $\mathcal{M}$ of a certain type, but it is still quite instructive. The second proof uses the Gauss-Bonnet theorem and is completely self-contained: We push the triangulation question back to the corresponding difficulty in the proof of the Gauss-Bonnet theorem.

Proof I of Lemma 8.7.17. Choose a triangulation of $\mathcal{M}$ such that the singular orbits $\mathcal{O}_1, \ldots, \mathcal{O}_m$ (thought of as points of $\mathcal{M}$) occur as vertices of the triangulation. This triangulation pulls back to a triangulation of M. In this triangulation of M, every edge of the $\mathcal{M}$-triangulation gives rise to $|\Gamma|$ edges and every face from $\mathcal{M}$ to $|\Gamma|$ faces. Let F_M denote the number of faces, E_M the number of edges, and V_M the number of vertices for the M-triangulation. We have a similar notation for the $\mathcal{M}$-triangulation. Then

$$E_M = |\Gamma| \cdot E_{\mathcal{M}}$$

and

$$F_M = |\Gamma| \cdot F_{\mathcal{M}}.$$

However, the pre-image $\pi^{-1}\left(\mathcal{O}_j\right)$ contains not $|\Gamma|$ points but rather $|\Gamma|/k_j$ points. Of course pre-images of vertices of the $\mathcal{M}$-triangulation that are nonsingular still contain $|\Gamma|$ points. Clearly the errors arising from the singular orbits are accounted for correctly by

$$V_M = |\Gamma| \cdot V_{\mathcal{M}} - \sum_{j=1}^{m} \left(|\Gamma| - \frac{|\Gamma|}{k_j} \right).$$

Thus the Euler characteristic $\chi(M) = V_M - E_M + F_M$ is related to $\chi(\mathcal{M}) = V_{\mathcal{M}} - E_{\mathcal{M}} + F_{\mathcal{M}}$ by

$$\chi(M) = |\Gamma| \cdot \chi(\mathcal{M}) - \sum_{j=1}^{m} \left(|\Gamma| - \frac{|\Gamma|}{k_j} \right)$$

or

$$2 - 2g = |\Gamma| \cdot (2 - 2g^*) - |\Gamma| \cdot \sum_{j=1}^{m} \left(1 - \frac{1}{k_j} \right).$$

Proof II of Lemma 8.7.17. We cannot apply the Gauss-Bonnet theorem directly to $\mathcal{M}$ because of the "cone point" singularities of the metric at $\mathcal{O}_1, \ldots, \mathcal{O}_m$. We instead set $\mathcal{M}^\epsilon$ equal to the image under the mapping $M \to M/\Gamma$ of

$M \setminus \bigcup_{p \text{ singular}} \mathbb{B}(p, \epsilon)$. Then $\mathcal{M}^\epsilon$ is, for $\epsilon > 0$ small enough, an orientable, compact, $\mathcal{C}^\infty$-manifold with boundary having a $\mathcal{C}^\infty$ metric inherited from M (we use here the fact that the elements of Γ are isometries). The manifold with boundary $\mathcal{M}^\epsilon$ is in effect $\mathcal{M}$ with the cone points and ϵ-discs around them removed. The Gauss-Bonnet theorem does apply to $\mathcal{M}^\epsilon$ as a manifold with boundary. Thus

$$2\pi \chi \left(\mathcal{M}^\epsilon\right) = \int_{\mathcal{M}^\epsilon} K dA - \int_{\partial \mathcal{M}^\epsilon} K_g ds,$$

where K_g is the geodesic curvature of the boundary (with $K_g > 0$ in our situation). Now $\chi\left(\mathcal{M}^\epsilon\right) = \chi(\mathcal{M}) - m$ by standard topology. Also write C^j_ϵ for the component of $\partial\mathcal{M}^\epsilon$ at $\mathcal{O}_j$ and $\widehat{C}^j_\epsilon$ for its pre-image in M at a particular point $p_j \in \mathcal{O}_j$; then

$$\lim_{\epsilon \to 0^+} \int_C K_g ds = \frac{1}{k_j} \lim_{\epsilon \to 0^+} \int_{\widehat{C}} K_g ds = 2\pi \cdot \frac{1}{k_j}.$$

Here, for typographical convenience, $C = C^j_\epsilon$ and $\widehat{C} = \widehat{C}^j_\epsilon$. Combining these results gives

$$\chi(\mathcal{M}) = m - \sum_{j=1}^{m} \frac{1}{k_j} + \lim_{\epsilon \to 0^+} \frac{1}{2\pi} \int_{\mathcal{M}^\epsilon} K dA$$

or

$$\chi(\mathcal{M}) = \sum_{j=1}^{m} \left(1 - \frac{1}{k_j}\right) + \lim_{\epsilon \to 0^+} \frac{1}{2\pi} \int_{\mathcal{M}^\epsilon} K dA.$$

But

$$\lim_{\epsilon \to 0^+} \frac{1}{2\pi} \int_{\mathcal{M}^\epsilon} K dA = \frac{1}{2\pi |\Gamma|} \int_M K dA = \frac{1}{|\Gamma|} \cdot \chi(M).$$

Thus

$$\chi(M) = |\Gamma| \cdot \chi(\mathcal{M}) - |\Gamma| \sum_{j=1}^{m} \left(1 - \frac{1}{k_j}\right).$$

With Lemma 8.7.17 now in hand, we are ready to prove Hurwitz's theorem.

Proof of Hurwitz's Theorem 8.7.14

$$\frac{2 - 2g}{|\Gamma|} = (2 - 2g^*) - \sum_{j=1}^{m} \left(1 - \frac{1}{k_j}\right)$$

or

$$\frac{2g-2}{|\Gamma|} = (2g^* - 2) + \sum_{j=1}^{m} \left(1 - \frac{1}{k_j}\right).$$

To get an upper bound for $|\Gamma|$, we need a lower bound for the right-hand side of this last equation (for g fixed). Now $2g-2 > 0$ by hypothesis. So the right-hand side is required to be positive. If $g^* \geq 2$, then the right-hand side is at least 2. If $g^* = 1$, then $2g^* - 2 = 0$, and there must be at least one term of the form $1 - 1/k_j$, with k_j a positive integer greater than 1. Hence in this case, the right-hand side is at least $1/2$. Finally, if $g^* = 0$, so that $2g^* - 2 = -2$, then there must be at least three terms of the form $1 - 1/k_j$ with $k_j \geq 2$. If there are five or more of these, then the right-hand side would be at least $-2 + 5(1/2) = 1/2$. If there are four, then the right-hand side—in order to be positive—must be at least $-2 + 3(1/2) + 2/3 = 1/6$. Finally if there are three terms of the form $1 - 1/k_j$, then only one of these can have $k_j = 1/2$. A little further experimentation shows that $-2 + 3 - (1/k_1 + 1/k_2 + 1/k_3)$ has the minimum possible value $1/42$, obtained when $k_1 = 2$, $k_2 = 3$, and $k_3 = 7$.

Thus the minimum possible positive value in all cases for the right-hand side is $1/42$, from which

$$|\Gamma| \leq 42(2g-2) = 84(g-1).$$

□

As already noted earlier in this section, the number 84 is sharp: The Klein quartic is a Riemann surfaces of genus 3 with $168 = 84(3-1)$ automorphisms. There is a considerable literature on the subject of when the $84(g-1)$ bound is attained (cf. [364] and [530]). Since the automorphism group of a Riemann surface of genus $g > 1$ is precisely the group of orientation-preserving automorphisms of the "push-down" metric from its uniformization by the unit disc (see [190, Proposition 2.3.1]), the examples where the order of the automorphism group has order $84(g-1)$ also serve as examples for the sharpness of the bound for the order of the group of orientation-preserving isometries.

Another application of the idea that geometric quotients give rise to holomorphic ones is to the question of which compact Riemann surfaces cover others in the sense of covering spaces. This corresponds essentially to Lemma 8.7.17 without branch points. Specifically, one can see either by the triangular argument of Proof I of Lemma 8.7.17 or via the Gauss-Bonnet argument used in Proof II that the following holds:

If $\pi : M \to M'$ is a covering map from one compact Riemann surface to another, then, with g denoting the genus of M and g' denoting the genus of M',

$$2 - 2g = k \cdot (2 - 2g').$$

Here k is a positive integer, and indeed k is the "sheeting number" of π, i.e., the number of points in $\pi^{-1}(\{p\})$ for each $p \in M$.

The necessary condition that when $g' \geq 2$, then $g'-1$ is a divisor of $g-1$ is in fact also sufficient. This is clear geometrically: Consider M as a torus of revolution in $\mathbb{R}^3$ with $g-1$ tubes smoothly attached in such a way that rotation by $2\pi/(g-1)$ takes each one to the next. Let Γ be the group of rotations by multiples of $2\pi/[(g-1)/k]$. Then $M \to M/\Gamma$ is a covering onto a surface with $(g-1)/k = g^* - 1$ tubes attached. That is a Riemann surface of genus $1 + (g^* - 1) = g^*$. This geometric construction is then, according to the principle of associating complex structures with metrics, holomorphic in suitable complex structures. First, there is a unique complex structure for which the metric of M as a surface in $\mathbb{R}^3$ is Hermitian. Since Γ acts as orientation-preserving isometries on M, it acts holomorphically for this complex structure. And thus $M \to M/\Gamma$ is a holomorphic covering space.

Notes to This Chapter

It will come as no surprise that the subject of Riemann surfaces goes back to Riemann [436]; Klein's exposition [288] followed in the last century. The first modern treatment of Riemann surfaces dates from Weyl's landmark text [525], first published in 1913. The third edition, published in 1955, was reworked considerably, and Weyl's approach there was foreshadowed by Chevalley a few years earlier [112].

The literature on Riemann surfaces is often referred to as "vast," but this word is almost an understatement. For the basic definitions [82, 413, 481], and [23] are still useful; these are in a slightly older style but have especially strong treatments of the topological issues. More recent are [101, 148, 158, 207, 215–217, 431, 457, 536], and [395], all of which are solid and relatively complete in what they do. As an excellent survey the reader may wish to consult [469]. The texts [46, 261], and [285] are somewhat more elementary. Especially delightful is Clemens' scrapbook [117], which is informal yet substantial.

We have downplayed the topological questions that arise in the study; the reader could consult any number of good texts for the basic material on manifolds. In the text are mentioned [31, 371, 373, 388, 389], and [475]; the analysis on manifolds is very well done in [389], and [101] has a solid discussion of the topological questions arising specifically for Riemann surfaces.

As to preliminary material, namely the basics on functions of one complex variable, the authors have taught or taken courses using [15, 118, 202, 349], and [86] as texts, and all have their strengths.

Complex manifolds of higher dimension are the subject of [385].

Chapter 9
The Uniformization Theorem

Classical uniformization theory, developed mainly during the last two decades of the nineteenth century and the first decade of the twentieth, was concerned with proving that every algebraic or analytic curve can be uniformized, that is, represented parametrically by single-valued (or "uniform") functions.

The basic ideas of the theory of uniformization are due to Klein and Poincaré who developed them independently in the 1880s without giving rigorous proofs. At about that time Schwarz conceived the fruitful idea of the covering surface of a Riemann surface. A complete proof of the uniformization theorem for algebraic functions and further generalizations were given in 1907 by Koebe and, at about the same time, by Poincaré.

Research in uniformization theory has gone through several dormant periods since the concept of a (global) uniformization was introduced by Klein in 1882. The original papers by F. Klein are collected in [290]; this volume contains historical notes by Klein and his correspondence with Poincaré. Concerning Poincaré, see his Oeuvres, especially Vol.2 [420].

At about that time Schwarz conceived the fruitful idea of the covering surface of a Riemann surface. A complete proof of the uniformization theorem for algebraic functions and further generalizations were given in 1907 by Koebe and, at about the same time, by Poincaré.

The story is fascinating, first of all because of the wealth of mathematical ideas that arose out of efforts to find a proof: manifolds, topological dimension, universal covering space, methods for solving nonlinear elliptic partial differential equations, distortion theorems for conformal mappings, and many more.

The personal element is also fascinating. It involves some of the most illustrious mathematicians of that time: Schottky, about to conjecture, in 1875, a fairly general uniformization theorem, but deflected by the authority of Weierstrass (according to Klein); the rivalry between Klein, then at the height of fame and productivity, and the yet unknown Poincaré (see their correspondence from the years 1881–1882); Schwarz, suggesting, in a private communication, two methods for proving the main

P. V. Dovbush, S. G. Krantz, *The Geometric Theory of Complex Variables*,
https://doi.org/10.1007/978-3-031-77204-7_9

uniformization theorem (one using universal covering surfaces, the other involving the partial differential equation $\Delta u = e^{2u}$); Hilbert, reviving interest in the problem by his Paris lecture, and several years later creating a new tool for uniformization by "rehabilitating" the Dirichlet principle; Brouwer, embarking on his epoch-making topological investigations in order to put the original "continuity method" of Klein and Poincaré on a firm foundation; Poincaré, returning to the uniformization problem after a quarter of a century and finally achieving a full solution, but having to share this honor with Koebe. The full story of the uniformization theorem from its formulation by Poincaré in 1883 to the first rigorous proofs published by Poincaré and the young, ambitious, Paul Koebe independently in 1907 is given in the book of Umberto Bottazzini and Jeremy Gray: *Hidden Harmony—Geometric Fantasies. The Rise of Complex Function Theory* [91].

While such theorems had earlier been proved in special cases, no one had dared even conjecture that every compact Riemann surface could be parametrized by a variable whose domain of definition lay in the Riemann sphere. Uniformization theorems of power unimaginable to the classical masters have been proved in the past hundred years and no end is in sight.

Accounts of the classical uniformization theory will be found in the second volume of Fricke-Klein [289] (monumental and hard to read), in the second volume of Appel-Goursat [29] (written by Fatou), in the treatises by Ford [155], and by Nevanlinna [401], and in most standard texts on Riemann surfaces (Weyl [524], Pfluger [413], Springer [481], Ahlfors-Sario [23]). It is of interest to consult also the original works by Klein (see [290] that contain also the amusing correspondence with Poincaré), Poincaré [419, 420], and Koebe [300, 302, 305, 311].

9.1 Double Periodicity of Elliptic Functions

The view of complex integration provided by Cauchy's theorem is one step toward understanding elliptic integrals such as $\int_0^z dt/\sqrt{t(t-\alpha)(t-\beta)}$. The other important step is the idea of a Riemann surface (Sect. 8.1.1), which enables us to visualize the possible paths of integration from 0 to z. The "function" $1/\sqrt{t(t-\alpha)(t-\beta)}$ is of course two-valued and, by an argument like that in Sect. 8.1.1, is represented by a two-sheeted covering of the t-sphere, with branch points at $0, \alpha, \beta, \infty$. Thus the paths of integration, correctly viewed, are curves on this surface, which is topologically a torus (again, as in Sect. 8.1.1).

Now a torus contains certain closed curves that do not bound a piece of the surface, such as the curves $\mathcal{E}_1$ and $\mathcal{E}_2$ shown in Fig. 9.1. There is no region $\mathcal{R}$ bounded by $\mathcal{E}_1$ or $\mathcal{E}_2$; hence, Green's theorem does not apply, and we in fact obtain nonzero values

$$\omega_1 = \int_{\mathcal{E}_1} \frac{dt}{\sqrt{t(t-\alpha)(t-\beta)}},$$

$$\omega_2 = \int_{\mathcal{E}_2} \frac{dt}{\sqrt{t(t-\alpha)(t-\beta)}}.$$

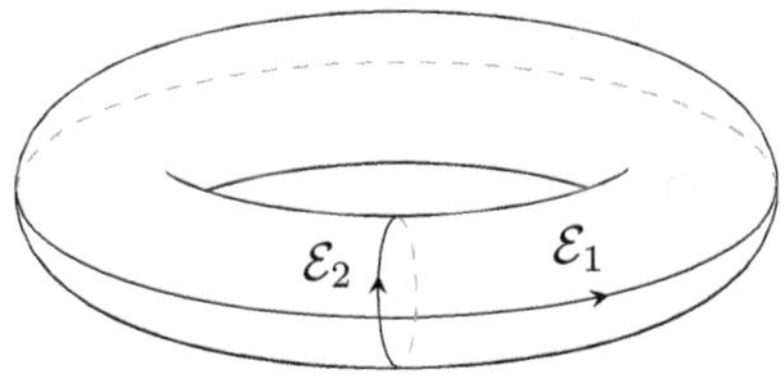

Fig. 9.1 Nonbounding curves on the torus

Consequently, the integral

$$\Phi^{-1}(z) = \int_0^z \frac{dt}{\sqrt{t(t-\alpha)(t-\beta)}}$$

will be ambiguous: For each value $\Phi^{-1}(z) = w$ obtained for a certain path $\mathcal{E}$ from 0 to z, we also obtain the values $w + m\omega_1 + n\omega_2$ by adding to $\mathcal{E}$ a detour that winds m times around $\mathcal{E}_1$ and n times around $\mathcal{E}_2$. (For topological reasons, this is essentially the most general path of integration.)

It follows that the inverse relation $\Phi(w) = z$, the elliptic function corresponding to the integral, satisfies

$$\Phi(w) = \Phi\,(w + m\omega_1 + n\omega_2)$$

for any integers m, n. That is, Φ is doubly periodic, with periods ω_1, ω_2. This intuitive explanation of double periodicity is due to Riemann, who later developed the theory of elliptic functions from this standpoint.

Eisenstein showed that doubly periodic functions could be obtained by analogous expressions, such as

$$\sum_{m,n=-\infty}^{\infty} \frac{1}{(z + m\omega_1 + n\omega_2)^2},$$

which again (with suitable interpretation to ensure convergence) are obviously unchanged when z is replaced by $z + \omega_1$ or $z + \omega_2$. Hence we obtain a function with periods ω_1, ω_2. The function above is in fact identical (up to a constant) with the Weierstrass $\wp$-function, as the inverse to the integral $\int dt/\sqrt{4t^3 - g_2 t - g_3}$. Weierstrass [521, p. 121] found the relations between g_2, g_3 and the periods ω_1, ω_2:

$$g_2 = 60 \sum \frac{1}{(m\omega_1 + n\omega_2)^4},$$

$$g_3 = 140 \sum \frac{1}{(m\omega_1 + n\omega_2)^6},$$

where the sums are over all pairs $(m, n) \neq (0, 0)$. Elegant modern accounts of the Eisenstein and Weierstrass theories may be found in Weil [522] and Robert [437].

9.2 Elliptic Curves

Jacobi (1834) saw, at least implicitly, that the nonsingular cubic curve of the form

$$y^2 = ax^3 + bx^2 + cx + d \tag{9.1}$$

could be parametrized as

$$x = f(z), \quad y = f'(z), \tag{9.2}$$

where f and its derivative f' are elliptic functions. Knowing that f and f' were doubly periodic, with the same periods ω_1, ω_2, say, he would have seen that this gave a map of the z plane $\mathbb{C}$ onto the curve (9.1) for which the preimage of a given point on (9.1) is a set of points in $\mathbb{C}$ of the form

$$z + \Lambda = \{z + m\omega_1 + n\omega_2 : m, n \in \mathbb{Z}\}\ ,$$

where

$$\Lambda = \{m\omega_1 + n\omega_2 : m, n \in \mathbb{Z}\}$$

is called the *lattice of periods* of f. The numbers $z + m\omega_1 + n\omega_2$ in $z + \Lambda$ are also called "equivalent with respect to Λ." One such equivalence class is shown by asterisks in Fig. 9.2.

The parametrization (9.2) means that there is a one-to-one correspondence between the points $\left(f(z), f'(z)\right)$ of the curve and the equivalence classes $z + \Lambda$. Today we express this relation by saying that the curve is isomorphic to the space $\mathbb{C}/\Lambda$ of these equivalence classes. Jacobi might have seen, though it was probably not of interest to him, that $\mathbb{C}/\Lambda$ is a torus. One sees this by taking one parallelogram in $\mathbb{C}$, which includes a representative of each equivalence class, and identifying the equivalent points on its boundary (that is, pasting opposite sides together, as in Fig. 9.3). Of course, the torus form of (9.1) eventually came to light through the Riemann surface construction given in Sect. 8.4.

An elegant way of demonstrating both the double periodicity of elliptic functions and the parametrization of cubic curves was given by Weierstrass (1863). Beginning

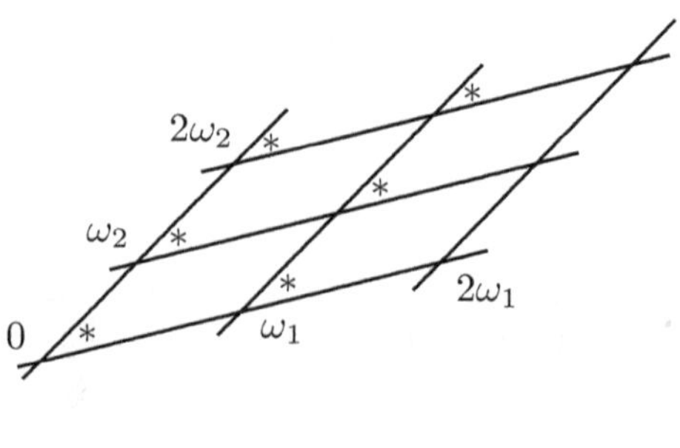

Fig. 9.2 Lattice-equivalent points

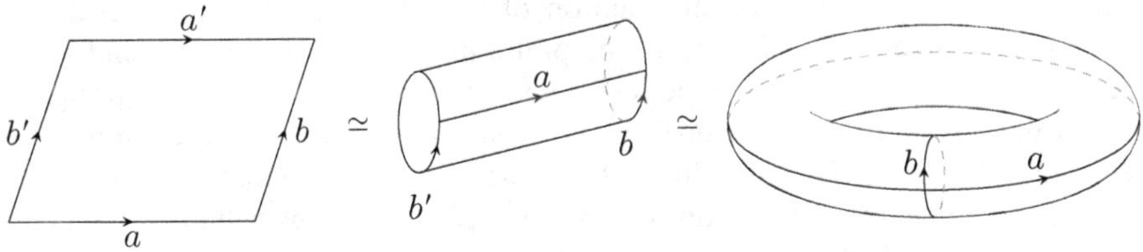

Fig. 9.3 Construction of a torus by pasting

with the function

$$\sum_{m,n=-\infty}^{\infty} \frac{1}{(z+m\omega_1+n\omega_2)^2},$$

which makes the double periodicity evident, Weierstrass defined the function

$$\wp(z) = \frac{1}{z^2} + \sum_{m,n\neq 0,0}^{\infty} \left(\frac{1}{(z+m\omega_1+n\omega_2)^2} - \frac{1}{(m\omega_1+n\omega_2)^2} \right),$$

which has better convergence properties and is also doubly periodic. He then showed by simple computations with series that

$$\wp'(z)^2 = 4\wp(z)^3 - g_2\wp(z) - g_3\,,$$

where g_2, g_3 are the constants depending on ω_1, ω_2, which were defined in Sect. 9.1. It follows that the point $(\wp(z), \wp'(z))$ lies on the curve

$$y^2 = 4x^3 - g_2 x - g_3, \tag{9.3}$$

and a little further checking shows that (9.3) is in fact isomorphic to $\mathbb{C}/\Lambda$, where Λ is the lattice of periods of $\wp$. The parametrization of all curves (9.1) by elliptic functions follows by making a linear transformation.

A representation of the x and y values on a curve by simultaneous functions of a single parameter z is sometimes called a *uniform representation*, and so the problem of parametrizing all algebraic curves in this way came to be known as the uniformization problem. Once the elliptic case was understood, it became clear that a solution of the uniformization problem for arbitrary algebraic curves would depend on a better understanding of surfaces: their topology, the periodicities associated with their closed curves, and the way these periodicities could be reflected in $\mathbb{C}$. These problems were first attacked by Poincaré and Klein in the 1880s, and their work led to the eventual positive solution of the uniformization problem by Poincaré [418] (1907) and Koebe [300] (1907).

Even more important than the solution of this single problem, however, was the amazing convergence of ideas in the preliminary work of Poincaré and Klein. They discovered that multiple periodicities were reflected in $\mathbb{C}$ by groups of transformations and that the transformations in question were of the simple type $z \mapsto (az + b)/(cz + d)$, called linear fractional. Linear fractional transformations generalize the linear transformations $z \mapsto z + \omega_1$, $z \mapsto z + \omega_2$ naturally associated with the periods of elliptic functions. However, while the transformations $z \mapsto z+\omega_1$, $z \mapsto z+\omega_2$ are algebraically and geometrically transparent—they commute, and they generate the general transformations $z \mapsto z + m\omega_1 + n\omega_2$, which are simply translations of the plane—the more general linear fractional transformations are not as easily understood. Linear fractional transformations do not normally commute, and their mastery requires a simultaneous grasp of algebraic, geometric, and topological aspects.

The simultaneous view proved to be enormously fruitful in the development of group theory and topology. Geometry was also given a new lease of life when Poincaré (1882) discovered that linear fractional transformations give a natural interpretation of non-Euclidean geometry, a field that until then had been a curiosity on the fringes of mathematics. In Chap. 13 we look at the origins of non-Euclidean geometry and see how the subject was transformed by Poincaré's discovery.

9.3 How to Look at the Problem

To return to the uniformization problem, the question changed as the notion of Riemann surfaces changed. Klein's original claim is to have proved the theorem for compact Riemann surfaces, possibly missing a finite number of points, in the sense of Riemann. All proofs until the 1920s assumed that the surfaces were first countable; this latter assumption was made by Weyl. The assumption was removed by Radó but now may be derived as a simple consequence of the uniformization theorem.

We may now state the *general uniformization problem*. Let S be a Riemann surface. Find all domains $U \subset \widehat{\mathbb{C}}$ and holomorphic functions $t : U \to S$ so that at each point $p \in S$, t is a local uniformizing variable at p.

Equivalently, there is a topological disc $B \subset S$ with center p so that the restriction of t to each component of $t^{-1}(B)$ is a homeomorphism. The reader should notice that the required conditions on t and U say precisely that the triple $(\mathbb{D}, S, t)$ is a (smooth) covering space with base S, total space U, and holomorphic projection t.

The early problem was less ambitious and discussed in Chap. 2. It was simply to find one uniformization, but where U is simply connected. This theorem was proved independently by Koebe and Poincaré in 1907. Until recently the most common proof was that offered by Hilbert in 1909. We give the modern statement of the Uniformization Theorem in the next subsection following the preliminaries that tie it to the uniformization problem.

9.3.1 Covering Surfaces and Classical Plane Geometries

Let S and S' denote not necessarily compact Riemann surfaces. We study continuous maps $f : S \to S'$, and we will see that if we make some assumptions on the structure of f, then there are not many of them.

Definition 9.3.1

(a) $f : S \to S'$ is called a *covering* of S' if every point $x \in S'$ admits a neighborhood U in such a way that

$$f^{-1}(U) = \bigcup_j V_j, \quad V_j \cap V_k = \emptyset \quad \text{for } j \neq k$$

and

$$f_{|V_j} : V_j \cong U.$$

(b) f is called a *ramified covering* of S' if there is a discrete set of points A in S' such that the restriction

$$f_| : S \backslash f^{-1}(A) \to S' \backslash A$$

of f to $S \backslash f^{-1}(A)$ is a covering.

(c) A covering $f : S \to S'$ is called a *universal covering* of S' if S is simply connected.

By definition coverings are always surjective.

The reason for calling a simply connected covering universal is given by the following:

Proposition 9.3.2 *Let $f : S \to S'$ be a universal covering. Let $g : S'' \to S'$ be another covering. If we choose a point $x_0 \in S'$ and a pair of points $y_0 \in f^{-1}(x_0)$, $z_0 \in g^{-1}(x_0)$, then there is exactly one map $k : S \to S''$ with*

$$f = g \circ k \quad \text{and } z_0 = k(y_0).$$

The map k is a covering of S''.

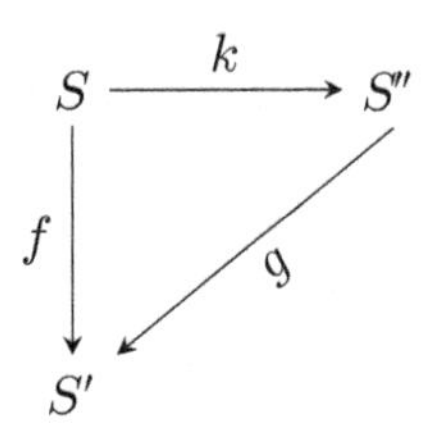

From this it follows easily that, up to canonical isomorphy, the universal covering is unique.

Theorem 9.3.3 *Every compact Riemann surface admits a universal covering. In case of genus* $0\ (= \mathbb{S}^2)$*, it is its own universal covering. In case of genus* 1 *its universal covering is the complex number plane* $\mathbb{C}$*. In case of higher genus it is the upper half plane* $\mathbb{H}$*.*

Strictly speaking we have now left the field of topology because in this theorem we speak of Riemann surfaces with their complex structures. In fact these coverings are even analytic maps.

In the book by Forster [157] or its English translation [158], one will find an extensive treatment of the topology of Riemann surfaces from the viewpoint of fundamental groups and covering spaces. The simplicial homology is more or less skipped. For this, you can consult [465]. You find there the explicit reduction of every two-dimensional manifold to the standard polygon model. You will also find this in a textbook [489].

In fact for arbitrary spaces one uses today other homology theories. One example is the so-called singular homology. It is not so easy to visualize but much easier to handle in general constructions. Fortunately, there is a theorem that says that in the cases that are of interest to us it coincides with the simplicial homology. You can find the singular homology in every modern textbook on algebraic topology, see for example [372].

There is another important generalization: sheaf cohomology. The definition of it is even more technical, but one is also able to cover analytic or algebraic aspects with such methods. In the book of Forster, you find one representative of these theories, the so-called Čech cohomology. There are a lot of books on the subject. But most of them are rather voluminous and abstract. A standard reference is [177].

Let S_1 and S_2 be two Riemann surfaces and $\pi : S_1 \to S_2$ be a local homeomorphism so that each point on S_2 has an evenly covered neighborhood. We then say that (S_1, S_2, π) is a covering surface or, more commonly, that S_1 is a *covering surface* of S_2 with *cover map* or projection π. The number of points in $\pi^{-1}(p)$ is independent of $p \in S_2$ and is called the *number of sheets of the covering*.

Covering surfaces occur classically in the study of algebraic equations with symmetries. For example, the equation $w^2 - z = 0$ admits the symmetry obtained by replacing w by $-w$. In this context the symmetry is called sheet interchange. The Riemann surface S of $w^2 - z = 0$ over $\mathbb{C}\backslash\{0\}$ is a covering surface of $\mathbb{C}\backslash\{0\}$. Notice that the sheet interchange cannot be defined to be a covering surface at $z = 0$. Points where the map topologically looks like $z \mapsto z^n$ are called *ramification points*.

Covering surfaces with isolated singularities, all of which are ramification points, are called *ramified covers*.

A method for constructing a covering surface is to take two copies S' and S'' of a given surface S, slice them along corresponding curves, and glue as indicated in Fig. 9.4. The resulting surface is a two-sheeted cover of S with the projection taking z' and z'' to z. Notice that there is a simple closed curve that covers β twice. In a sense that may be made quite precise, passing to the covering surface has replaced

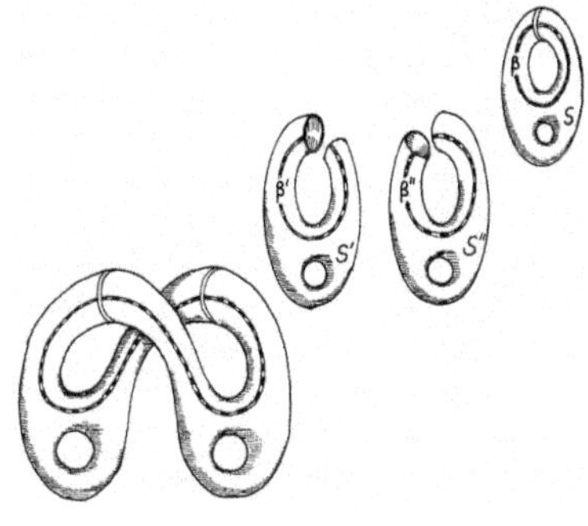

Fig. 9.4 A method for constructing a covering surface for a given surface S

β by a curve of twice the length of β. If we repeat this process infinitely often, β will have been replaced by an infinitely long, simple curve; i.e., β will no longer be an obstruction to simple connectivity. We may repeat this process for sufficiently many curves so that the resulting covering surface $\widetilde{S}$ has only homotopically trivial curves. $\widetilde{S}$ is called the *universal covering surface.*

This approach is the most classical one; the modern approach (see, e.g., Greenberg [188]) is an abstract reformulation of the basic idea of opening up all homotopically nontrivial closed curves. Klein states that the construction above is due to Schwarz.

As we have noted in the previous section, if S is the Riemann surface of some multivalued function Y of X, then both X and Y are functions of ζ if ζ varies over the points of S. If S_1 is a covering surface of S, and ζ_1 varies over S_1, then $X \cdot \pi$ and $Y \cdot \pi$ parametrize the variety in $\mathbb{C}^2$ defined by the functional relationship between X and Y. Thus if S_1 lies in $\widehat{\mathbb{C}}$, we obtain a solution to the general uniformization problem. We will return to this question in §6 but now restrict our attention to $S_1 = \widetilde{S}$.

$\widetilde{S}$ inherits the structure of a Riemann surface from the conformal structure on S. $\widetilde{S}$ is a simply connected Riemann surface because any homotopically nontrivial closed curve has been opened. It follows immediately that we may obtain a solution to the uniformization problem by showing that every simply connected Riemann surface is biholomorphically equivalent to a subdomain of $\widehat{\mathbb{C}}$; this is merely a rephrasing of the Riemann mapping theorem. So a solution to the uniformization problem follows from the statement now known as:

Theorem 9.3.4 (The Uniformization Theorem) *Every simply connected Riemann surface is biholomorphically equivalent to either* $\mathbb{C}$, $\widehat{\mathbb{C}}$, *or the unit disc* $\mathbb{D}$.

Of these, only the Riemann sphere is compact. In particular, it is not homeomorphic to $\mathbb{C}$ or $\mathbb{D}$. There is no isomorphism $f : \mathbb{C} \to \mathbb{D}$ because any such f would be a bounded holomorphic function on $\mathbb{C}$, and hence constant by Liouville's theorem. Thus, the three are distinct. A special case of the theorem says that every simply connected open subset of C different from $\mathbb{C}$ is isomorphic to $\mathbb{D}$.

The universal cover has several very strong properties. Since it is simply connected, every multivalued locally meromorphic function on S lifts to a meromorphic

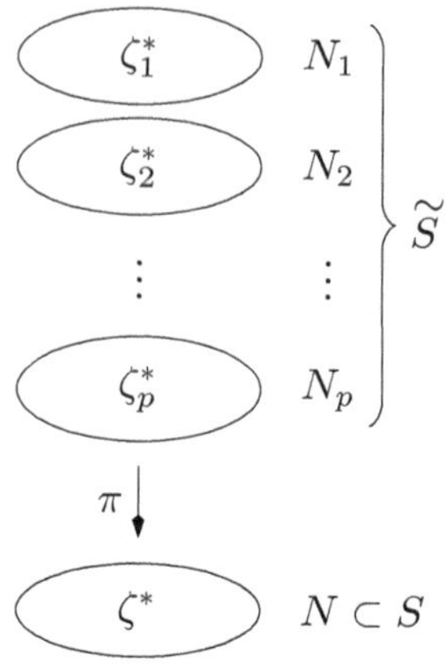

Fig. 9.5 The covering map $\pi : \widetilde{S} \to S$

function on $\widetilde{S}$; here we are restating the monodromy theorem. The universal cover is the only covering surface with this property.

Another important property is the following. Suppose $\zeta \in S$ and N is an evenly covered neighborhood of ζ. The cover map $\pi : \widetilde{S} \to S$ defines a map γ from one component N_1 of $\pi^{-1}(N)$ to another, say N_2, by the rule $\zeta_1 \mapsto \zeta_2$ where ζ_2 is the unique point in N_2 for which $\pi(\zeta_1) = \pi(\zeta_2)$ (see Fig. 9.5).

Using the evenly covered character of N_1, one shows that γ extends to a biholomorphic self-map of $\widetilde{S}$. These maps form a group G called the *group of the covering* $(\widetilde{S}, S, \pi)$ or *cover group* or *group of deck (cover) transformations*. Clearly, G acts without fixed points. G also has the stronger property of being properly discontinuous, which means that each point $\widetilde{\zeta} \in \widetilde{S}$ has a neighborhood $\widetilde{N}$ so that $\gamma(\widetilde{N}) \cap \widetilde{N} \neq \varnothing$ if and only if $\gamma =$ identity. Since the sheet interchange maps $\gamma \in G$ are in one-to-one correspondence with the free homotopy classes of closed curves on S, the group G is isomorphic to the fundamental group $\pi_1 S$ of S based at any point. The orbit space $\widetilde{S}/G$ is the set of equivalence classes in $\widetilde{S}$ where $\zeta_1 \sim \zeta_2$ if there exists $\gamma \in G$ so that $\zeta_2 = \gamma(\zeta_1)$. It is not difficult to show that $\widetilde{S}/G$ inherits a conformal structure from $\widetilde{S}$ and that S is biholomorphically equivalent to $\widetilde{S}/G$.

The line of reasoning developed above is, more or less, historically accurate when $\widetilde{S}$ is $\widehat{\mathbb{C}}$ or $\mathbb{C}$ (we shall no longer distinguish between biholomorphically equivalent surfaces except when necessary). It shows immediately that $\widetilde{S} = \widehat{\mathbb{C}}$ only if $S = \widehat{\mathbb{C}}$. If S is compact and $\widetilde{S} = \mathbb{C}$, π is the inverse to an elliptic integral (see Siegel [474, Chapter 1]). The problem of finding inverse functions to algebraic integrals, i.e., cover maps, is called the Jacobi inversion problem, the rich history of which leads directly to Riemann surfaces and uniformization (cf. Weyl [525, p. 144]).

When $\widetilde{S} = \mathbb{C}$, G must be a properly discontinuous subgroup of the group Aut $\mathbb{C}$ of biholomorphic self-maps of $\mathbb{C}$. There are precisely three types. The first is when G is trivial. The Riemann surface is then $\mathbb{C}$. The second are the cyclic groups $\{z \mapsto z + nz_0 \mid n \in \mathbb{Z}\}$. All are conjugate in Aut $\mathbb{C}$ to $G = \{z \mapsto z + n \mid n \in \mathbb{Z}\}$ and conjugate groups determine biholomorphically equivalent surfaces. The surface is $\mathbb{C}\backslash\{0\}$ and $\pi : \mathbb{C} \to S$ is the map $\exp(2\pi i z)$. The last is a class of groups, the

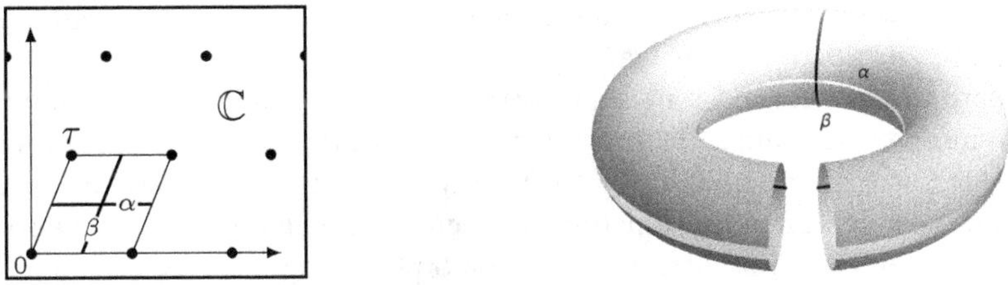

Fig. 9.6 A free abelian group on two generators

lattices, which are, up to conjugation, generated by $\gamma_1 : z \mapsto z + 1$ and $\gamma_2 : z \mapsto z + \tau$, where $\operatorname{Im} \tau > 0$. In this case $\pi_1 S$ is free abelian on two generators and S is topologically a torus. The two generators may be chosen as in Fig. 9.6.

We have thus "found" all Riemann surfaces with $\widetilde{S} = \mathbb{C}$ or $\widetilde{S} = \widehat{\mathbb{C}}$. The uniformization theorem tells us that every other Riemann surface S has $\widetilde{S} = \mathbb{D}$. Here the history proceeded in the opposite direction. Klein knew in 1882 that if G is a properly discontinuous subgroup of Aut $\mathbb{D}$, then $\mathbb{D}/G$ looks like a Riemann surface. However in 1882, a Riemann surface required a defining multivalued function. While it seems that Klein had a method for obtaining such a function, Poincaré claimed two methods, one of which remains the standard method for obtaining automorphic functions for discrete groups acting on bounded domains. The existence of this function shows that all the definitions of Riemann surfaces coincide.

The Riemann sphere has a metric of constant positive curvature induced by the usual embedding in $\mathbb{R}^3$. The space $\mathbb{C}$, hence by projection $\mathbb{C}\backslash\{0\}$ and tori, has Euclidean or flat metrics, i.e., of zero curvature. Using the uniformization theorem, we see that all other Riemann surfaces inherit any Aut $\mathbb{D}$ invariant metric from $\mathbb{D}$. Up to a scale factor, there is only one; it is the Poincaré metric on the hyperbolic plane. It is a metric of constant negative curvature. The universal cover of a surface of constant negative curvature may be conformally and injectively developed on a hyperbolic plane. This is one of many successful approaches to proving the uniformization theorem. The relevant differential equation is $\Delta u = e^{2u}$. Studies of this equation were among the earliest in nonlinear partial differential equations.

We also recall that the hyperbolic plane, in the Poincaré model, is the unit disc $\mathbb{D}$. When we map $\widetilde{S}$ to $\mathbb{D}$, we give natural boundary to $\widetilde{S}$, namely $\partial\mathbb{D}$. To Klein the edge was a limiting circle and he called his uniformization theorem the limit circle theorem.

9.3.2 The Uniformization Theorem for Riemann Surfaces

The concept of uniformization, which is one of the main concepts in complex analysis and other areas of mathematics, was introduced in the classical literature as far back as the second half of the eighteenth century when a detailed study of multivalued analytic functions was beginning.

To uniformize means to represent parametrically by single-valued holomorphic or meromorphic functions. In its original understanding, to uniformize a given multivalued analytic function $w = F(z)$ (or the corresponding analytic expression or set) means to represent this function parametrically with the help of single-valued (uniform) holomorphic or, in general, meromorphic functions $z = z(t)$, $w = w(t)$ such that $w(t) = F[z(t)]$.

More precisely, let Σ be a set in $\mathbb{C}^n$, the complex n-space, and let $f_1, \ldots, f_n$ be n meromorphic functions defined in some domain Ω of $\mathbb{C}$ (or of $\mathbb{C}^m$, for some $m > 1$). We say that the functions $f_1, \ldots, f_n$ *uniformize* Σ if the following conditions are satisfied:

(i) Let $\Omega_0 \subset \Omega$ be the set on which $f = (f_1, \ldots, f_n)$ is defined as an n-tuple of complex numbers; the set Σ is the closure of $f(\Omega_0)$.

(ii) There is a discrete group G of holomorphic self-mappings of Ω such that $f(t_1) = f(t_2)$ if and only if $t_2 = g(t_1)$ for some $g \in G$. (There is a similar definition for subsets of a projective space $\mathbb{P}_n$.)

For instance, the circle ${\zeta_1}^2 + {\zeta_2}^2 = 1$ can be uniformized by the transcendental functions $\zeta_1 = \cos\theta$, $\zeta_2 = \sin\theta$ and also, which is more interesting, by the rational functions

$$\zeta_1 = \frac{1-t^2}{1+t^2}, \qquad \zeta_2 = \frac{2t}{1+t^2}.$$

In the first case, $\Omega_0 = \mathbb{C}$ and G is generated by the translation $z \mapsto z + 2\pi$. In the second case, $\Omega_0 = \mathbb{C} - \{i, -i\}$ and $G = \{1\}$, the trivial group.

A cubic curve

$$\zeta_2^2 = 4(\zeta_1 - c_1)(\zeta_1 - c_2)(\zeta_1 - c_3),$$

where c_1, c_2, c_3 are distinct numbers, with $c_1 + c_2 + c_3 = 0$, cannot be uniformized by rational functions. But it admits a parametric representation $\zeta_1 = \wp(z)$, $\zeta_2 = \wp'(z)$, where $\wp$ is the Weierstrass $\wp$-function with periods w_1 and w_2:

$$\wp(z) = z^{-2} + \Sigma\left\{(z - mw_1 - nw_2)^{-2} - (mw_1 - nw_2)^{-2}\right\}.$$

Here w_1 and w_2 are appropriately chosen complex numbers, $\tau = w_1/w_2$ has a positive imaginary part, and the summation is extended over all $(m, n) \in Z^2$, $(m, n) \neq (0, 0)$. In this case, G is the elliptic group generated by the two translations $z \mapsto z + w_1$, $z \mapsto z + w_2$—see Sect. 9.2.

The uniformization problem for arbitrary plane algebraic curves reads: Represent all pairs of complex numbers (ζ_1, ζ_2) satisfying the equation

$$P(\zeta_1, \zeta_2) = 0$$

as single-valued functions of a complex variable. Here P is an irreducible polynomial with complex coefficients.

With an algebraic curve, there is associated a nonnegative integer p, its genus, and the solution of the uniformization problem depends on this number. If $p = 0$, the curve can be uniformized by rational functions; if $p = 1$, the curve can be uniformized by elliptic functions. If $p > 1$, uniformization necessarily requires functions defined only on proper subsets of $\mathbb{C}$. Indeed, all algebraic curves can be uniformized by functions defined in the upper half plane $\mathbb{H} = \{z \mid z = x + iy \in \mathbb{C}, y > 0\}$.

The solution set of every polynomial equation in two unknowns can be uniformized by meromorphic functions defined in U or is a finite union of sets with this property. (The second case occurs when $P(\zeta_1, \zeta_2)$ is not irreducible but a product of distinct factors.)

A basic tool in establishing any uniformization theorem is the concept of a *Riemann surface*.

We may now state the general uniformization problem. Let S be a Riemann surface. Find all domains $\Omega \subset \widehat{\mathbb{C}}$ and holomorphic functions $t : \Omega \to S$ so that at each point $p \in S$, t is a local uniformizing variable at p.

Equivalently, there is a topological disc $B \subset S$ with center p so that the restriction of t to each component of $t^{-1}(B)$ is a homeomorphism. The reader should notice that the required conditions on t and Ω say precisely that the triple (Ω, S, t) is a (smooth) covering space with base S, total space Ω, and holomorphic projection t.

The early problem was less ambitious. It was simply to find one uniformization, but where $\Omega \subset \mathbb{C}$ is simply connected.

We have seen that every domain of a Riemann surface is a Riemann surface. We know from Chap. 8 that every domain in Ω is a Riemann surface: For example, the unit disc (or the conformally equivalent under $\phi(z) = (z - i)/(z + i)$ upper half plane) is a Riemann surface.

Next, the Riemann sphere S^2 is a Riemann surface. The basis for the topology of the space S^2 will consist of neighborhoods of all points of $\mathbb{C}$ plus the neighborhoods of the point ∞; a "neighborhood of the point ∞" is defined as the complement of any compact set in $\mathbb{C}$, plus the point ∞. A function $f(z)$ is analytic at a point z ($z \neq \infty$) of S^2 if $f(z)$ is holomorphic in $\mathbb{C}$; and $f(z)$ is analytic at $z = \infty$ if $f(1/z)$ is holomorphic at $z = 0$. With this topology and complex structure, the Riemann sphere S^2 is a Riemann surface, and, as is well known (see Example 7.5.11), homeomorphic (under stereographic projection) to the sphere ($\widehat{\mathbb{C}}$ is also called the Riemann sphere).

Now we might ask whether or not these three Riemann surfaces are really different.

Notice first that $\mathbb{C}$, $\mathbb{D}$ are homeomorphic (see Example 2.1.3), but neither is homeomorphic to $\widehat{\mathbb{C}}$ (which unlike $\mathbb{C}$, $\mathbb{D}$ is compact). Next, the classes of analytic functions on the spaces $\mathbb{C}$, $\mathbb{D}$, $\widehat{\mathbb{C}}$ differ: The holomorphic functions on $\mathbb{C}$ are the entire functions, those on $\mathbb{D}$ may be the restrictions of functions meromorphic on $\mathbb{C}$, and finally the only bounded functions holomorphic on $\mathbb{C}$ are constants.

Although $\mathbb{C}$, $\mathbb{D}$, $\widehat{\mathbb{C}}$ differ as Riemann surfaces, all three are simply connected.

It is easy to see that these three Riemann surfaces are non-isomorphic: that $\mathbb{C}$ and $\mathbb{D}$ are non-isomorphic is an immediate consequence of Liouville's theorem and $\widehat{\mathbb{C}}$ is not even homeomorphic to both of them.

It is the opinion of many experts that Riemann surface theory as a subject distinct from the study of complex manifolds consists of two theorems and their consequences. The second is the subject of this chapter; the first is the uniformization theorem whose statement follows.

Theorem 9.3.5 (Uniformization theorem for Riemann surfaces) *If S is a Riemann surface, then S is holomorphically equivalent to either:*

(i) $\mathbb{C}$.
(ii) $\mathbb{C}\backslash\{0\}$.
(iii) $\mathbb{C}/\mathrm{L}$, *where L is a lattice.*
(iv) $\widehat{\mathbb{C}} = \mathbb{C}P^1 = \mathbb{C} \cup \{\infty\}$.
(v) H/G, *where* $H = \{\mathrm{Im}\, z > 0\}$ *and G is a properly discontinuous group of holomorphic self-maps of H.*

Proof See Ahlfors [17]. □

Theorem 9.3.6 (Uniformization theorem) *Each simply connected Riemann surface is conformally equivalent to either the open unit disc $\mathbb{D}$, the complex plane $\mathbb{C}$, or the Riemann sphere $\widehat{\mathbb{C}}$.*

To illustrate the use of the uniformization theorem, we note:

Corollary 9.3.7 *Every Riemann surface is second countable and separable.*

Proof These properties project from the universal cover. Certainly $\mathbb{D}$, $\mathbb{C}$, and $\widehat{\mathbb{C}}$ have these properties. □

Corollary 9.3.8 (Picard's Theorem) *Let f be a meromorphic function in $\mathbb{C}$. If $\widehat{\mathbb{C}}\backslash f(\mathbb{C})$ contains at least three points, then f is constant.*

Proof Let $\Omega = f(\mathbb{C})$. By the uniformization theorem, the universal cover $\widetilde{\Omega}$ of Ω is conformally equivalent to U. Let π denote the universal cover map, $\pi : \mathbb{D} \to \Omega$. Then, locally, π^{-1} exists and $\pi^{-1} \cdot f$ may be continued along all paths in $\mathbb{C}$ to define a map $F : \mathbb{C} \to \mathbb{D}$. Since F is holomorphic, Liouville's theorem says F is constant; hence so is f. □

Theorem 9.3.4 was proved independently by Koebe and Poincaré in 1907. Poincaré also generalized the theorem to arbitrary analytic curves, that is, to the case where P is not a polynomial, but a convergent power series in two variables, together with all its analytic continuations. It took some of the best mathematicians of the world 25 years to give rigorous proofs of this and some related theorems; the first complete proofs were found by Poincaré and by Koebe in 1907. Many basic concepts and methods of contemporary mathematics were developed explicitly for establishing uniformization. The list includes covering spaces, methods for solving nonlinear elliptic partial differential equations, existence and distortion theorems for conformal mappings, topological dimension, and invariance of domain.

Ahlfors has hailed the famous uniformization theorem of Koebe as "perhaps the single most important theorem in the whole theory of analytic functions of one variable" [21, Chapter 10, p. 136]. As he went on to remark: the uniformization theorem "does for Riemann surfaces what Riemann's mapping theorem does for plane regions."

Despite their importance, it remains the case, as Lipman Bers commented that "a scholarly history of that period is yet to be written" [81, p. 559].

9.4 Proof of the Uniformization Theorem

The uniformization theorem of Koebe and Klein asserts a more general result than Riemann mapping theorem by identifying the simply connected Riemann surfaces. Our aim in the rest of this chapter is to prove theorem of Koebe and Poincaré. The uniformization theorem, even today, commands a nontrivial proof. Here we will sketch one style of proof with some details omitted. There are a number of different proofs of the uniformization theorem. Lipman Bers's book on Riemann surfaces [76] has a nice presentation of such a proof.

9.4.1 Green's Functions

The goal of this section is to develop the machinery of Green's functions on a Riemann surface. This is in fact the key to the traditional proof of the uniformization theorem.

In complex analytic terms, the Poincaré–Koebe uniformization theorem (a generalization of the Riemann mapping theorem) states that every simply connected Riemann surface is conformally equivalent to one of the following:

- The Riemann sphere $\widehat{\mathbb{C}} := \mathbb{C} \cup \{\infty\}$, which is isomorphic to the complex projective line $\mathbb{CP}^1$.
- The complex plane $\mathbb{C}$.
- The open disc $\mathbb{D} = \{z \in \mathbb{C} : |z| < 1\}$, which is isomorphic to the upper half plane $\mathbb{H} := \{z \in \mathbb{C} : \mathrm{Im}(z) > 0\}$.

A Riemann surface is *elliptic*, *parabolic*, or *hyperbolic* according to whether its universal cover is isomorphic to $\mathbb{CP}^1$, $\mathbb{C}$ or U.

The classification scheme above is typically used by geometers. There is a different classification for Riemann surfaces that is typically used by complex analysts. It employs a different definition for "parabolic" and "hyperbolic." In this alternative classification scheme, a Riemann surface is called parabolic if there are no nonconstant negative subharmonic functions on the surface and is otherwise called hyperbolic [23].

The complex plane does not carry any negative nonconstant subharmonic function. This property could be used to extend the definition of parabolicity to arbitrary open surfaces. Traditionally, however, the name of parabolic surfaces has been reserved for surfaces that have no Green's function.

Definition 9.4.1 A Riemann surface S is called *hyperbolic* if there exists a nonconstant negative subharmonic function on S.

Hyperbolic Riemann surfaces are also said to have a "positive ideal boundary." We now give a definition of the Green's function using Perron's method.

Definition 9.4.2 Let S be a Riemann surface and $p \in S$. A function g is called a *Green's function* for S with singularity at p provided:

$$g \text{ is harmonic in } S \backslash \{p\}. \tag{9.4}$$

$$g > 0 \text{ in } S \backslash \{p\}. \tag{9.5}$$

If z is a local parameter vanishing at p, then

$$g(z) + \log |z| \text{ is harmonic in a neighborhood of } p. \tag{9.6}$$

$$\text{If } \widehat{g} \text{ is another function satisfying (9.4)–(9.6), then } \widehat{g} \geq g. \tag{9.7}$$

Remark 9.4.3 Condition (9.6) is independent of the choice of local parameter vanishing at p (that is, the condition makes sense on S). For if ζ is another such parameter, then (with $a_1 \neq 0$)

$$\begin{aligned} \zeta(z) &= a_1 z + a_2 z^2 + \cdots \\ &= a_1 z \left(1 + \frac{a_2}{a_1} z + \cdots\right) = a_1 z f(z), \end{aligned}$$

where f is holomorphic near $z = 0$ and $f(0) \neq 0$. We write $f(z) = e^{h(z)}$, with h holomorphic near $z = 0$, and conclude that

$$\log |\zeta| = \log |a_1| + \log |z| + \operatorname{Re} h(z).$$

Since $\operatorname{Re} h(z)$ is harmonic, we see that $g(z) + \log |z|$ is harmonic if and only if $g(\zeta) + \log |\zeta|$ is.

The nonexistence of a Green's function is undoubtedly the most important characteristic of a parabolic surface, but as a definition it has the disadvantage that it refers to a fixed choice of the pole. In order to justify the definition, it is necessary to prove that the existence of a Green's function does not depend on the location of the pole.

Theorem 9.4.4 *Let S be a Riemann surface. There exists a Green's function on S (with singularity at some point $p \in S$) if and only if S is hyperbolic.*

Proof Let g be a Green's function on S with singularity at p. Let $m > 0$. Set $f = \min\{m, g\}$. Then f is clearly superharmonic in $S\backslash\{p\}$. Since $g(q) \to +\infty$ as $q \to p$, $f \equiv m$ in a neighborhood of p. Thus f is superharmonic on S. Also $f > 0$. If f were constant, then $g \geq m$, and thus $g - m$ would be another candidate for the Green's function with $g - m < g$. This contradiction shows that f cannot be constant and hence S is hyperbolic.

Conversely, assume that S is hyperbolic. Let $p \in S$. Let K be a conformal disc about p with local coordinate z. Set

$$\begin{aligned} \mathcal{F} = \{v;\quad & v \text{ is subharmonic in } S\backslash\{p\}, \\ & v \geq 0,\ v \text{ has compact support, and} \\ & v(z) + \log|z| \text{ is subharmonic in } |z| < 1\}. \end{aligned}$$

To note that $\mathcal{F}$ is non-empty, we define

$$v_0(z) = \begin{cases} -\log|z|, & 0 < |z| < 1 \\ 0, & |z| \geq 1 \end{cases}$$

and observe that $v_0 \in \mathcal{F}$. It is easy to see that $\mathcal{F}$ is a Perron family (on $S\backslash\{p\}$). □

Definition 9.4.5 We shall call a function f_p *admissible* if:

1. $f_p(p) = 0$.
2. $|f_p| < 1$.
3. f_p holomorphic on S.
4. If h_p has properties **1**, **2**, **3**, then $|h_p| \leq |f_p|$.

Lemma 9.4.6 *If p is any point on a simply connected hyperbolic Riemann surface S, then there exists an admissible function f_p such that f_p maps S to the unit disc.*

Proof Consider the Green's function $G(p, q)$. Near p, $G(p, q) = -\log|z| + u$, where u is harmonic. In some neighborhood of p, u has a conjugate harmonic function v. Form the function $f_p = ze^{-(u+iv)}$. Then $\log|f_p| = \log|z| - u = -G$. Since G has a conjugate H at every point of $S \setminus \{p\}$, we may analytically continue f_p. At any point $q \in S \setminus \{p\}$,

$$f_p = e^{-(G+iH)}.$$

It is clear that $f_p(q) = 0$ if and only if $p = q$. Furthermore f_p is analytic on S, and $|f_p| < 1$ since $G > 0$. Now suppose that h_p is a function satisfying properties 1, 2, 3. Consider $-\log|h_p|$. Since h_p has a zero at p, $-\log|h_p| \geq G = -\log|r_p|$ from which it follows that $|h_p| \leq |f_p|$. The first inequality results from the fact that $-\log|h_p|$ satisfies the first three conditions of a Green's function.

Consider

$$\phi(p) = \frac{f_{q_0}(p) - f_{q_0}(q)}{1 - \overline{f_{q_0}(q)} f_{q_0}(p)}.$$

It is easily seen that $|\phi(p)| < 1$, $\phi(q) = 0$. Hence $|\phi(p)| \le |f_q(p)|$. If we let $p = q_0$, we obtain

$$|f_{q_0}(q)| \le |f_q(q_0)|.$$

Hence

$$|f_{q_0}(q)| \le |f_q(q_0)|$$

and

$$|\phi(q_0)| = |r_q(q_0)|.$$

That is, $\frac{|\phi|}{|f_q|} \equiv 1$, so that

$$\phi = e^{-i\theta f_q}.$$

Therefore $f_{q_0}(p) = f_{q_0}(q)$ implies that

$$p = q,$$

and f_p maps S $1-1$ conformally into the unit disc. All that is left to show is that f_p maps S onto the unit disc.

Let a be a boundary point of $f_p(S)$ in the w plane. We can select a single-valued branch of $\log(w - a) = \zeta$. This maps S into a domain contained in $\mathrm{Re}\zeta < \log 2$. Let the image of a be ζ_0. Now we map the half plane $Re\zeta < \log 2$ conformally into the unit disc taking ζ_0 into 0. Call this composed mapping of S into $|\zeta| < 1$ by the name g. Since $|g| < 1$ and $g(p) = 0$, it follows that $|g| \le |f_p|$. Let p_n correspond to a_n a sequence of points in $f_p(S)$ that converge to a. It follows that $|g(p_n)| \to 1$, hence $|f_p(p_n)| \to 1$. Hence every boundary point a has modulus 1. $f_p(S)$ is conformally equivalent to the open unit disc, and then so is S. □

9.4.2 *Bipolar Green's Functions*

By Theorem 9.4.4 not every Riemann surface has a Green's function. However, every Riemann surface does have a bipolar Green's function. In this subsection we will prove the existence of the bipolar Green's function, and in the next we will use it to prove the celebrated uniformization theorem (see [166]).

Let q_1, q_2 be distinct points of the Riemann surface S, and let Δ_1 and Δ_2 be disjoint coordinate discs containing q_1 and q_2, respectively, with coordinate maps $z_1(p)$ and $z_2(p)$ satisfying $z_1(q_1) = 0$ and $z_2(q_2) = 0$.

We define the *bipolar Green's function* with poles at q_1 and q_2 to be any harmonic function $G(p, q_1, q_2)$ on $S \setminus \{q_1, q_2\}$ such that:

1. $G(p, q_1, q_2) + \log |z_1(q)|$ is harmonic at q_1.
2. $G(p, q_1, q_2) - \log |z_2(q)|$ is harmonic at q_2.
3. $G(p, q_1, q_2)$ is bounded on $S \setminus (\Delta_1 \cup \Delta_2)$.

Thus $G(p, q_1, q_2)$ has a logarithmic pole at q_1, and $-G(p, q_1, q_2)$ has a logarithmic pole at q_2. Note that $G(p, q_1, q_2)$ is not uniquely determined. It is unique up to adding a bounded harmonic function.

Example 9.4.7 The bipolar Green's function for the Riemann sphere $\widehat{\mathbb{C}}$ with poles at $q_1 = 0$ and $q_2 = \infty$ is given by $G(z, 0, \infty) = -\log |z|$. In this case, $G(z, 0, \infty)$ is uniquely determined up to an additive constant, since a bounded harmonic function on $\widehat{\mathbb{C}}$ is constant.

Example 9.4.8 If the Green's function $g(p, q)$ exists for the Riemann surface S, then

$$G(p, q_1, q_2) = g(p, q_1) - g(p, q_2)$$

is a bipolar Green's function for S.

Theorem 9.4.9 *For each pair of distinct points q_1, q_2 on a Riemann surface, there is a bipolar Green's function $G(p, q_1, q_2)$.*

The technical part of the proof is contained in the following lemma.

Lemma 9.4.10 *Let S be a finite bordered Riemann surface, and let q_1, q_2 be distinct points of S. Let $B_1 = \{|z_1(p)| \le \sigma\}$ and $B_2 = \{|z_2(p)| \le \sigma\}$ be disjoint closed coordinate discs, where $z_1(q_1) = 0$, $z_2(q_2) = 0$. There is a constant $C > 0$ such that*

$$|g_S(p, q_1) - g_S(p, q_2)| \le C, \quad p \in S \setminus (B_1 \cup B_2),$$

for all Riemann surfaces S containing $S \cup \partial S$ for which the Green's function $g_S(p, q)$ exists.

Proof It suffices to establish the lemma for finite bordered Riemann surfaces S containing $S \cup \partial S$. Let $\rho > 0$ satisfy $\rho < \sigma$. For $j = 1, 2$, let A_j be the closed coordinate disc $\{|z_j(p)| \le \rho\}$, and let $M_j = M_j(S)$ be the maximum of $g_S(p, q_j)$ on ∂A_j. For $p \in \partial B_j$ we have

$$g_S(p, q_j) + \log |z_j(p)| \le \max \{g_S(q, q_j) : q \in \partial B_j\} + \log \sigma$$

and $g_S(p, q_j) + \log |z_j(p)|$ is harmonic on B_j. By the maximum principle, the estimate persists for all $p \in B_j$. Taking the supremum over $p \in \partial A_j$, we obtain

$$M_j + \log \rho \leq \max \left\{ g_S\left(q, q_j\right) : q \in \partial B_j \right\} + \log \sigma.$$

Thus there is $p_j \in \partial B_j$ such that $M_j + \log \rho \leq g_R\left(p_j, q_j\right) + \log \sigma$, or

$$M_j - g_S\left(p_j, q_j\right) \leq \log(\sigma/\rho).$$

Now, $M_j - g_S\left(p, q_j\right)$ is a positive harmonic function on $S \backslash \left(A_1 \cup A_2\right)$. If we apply the Harnack estimate to the Riemann surface $S \backslash \left(A_1 \cup A_2\right)$ and the compact subset $\partial B_1 \cup \partial B_2$, we obtain a constant C_0, independent of S, such that $M_j - g_S\left(p, q_j\right) \leq C_0$ for $p \in \partial B_1 \cup \partial B_2$. Thus

$$M_j - C_0 \leq g_R\left(p, q_j\right) \leq M_j, \quad p \in \partial B_1 \cup \partial B_2. \tag{9.8}$$

Since $g_R\left(p, q_1\right)$ is harmonic for $p \in B_2$ and satisfies (9.8) on ∂B_2, it must satisfy (9.8) for all $p \in B_2$. In particular,

$$M_1 - C_0 \leq g_R\left(q_2, q_1\right) \leq M_1.$$

Similarly,

$$M_2 - C_0 \leq g_R\left(q_1, q_2\right) \leq M_2.$$

Since $g_R\left(q_2, q_1\right) = g_R\left(q_1, q_2\right)$, these estimates imply that $\left|M_1 - M_2\right| \leq C_0$. From (9.8), we then obtain

$$\left|g_R\left(p, q_1\right) - g_R\left(p, q_2\right)\right| \leq 2C_0, \quad p \in \partial B_1 \cup \partial B_2.$$

Since the Green's function vanishes on ∂R, this estimate persists on $R \backslash \left(B_1 \cup B_2\right)$, by the maximum principle. This establishes the lemma, with $C = 2C_0$.

Now we complete the proof of Theorem 9.4.9 and approximate S by surfaces $S \backslash B_\varepsilon$, where B_ε is a closed coordinate disc $\{|z_0(p)| \leq \varepsilon\}$ centered at some arbitrary point $p_0 \in S$. Green's function $g_\varepsilon(p, q)$ exists for $S \backslash B_\varepsilon$, and we can form bipolar Green's function $g_\varepsilon\left(p, q_1\right) - g_\varepsilon\left(p, q_2\right)$. Lemma 9.4.10 and compactness results allow us to pass to a normal limit through some sequence $\varepsilon_j \to 0$, to obtain a bipolar Green's function $G\left(p, q_1, q_2\right)$ for $S \backslash \{p_0\}$. Since $G\left(p, q_1, q_2\right)$ is bounded near the isolated singularity p_0, it extends harmonically to p_0 and is a bipolar Green's function for S. □

9.4.3 *Proof of the Uniformization Theorem 9.3.4*

Our strategy is to split the proof into two cases. First we show that if S is hyperbolic Riemann surface, then the Green's function for S exists, and it can be used to map S conformally to the open unit disc (see Lemma 9.4.6).

Second if the Green's function for S does not exist (i.e., S is not a hyperbolic Riemann surface), we use the bipolar Green's function to map S to the sphere or the punctured sphere, which is conformally equivalent to $\mathbb{C}$.

If S is a hyperbolic Riemann surface, then, by Theorem 9.4.4, at every point p there exists an admissible function f_p. By Lemma 9.4.6, f_p maps S to the unit disc.

Now suppose that the Green's function for S does not exist.

By Theorem 9.4.9 for each pair of distinct points q_1, q_2 on S, there is a bipolar Green's function $G(p, q_1, q_2)$.

Since S is simply connected, we can then find by analytic continuation a meromorphic function $\psi : S \to \overline{\mathbb{C}}$ on S such that

$$|\psi(p)| = e^{-G(p,q_1,q_2)}, \quad p \in S.$$

Notice that ψ has one simple zero at q_1 and one simple pole at q_2.

We claim that ψ is one-to-one. To see this, let q_0 be any point of S distinct from q_1 and q_2, select a bipolar Green's function $G(p, q_0, q_2)$, and let $\varphi : S \to \widehat{\mathbb{C}}$ be a meromorphic function such that

$$|\varphi(p)| = e^{-G(p,q_0,q_2)}, \quad p \in S.$$

Consider the quotient

$$\phi(p) = \frac{\psi(p) - \psi(q_0)}{\varphi(p)}, \quad p \in S.$$

From the definition of bipolar Green's function, ψ is bounded away from q_1, q_2 and φ is bounded away from q_0, q_2. Also, we have $\psi(q_1) = \varphi(q_2) = 0$ and $\psi(q_2) = \varphi(q_2) = \infty$. Since all poles and zeros are simple, the function $\phi : S \to \mathbb{C}$ defined above is well defined, as well as holomorphic.

The poles at q_1 and q_2 cancel, so ϕ is a bounded holomorphic function on S. Then the function $\log(|\phi(z)|/|\phi|_\infty)$ is nonnegative and subharmonic on S. Since the Green's function does not exist, Theorem 9.4.4 shows that $\log(|\phi(z)|/|\phi|_\infty)$ must be constant. Hence ϕ is constant.

Since $\phi(q_1) = -\psi(q_0)/\varphi(q_1) \neq 0$, we see that ϕ is nonzero. Hence $\psi(p) = \psi(q_0)$ if and only if $p = q_0$. Since q_0 was arbitrary, we have proved that ψ is injective.

Now, ψ maps S conformally onto a simply connected domain $\psi(S)$ on the Riemann sphere $\widehat{\mathbb{C}}$. If $\widehat{\mathbb{C}} \backslash \psi(S)$ has more than one point, then by the Riemann mapping theorem $\psi(S)$ can be mapped conformally onto the unit disc, and consequently, Green's functions for $\psi(S)$ and hence for S exist.

We conclude that $\widehat{\mathbb{C}} \backslash \varphi(S)$ has at most one point. Thus either $\varphi(S) = \widehat{\mathbb{C}}$, or else $\varphi(S)$ is the punctured sphere, which is conformally equivalent to $\mathbb{C}$. The proof is complete.

9.4.4 *Another Proof of the Uniformization Theorem 9.3.4*

As we mentioned above, there are a number of different proofs of the uniformization theorem. We present here another one.

Theorem 9.3.4 asserts a more general result than the Riemann mapping theorem by identifying the simply connected Riemann surfaces. Now it is a basic fact of topology, more precisely covering space theory [189], that any Riemann surface admits a universal cover $\widetilde{\Sigma}$, a simply connected Riemann surface, for which the deck transformations (the fundamental group) will act as a discrete group Γ of conformal transformations without fixed points. The discrete groups of conformal transformations of $\mathbb{C}$ and $\widehat{\mathbb{C}}$ are easily identified. Any homeomorphism of $\widehat{\mathbb{C}}$ has a fixed point. A conformal transformation of $\mathbb{C}$ is a similarity. A similarity without fixed points is a translation. It follows with a little work that $\mathbb{C}$ is the universal cover only for Tori $\approx \mathbb{S} \times \mathbb{S}$ and cylinders $\mathbb{C}\backslash\{0\} \approx \mathbb{S} \times \mathbb{R}$. Every other Riemann surface therefore admits the disc as its universal cover.

Notes to This Chapter

A complete bibliography on uniformization would require a volume. We list only some books and papers.

Accounts of the classical uniformization theory will be found in the second volume of Fricke-Klein [289] (monumental and hard to read), in the second volume of Appel-Goursat [29] (written by Fatou), in the treatises by Ford [155], and by Nevanlinna [401], and in most standard texts on Riemann surfaces (Weyl [526], Pfluger [413], Springer [481], Ahlfors-Sario [23]). It is of interest to consult also the original works by Klein (see [290] that contain also the amusing correspondence with Poincaré), Poincaré [419, 420], and Koebe [300, 301, 303–305, 305, 306, 306, 307, 310].

Standard reference texts for classical uniformization theory are H. Weyl [524], available also in an English translation [525], R. Fricke and F. Klein [160], P. Appell and E. Goursat [29], L. Ford, [155], R. Nevanlinna [401]. Of more recent texts we mention only Springer [481] and L.V. Ahlfors [17].

Chapter 10
Automorphism Groups

10.1 Introduction

Every mathematical subject is shaped by its isomorphisms. In point set topology, these are *homeomorphisms*, in Riemannian geometry these are *isometries*, in group theory these are *group isomorphisms*, and so forth. This point of view is part of Felix Klein's *Erlangen Program*, and it has exerted a strong influence over mathematics for the past hundred years.

In complex analysis, the isomorphisms are the *bijective holomorphic mappings*, which are often called *conformal mappings* (in one complex variable) or *biholomorphic mappings* (in one and several complex variables). In this book, a *domain* is a non-empty, connected, open set. If Ω_1 and Ω_2 are domains in $\mathbb{C}^n$ with $n \geq 1$ and if $\varphi : \Omega_1 \to \Omega_2$ is a biholomorphic mapping, then we may transfer the function theory of one domain to the other by way of composition with the mapping φ or φ^{-1}.

In this discussion we are particularly interested in the *automorphism group* of a given domain. For a domain Ω in $\mathbb{C}^n$, its automorphism group is the collection of biholomorphic self-mappings of Ω. This collection forms a *group* under the binary operation of composition of mappings; we denote it by $\mathrm{Aut}(\Omega)$. The automorphism group comes with the natural *topology of uniform convergence on compact subsets of* Ω (i.e., what topologists call the compact open topology in Ω). Equipped with this topology, the mapping $(\varphi, \psi) \to \varphi \circ \psi^{-1}$ from $\mathrm{Aut}(\Omega) \times \mathrm{Aut}(\Omega)$ to $\mathrm{Aut}(\Omega)$ turns out to be continuous. Thus every automorphism group is a *topological group*.

We observe that the automorphism group is an *invariant*, in the sense that it is preserved by biholomorphic mappings. More precisely, for two domains Ω_1 and Ω_2 that admit a biholomorphic mapping $f : \Omega_1 \to \Omega_2$, one can easily check that the map $f_* : \mathrm{Aut}(\Omega_1) \to \mathrm{Aut}(\Omega_2)$ given by $\varphi \mapsto f \circ \varphi \circ f^{-1}$ is a topological group isomorphism, meaning a continuous group isomorphism with a continuous inverse. Thus it is natural to investigate the relationship between the biholomorphic

P. V. Dovbush, S. G. Krantz, *The Geometric Theory of Complex Variables*,
https://doi.org/10.1007/978-3-031-77204-7_10

equivalence of domains and the isomorphic equivalence of their automorphism groups.

As this exposition progresses, we shall give several references for further study. This reference section is far from being complete. See [198] and [254] for more extensive bibliographies.

10.2 Automorphism Groups of Planar Domains

We begin with some examples. The automorphism group of the open unit disc $\mathbb{D}$ in the complex plane is noncompact. In fact any textbook will tell us (see, for instance, [202]) that the automorphism group of the unit disc is generated by the rotations about the origin and the Möbius transforms of type $z \mapsto (z-a)/(1-\overline{a}z)$ for a in $\mathbb{C}$ with $|a| < 1$.

If one considers the point sequence $a_j = (-j+1)/j$ $(j = 1, 2, \ldots)$ in the unit disc, then the maps

$$\varphi_j(z) = \frac{z - a_j}{1 - a_j z} \quad (j = 1, 2, \ldots)$$

form a sequence in $\text{Aut}(\mathbb{D})$. However, it is immediately clear that the limit of this sequence in the compact open topology is the constant map $\widehat{\varphi}(z) \equiv 1$. Since $\widehat{\varphi}$ is not an automorphism of the disc, we conclude that $\text{Aut}(\mathbb{D})$ is (sequentially) noncompact.

Now consider the annulus $A = \{z \in \mathbb{C} : r < |z| < R\}$ for some positive constants r and R with $0 < r < R$. The automorphism group of A is compact. In fact its automorphism group is generated by the rotations about the origin and the inversion mapping $z \mapsto Rr/z$.

It is a classical fact due to Heins (see [228] or [148]) that every planar domain with at least two, but finitely many, holes has a *finite* automorphism group (i.e., a group with finitely many elements).

In fact if the domain is multiply connected with connectivity $p > 2$, then the number of automorphisms does not exceed $p(p-1)(p-2)$, an estimate that can be sharpened in a number of cases. Thus the automorphism groups of such domains are certainly compact.

The discussion thus far suggests that any domain with a noncompact automorphism group must be either simply connected or have infinitely many holes. We now present an example to illustrate the problem.

Example 10.2.1 Let $\mathbb{D} = \{z \in \mathbb{C} : |z| < 1\}$ as before. Let $\overline{K} = \{\zeta \in \mathbb{C} : |\zeta| \leq 1/10\}$. Define $\varphi : \mathbb{D} \to \mathbb{D}$ by

$$\varphi(\zeta) = \frac{\zeta - 1/2}{1 - (1/2)\zeta}.$$

For $j = 1, 2, \ldots$, we let φ^j denote the j-fold iterate of φ,

$$\varphi^j \equiv \underbrace{\varphi \circ \varphi \circ \cdots \circ \varphi}_{j\ times};$$

if j is a negative integer, we let

$$\varphi^j \equiv [\varphi^{-1}]^{|j|};$$

and, finally, we interpret φ^0 to mean the identity map of $\mathbb{D}$,

$$\varphi^0 = \mathrm{id}.$$

With this language, we define

$$U = \mathbb{D} \setminus \bigcup_{j=-\infty}^{\infty} \varphi^j(\overline{K}).$$

Figure 10.1 exhibits this domain. It is obviously infinitely connected. What is its automorphism group?

Certainly any φ^k is an automorphism of U, just because $\varphi^k \circ \varphi^j = \varphi^{j+k}$. Also the mapping $\zeta \mapsto -\zeta$ is an automorphism of U. It is not difficult to show, though we shall not provide the details, that these are all the automorphisms of U.

It is easy to prove by induction that the explicit formula for φ^j is

$$\varphi^j(\zeta) = \frac{\zeta - \frac{3^j-1}{3^j+1}}{1 - \frac{3^j-1}{3^j+1}\zeta}.$$

Fig. 10.1 The domain U

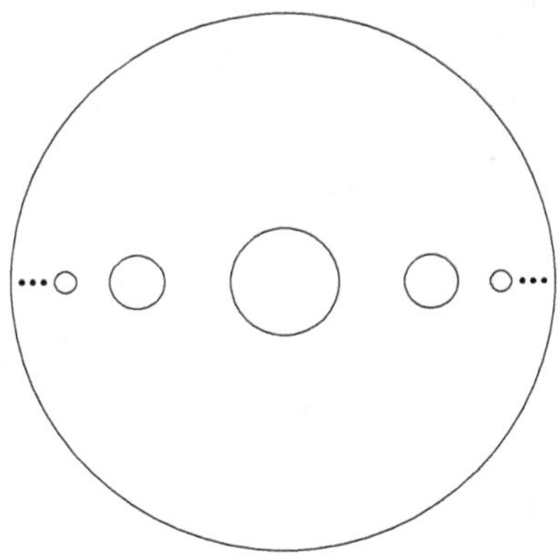

We see, therefore, that $\varphi^j \to -1$ uniformly on compact sets in $\mathbb{D}$ as $j \to +\infty$ (and, likewise, $\varphi^j \to +1$ uniformly on compact sets in $\mathbb{D}$ as $j \to -\infty$). In conclusion, Aut(U) is noncompact. (However, note that the boundary of U is not C^1-smooth at the points 1 and -1—in fact it is not a manifold.)

Keeping this example in mind, we present the following one-dimensional version of the Wong/Rosay theorem (see [441, 533]).

Theorem 10.2.2 *Let U be a planar domain with C^1-boundary (i.e., ∂U is the union of simple, closed, continuously differentiable curves, and each point of ∂U has a neighborhood that intersects exactly one of those curves). If there exists a point Q in U, a point P of ∂U, and a sequence of mappings φ_j in* Aut(U) *such that $\lim_{j\to\infty} \varphi_j(Q) = P$, then U is conformally equivalent to the unit disc.*

Such a special boundary point P is called a *boundary orbit accumulation point* (or, simply, an *orbit accumulation point*) throughout this treatment.

Before giving a proof of Theorem 10.2.2, we present a discussion concerning the noncompactness of the automorphism group in relation to the existence of boundary orbit accumulation points. In case the domain U in the theorem is a bounded domain with C^1-boundary, there are (by definition) at most finitely many holes in the domain. There obviously cannot be just one hole; if that were so, then the domain would be conformally equivalent to an annulus (see [15]), hence would have a compact automorphism group.

Of course the domain we are discussing is not the plane. Thus the preceding discussion implies that the domain U in the theorem cannot be multiply connected unless it has connectivity at least three, in which event Aut(U) is compact. On the other hand, *the existence of the orbit accumulation point P in the hypothesis implies that* Aut(U) *is noncompact*, since every subsequential limit (if it exists) of the sequence $\{\varphi_j\}$ has to be constant (as we see from the open mapping theorem). We conclude that U is simply connected. Since the domain certainly has a boundary point, it is then conformally equivalent to the unit disc by the Riemann mapping theorem.

We also remark that when the domain U is bounded, the existence of a boundary orbit accumulation point is actually equivalent to the noncompactness of Aut(U) (see [393]). To see this, suppose that U is a bounded domain and that Aut(U) has the property that, for each Q in U, the *orbit set*

$$\text{Orbit}(Q) = \{\varphi(Q) : \varphi \in \text{Aut}(U)\}$$

is bounded away from the boundary of U, so that the closure of Orbit(Q) is a compact subset of U.

Let $\{\varphi_j\}$ be a sequence of automorphisms and fix Q in U. Then, taking a subsequence if need be, we may assume that $\lim_{j\to\infty} \varphi_j(Q) = Q_0$ for some Q_0 in U. By Montel's theorem, we may pass to a further subsequence and suppose that the sequence φ_j converges uniformly on compact subsets of U to a holomorphic self-mapping $\varphi : U \to \overline{U}$. Since $\overline{\text{Orbit}(Q)}$ is a compact subset in U, it is clear that $\varphi(U) \subset U$. Moreover, considering the sequence $\{\varphi_j^{-1}\}$ of the inverse mappings of the last subsequence, one sees that $(\varphi_j^{-1})'$ is defined at all points. Hence $\varphi'(Q) \neq 0$. Now Hurwitz's theorem implies that φ' is nowhere zero on U, since φ_j' converges to φ' uniformly on compact subsets by the Cauchy estimates. Hence the limit map φ is locally one-to-one.

Finally, using the uniform convergence of φ_j on compact subsets, it is not hard to show (use the argument principle, for instance) that φ is globally one-to-one. Applying the same argument to the sequence of inverse mappings, one can show that φ is in fact an automorphism of U. The upshot of this argument is that, if U is bounded and if there are no boundary orbit accumulation points, then $\text{Aut}(U)$ is sequentially compact. But then one can indeed conclude that $\text{Aut}(U)$ is compact, since the automorphism group admits the following metric $\mathbb{D}$ that is compatible with the compact open topology: For φ and ψ in $\text{Aut}(U)$,

$$d(\varphi, \psi) = \sum_{m=1}^{\infty} 2^{-m} \cdot \frac{\sup\limits_{z \in K_m} |\varphi(z) - \psi(z)|}{1 + \sup\limits_{z \in K_m} |\varphi(z) - \psi(z)|},$$

where $\{K_m : m = 1, 2, \ldots\}$ is a compact exhaustion of U (i.e., each K_j is compact, $K_1 \subset K_2 \subset \cdots$, and $\cup_j K_j = U$).

Now we present the proof of our theorem.

Proof of Theorem 10.2.2 By the Riemann mapping theorem, it suffices to prove that U is simply connected. We establish this fact in several steps:

(1) There is a peaking function for U at the point P of ∂U. A *peaking function* is a function $\mu : \overline{U} \to \overline{\mathbb{D}}$ such that:

- μ is continuous on $\overline{U}$.
- μ is holomorphic on U.
- $\mu(P) = 1$.
- $|\mu(\zeta)| < 1$ for all ζ in $\overline{U} \setminus \{P\}$.

We can construct a peaking function as follows. Since the boundary ∂U is a C^1-curve, we can choose a point P' of $\mathbb{C} \setminus \overline{U}$ such that the line segment $\overline{PP'}$ joining P and P' is perpendicular to ∂U and is contained in $\mathbb{C} \setminus U$. Now let ψ_1 be a linear fractional transformation such that $\psi_1(P) = 0$, $\psi_1(P') = \infty$, and $\psi_1(\overline{PP'})$ is the negative real axis. Let $\psi_2(z) = \sqrt{z}$ (taking the principal branch), $\psi_3(z) = (1 - z)/(1 + z)$, and $\psi_4(z) = (z + 1)/2$. Then one can check easily that $\mu = \psi_4 \circ \psi_3 \circ \psi_2 \circ \psi_1$ is the peaking function that we seek (see Fig. 10.2).

(2) Let K be any compact subset of U. If φ_j, P, and Q are as in the hypothesis of the theorem, then there is a subsequence φ_{j_k} such that $\varphi_{j_k}(z) \to P$ uniformly for z in K. To see this, let μ be the peaking function as in part (1). Consider the maps $g_j \equiv \mu \circ \varphi_j$. Observe that each g_j is holomorphic and bounded by 1.

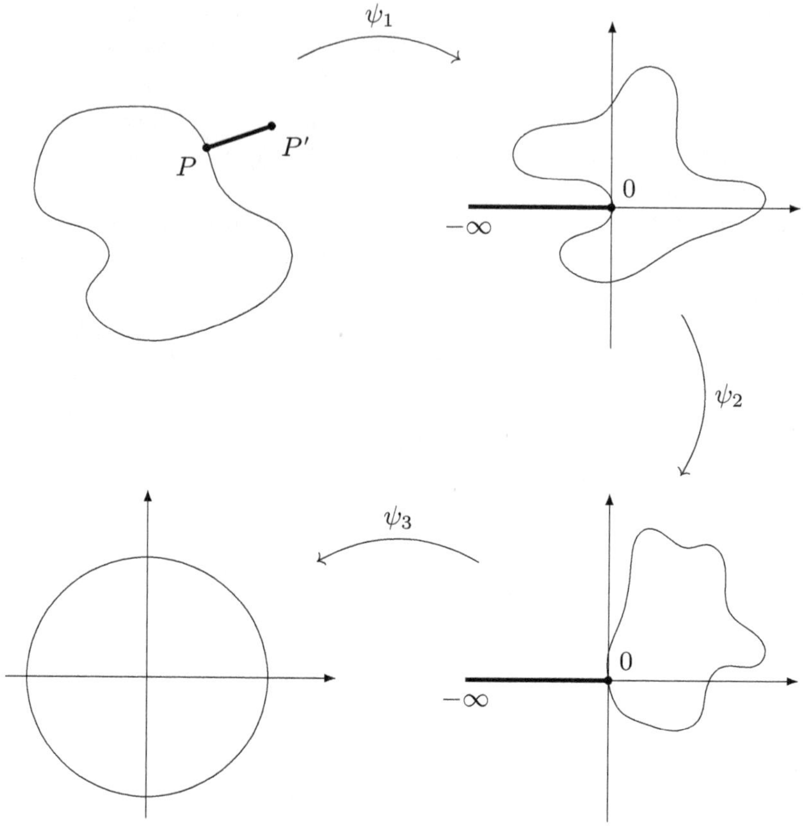

Fig. 10.2 The peaking function that we seek

By Montel's theorem, there is a subsequence $\{g_{j_k}\}$ that converges uniformly on compact sets to a limit function g_0.

Of course $|g_0(\zeta)| \leq 1$ for all ζ in U. On the other hand, observe that

$$g_0(Q) = \lim_{k\to\infty} g_{j_k}(Q) = \lim_{k\to\infty} \mu(\varphi_{j_k}(Q)) = \mu(\lim_{k\to\infty} \varphi_{j_k}(Q)) = \mu(P) = 1\,.$$

We have thereby contradicted the maximum modulus principle (see [201]) unless $g_0 \equiv 1$. But this means that the φ_{j_k} are converging uniformly on compact sets to the constant function with value P. That is what was claimed.

(3) There is a small open disc $\mathcal{D}$ centered at P such that $U \cap \mathcal{D}$ is simply connected. In fact this is a simple application of the implicit function theorem (see [336]), and we leave the details to the reader. *Notice that this is the only step in the proof where we use the fact that the boundary curve is continuously differentiable.*

(4) The domain U is simply connected. Consider a closed loop $\gamma : [0, 1] \to U$. The image curve of γ is a compact set Γ. Let $\mathcal{D}$ be the disc that we found in

step (3). By step (2), there is a k so large that $\varphi_{j_k}(\Gamma) \subseteq \mathcal{D} \cap U$. This ensures that $\varphi_{j_k} \circ \gamma$ is a closed curve that lies entirely in $\mathcal{D} \cap U$. Hence the image of $\varphi_{j_k} \circ \gamma$ can certainly be deformed to a point inside $\mathcal{D} \cap U$. Because the map φ_{j_k} is a homeomorphism, the image Γ of γ itself can be deformed to a point in U. Thus γ is homotopically trivial.

(5) And now our proof is complete, and for the simply connected domain U is conformally equivalent to the disc. That is what was claimed.

□

It should be noted that the possession of holes is not the sole determinant of whether a planar domain has compact automorphism group. We point out that Example 10.2.1 with just one more small closed disc centered at $(1 + 4i)/5$ excised will have no automorphism other than the identity map and hence will have compact automorphism group. Part of what makes this subject fascinating is that the theorems require a combination of algebra, complex function theory, topology, and differential geometry. The arguments in the preceding proof illustrate this cross-fertilization of techniques. We note in passing that the argument just presented can be used to give a proof of the Riemann mapping theorem (see [334]).

We now have a fairly complete understanding of which domains in the complex plane have compact automorphism group. We will next see that matters in the function theory of several complex variables are considerably more subtle and more beautiful.

10.3 Holomorphic Functions and Mappings of Several Variables

Let Ω be a domain in $\mathbb{C}^n$. A point in Ω has the form $z = (z_1, \ldots, z_n)$. A *holomorphic function* on Ω is a function $f : \Omega \to \mathbb{C}$ that is continuously differentiable and that satisfies the Cauchy-Riemann equations in each complex variable separately. This is the requirement that

$$\frac{\partial f}{\partial \overline{z}_j} \equiv 0$$

on U for $j = 1, \ldots, n$. Here $z_j = x_j + iy_j$ and

$$\frac{\partial}{\partial z_j} = \frac{1}{2}\left(\frac{\partial}{\partial x_j} - i\frac{\partial}{\partial y_j}\right) \quad , \quad \frac{\partial}{\partial \overline{z}_j} = \frac{1}{2}\left(\frac{\partial}{\partial x_j} + i\frac{\partial}{\partial y_j}\right) .$$

This definition is convenient and is equivalent to any other reasonable definition:

(a) That f is holomorphic in each variable separately.
(b) That f is holomorphic on any complex line in Ω.
(c) That f has a local power series representation at each point of Ω.

The reference [331] has a thorough treatment of all the different ways to think about holomorphic functions of several complex variables in $\mathbb{C}^n$.

Now let Ω_1 and Ω_2 be domains in $\mathbb{C}^n$. Then a *holomorphic mapping* $\varphi : \Omega_1 \to \Omega_2$ is a mapping

$$F(z_1, \ldots, z_n) = (f_1(z_1, \ldots, z_n), \ldots, f_n(z_1, \ldots, z_n))$$

such that each f_j is holomorphic on Ω in the sense of **(a)** or **(b)** or **(c)**.

Example 10.3.1 Let

$$\mathbb{B} = \{(z_1, z_2) \in \mathbb{C}^2 : |z_1|^2 + |z_2|^2 < 1\}.$$

This is the unit ball in complex two-space. If a is a complex number of modulus less than 1, then the transformation

$$\varphi(z_1, z_2) = \left(\frac{z_1 - a}{1 - \bar{a}z_1}, \frac{\sqrt{1 - |a|^2} z_2}{1 - \bar{a}z_1} \right) \tag{10.1}$$

is a holomorphic mapping from $\mathbb{B}$ to $\mathbb{B}$ that is both one-to-one and onto (see [451] and [331] for further discussion of this mapping). We note that unitary transformations are also automorphisms of $\mathbb{B}$. In fact, the automorphism group of $\mathbb{B}$ is generated by the transformations of type (10.1) (which, for want of a better name, we call complex two-dimensional Möbius transformations) and unitary mappings.

10.4 Failure of the Riemann Mapping Theorem in Several Complex Variables

We have already, in Example 10.3.1, defined the unit ball $\mathbb{B}$ in $\mathbb{C}^2$. The *unit bidisc* is the domain

$$\mathbb{D}^2 = \{(z_1, z_2) \in \mathbb{C}^2 : |z_1| < 1, |z_2| < 1\}.$$

Both these domains are topologically trivial, and they are certainly topologically (indeed, diffeomorphically) equivalent. Thus it was quite a surprise, and a blow to the development of the theory of several complex variables, when H. Poincaré proved in 1906 that $\mathbb{B}$ and $\mathbb{D}^2$ are *not* biholomorphically equivalent. We now sketch Poincaré's proof:

Proof of Poincaré's Theorem Seeking a contradiction, we assume that $\mathbb{B}$ and $\mathbb{D}^2$ *are* in fact biholomorphically equivalent via a mapping $\varphi : \mathbb{B} \to \mathbb{D}^2$. After composition with a unitary mapping and a Möbius transformation as in Example 10.3.1, we may assume that $\varphi(0) = 0$. Let the *isotropy group* $I^{\mathbb{B}}(0)$ of 0 be those elements of the automorphism group of $\mathbb{B}$ that fix 0. Likewise, let the isotropy group $I^{\mathbb{D}^2}(0)$ of 0 be those elements of the automorphism group of $\mathbb{D}^2$ that fix 0. It is easy to check that conjugation by the biholomorphism φ induces an isomorphism of $I^{\mathbb{B}}(0)$ with $I^{\mathbb{D}^2}(0)$. Thus it induces an isomorphism of the *connected component E of the identity mapping* in $I^{\mathbb{B}}(0)$ with the *connected component F of the identity mapping in* $I^{\mathbb{D}^2}(0)$. Now elementary but tricky calculations (see [331]) show that F consists of mappings that are rotations $(z_1, z_2) \mapsto (e^{i\theta_1} z_1, e^{i\theta_2} z_2)$ in each variable separately. We see, therefore, that F is Abelian. But E contains the unitary maps in two complex variables and hence is *not* Abelian. Hence E *cannot* be isomorphic to F. We have a contradiction. □

There are many other ways to prove Poincaré's theorem, several of them more geometric. But Poincaré's original proof, which we have just presented, serves once again to showcase the primacy of automorphism groups.

There are many modern results, due to Burns/Shnider/Wells [99], Chern/Moser [111], Greene/Krantz [194], and others, that show that the Riemann mapping theorem fails generically in several complex variables. Stated informally, if one considers two topologically trivial domains at random in $\mathbb{C}^n$, then with probability 1 (using a suitable probability measure on the collection of all domains) they will not be biholomorphically equivalent. See the paper [198] for a detailed discussion of this matter with references to the primary sources.

The next section begins to treat positive results, in the vein of the Riemann mapping theorem, in the several complex variables context. These are in part inspired by the material treated in Sect. 10.2.

10.5 Convexity in Complex Analysis

The Riemann mapping theorem clearly need not imply that biholomorphic mappings in one-complex variable preserve convexity. However, in several complex variables, the concept called *pseudoconvexity*, which can be viewed as a complex analytic version of convexity, plays a primary role because it *is* preserved by biholomorphic mappings.

The concept of pseudoconvexity is rather subtle, involving ideas from differential geometry, contact geometry, and the theory of quadratic forms. We cannot give it a complete treatment here, but the reader may consult the reference [328] for a thorough discussion of the notion. For our current purposes, we isolate and give a very informal (but completely correct) treatment of just one approach to pseudoconvexity theory, beginning with the restricted but in some sense more intuitive concept of *strong pseudoconvexity*.

Let Ω be a domain in $\mathbb{C}^n$ as usual, and let P belong to Ω. We assume that Ω has C^2-boundary. What does this mean? One way to describe the idea—for a bounded domain—is that the boundary has finitely many components and each component is a regularly imbedded, codimension-one C^2-manifold (i.e., a C^2-hypersurface) in Euclidean space. For an unbounded domain we would allow infinitely many boundary components.

For our current purposes, there is a more technical but more useful way to think about this idea. We think of our domain as specified in the following manner:

$$\Omega = \{z \in \mathbb{C}^n : \rho(z) < 0\},$$

where ρ is a C^2-function defined in an open neighborhood of $\overline{\Omega}$ such that $\nabla\rho$ vanishes at no point of $\partial\Omega$. The function ρ is called a *defining function* for Ω. Its existence is an easy consequence of the implicit function theorem. Observe that the defining function is not unique. For if h is any positive, smooth function on a neighborhood of $\overline{\Omega}$, then $h \cdot \rho$ is another defining function.

Definition 10.5.1 We say that the domain Ω is *strongly pseudoconvex* at a point P in $\partial\Omega$ if there exists a defining function ρ for Ω such that

$$\sum_{j,k=1}^{n} \frac{\partial^2 \rho}{\partial z_j \partial \overline{z}_k}(P) w_j \overline{w}_k \geq C \cdot |w|^2 \tag{10.2}$$

for every complex vector $w = (w_1, \ldots, w_n)$ in $\mathbb{C}^n$.

Thus strong pseudoconvexity is specified by the positive definiteness of a certain quadratic form. In fact the quadratic form is given by the complex Hessian. The reader may recall that positive semi-definiteness (respectively, positive definiteness) of the $2n \times 2n$ *real Hessian* matrix

$$H \equiv \left(\frac{\partial^2 \rho}{\partial x_j \partial x_k}\right)_{j,k=1}^{2n},$$

where we think of $\mathbb{C}^n$ as R^{2n} and write $x_1, \ldots, x_{2n}$ for the coordinates, characterizes classical convexity (respectively, strict convexity) of the boundary (see [357] for the details). The condition that we have specified for strong pseudoconvexity is somewhat more subtle. The following result of Narasimhan (see [331]) puts the concept of strong pseudoconvexity on a more concrete and intuitive footing:

Proposition 10.5.2 *Let $\Omega = \{z \in \mathbb{C}^n : \rho(z) < 0\}$ be a domain with C^2-boundary, and let P belong to $\partial\Omega$. Then Ω is strongly pseudoconvex at P if and only if there is a neighborhood Ω' of P and a biholomorphic mapping $\Psi : \Omega' \to \mathbb{C}^n$ such that $\Psi(\Omega' \cap \partial\Omega)$ is strongly convex in the sense of classical geometry.*

We can now give a definition of the concept of (weak) *pseudoconvexity*. A domain Ω in $\mathbb{C}^n$ is called (weakly) *pseudoconvex* if $\Omega = \cup_{j=1}^{\infty} \Omega_j$, where $\{\Omega_j\}$ is an increasing sequence of strongly pseudoconvex domains. (There is an intrinsic definition of pseudoconvexity in terms of the Levi form (10.2), but we shall not need it here.)

The other type of domain that plays an important role in the theory of several complex variables goes under the heading of "domain of holomorphy." A subdomain W of $\mathbb{C}^n$ is called a *domain of holomorphy* if, roughly speaking, there is a holomorphic function f on W that cannot be analytically continued past any boundary point. (The precise definition is rather delicate; see [331].) In one complex variable, *every* domain is a domain of holomorphy. In several complex variables, some domains are domains of holomorphy and some are not. Pseudoconvex domains are important because of the following equivalence (known as *the solution of the Levi problem*): A domain W is a domain of holomorphy if and only if it is pseudoconvex.

10.6 The Theorem of Bun Wong and Jean-Pierre Rosay

We now develop further the concept of noncompact automorphism group. The following examples show that it is *not* the case that a bounded domain in $\mathbb{C}^n$ ($n \geq 2$) with an orbit accumulation point at which the boundary is smooth is always biholomorphic to the ball (in contrast with Theorem 10.2.2 of Sect. 10.2). Consider the following two examples:

Example 10.6.1 The automorphism group of the bidisc

$$\mathbb{D}^2 = \{(z_1, z_2) \in \mathbb{C}^2 : |z_1| < 1, |z_2| < 1\}$$

acts transitively. This means that, for arbitrary points z and w of $\mathbb{D}^2$, there is an automorphism φ of $\mathbb{D}^2$ such that $\varphi(z) = w$. But we saw in the theorem of Poincaré that this domain is not biholomorphic to the ball. On the other hand, in this example the boundary orbit accumulation points are never strongly pseudoconvex.

Example 10.6.2 Consider

$$E_{1,2} = \{(z_1, z_2) \in \mathbb{C}^2 : |z_1|^2 + |z_2|^4 < 1\}.$$

The mappings

$$(z_1, z_2) \mapsto \left(\frac{z_1 - a}{1 - \bar{a}z_1}, \frac{\sqrt[4]{1 - |a|^2} z_2}{\sqrt{1 - \bar{a}z_1}} \right)$$

for a in $\mathbb{C}$ with $|a| < 1$ are automorphisms of $E_{1,2}$. Taking $a = (j - 1)/j$, we can see that the automorphism group is noncompact. But it is known that this domain is not biholomorphic to the ball. (If there were a biholomorphism $f : E_{1,2} \to \mathbb{B}$, then, without loss of generality, we could assume, after composition with an automorphism of $\mathbb{B}$, that f maps the origin to itself. A theorem of H. Cartan (see [393]) implies that f must be linear. Now $E_{1,2}$ is *not* strongly convex at the point $(1, 0)$, whereas $\mathbb{B}$ *is* strongly convex at every boundary point. Since a linear isomorphism must preserve strong convexity, we arrive at a contradiction.) Notice again that the boundary of $E_{1,2}$ at the orbit accumulation point $(1, 0)$ is not strongly pseudoconvex.

In order to obtain a positive result, some additional conditions must be imposed. We now present the basic theorem of Bun Wong [533] and Jean-Pierre Rosay [441].

Theorem 10.6.3 *Let Ω be a domain in $\mathbb{C}^n$. Assume that Ω admits a boundary orbit accumulation point P that is strongly pseudoconvex and that $\partial\Omega$ is C^2-smooth in a neighborhood of P. Then Ω is biholomorphic to the unit ball.*

We see that the new ingredient that has been added here is that, in addition to assuming that the automorphism group is noncompact, we must also assume that the resulting boundary orbit accumulation point is (smooth and) strongly pseudoconvex. A prevailing theme in this subject, which originally arose in the paper [441] and was further developed in the work [201] of Greene and Krantz, is that the geometry of the boundary orbit accumulation point determines the global characteristics of the domain. The Wong/Rosay theorem is *the* classical example of this phenomenon. For the interested reader, we point out the recent surveys [254] and [534].

Sketch of Proof of the Wong/Rosay Theorem This discussion will rely heavily on Fig. 10.3. We assume as usual that we are working in $\mathbb{C}^2$. First, the orbit accumulation point hypothesis implies that there is an interior point Q of Ω, a point P of $\partial\Omega$ at which the boundary is strongly pseudoconvex, and automorphisms φ_j of Ω such that $\varphi_j(Q) \to P$. By Proposition 10.5.2, the strong pseudoconvexity at P allows us to have a (local) peaking function for Ω at P. So, if K is a large compact subset of Ω as indicated in the figure and if j is sufficiently large, then $\varphi_j(K)$ will lie in a small neighborhood of the boundary point P. Now, by hypothesis, the quadratic form in (10.2) is positive definite. We may thus perform elementary complex linear transformations (and these are certainly biholomorphic mappings) to diagonalize this quadratic form and then to arrange for all its eigenvalues to equal 1. This means that, modulo this complex linear transformation, the boundary of Ω in the vicinity of P very closely resembles the unit ball. In fact it can be shown that the set $\partial\Omega$ near P agrees with (a biholomorphic image of) the unit sphere up to third order.

Thus it is possible to map this normalized neighborhood near the boundary at P into a cognate neighborhood in the unit ball near $(1, 0)$. We call this mapping η_j, again referring to Fig. 10.3. Now we apply a two-dimensional Möbius transformation λ_j (as described in Example 10.2.1) to map this image of K back into a large compact subset L of the ball $\mathbb{B}$.

The composition φ_j of these three mappings sends a large compact subset K of Ω to a large compact subset L of $\mathbb{B}$. We can repeat this construction for an increasing sequence of compact sets K_ℓ exhausting Ω to generate a subsequence $\{\varphi_{j_\ell}\}$ of $\{\varphi_j\}$. If this argument is carried out with sufficient care, then it can be seen that the sets $L_\ell = \varphi_{j_\ell}(K_\ell)$ exhaust $\mathbb{B}$ and the mappings φ_{j_ℓ} converge to a biholomorphic mapping of Ω to $\mathbb{B}$. That is what we wished to establish. □

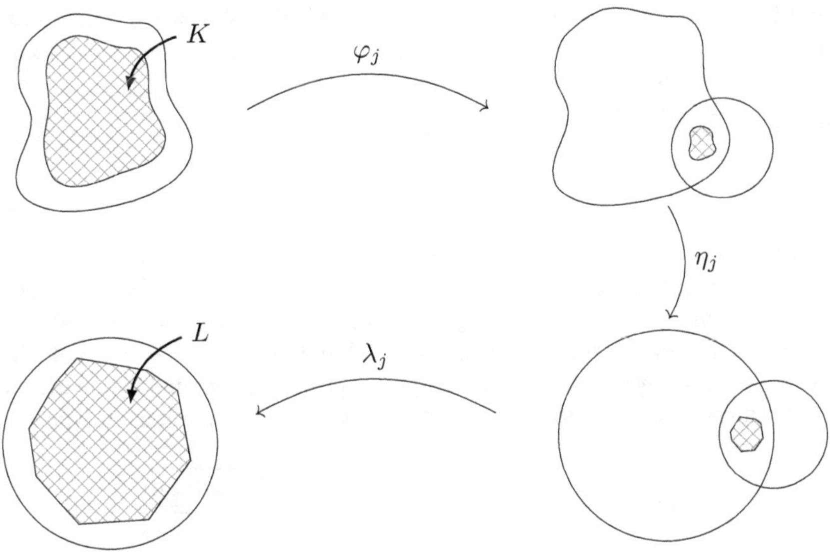

Fig. 10.3 The Wong-Rosay theorem

One might wonder what ideas would be required to put the automorphism group results for the bidisc and the complex ellipsoids into perspective. Since such results require further techniques, we shall introduce some of them in Sect. 10.7. Before leaving this section, however, we mention that there are bidisc analogues of the Wong/Rosay theorem. We refer the interested reader to [280, 284], or [283] for the details.

10.7 Some Startling Examples

Next we present some very simple examples that illustrate how truly complicated life is in the world of several complex variables. We note in advance that the domains we discuss are all pseudoconvex and topologically trivial. All but one has real analytic boundary (i.e., boundary that is locally the graph of a function defined by power series in the real coordinates), and the fourth has $\mathcal{C}^\infty$-boundary. And all four domains are diffeomorphic to each other. Nevertheless, their automorphism group characteristics are quite different.

In what follows, we let

$$E_{j,k} = \{(z_1, z_2) \in \mathbb{C}^2 : |z_1|^{2j} + |z_2|^{2k} < 1\}$$

for positive integers j and k. These domains are sometimes called "(complex) ellipsoids."

Example 10.7.1 The domain $E_{1,1}$ is the unit ball. Recall that the automorphism group acts *transitively* on $\mathbb{B}$ (i.e., if z and w belong to $\mathbb{B}$, then there is some automorphism φ of $\mathbb{B}$ such that $\varphi(z) = w$).

Example 10.7.2 The second example is

$$E_{1,2} = \{(z_1, z_2) \in \mathbb{C}^2 : |z_1|^2 + |z_2|^4 < 1\}.$$

This repeats an example given in the preceding section. But now we emphasize that the automorphism group is *not* transitive. If it were, then there would be an orbit that accumulated at the strongly pseudoconvex boundary point (0, 1). By Theorem 10.6.3, the domain would then be biholomorphic to the unit ball. But it is not. In fact, it is known that its automorphism group is generated by the mappings that are rotations in each variable separately and the transformations of type

$$\Phi_a(z_1, z_2) \equiv (z_1, z_2) \mapsto \left(\frac{z_1 - a}{1 - \bar{a} z_1}, \frac{\sqrt[4]{1 - |a|^2} z_2}{\sqrt{1 - \bar{a} z_1}} \right)$$

for complex numbers a with $|a| < 1$. The unitary group does *not* act on $E_{1,2}$. We see that the automorphism group of $E_{1,2}$ is noncompact, as the automorphisms Φ_a with $a \to 1$ illustrate.

Example 10.7.3 Consider the domain

$$E_{2,2} = \{(z_1, z_2) \in \mathbb{C}^2 : |z_1|^4 + |z_2|^4 < 1\}.$$

It is known that this domain does not have any automorphisms other than those generated by the rotations in each variable separately and the maps $(z_1, z_2) \mapsto (z_2, z_1)$ and $(z_1, z_2) \mapsto (-z_2, -z_1)$. Why? Here is a brief indication of the reason. First, we claim that the automorphism group is compact. If the automorphism group were noncompact, then there would be a boundary orbit accumulation point. Where could it be? It could not be any of the boundary points at which $E_{2,2}$ is strongly pseudoconvex (i.e., points with $z_1 z_2 \neq 0$) because then the Wong/Rosay theorem would tell us that the domain is the ball (which it is not). If a point of the form $(e^{i\theta}, 0)$ or of the form $(0, e^{i\theta})$ were an orbit accumulation point, then we could exploit the deep results on the extension of biholomorphic mappings to the boundary (see, for instance, [66]), which imply in particular that every automorphism of $E_{2,2}$ extends to a diffeomorphism of its closure. Then it can be shown that any sequence of automorphisms ψ_j that drags an interior point out to this orbit accumulation point will drag the boundary points out there as well. This would force strongly pseudoconvex points to be moved to weakly pseudoconvex points (see Fig. 10.4) by the action of a biholomorphic mapping. That is impossible (we again omit the details). This contradiction implies that $E_{2,2}$ has compact automorphism group.

Now we observe further in the preceding arguments concerning $E_{2,2}$ that the union of the two unit circles in each of the complex coordinate planes must be

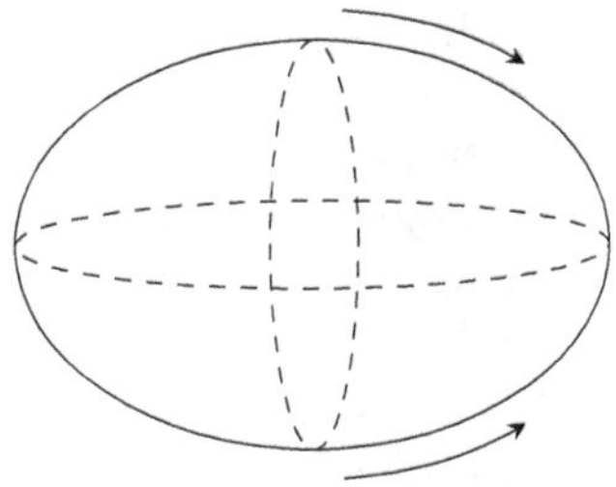

Fig. 10.4 Strongly pseudoconvex points moved to weakly pseudoconvex points

preserved by the (boundary extension of the) automorphisms. Hence the union of the two discs in the complex coordinate planes must be preserved. In addition, the intersection of these two discs must be preserved as well. Thus every automorphism of $E_{2,2}$ preserves the origin (which is the sole element in the intersection of the two discs!). Now a theorem of H. Cartan (see [393, Proposition 2, p. 67], for instance) implies that every element of $\text{Aut}(E_{2,2})$ is linear. It is then easy to deduce that the automorphism group is in fact generated by the mappings that are rotations in each variable separately and the mapping that transposes the variables.

Example 10.7.4 Consider

$$E_{1,\infty} = \{(z_1, z_2) \in \mathbb{C}^2 : |z_1|^2 + 2e^{-1/|z_2|^2} < 1\}.$$

The idea for this domain is due to range.

Every boundary point of this domain is strongly pseudoconvex except for those of the form $(e^{i\theta}, 0)$ for real θ. One might expect, therefore, that the automorphism group's behavior would be similar to that of the automorphism group for the domain $E_{1,2}$ in Example 10.7.2. But such is not the case. The infinite flatness of the boundary points $(e^{i\theta}, 0)$ turns out to be incompatible with any orbits that might be accumulating there. Details may be found in [201]. It can be shown that the only automorphisms of $E_{1,\infty}$ are the rotations in each variable separately.

In fact Example 10.7.4 has turned out to be quite important in the subject. There is an influential conjecture that a boundary orbit accumulation point cannot be infinitely flat in a precise, complex analytic sense. We shall use the next, and last, section of this article to discuss the attendant ideas and give a precise formulation of the conjecture.

10.8 Boundary Flatness and an Influential Conjecture

10.8.1 More on the Preceding Examples

One of the critical features of complex analysis in higher dimensions is the role of nontrivial *analytic varieties*. Here the term "variety" means a set that is locally the common zero set of finitely many holomorphic functions. We are particularly interested in one-dimensional varieties, and such a variety can usually be parametrized by a holomorphic mapping $\varphi : \mathbb{D} \to \mathbb{C}^n$.

Consider the domains $E_{1,1}$, $E_{1,2}$, $E_{2,2}$, and $E_{1,\infty}$ in $\mathbb{C}^2$. Each of these has the point $(1, 0)$ as a boundary point, and each has the image of the complex curve $\varphi : \mathbb{D} \to \mathbb{C}^2$ given by $\zeta \mapsto (1, \zeta)$ as a variety that is tangent to the boundary of the domain at $(1, 0)$. What is of particular interest is the *order of contact* of the variety with the boundary at the point of contact.

To be specific and concrete, we restrict attention to domains in two complex dimensions for the remainder of this section. Consider the defining function $\rho(z_1, z_2)$ for the domain Ω:

$$U = \{z = (z_1, z_2) \in \mathbb{C}^2 : \rho(z) < 0\} .$$

The order of contact between the boundary $\partial\Omega$ and the variety determined by φ is defined to be the vanishing order of the function $\rho \circ \varphi(\zeta)$ at $\zeta = 0$.

Example 10.8.1 Let

$$E_{1,1} = \{(z_1, z_2) : \rho(z_1, z_2) = |z_1|^2 + |z_2|^2 - 1 < 0\} ,$$

which is just the unit ball in $\mathbb{C}^2$. Let $\varphi(\zeta) = (1, \zeta)$ for ζ in $\mathbb{D}$. Then

$$\rho \circ \varphi(\zeta) = |\zeta|^2 .$$

Clearly $\rho \circ \varphi$ vanishes to order 2 at the origin. Thus the variety $\varphi(\mathbb{D})$ has order of contact 2 with $\partial E_{1,1}$ at the boundary point $(1, 0)$.

Now let

$$E_{1,2} = \{(z_1, z_2) : \rho(z_1, z_2) = |z_1|^2 + |z_2|^4 - 1 < 0\} .$$

This is one of our ellipsoid domains. If again $\varphi(\zeta) = (1, \zeta)$ for ζ in $\mathbb{D}$, then

$$\rho \circ \varphi(\zeta) = |\zeta|^4 .$$

In this instance $\rho \circ \varphi$ vanishes to order 4 at the origin, so the variety $\varphi(\mathbb{D})$ has order of contact 4 with $\partial E_{1,2}$ at the boundary point $(1, 0)$.

More generally, for any positive integer m, we let

$$E_{1,m} = \{(z_1, z_2) : \rho(z_1, z_2) = |z_1|^2 + |z_2|^{2m} - 1 < 0\} .$$

Here we have

$$\rho \circ \varphi(\zeta) = |\zeta|^{2m},$$

and the variety $\varphi(\mathbb{D})$ has order of contact $2m$ at the boundary point $(1, 0)$.

Finally, let

$$E_{1,\infty} = \{(z_1, z_2) : \rho(z_1, z_2) = |z_1|^2 + 2\exp^{-1/|z_2|^2} - 1 < 0\}\,.$$

Proceeding as earlier, we find that

$$\rho \circ \varphi(\zeta) = 2\exp^{-1/|\zeta|^2}\,,$$

which vanishes to infinite order at the origin. In this case the variety $\varphi(\mathbb{D})$ has order of contact ∞ at the boundary point $(1, 0)$.

Now consider the explicit form of automorphism orbits accumulating at $(1, 0)$ for the domain $E_{1,m}$. We observed earlier that those automorphisms take the form

$$(z_1, z_2) \mapsto \left(\frac{z_1 + r_j}{1 + r_j z_1}, \left(\frac{\sqrt{1 - r_j^2}}{1 + r_j z_1} \right)^{\frac{1}{m}} z_2 \right),$$

where $\{r_j\}$ is an increasing sequence of positive real numbers such that $r_j \to 1$ and where appropriate branches are chosen for the mth roots. One cannot miss the close relationship between the order of contact of $\varphi(\mathbb{D})$ with $\partial E_{1,m}$ and the mth root in the second component of the expression for the automorphisms.

In the case of $E_{1,\infty}$ there are no obvious ways of creating a sequence of automorphisms that sends $(0, 0)$ successively to the boundary point $(1, 0)$. In fact none exists, as demonstrated in [201]. Bearing these considerations in mind, the reader may perhaps guess what the correct insight about domains like $E_{1,\infty}$ should be.

It is therefore natural to ask whether the boundary of a domain must have finite order of contact with analytic varieties at a boundary orbit accumulation point. We shall eventually formulate a general conjecture about this question. We shall also introduce some partial results and make several forward-looking remarks.

10.8.2 The Contact Type of D'Angelo

With the preceding discussion in mind, we now introduce the concept of the "contact type" for a boundary point of a given domain, following John P. D'Angelo's celebrated work (see [125–127] for further details and a survey). Since this is only an introduction intended for a broad audience, we confine our attention to complex two-space and to nonsingular complex varieties. We refer the reader to [126] or [127] for a more thoroughgoing consideration of all these matters.

Let Ω be a domain in $\mathbb{C}^2$ with defining function $\rho : \mathbb{C}^2 \to \mathbb{R}$ of class C^∞. We first clarify the notion of the order of vanishing. As earlier, let $\mathbb{D} \subseteq \mathbb{C}$ be the unit disc in $\mathbb{C}$. If $\psi : \mathbb{D} \to \mathbb{C}^2$ is a C^∞-smooth function with $\psi(0) = 0$, then we consider its associated Taylor expansion in real coordinates. To be more precise, writing $\psi = (\psi_1, \psi_2)$ in components, we consider the Taylor expansions corresponding to ψ_1, ψ_2 at 0. If both Taylor series are identically zero, then we say that ψ vanishes to infinite order at 0. If not, then we can seek the lowest order, say m, of nonvanishing terms in either of the series. In this case, we say that ψ vanishes to order m at 0 and express this by writing $\nu(\psi) = m$.

Let P be a point of $\partial\Omega$. We consider the family of one-dimensional complex analytic varieties in $\mathbb{C}^2$ that pass through P and are parametrized over $\mathbb{D}$. Denote by $\mathcal{F}_P$ the set of all nonconstant holomorphic mappings $\varphi : \mathbb{D} \to \mathbb{C}^2$ such that $\varphi(0) = P$ and $\varphi'(0) \neq 0$ (this last condition forces the complex curve $\varphi(\mathbb{D})$ to be nonsingular near P). Then the *D'Angelo type* $\tau(\partial U, P)$ *of* $\partial\Omega$ *at* P is defined to be

$$\tau(\partial U, P) \equiv \sup_{\varphi \in \mathcal{F}_P} \nu(\rho \circ \varphi) . \tag{10.3}$$

Roughly speaking, this concept detects the maximum possible order of contact between $\partial\Omega$ and the germs of analytic curves at P. We remark that τ is independent of the choice for the defining function of Ω. In case this number is finite, then we say that the point P has *finite D'Angelo type*. Of course calculating the supremum in line (10.3) can be challenging. McNeal [393] has proved a remarkable result asserting that, for a convex domain, the supremum is always realized by a *complex linear* variety. In particular, for the domains $E_{1,1}$, $E_{1,m}$, $E_{2,2}$, and $E_{1,\infty}$, the variety $\varphi(\mathbb{D})$ with $\varphi(\zeta) = (1, \zeta)$ is the one that realizes the supremum in (10.3).

In many cases, the first nonvanishing terms in the Taylor series associated with the defining function of the domain provide the D'Angelo type. Such simplicity is the norm in complex dimension two. But matters become much more subtle in higher dimensions.

10.8.3 The Greene/Krantz Conjecture

Despite their limitations, the examples of domains that we have considered and the discussions thereof lead us very naturally to the following conjecture:

> **Conjecture:** (Greene/Krantz) Let Ω be a bounded domain in $\mathbb{C}^2$ with C^∞-boundary. Then every boundary orbit accumulation point for the automorphism group of Ω is of finite D'Angelo type.

The domain $E_{1,\infty}$ was presented as the first evidence for the truth of this conjecture in [201]. It was further analyzed in [282]. Special cases of the conjecture are known to hold, but the full conjecture itself is still open. We now give some partial results related to the conjecture; these also serve to indicate its importance.

Theorem 10.8.2 (Bedford/Pinchuk) *Let Ω be a bounded pseudoconvex domain in $\mathbb{C}^2$ with real analytic boundary. If* $\mathrm{Aut}(\Omega)$ *is noncompact, then Ω is biholomorphic to $E_{1,k}$ for some positive integer k.*

We remark that, under the hypotheses of Theorem 10.8.2, every boundary point of Ω is of finite D'Angelo type (see [134]). Such a global assumption is necessary, because the methods of Bedford and Pinchuk require that the boundary orbit accumulation point be of finite D'Angelo type. So, if the Greene-Krantz conjecture is true, one might hope to sharpen this theorem.

The next theorem settles a special case of the conjecture. We state only a simplified version here.

Theorem 10.8.3 (Kim) *Let Ω be a bounded convex domain in $\mathbb{C}^2$ with C^∞-boundary. If a boundary point P of Ω has an open neighborhood Ω such that $\partial\Omega$ is of infinite D'Angelo type at each point of $\Omega \cap \partial\Omega$, then P is not an orbit accumulation point.*

At this juncture, it may be relevant to state the following special (but still significant) instance of the conjecture. We formulate the idea as an open problem.

> Let Ω be a bounded convex domain in $\mathbb{C}^2$ with C^∞-boundary. Assume that $\mathrm{Aut}(\Omega)$ admits at least one boundary orbit accumulation point. Show that $\mathrm{Aut}(\Omega)$ must have a boundary orbit accumulation point of finite D'Angelo type.

There are several recent results that are relevant to the study of the Greene/Krantz conjecture. See, for example, [102, 280–282], and [162] and the references therein.

10.8.4 Closing Remarks

One of the beautiful features of the subject of automorphism groups is that it gives rise so naturally to an interaction of complex function theory, geometry, and analysis. We hope that the results presented here illustrate this symbiosis and provide some motivation for the reader to delve more deeply into complex geometry.

The subject of automorphism groups is sophisticated modern mathematics. But some of the most interesting questions in the field are quite accessible. For example: *Which finite groups can occur as automorphism groups of domains in the plane?* This problem has been open for three quarters of a century. (The challenging part is that we are restricting ourselves to planar domains. If one is allowed to go to higher dimensions, then there is a remarkable theorem by Bedford and Dadok and by Saerens and Zame that asserts: *For every compact Lie group G, there is a bounded strongly pseudoconvex domain Ω in some $\mathbb{C}^n$ whose automorphism group is isomorphic to G*; see [49, 453].) In fact it is known that the domain Ω may be taken to have real analytic boundary—see [198] for the details.

Another interesting question is to determine the optimal boundary smoothness condition in Theorem 10.2.2. The condition of C^1-boundary is used only to apply the implicit function theorem, but it is clear that a considerably weaker hypothesis will suffice.

It is important to generate "toy problems" that will entice students to study mathematical subject areas of current interest, and on which modern research focuses. Certainly computer calculations can be used in the determination of automorphism groups. Automorphisms are a wonderful, tactile subject area for experimentation. They should be a fruitful source of questions for many years to come.

Chapter 11
Rigidity of Holomorphic Mappings and a New Schwarz Lemma at the Boundary

11.1 Introduction

The uniqueness portion of the classical Schwarz lemma has played an interesting historical role. Usually first encountered in the classification of the conformal self-maps of the disc, it arises in more general considerations of automorphism groups, in the construction of the Carathéodory and Kobayashi/Royden metrics, and in a variety of contexts in analysis on manifolds. Useful references for this material are [17, 192, 194, 294, 321, 537].

Generalizations of the disc Schwarz lemma to multiply connected domains in one complex variable are generated naturally and easily using the uniformization theorem. The generalization to several complex variables requires insights of a different nature. For example, the so-called Carathéodory-Cartan-Kaup-Wu theorem [535] says in part that if Ω is a bounded domain in $\mathbb{C}^n$ and if Φ is a holomorphic self-map of Ω that fixes a point $P \in \Omega$, then the holomorphic Jacobian determinant $\det \operatorname{Jac} \Phi(P)$ has modulus less than or equal to 1, and equals 1 if and only if Φ is a biholomorphism of Ω. An interesting and important feature of this result is that it has a global hypothesis (that Ω be mapped to itself) and a local hypothesis (the condition on the behavior of the mapping at P). The conclusion is then a strong global one. We will see this paradigm repeated in the work that follows.

The purpose of this paper is to seek versions of the last stated result when the point P lies in the boundary of Ω. (A primitive version of such a result appears in [324].) This problem, while of considerable intrinsic interest, is also related to a variety of other work in the literature. We now indicate some of these connections.

Our interest in this problem arose originally from a question of Warren Wogen: Does there exist a holomorphic self-mapping Φ of the ball $\mathbb{B}$ in $\mathbb{C}^2$ (i) that takes the point $\mathbf{1} = (1, 0) \in \partial\mathbb{B}$ to itself (in a suitable sense) and (ii) for which the set $\partial(\Phi(\mathbb{B}))$ has "high order of contact," as a set, with the boundary $\partial\mathbb{B}$ of the target ball? Here part of the problem is to find a suitable interpretation for the phrase "high order of contact." The mappings $\Phi_\varepsilon(z_1, z_2) = ((1-\varepsilon)z_1 + \varepsilon, (1-\varepsilon)z_2)$, $\varepsilon > 0$,

P. V. Dovbush, S. G. Krantz, *The Geometric Theory of Complex Variables*,
https://doi.org/10.1007/978-3-031-77204-7_11

satisfy (i) and (ii) with the boundary of the image ball having order of contact 1 with the boundary of the target ball (the tangent planes agree, but the second fundamental forms do not). The mapping Φ_0 is the identity and the order of contact is infinite. Wogen's question may be interpreted as asking whether there are maps that are intermediate to the $\Phi_\varepsilon, \varepsilon > 0$, and id $= \Phi_0$. The original interest in constructing such maps was in finding counterexamples to certain assertions about composition operators (see [531]).

As our understanding of the problem developed, we also saw that it relates to a result of H. Alexander [26]: If $U \cap \mathbb{B}$ and $U' \cap \mathbb{B}$ are boundary neighborhoods in the ball and if Φ is a biholomorphic mapping of these neighborhoods that extends C^2 to the boundary, then Φ must be the restriction to $U \cap \mathbb{B}$ of a biholomorphism of the entire ball. S. Pincuk [414] generalized this result to bounded strictly pseudoconvex domains with real-analytic boundaries and W. Rudin [452] reduced the hypothesis of C^2 to the boundary to an assumption that is even weaker than continuity at a point. One interpretation of the main result of the present paper is that the hypothesis of direct coincidence of boundary neighborhoods in the results of Alexander, Pincuk, and Rudin may be weakened to high order of contact of image boundary and target boundary. (See also [201] for other more general versions of the Alexander phenomenon—in particular, that paper characterizes not just the biholomorphic mappings of the ball but mappings that are "approximate" biholomorphic mappings of the ball.)

11.2 A Result on the Disc About Boundary Derivatives

The famous theorem of Daniel M. Burns and Steven G. Krantz is:

Theorem 11.2.1 (Burns-Krantz Theorem [100]) *Let $\phi : \mathbb{D} \to \mathbb{D}$ be a holomorphic function from the unit disc to itself such that*

$$\phi(\zeta) = 1 + (\zeta - 1) + \mathcal{O}\left((\zeta - 1)^4\right)$$

as $\zeta \to 1$. Then $\phi(\zeta) \equiv \zeta$ on the disc.

Remark 11.2.2 Results such as this one appear in the literature of conformal mappings (see, for instance, [516]) with the additional hypothesis that ϕ be univalent (and often the function is assumed to be quite smooth—even analytic—in a neighborhood of 1). The theorem presented here has no such hypothesis, so far as we know it is new. Even more surprising is that the exponent 4 is sharp: Simple geometric arguments show that the function

$$\phi(\zeta) = \zeta - \frac{1}{10}(\zeta - 1)^3$$

satisfies the hypotheses of the theorem with 4 replaced by 3. Note also in the proof that $O\left((z-1)^4\right)$ can be replaced by $o\left((z-1)^3\right)$. A similar remark applies to Sects. 11.4, 11.5, and 11.6.

X. Huang, in the papers [245–247], has explored additional conditions under which the exponent in the error term may be decreased.

Proof of Theorem 11.2.1 Consider the holomorphic function

$$g(\zeta) = \frac{1+\phi(\zeta)}{1-\phi(\zeta)}.$$

Then g maps the disc $\mathbb{D}$ to the right half plane. By the Herglotz representation (this is just an application of the Banach-Alaoglu theorem; see [17]), there must be a positive measure μ on the interval $[0, 2\pi)$ and an imaginary constant C such that

$$g(\zeta) = \frac{1}{2\pi}\int_0^{2\pi} \frac{e^{i\theta}+\zeta}{e^{i\theta}-\zeta} d\mu(\theta) + C. \tag{11.1}$$

We use the hypothesis on ϕ to analyze the structure of g and hence that of μ. To wit,

$$g(\zeta) = \frac{1+\zeta+O(\zeta-1)^4}{1-\zeta-O(\zeta-1)^4} = \frac{1+\zeta}{1-\zeta} + O(\zeta-1)^2.$$

From this and equation (11.1) we easily conclude that the measure μ has the form $\mu = \delta_0 + \nu$, where δ_0 is (2π times) the Dirac mass at the origin and ν is another positive measure on $[0, 2\pi)$. In fact, a nice way to verify the positivity of ν is to use the equation

$$\frac{1+\zeta}{1-\zeta} + O(\zeta-1)^2 = \frac{1}{2\pi}\int_0^{2\pi} \frac{e^{i\theta}+\zeta}{e^{i\theta}-\zeta} d\left(\delta_0+\nu\right)(\theta) + C \tag{11.2}$$

to derive a Fourier-Stieltjes expansion of $\delta_0 + \nu$ and then to apply the Herglotz criterion [265, p. 38].

We may simplify equation (11.2) to

$$O(\zeta-1)^2 = \frac{1}{2\pi}\int_0^{2\pi} \frac{e^{i\theta}+\zeta}{e^{i\theta}-\zeta} d\nu(\theta) + C.$$

Pass to the real part of the last equation, thus eliminating the constant C. Since ν is a positive measure, we thus see that the real part of the integral on the right of this last equation represents a positive harmonic function h on the disc that satisfies

$$h(\zeta) = O(\zeta-1)^2.$$

In particular, h takes a minimum at the point $\zeta = 1$ and is $O\left(|z-1|^2\right)$ there as well. This contradicts Hopf's lemma [331] unless $h \equiv 0$. But $h \equiv 0$ implies that $\nu \equiv 0$, hence that

$$g(\zeta) = \frac{1+\zeta}{1-\zeta}.$$

Therefore $\phi(\zeta) \equiv \zeta$, and the theorem is proved. □

Remark 11.2.3 Notice that the formulation of Theorem 11.1 is a boundary uniqueness theorem in the vein of the uniqueness part of the classical Schwarz lemma. However, the result sidesteps the original questions about "order of contact." In fact the Riemann Mapping Theorem guarantees that, with any reasonable definition of the phrase, there are univalent holomorphic maps of the disc to the disc so that the image boundary has arbitrarily high (even infinite order) contact with the target boundary. Thus Theorem 11.1 addresses some more rigid structural phenomenon.

Theorem 11.2.1 was improved by T.L. Kriete and B.D. MacCluer [337], who replaced F with its real part and considered the radial limit in $o\left((z-1)^3\right)$ instead of the unrestricted limit. Here is a more precise statement of their result.

Theorem 11.2.4 *Let $F \in \mathcal{O}(\mathbb{D})$ with radial limit $F(1) = 1$ and angular derivative $F'(1) = 1$. If*

$$\liminf_{r\to 1^-} \frac{\operatorname{Re}(F(r)-r)}{(1-r)^3} = 0,$$

then $F(z) \equiv z$.

Corollary 11.2.5 *Suppose $F \in \mathcal{O}(\mathbb{D})$ with angular derivative at some $\zeta_0 \in \partial\mathbb{D}$ satisfying $\left|F'(\zeta_0)\right| = 1$. If the maximum modulus function $M(r) := \max_{\theta\in[0,2\pi]}\left|F\left(re^{i\theta}\right)\right|$ of F satisfies*

$$\liminf_{r\to 1^-} \frac{M(r)-r}{(1-r)^3} = 0,$$

then $M(r) \equiv r$ and F is a rotation.

L. Baracco, D. Zaitsev, and G. Zampieri [40] reformulated the Burns-Krantz Theorem 11.2.1 and gave a simple proof that its conclusion remains valid even if the hypothesis is satisfied only on a sequence in $\mathbb{D}$ that converges nontangentially to 1.

Theorem 11.2.6 *Let $F \in \mathcal{O}(\mathbb{D})$ and $z_k \in \mathbb{D}$ be a sequence converging nontangentially to 1 such that $F(z_k) = z_k + o\left(|z_k - 1|^3\right)$ as $k \to \infty$. Then $F(z) \equiv z$.*

The next result, by R. Tauraso and F. Vlacci [496], gives sufficient conditions for a holomorphic self-mapping F of $\mathbb{D}$ to coincide with an arbitrary function $G \in$

$\mathcal{O}(\mathbb{D})$ if they agree up to the third order at a boundary contact point and one of the functions is "subordinated," in a sense, to the other one.

Theorem 11.2.7 *Let $F, G \in \mathcal{O}(\mathbb{D})$ and $\zeta \in \partial\mathbb{D}$ be such that*

$$\lim_{r \to 1^-} \frac{F(r\zeta) - G(r\zeta)}{(r-1)^3} = \ell$$

for some $\ell \in \mathbb{C}$, and

$$\frac{1 - |F(z)|^2}{|w - F(z)|^2} \geq \frac{1 - |G(z)|^2}{|w - G(z)|^2}, \quad z \in \mathbb{D}$$

for some $w \in \partial\mathbb{D}$. Then $F \equiv G$ if and only if $\ell = 0$. Moreover, $\overline{w}\ell$ is a nonpositive real number.

The Burns-Krantz Theorem 11.2.1 is a particular case of this result for $\zeta = 1$ and $G = I$.

11.3 A Result for the Ball

Theorem 11.3.1 *Let $\mathbb{B} \subset \mathbb{C}^n$ be the unit ball. Let $\Phi : \mathbb{B} \to \mathbb{B}$ be a holomorphic mapping of the ball to itself such that*

$$\Phi(z) = \mathbf{1} + (z - \mathbf{1}) + O\left(|z - \mathbf{1}|^4\right)$$

as $z \to \mathbf{1}$. (Here $\mathbf{1}$ denotes the distinguished boundary point $\mathbf{1} = (1, 0, \ldots, 0)$ of the ball.) Then $\phi(z) \equiv z$ on the ball.

Proof There is no useful Herglotz representation on the ball (however, see [25] for related ideas): This is a deep fact that cannot be circumvented. Thus we present a new argument that reduces the ball case to the disc case. For simplicity we restrict attention to dimension two.

For each point $a \in \mathbb{B}$ let $\mathcal{L}_a$ be the complex line joining a and $\mathbf{1}$. Let d_a be the complex disc given by $\mathcal{L}_a \cap \mathbb{B}$. Now for fixed a consider the holomorphic function

$$\begin{aligned} \psi : \mathbb{D} &\longrightarrow \mathbb{B} \\ \zeta &\longmapsto (\zeta, 0). \end{aligned}$$

Define also the mapping $\phi_a : \mathbb{B} \longrightarrow \mathbb{B}$, which is an automorphism of the ball mapping d_0 onto d_a and fixing $\mathbf{1}$. (That such maps exist follows from elementary geometric considerations, or see [451] for an explicit formula.) Finally define

$$\pi_1 : \mathbb{B} \longrightarrow \mathbb{B},$$
$$\pi_1 : (z_1, z_2) \longmapsto (z_1, 0) ;$$

and

$$\eta : d_0 \longrightarrow \mathbb{D},$$
$$\eta : (z_1, 0) \longmapsto z_1 .$$

Then the function

$$H_a : \mathbb{D} \longrightarrow \mathbb{D}$$
$$\zeta \longmapsto \eta \circ \pi_1 \circ (\phi_a)^{-1} \circ \Phi \circ \phi_a \circ \psi(\zeta)$$

is well defined. Moreover, it is straightforward to check that H_a satisfies the hypotheses of Theorem 11.1 (since Φ agrees with the identity to high order, the composition $(\phi_a)^{-1} \circ \Phi \circ \phi_a$ is the identity to high order). It follows that $H_a(\zeta) \equiv \zeta$.

Now set

$$G_a = (\phi_a)^{-1} \circ \Phi \circ \phi_a \circ \psi(\zeta) \equiv \left(g_a^1(\zeta), g_a^2(\zeta)\right) .$$

The statement that $H_a(\zeta) \equiv \zeta$ means that $g_a^1(\zeta) \equiv \zeta$. But

$$\left|g_a^1(\zeta)\right|^2 + \left|g_a^2(\zeta)\right|^2 < 1$$

for $\zeta \in \mathbb{D}$. Letting $|\zeta| \to 1$ yields then that $\left|g_a^2(\zeta)\right| \to 0$. Thus $g_a^2 \equiv 0$. It now follows that the image of G_a already lies in d_0. As a result, it must be that Φ preserves d_a. But this can only hold for every choice of a if Φ is the identity mapping.

The proof is now complete. □

11.4 Generalization of the Result to Strongly Convex and to Strongly Pseudoconvex Domains

It is clear that the methods of the last section will not apply to strongly pseudoconvex domains. Indeed most strongly pseudoconvex domains have no automorphisms except the identity (see [192]), so we are missing a major tool.

However, an inspection of the proof of Theorem 11.3.1 shows that the principal geometric construct is of a family of mappings of the disc into the ball such that the image analytic discs are holomorphic retracts of the entire domain (this is where the global nature of the result, alluded to in the introduction, comes in). Thanks to work of Lempert [357], we know a large family of domains for which such special analytic discs exist:

Proposition 11.4.1 (Lempert) *Let $\Omega \subset \mathbb{C}^n$ be a smoothly bounded, strictly convex (in the real sense) domain with C^k-boundary ($k \geq 6$). Let $Q \in \Omega$ and $P \in \partial\Omega$. Then there exist unique holomorphic mappings*

$$\varphi_{P,Q} : \overline{\mathbb{D}} \longrightarrow \overline{\Omega}, \quad \psi_{P,Q} : \overline{\Omega} \longrightarrow \overline{\mathbb{D}}$$

that are C^{k-4} up to the boundary, and such that:

(1) $\varphi_{P,Q}(0) = Q$, $\varphi_{P,Q}(1) = P$.
(2) $\psi_{P,Q} \circ \varphi_{P,Q}(\zeta) = \zeta$, *for every* $\zeta \in \overline{\mathbb{D}}$.
(3) If $\varphi : \mathbb{D} \to \Omega$ *is holomorphic, and if* $\psi_{P,Q} \circ \varphi(\zeta) = \zeta$ *for* $\zeta \in \mathbb{D}$, *then* $\varphi(\zeta) \equiv \varphi_{P,Q}(\zeta)$.

Note that **(2)** implies that $\varphi_{P,Q}$ is extremal for the infinitesimal Kobayashi metric on Ω in the direction $\xi := \varphi'_{P,Q}(0) \in T_Q(\Omega)$, while $\psi_{P,Q}$ is the extremal for the "dual" infinitesimal Carathéodory metric at ξ.

Proof The statements of the proposition follow from [357, Théorème 3] and [358]. □

Given Proposition 11.4.1, the proof of Theorem 11.3.1 gives the following result.

Theorem 11.4.2 *Let Ω be a smoothly bounded, strictly convex domain in $\mathbb{C}^n$ with C^6 boundary, and let $P \in \partial\Omega$ be fixed. If $\Phi : \Omega \to \Omega$ is a holomorphic mapping such that*

$$\Phi(z) = z + O\left(|z - P|^4\right),$$

then $\Phi(z) \equiv z$.

Proof Indeed, the maps $\varphi_{P,Q}$ above replace $\phi_a \circ \psi$ in the proof of Theorem 11.3.1, and the $\psi_{P,Q}$ play the role of $\eta \circ \pi_1 \circ (\phi_a)^{-1}$ there. The uniqueness argument at the end of the proof of Theorem 11.3.1 is here achieved by **(3)** of Proposition 11.4.1. □

Unfortunately, Lempert's results fail for a general strictly pseudoconvex domain (see [473]), but they remain true in a neighborhood of P, in directions approximately $\mathbb{C}$-tangent to $\partial\Omega$ at P. More precisely, we have the following.

Proposition 11.4.3 *Let Ω be a smoothly bounded, strongly pseudoconvex domain in $\mathbb{C}^n$ with C^6 boundary, and let $P \in \partial\Omega$ be fixed. Then there exists $Q \in \Omega$, an open neighborhood $V \subset \Omega$ of Q, and holomorphic mappings*

$$\varphi_{P,Q'} : \overline{\mathbb{D}} \longrightarrow \overline{\Omega}, \quad \psi_{P,Q'} : \overline{\Omega} \longrightarrow \overline{\mathbb{D}}$$

as in Proposition 11.4.1 above, for every $Q' \in V$. Furthermore, given any neighborhood $U \subset \mathbb{C}^n$ of P, one can assume that both Q and the images $\varphi_{P,Q'}(\overline{\mathbb{D}})$ lie in $\overline{\Omega} \cap U$.

Proof Given U, that the stationary disc $\varphi_{P,Q}$ exists for some $Q \in U \cap \Omega$ is shown in [357, p. 468]. The existence of the open set $V \ni Q$, and of the corresponding $\varphi_{P,Q'}$, $Q' \in V$, follows from [357, Proposition 10, p. 446]. □

To get the dual functions $\psi_{P,Q'}$, we need the following preparatory geometric lemma.

Lemma 11.4.4 *Let Ω and P be as in Proposition 11.4.3 above. Then there exist a neighborhood Ω' of $\overline{\Omega}$, a bounded, strictly convex domain $\Omega'' \subset \mathbb{C}^n$, and a holomorphic map $F : \Omega' \to \mathbb{C}^n$ such that:*

(1) For a suitable open neighborhood W of $F(P)$, F is biholomorphic, $F : F^{-1}(W) \to W$.
(2) $F(\Omega) \subset \Omega''$.
(3) $F(\partial\Omega) \cap W = \partial\Omega'' \cap W$.

Admitting the lemma for the moment, we return to the proof of Proposition 11.4.3. We now choose the Q and V more carefully, namely so that $V \subset U \cap F^{-1}(W) \cap \overline{\Omega}$, where W, F are as in the lemma. Then the mappings $F \circ \varphi_{P,Q'}$ are stationary discs for the strictly convex domain $\Omega'' \subset \mathbb{C}^n$. Thus there exist dual mappings

$$\psi^0_{F(P),F(Q')} : \overline{\Omega}'' \longrightarrow \overline{\mathbb{D}},$$

and the mappings $\psi_{P,Q'} = \psi^0_{F(P),\, F(Q')} \circ F$ will have all the desired properties in the statement of Proposition 11.4.3, as follows directly from Proposition 11.4.1.

Using Proposition 11.4.3, the proof of Theorem 11.4.2 extends directly to give:

Theorem 11.4.5 *Let Ω be a connected, smoothly bounded, strongly pseudoconvex domain. Let $P \in \partial\Omega$, and let $\Phi : \Omega \to \Omega$ be a holomorphic map such that $\Phi(z) = z + O\left(|z - P|^4\right)$. Then $\Phi(z) \equiv z$.*

Proof Find Q, V as in Proposition 11.4.3 above. The proof as in Theorems 11.3.1 and 11.4.2 shows that Φ is the identity on each disc $\varphi_{P,Q'}(\mathbb{D})$; in particular, $\Phi\left(Q'\right) = Q'$ for all $Q' \in V$. By analytic continuation, $\Phi(z) \equiv z$. □

We finish this section with the proof of Lemma 11.4.4 We start with Fornaess's embedding theorem [156], which gives a neighborhood Ω' of $\overline{\Omega}$, a holomorphic embedding $G : \Omega' \hookrightarrow \mathbb{C}^N$ (for some large N), a smoothly bounded strictly convex set $\Omega''_0 \subset \mathbb{C}^N$ such that $\partial\Omega''_0$ intersects $G\left(\Omega'\right)$ transversally, and $G(\Omega) = \Omega''_0 \cap G\left(\Omega'\right)$. Let T be the tangent space to $G\left(\Omega'\right)$ at $G(P)$; T is isomorphic to $\mathbb{C}^n$, say $E : T \xrightarrow{\sim} \mathbb{C}^n$ linearly. Let $\Pi : \mathbb{C}^N \to T$ be the Hermitian orthogonal projection onto T (with respect to the usual Euclidean metric on $\mathbb{C}^N$), so that Π is $\mathbb{C}$-linear. We claim that the composition

$$F : \Omega' \xrightarrow{G} \mathbb{C}^N \xrightarrow{\Pi} T \xrightarrow{E} \mathbb{C}^n$$

will be the desired map, if we shrink Ω' about $\overline{\Omega}$ as necessary. To see this, notice first that property **(1)** of Lemma 11.4.4 follows from the strict convexity of Ω_0'' and the fact that G is an embedding. (We may be required to shrink Ω' here.) It is an easy exercise in the implicit function theorem to see that, in a neighborhood of P, F is a local diffeomorphism of $\partial\Omega$ onto a strictly convex hypersurface in $\mathbb{C}^n$. Since $E \circ \Pi$ is linear, the compact set $E \circ \Pi\left(\overline{\Omega}_0''\right) := \overline{\Omega}_1''$ is convex and $F(\overline{\Omega}) \subset \Omega_1''$. It is again a simple exercise in the implicit function theorem to see that the boundary of $\overline{\Omega}_1''$ is smooth and strictly convex near $F(P)$. Thus there is a closed Euclidean ball $\overline{B}$ that is tangent to $\partial\overline{\Omega}_1''$ at $F(P)$ and that contains $\overline{\Omega}_1''$. Translate $\overline{B}$ a distance $\varepsilon > 0$ in the direction of the inward Euclidean normal to $\partial\Omega_1''$ at $F(P)$, to a new ball called $\overline{B}_\varepsilon$. For ε sufficiently small, let H_ε denote the (real) convex hull of $\overline{B}_\varepsilon \cup F(\overline{\Omega})$ in $\mathbb{C}^n$, so that the boundary of H_ε coincides with $F(\partial\Omega)$ in a neighborhood of $F(P)$. The domain H_ε is convex, but not smooth and not strictly convex. We can enlarge H_ε slightly to a smoothly bounded, strictly convex domain $\Omega'' \subset \mathbb{C}^n$ whose boundary still agrees with $F(\partial\Omega)$ in a neighborhood of $F(P)$. That this last is possible is clear from a picture but is elementary and tedious to write down explicitly, so we omit this verification. The Ω'' just constructed satisfies all the claims of Proposition 11.4.3.

□

We close this section with the remark that Lemma 11.4.4 is an attempt to re-embed Ω in $\mathbb{C}^n$ as a strictly convex domain, which is certainly impossible. The point here is that F is globally defined, and does re-embed Ω as a strictly convex domain, at least in a neighborhood of P.

11.5 Another Direct Extension

For $P \neq Q$ it is interesting to consider the setting of Theorem 11.4.2 with the modified hypothesis that

$$\Phi(z) = Q + \mu(z - P) + O\left(|z - P|^4\right);$$

here μ is some local holomorphic transformation that takes the tangent plane to $\partial\Omega$ at P to the tangent plane to $\partial\Omega$ at Q. Of course we cannot hope to conclude under these circumstances that Φ is the identity. To see what the correct conclusion might be, consider two special cases:

(1) If Ω is the ball, then Φ may be composed with a unitary rotation taking Q to P. This reduces the situation to the result of Sect. 11.4. We conclude that Φ is a rotation.

(2) If Ω is strongly pseudoconvex and there is a biholomorphism τ of Ω whose extension to the boundary takes Q to P, then $\tau \circ \Phi$ satisfies the hypotheses of Theorem 11.4.5. We conclude that Φ is a biholomorphism of Ω.

A generic strongly pseudoconvex Ω has no nontrivial biholomorphisms (see [192]); so the second case above can be considered descriptive but not prescriptive. Moreover, for any positive integer m, it is a simple matter to construct a strongly pseudoconvex domain Ω with boundary points P, Q such that $\partial\Omega$ near Q is (after a rigid motion) a $2m$-order perturbation of $\partial\Omega$ near P—thus it can be arranged that a rigid motion of Ω maps Ω into Ω, maps P to Q, and agrees with the identity to order $2m$, yet the mapping is not biholomorphic. In summary, a generalization of our results in the direction we have been discussing must take a more restrictive form. We have the following:

Theorem 11.5.1 *Let $\Omega \subset \mathbb{C}^n$ be a smoothly bounded, strongly pseudoconvex domain with C^ω boundary. Suppose $P, Q \in \partial\Omega$ and Φ is a local biholomorphic equivalence of $\partial\Omega$ sending P to Q. If $F : \Omega \to \Omega$ is a holomorphic map such that*

$$F(z) = \Phi(z) + O\left(|z - P|^4\right)$$

near P, then Φ continues analytically to a biholomorphism of Ω and $F(z) \equiv \Phi(z)$ on Ω.

Proof We first note that the localized argument in the proof of Theorem 11.4.5 can be applied to $\Phi^{-1} \circ F$ in a neighborhood of P to show that $F \equiv \Phi$ near P in $\overline{\Omega}$. We would like to use analytic continuation now but must distinguish between two cases.

(1) $\partial\Omega$ is not spherical, i.e., is nowhere locally CR-equivalent to $\partial\mathbb{B}^n$. In this case, the theorem of Pinchuk [415] says that the local holomorphic equivalence Φ can be analytically continued along arbitrary continuous curves in $\partial\Omega$, as a local biholomorphic equivalence sending $\partial\Omega$ to itself. By analytic continuation, then, we conclude that F extends C^ω to $\partial\Omega$ and has nonvanishing Jacobian determinant. The mapping F is therefore a biholomorphism, since it is a self-map of Ω.
(2) $\partial\Omega$ is spherical, i.e., locally CR-equivalent to $\partial\mathbb{B}^n$ at some (and hence every, by analytic continuation) point. Then, by a result of [399, 400], the universal cover $\widetilde{\Omega}$ is biholomorphic to $\mathbb{B}^n$.

The universal covering $\widetilde{\Omega}'$ of a sufficiently small open neighborhood Ω' of Ω is also realizable as an open set $\mathbb{B}^n \subset \widetilde{\Omega}' \subset \mathbb{C}^n$, and we let $\widetilde{P} \in \partial\mathbb{B}^n$ be an inverse image of P for this extended covering map. The mapping F lifts to $\widetilde{F} : \mathbb{B}^n \to \mathbb{B}^n$, and what we have proved above now shows that in a neighborhood of $\widetilde{P}$, $\widetilde{F}$ gives a local biholomorphic equivalence of $\partial\mathbb{B}^n$ to itself. But it is known from [111] that such a local equivalence is the restriction of a global automorphism $\widetilde{\Phi}$ of $\mathbb{B}^n$. By analytic continuation, $\widetilde{F} \equiv \widetilde{\Phi}$ on $\mathbb{B}^n$, and $\widetilde{\Phi}$ induces the claimed analytic continuation of Φ to Ω. □

Remark 11.5.2 In conclusion, we note that these two cases behave differently with respect to analytic continuation of boundary equivalences, namely, the theorem of Pinchuk is false if $\partial\Omega$ is spherical. We do not know, incidentally, whether in general the analytic continuation given by the Pinchuk theorem is single valued.

11.6 Geometric Interpretations and Return to the Question of Order of Contact

An elegant theorem of Ian Graham [182] calculates the asymptotic boundary behavior of the infinitesimal Carathéodory and Kobayashi metrics near a strongly pseudoconvex boundary point. The work of Fefferman in [150] calculates the asymptotic boundary behavior of the infinitesimal Bergman metric near a strongly pseudoconvex boundary point. In both papers the proofs make it clear that if two strongly pseudoconvex domains have local defining functions that agree to fourth order at a strongly pseudoconvex boundary point P, then the Carathéodory (resp. Kobayashi, resp. Bergman) metric has local boundary behavior at P that is the same for both domains. Thus we see that the hypotheses of Theorem 11.4.5 are precisely what are needed to guarantee that the Carathéodory (resp. Kobayashi, resp. Bergman) metric near the distinguished boundary point P has asymptotic boundary behavior that is identical to that for the pullback metric (under the mapping Φ) near P. To what extent can this reasoning be reversed? That is, if Φ is a self-map of Ω, taking a boundary point P to itself and such that Φ preserves asymptotically some invariant metric near P, then must Φ be an automorphism of Ω?

Interestingly, the problem is more subtle than this discussion suggests. For the fourth order contact that we have been discussing is enough to guarantee identical local behavior of the metrics (for instance, it is enough to pick up the first term of the asymptotic expansion for the Bergman kernel at P), but it does not suffice to capture the global information that is necessary for the truth of the theorems we have been considering. Thus at this time we do not have a purely geometric method for proving the theorems of Sects. 11.4 and 11.5.

Now let us return to our original question concerning "geometric order of contact" of the image of a mapping with its target boundary. To give a more natural geometrical notion of contact here for our problem, we consider an anisotropic notion appropriate to the strong pseudoconvexity of our domains. Let Ω as before be a smoothly bounded strongly pseudoconvex domain, and fix $P \in \partial\Omega$. Let ρ be a defining function for $\partial\Omega$ near P. Normalize the coordinate system at P so that $z = 0$ at P, the z_1-direction is the complex normal direction, and $z' := (z_2, \ldots, z_n)$ are the complex tangential directions. Define $\delta(z) = \left(|z_1|^2 + |z'|^4\right)^{1/4}$. We say that a function $W_N = W_N(z)$ is of weight N at P if $|W_N(z)| \le C\delta(z)^N$ near P, for some $C > 0$. Similarly, in analogy with Landau's notation, we will denote by $w_N(z)$ a function such that $\lim_{z\to 0} |w_N(z)| / \delta^N(z) = 0$. These notions are independent of the coordinate normalizations. Now let Φ be a holomorphic map that is C^k at the boundary near $P \in \partial\Omega$, $\Phi(\Omega) \subset \Omega'$, $\Phi(P) = Q \in \partial\Omega'$, and let ρ' be a defining function for $\partial\Omega'$ at Q.

Definition 11.6.1 The set $\Phi(\partial\Omega)$ is said to have geometric contact with $\partial\Omega'$ of weight N at $\Phi(P) = Q$ if

$$\rho' \circ \Phi = h \cdot \rho + w_N , \qquad (*),$$

where h is a C^{k-1} positive function on $\overline{\Omega}$ near P and w_N is as above.

Remark 11.6.2 In this definition, k must be at least N for the definition to make intrinsic sense. See below for some remarks on more efficient measures of differentiability here.

Now let us relate this geometric notion of contact with the analytic osculation discussed earlier.

Proposition 11.6.3 *Let $\mathbb{B}^n$ be the unit ball in $\mathbb{C}^n$, $n \geq 2$, and Φ a holomorphic mapping, $\Phi : \mathbb{B}^n \to \mathbb{B}^n$. Suppose Φ is C^6 to the boundary near $P \in \partial\mathbb{B}^n$ and $Q = \Phi(P) \in \partial\mathbb{B}^n$. If $\Phi(\partial\mathbb{B}^n)$ has geometric contact of weight* 6 *with $\partial\mathbb{B}^n$ at Q, then there is a global biholomorphism $\Phi : \mathbb{B}^n \to \mathbb{B}^n$ such that*

$$\Phi(z) = \Psi(z) + o\left(|z - P|^3\right)$$

near P.

Remark 11.6.4 The Riemann Mapping Theorem shows that the proposition is false when $n = 1$. Concerning the sharpness of the proposition, and of Theorem 11.3.1, see Example 11.6.6 below.

Proof of Proposition 11.6.3 Without loss of generality, by composing Φ with automorphisms of $\mathbb{B}^n$, we can assume that $P = Q = (1, 0, \ldots, 0)$. Conjugating with a Cayley transform, we can also replace $\mathbb{B}^n$ with $\mathcal{U}_n = \{(z, w) \in \mathbb{C}^{n-1} \times \mathbb{C} \mid \operatorname{Im} w > |z|^2\}$, and P with 0. The correspondence can be given explicitly by

$$w = -i\frac{z_1 - 1}{z_1 + 1}, \qquad z_i = \frac{z_{j+1}}{z_1 + 1}, \qquad j = 1, \ldots, n - 1.$$

The coordinates (z, w) at 0 are, suitably renumbered, normalized as in Definition 11.6.1, and the proof of the proposition is an examination of our hypothesis $(*)$ with $N = 6$ in terms of homogeneous with respect to the dilations $T_\delta : (z, w) \to (\delta z, \delta^2 w)$, $\delta > 0$.

Again, composing with automorphisms of $\mathcal{U}_n$, we can assume without loss of generality that Φ has an expansion near 0:

$$\Phi(z, w) = (z + A(z, w), w + B(z, w)),$$

where

$$A(z, w) = \sum_{\nu=2}^{N} a_\nu(z, w) + w(N+1),$$

$$B(z, w) = \sum_{\nu=3}^{N} b_\nu(z, w) + w(N+1).$$

Here, a_ν, b_ν are homogeneous of degree ν for the T_δ-dilations. Also $W(k)$, as distinct from W_k, denotes an error term in an obvious sense. Similarly, h can be written

$$h = 1 + \sum_{\nu=2}^{N} h_\nu + w(N+1).$$

We will use $N \leq 6$ in these expansions.

Let a_{ν_0} be the first nonvanishing term in A; then $(*)$ shows that b_{ν_0+1} is the first nonvanishing term of $\mathbb{B}$. Thus

$$\Phi(z, w) = \left(z + a_{\nu_0} + \text{ h.w.t. }, w + b_{\nu_0+1} + \text{ h.w.t. }\right), \tag{$**$}$$

where "h.w.t." is "higher weight terms" and $\nu_0 \geq 2$. The part of equation $(*)$ homogeneous of weight $\nu_0 + 1$ (where $\nu_0 < N$), and with $P = Q = 0$, $\rho = \rho^1 = \operatorname{Im} w - |z|^2$, is just

$$h_{\nu_0-1}\left(\operatorname{Im} w - |z|^2\right) = -2 \operatorname{Re}\left\langle a_{\nu_0}, z\right\rangle + b_{\nu_0+1}, \tag{$(*)_{\nu_0+1}$}$$

where, for $v, v' \in \mathbb{C}^{n-1}$, $\left\langle v, v'\right\rangle = \sum_{i=1}^{n-1} v_i \overline{v}'_i$.

Now the operation

$$\left(a_{\nu_0}, b_{\nu_0+1}\right) \longmapsto -2 \operatorname{Re}\left\langle a_{\nu_0}, z\right\rangle + \operatorname{Im} b_{\nu_0+1},$$

evaluated at $\operatorname{Im} w = |z|^2$, is the operator "$L$" on formal power series that was defined in [111, formula (2.6)]. Equation $(*)_{\nu_0+1}$ says precisely that $L\left(a_{\nu_0}, b_{\nu_0+1}\right) = 0$. We shall consider $\nu_0 = 2, 3$ directly; the cases $\nu_0 = 4, 5$, which we shall also need, will then follow from [111, Lemma 2.1].

The case $\nu_0 = 2$. In this case we have

$$h_1 = \langle \alpha, z\rangle + \langle \overline{\alpha}, \overline{z}\rangle, \qquad \text{some } \alpha \in \mathbb{C}^n,$$

$$a_2 = Q(z) + \beta w, \quad Q \text{ quadratic in } z, \quad \beta \in \mathbb{C}^n,$$

$$b_3 = C(z) + w\langle z, ga\rangle, \quad C \text{ cubic, }, \gamma \in \mathbb{C}^n.$$

Then equation $(*)_3$ becomes

$$\left(\frac{\langle\alpha, z\rangle w}{2i} - \frac{\langle\overline{\alpha}, \overline{z}\rangle w}{2i}\right) - \left(\langle\alpha, z\rangle|z|^2 + \langle\overline{\alpha}, |z|\rangle|z|^2\right) + \left(\frac{\langle\overline{\alpha}, \overline{z}\rangle w}{2i} - \frac{\langle\alpha, z\rangle\overline{w}}{2i}\right)$$
$$= -(\langle Q(z), z\rangle + \overline{\langle Q(z), z\rangle}) - (w\langle\beta, z\rangle + \overline{w\langle\beta, z\rangle})$$
$$+\left(\frac{C(z)}{2i} - \frac{\overline{C(z)}}{2i}\right) + \left(\frac{w\langle z, \gamma\rangle}{2i} - \frac{\overline{w}\overline{\langle z, w\rangle}}{2i}\right).$$

Comparing like monomials on either side (let us label these with the notation (p, q)), we find that

$$(3, 0) \qquad 0 = \frac{C(z)}{2i} \text{ or } C(z) = 0$$

$$(2, 0) \qquad \frac{w\langle z, \alpha\rangle}{2i} = \frac{w\langle z, \gamma\rangle}{2i} \text{ or } \alpha = \gamma$$

$$(3, 1) \qquad -\langle z, \alpha\rangle\langle z, z\rangle = -\langle Q(z), z\rangle \text{ or } Q(z) = \langle z, \alpha\rangle z$$

$$(1, 1) \qquad \frac{\langle\alpha, z\rangle w}{2i} - \frac{\langle z, \alpha\rangle\overline{w}}{2i} = -w\langle z, \alpha\rangle - \overline{w}\overline{\langle\beta, z\rangle}$$

hence $\alpha = -2i\beta$, some $\beta \in \mathbb{C}^n$.

Before proceeding, we recall that the automorphisms of $\mathcal{U}_n$ with $0 \in \partial\mathcal{U}_n$ a fixed point and normalized as in $(**)$ have the form

$$(z, w) \longmapsto \left(\frac{z + w\beta}{1 - 2i\langle z, \beta\rangle + cw}, \frac{w}{1 - 2i\langle z, \beta\rangle + cw}\right),$$

where $\beta \in \mathbb{C}^n$ is arbitrary and $\operatorname{Im} c = -|\beta|^2$. The calculations performed above tell us that, after composing with one of these automorphisms of $\mathcal{U}_n$, we may assume that $(**)$ holds with $\nu_0 \geq 3$.

The case $\nu_0 = 3$. Now we have

$$h_2 = q(z) + \alpha w + \overline{\alpha w} + \overline{q}(z), \quad q \text{ quadratic}, \alpha \in \mathbb{C}^n;$$
$$a_3 = C(z) + wL(z), \quad C \text{ cubic}, L \text{ linear};$$
$$b_4 = Q(z) + wS(z) + \gamma w^2, \quad Q \text{ quartic}, S \text{ quadratic}, j \in \mathbb{C}.$$

Thus $(*)_4$ becomes

$$\operatorname{Im}\left(qw + \alpha w^2\right) - 2\operatorname{Re} q|z|^2 - \operatorname{Im}(q\overline{w}) - 2\operatorname{Re}\left(\alpha w|z|^2\right) - |w|^2 \operatorname{Im}\alpha$$
$$= \operatorname{Im}\left(Q + wS + \gamma w^2\right) - 2\operatorname{Re}[\langle C(z), z\rangle + w\langle L(z), z\rangle].$$

Comparing like terms now yields

$$\begin{aligned}
&(4,0) && Q \equiv 0;\\
&(3,0) && S \equiv q;\\
&(2,0) && \alpha = \gamma;\\
&(3,1) && C(z) = q(z);\\
&(1,1) && \operatorname{Im}\alpha = 0;\\
&(2,1) && -\alpha w\langle z, z\rangle - \frac{q\overline{w}}{2i} = -w\langle L(z), z\rangle;
\end{aligned}$$

hence $q \equiv 0$ and $L(z) = \alpha z$.

Note that we have now that $q = S = 0$ and $C = z \cdot q = 0$. The upshot of our calculation is that the mapping Φ agrees in form (up to order ν_0) with an automorphism of $\mathcal{U}_n$ that fixes 0. Composing Φ with the inverse of said automorphism (as we are obviously free to do), we have that Φ satisfies (**) with $\nu_0 = 4$.

The cases $\nu_0 = 4, 5$. Applying Lemma 2.1 of [111, p. 233], we find that

$$(\alpha_4, b_5) = (0, 0) \text{ and } (\alpha_5, b_6) = 0.$$

Thus we have proved that the geometric order of contact 6 implies that

$$\Phi(z, w) = (z, w) + (\text{terms of homogeneity at least six}).$$

This is almost the desired result. In terms of the usual notion of order, this result says that there is a constant $\alpha \in \mathbb{C}$ such that, without loss of generality,

$$\Phi(z, w) = \left(z + \alpha w^3, w\right) + O\left(|z - P|^4\right).$$

We can now use a global argument related to Theorem 11.3.1 above. Consider the mapping of the upper half plane $\mathcal{H} := \{\operatorname{Im} w > 0\}$ to itself given by $w \to w + B(0, w)$. By the above we know that $B(0, w)$ is $O\left(|w|^4\right)$ at $w = 0$ and so by the conjugated form of the boundary Schwarz lemma, $\mathbb{B}(0, w) \equiv 0$. Now consider the map of $\mathbb{D}$ into $\mathcal{U}_n$ given by

$$w \longrightarrow \Phi(0, w) = (A(0, w), w + B(0, w)) = (A(0, w), w).$$

By the uniqueness argument at the end of the proof of Theorem 11.3.1, it follows that $A(0, w) \equiv 0$. Hence, the coefficient α above is 0, which proves the proposition. □

Remark 11.6.5 Of course we could assume above that $\Phi(\partial\mathbb{B}^n)$ had contact of weight $N \geq 7$, and the proposition would follow from purely local calculations. It would certainly not be sharp then.

In the above proposition, we can weaken considerably the notion of differentiability on Φ and h as follows: Say that Φ is admissibly differentiable at P of weight N if, in coordinates normalized at $P = 0$ as above, there exists an approximation of $\mathcal{P}(z, w) = (A(z, w), B(z, w))$ where $A - \sum_{\nu=0}^{N-1} a_\nu(z, w) = w_N(z, w)$ and $B - \sum_{\nu=0}^{N} b_\nu(z, w) = w_N(z, w)$. Here the approximations are to hold on admissible approach regions at 0 in the sense of Koranyi. Note that the a_ν, b_ν are assumed to be T_δ-homogeneous as in the proof above. One can similarly define nontangential differentiability of order N, using nontangential approach regions. The corresponding concepts for the real function h we leave to the reader. Proposition 11.6.3 follows if Φ is admissibly differentiable of weight 6 and nontangentially differentiable of order 3 at P. Presumably one does not need all derivatives in z of Φ up to order 6 for the validity of Proposition 11.6.3. We have chosen to work with the notion of contact expressed in terms of weight because of its naturality with respect to CR geometry, as in [111]. Example 11.6.6 shows that, with this measure of contact, Theorem 11.3.1 and Proposition 11.6.3 are sharp.

Example 11.6.6 We will construct self-maps of the Siegel domain $\mathcal{U}_2$ to itself that describe with some precision how sharp the above results are. We start from the Riemann Mapping Theorem in $\mathbb{C}$ and take V to be a smoothly bounded, simply connected open set $\subset \mathcal{H}$ (the upper half plane in $\mathbb{C}$) such that $\overline{V} \cap \partial\mathcal{H}$ in $\mathbb{C}$ contains a neighborhood of 0 in the real axis. Let $\varphi : \mathcal{H} \to V$ be a Riemann mapping (i.e., biholomorphism). Since φ is one-to-one, $\varphi'(w) \neq 0$, for all $w \in \mathcal{H}$. Since $\mathcal{H}$ is simply connected, we can choose a branch of $\sqrt{\varphi'(w)}$ globally on $\mathcal{H}$. By the Schwarz-Pick lemma, applied to the map φ, we know that

$$\frac{|\varphi'(w)|^2}{|\operatorname{Im}\varphi(w)|^2} \leq \frac{1}{|\operatorname{Im} w|^2} \qquad (\#).$$

By (#), the mapping of $\mathbb{C} \times \mathcal{H} \subset \mathbb{C}^2$ to $\mathbb{C}^2$ given by

$$F : (z, w) \longrightarrow \left(\sqrt{\varphi'(w)}z, \varphi(w)\right)$$

sends $\mathcal{U}_2$ to itself. Without loss of generality, we may choose φ in such a way that

$$\varphi(w) = w + \lambda w^3 + O\left(|w|^4\right),$$

where λ is real, $\lambda > 0$. Then, in a neighborhood of 0, we have

$$F(z, w) = \left(z + \frac{3}{2}\lambda w^2 z, w + \lambda w^3\right) + O\left(|(z, w)|^4\right).$$

Here we have chosen the branch of $\sqrt{\varphi'(w)}$ that is 1 at $w = 0$. For this map, $F(\partial\mathcal{U}_2)$ has contact of weight 5 but not weight 6; the case $\nu_0 = 5$ above breaks down here. This example (extended to $\mathbb{C}^n$) also shows that Theorem 11.3.1 is sharp for all n. Note that the map F actually extends analytically past 0.

It is enlightening to attempt to carry out the proof of Proposition 11.6.3 for $n = 1$ to see that the coefficients of Φ are not determined by the contact equation, no matter what order of contact is assumed. Note also that the higher-dimensional example above has contact of $F(\partial\mathcal{U}_2)$ with $\partial\mathcal{U}_2$ along the entire curve $(0, u)$, $u = \operatorname{Re} w \in \mathbb{R}$, near $u = 0$. The contact weight is sharp at all points of this curve.

11.7 Discussion of the Weakly Pseudoconvex Case and Closing Remarks

The analysis we have been discussing does not carry over to the case of weakly pseudoconvex domains. A simple example serves to illustrate the wealth of questions and problems available: Consider the domains

$$\Omega_m = \left\{(z_1, z_2) : |z_1|^2 + |z_2|^{2m} < 1\right\},$$

for m a positive integer. What conditions near $\mathbf{1} = (1, 0)$ on a holomorphic mapping $\Phi : \Omega_m \to \Omega_m$ will force Ω to be the identity mapping? The key observation is that Ω_m covers the ball $\mathbb{B}$ via the mapping $\psi_m(z_1, z_2) = \left(z_1, (z_2)^m\right)$. For any $\alpha = (\alpha_1, \alpha_2) \in \mathbb{B}$ with $\alpha_2 \neq 0$, the analytic disc d_α (see §3) lifts to m extremal discs $\hat{d}_{\hat{\alpha}, j}$, $j = 1, \ldots, m$, in Ω_m. Then the analysis of §3 may be carried out with these extremal discs—provided that the mapping Φ has Taylor jet at P that is sufficiently restricted so that after it is pushed down with the function ψ_m off the branch locus, it satisfies the conditions of Theorem 11.3.1. Note in particular that the correct condition on the Taylor jet will be nonisotropic and will depend on m. X. Huang [245–247] has explored these matters in some detail. An argument similar to the one just outlined may be carried out near any weakly pseudoconvex boundary point that locally “covers” a strongly pseudoconvex boundary point. This is in fact a rather restrictive class and has recently been described by Barletta and Bedford [42] (however, see also [173]).

Chapter 12
The Schwarz Lemma and Its Generalizations

12.1 The Schwarz-Pick Lemma

Here we shall meet one of the most useful facts concerning holomorphic mappings.

Let $\mathbb{D}$ be the open unit disc in the complex plane $\mathbb{C}$, i.e.,

$$\mathbb{D} = \{z \in \mathbb{C}; |z| < 1\}.$$

Let $f : \mathbb{D} \to \mathbb{D}$ be a holomorphic mapping such that $f(0) = 0$. Then the classical Schwarz lemma states that

$$|f(z)| \leqq |z| \quad \text{for } z \in \mathbb{D}$$

and

$$\left|f'(0)\right| \leqq 1,$$

and the equality $\left|f'(0)\right| = 1$ or the equality $|f(z)| = |z|$ at a single point $z \neq 0$ implies

$$f(z) = \varepsilon z \quad \text{with } |\varepsilon| = 1.$$

Now we shall drop the assumption $f(0) = 0$.

If $f : \mathbb{D} \to \mathbb{D}$ is an arbitrary holomorphic mapping, we fix an arbitrarily chosen point $z \in \mathbb{D}$ and consider the automorphisms g and h of $\mathbb{D}$ defined by

P. V. Dovbush, S. G. Krantz, *The Geometric Theory of Complex Variables*,
https://doi.org/10.1007/978-3-031-77204-7_12

$$g(\zeta) = \frac{\zeta + z}{1 + \bar{z}\zeta} \quad \text{for } \zeta \in \mathbb{D}$$

$$h(\zeta) = \frac{\zeta - f(z)}{1 - \overline{f(z)}\zeta} \quad \text{for } \zeta \in \mathbb{D}.$$

Then the composed mapping $F = h \circ f \circ g$ is a holomorphic mapping of $\mathbb{D}$ into itself which sends 0 into itself. Since $F(0) = 0$ and $F'(0) = h'(f(z)) \cdot f'(z) \cdot g'(0)$, we obtain

$$F'(0) = \frac{1 - |z|^2}{1 - |f(z)|^2} f'(z).$$

Hence

$$\frac{1 - |z|^2}{1 - |f(z)|^2} \left| f'(z) \right| \leqq 1,$$

or

$$\frac{\left| f'(z) \right|}{1 - |f(z)|^2} \leqq \frac{1}{1 - |z|^2} \quad \text{for } z \in \mathbb{D}.$$

We may conclude the following:

Theorem 12.1.1 *Let f be a holomorphic mapping of the unit disc $\mathbb{D}$ into itself. Then*

$$\frac{|df|}{1 - |f|^2} \leqq \frac{|dz|}{1 - |z|^2} \quad \textit{for } z \in \mathbb{D},$$

and the equality at a single point z implies that f is an automorphism of $\mathbb{D}$.

This result, which was derived from the Schwarz lemma, is actually equivalent to the Schwarz lemma.

In fact, if $f : \mathbb{D} \to \mathbb{D}$ is a holomorphic mapping such that $f(0) = 0$, then by setting $z = 0$ in the inequality above, we obtain

$$\left| f'(0) \right| \leqq 1,$$

and if $\left| f'(0) \right| = 1$, then f is an automorphism of $\mathbb{D}$. Moreover,

$$\int_0^{|f(\zeta)|} \frac{|df|}{1 - |f|^2} \leqq \int_0^{|\zeta|} \frac{|dz|}{1 - |z|^2}.$$

Hence,

$$\log \frac{1+|f(\zeta)|}{1-|f(\zeta)|} \leqq \log \frac{1+|\zeta|}{1-|\zeta|},$$

which implies

$$\frac{2}{1-|f(\zeta)|} - 1 = \frac{1+|f(\zeta)|}{1-|f(\zeta)|} \leqq \frac{1+|\zeta|}{1-|\zeta|} = \frac{2}{1-|\zeta|} - 1.$$

It follows that $|f(\zeta)| \leqq |\zeta|$. The equality $|f(\zeta)| = |\zeta|$ at a single point $\zeta \neq 0$ implies the equality

$$\frac{|f'(z)|}{1-|f(z)|^2} = \frac{1}{1-|z|^2}$$

for all z lying on the line segment joining 0 and ζ. By Theorem 12.1.1, f is an automorphism of $\mathbb{D}$, and hence, $f(z) = \varepsilon z$ for some ε with $|\varepsilon| = 1$. This proves that Theorem 12.1.1 implies the classical Schwarz lemma. We shall now consider Theorem 12.1.1 from a differential geometric viewpoint. If we consider the Kähler metric $ds^2_{\mathbb{D}}$ on $\mathbb{D}$ given by

$$ds^2_{\mathbb{D}} = \frac{dzd\bar{z}}{\left(1-|z|^2\right)^2},$$

then the inequality in Theorem 12.1.1 may be written as follows:

$$f^*\left(ds^2_{\mathbb{D}}\right) \leqq ds^2_{\mathbb{D}}.$$

The metric $ds^2_{\mathbb{D}}$ is called the Poincaré metric or the Poincaré-Bergman metric of $\mathbb{D}$. Now Theorem 12.1.1 may be stated as follows:

Theorem 12.1.2 *Let $\mathbb{D}$ be the open unit disc in $\mathbb{C}$ with the Poincaré-Bergman metric $ds^2_{\mathbb{D}}$. Then every holomorphic mapping $f : \mathbb{D} \to \mathbb{D}$ is distance-decreasing, i.e., satisfies*

$$f^*\left(ds^2_{\mathbb{D}}\right) \leqq ds^2_{\mathbb{D}},$$

and the equality at a single point of $\mathbb{D}$ implies that f is an automorphism of $\mathbb{D}$.

If f is an automorphism of $\mathbb{D}$, then the Schwarz lemma applied to both f and f^{-1} implies that f is an isometry.

Let $G = g(z)dzd\bar{z}$ be a pseudo-Hermitian metric, on the unit disc $\mathbb{D}$, that is, $g \geq 0$ and g may vanish at an isolated set of points. The Gaussian curvature $K(G)$

is $-\frac{1}{g}\frac{\partial^2}{\partial z\partial\bar{z}}\log g$. It is easily verified that if $K(G_i) \leq -1$ for $i = 1, \ldots n$, then $K(G_1 + \cdots + G_n) \leq -\frac{1}{n}$.

We note that the Gaussian curvature of the metric $ds^2_{\mathbb{D}}$ is equal to -4 everywhere.

12.2 Curvature and the Schwarz Lemma Revisited

If $U \subseteq \mathbb{C}$ is a planar domain and ρ is a metric on U, then the curvature of the metric ρ at a point $z \in U$ (at which $\rho(z) \neq 0$) is defined to be

$$\kappa_{U,\rho}(z) = \kappa(z) \equiv \frac{-\Delta \log \rho(z)}{\rho(z)^2}.$$

(Here zeros of $\rho(z)$ will result in singularities of the curvature function $-\kappa$ which is undefined at such points.) We will study this quantity for a bit and then summarize and provide motivation afterward.

Since ρ is twice continuously differentiable, this definition makes sense. It assigns to each $z \in U$ a numerical quantity. The most important preliminary fact about κ is its conformal invariance:

Proposition 12.2.1 *Let U_1 and U_2 be planar domains and $h : U_1 \to U_2$ a conformal map (in particular, h' never vanishes). If ρ is a metric on U_2, then*

$$\kappa_{U_1,h^*\rho}(z) = \kappa_{U_2,\rho}(h(z)), \quad \forall z \in U_1.$$

Proof We need to calculate

$$\begin{aligned}
\kappa_{U_1,h^*\rho}(z) &\equiv \frac{-\Delta \log\left[\rho(h(z)) \cdot |h'(z)|\right]}{[\rho(h(z)) \cdot |h'(z)|]^2} \\
&= \frac{-\Delta \log[\rho(h(z))] - \Delta\left[\log|h'(z)|\right]}{[\rho(h(z)) \cdot |h'(z)|]^2}.
\end{aligned}$$

Now the second term in the numerator vanishes. Indeed, since h' never vanishes on U_1, it has a holomorphic logarithm. Then on U_1, we have

$$\begin{aligned}
\Delta\left(\log\left(|h'|^2\right)\right) &= \Delta(\log h' + \log \bar{h}') \\
&= 4\frac{\partial}{\partial z}\frac{\partial}{\partial \bar{z}}\log h' + 4\frac{\partial}{\partial \bar{z}}\frac{\partial}{\partial z}\log \bar{h}' \\
&= 0.
\end{aligned}$$

We may further simplify the numerator to obtain that the last line is

$$\begin{aligned}
&= \frac{-\left[(\Delta \log \rho)(h(z)) \left|h'(z)\right|^2\right]}{[\rho(h(z)) \cdot |h'(z)|]^2} \\
&= \frac{-\,\Delta \log \rho|_{h(z)}}{\rho(h(z))^2} \\
&= \kappa_{U_2,\rho}(h(z)).
\end{aligned}$$

□

Remark 12.2.2 In fact the proof gives the slightly more general fact that if U_1, U_2 are domains and $f : U_1 \to U_2$ is a holomorphic map (not necessarily one-to-one or onto), then the following holds. If ρ is a metric on U_2, then

$$\kappa_{f^*\rho}(z) = \kappa_\rho(f(z))$$

at every point $z \in U_1$ for which $f'(z) \neq 0$ and $\rho(f(z)) \neq 0$. Usually this is all points except for a discrete set.

From the point of view of geometry, any differential quantity that is invariant is automatically of great interest. But why do we call κ "curvature"? Why do not we call it "mass" or "humidity" or "color"? The answer is that a standard construction in Riemannian geometry assigns to any Riemannian metric a quantity called Gaussian curvature. In the setting of classical Euclidean geometry, this quantity is consistent with our intuitive notion of what curvature ought to be (i.e., the rate of change of the normal). In more abstract settings, the structural equations of Cartan lead to a quantity called curvature which is invariant but has somewhat less intuitive content. In the complex plane, the curvature calculation leads to the formula which we have used to define κ. One of the pleasant facts about the use of elementary differential geometry in complex function theory is that one does not require all the geometric machinery which leads to the formula for κ in order to make good use of κ.

Because of the considerations in the last paragraph, we shall simply accept the definition of κ, as given. The geometric calculations *that lead to the definition of* κ are provided in the Appendix (Chap. 22). Understanding of the material which follows in the text does not in any way depend on the material in this appendix.

Let us begin our study of curvature by calculating the curvature of the Euclidean metric.

Example 12.2.3 Let U be a planar domain equipped with the Euclidean metric $\rho(z) \equiv 1$. It follows from the definition that $\kappa(z) \equiv 0$. This should be expected, for the Euclidean metric does not change from point to point.

Example 12.2.4 The metric

$$\sigma(z) = \frac{2}{1 + |z|^2}$$

on $\mathbb{C}$ is called the *spherical metric*. A straightforward calculation shows that the curvature of σ is identically 1. Manifolds of constant curvature are a matter of great interest (see [532]).

Next we calculate the curvature of the Poincaré metric on the disc, which calculation will be of great utility for us. We will learn that the curvature is negative. Notice that since any point of the disc may be moved to any other by a Möbius transformation and since the curvature is a conformal invariant, we will expect the curvature function to be in fact a negative constant.

Proposition 12.2.5 *Consider the disc $\mathbb{D}$ equipped with the Poincaré metric ρ. For any point $z \in \mathbb{D}$, we have $\kappa(z) = -4$.*

Proof We notice that

$$-\Delta \log \rho(z) = \Delta \log \left(1 - |z|^2\right).$$

Now we write $\Delta = 4(\partial/\partial z)(\partial/\partial \bar{z})$ and $|z|^2 = z \cdot \bar{z}$ to obtain that the last line is

$$= -\frac{4}{\left(1 - |z|^2\right)^2}.$$

It follows that $\kappa(z) \equiv -4$. □

It was Ahlfors [10] who first realized that the Schwarz lemma is really an inequality about curvature. In his annotations to his collected works he modestly asserts that "This is an almost trivial fact and anybody who sees the need could prove it at once." However, he goes on to say (most correctly) that this point of view has been a decisive influence in modern function theory. It is a good place to begin our understanding of curvature. Here is Ahlfors's version of Schwarz's lemma:

Theorem 12.2.6 *Let the disc $\mathbb{D}$ be equipped with the Poincaré metric ρ, and let U be a planar domain equipped with a metric σ. Assume that, at all points of U, σ has curvature not exceeding -4. If $f : \mathbb{D} \to U$ is holomorphic, then we have*

$$f^*\sigma(z) \leq \rho(z), \quad \forall z \in \mathbb{D}.$$

Proof following [381] Let $0 < r < 1$. On the disc $\mathbb{D}(0, r)$, define the metric

$$\rho_r(z) = \frac{r}{r^2 - |z|^2}.$$

Then straightforward calculations (or change of variable) show that ρ_r is the analog of the Poincaré metric for $\mathbb{D}(0, r)$: It has constant curvature -4 and is invariant under conformal maps. Define

$$v = \frac{f^*\sigma}{\rho_r}.$$

Observe that v is continuous and nonnegative on $\mathbb{D}(0, r)$ and that $v \to 0$ when $|z| \to r$ (since $f^*\sigma$ is bounded above on $\bar{D}(0, r) \subset \mathbb{D}$ while $\rho_r \to \infty$). It follows that v attains a maximum value $M = M(r)$ at some point $\tau \in \mathbb{D}(0, r)$. We will show that $M \leq 1$; hence $v \leq 1$ on $\mathbb{D}(0, r)$. Letting $r \to 1^-$then finishes the proof.

If $f^*\sigma(\tau) = 0$, then $v \equiv 0$. So we may suppose that $f^*\sigma(\tau) > 0$. Therefore $\kappa_{f^*\sigma}$ is defined at τ. By hypothesis,

$$\kappa_{f*\sigma} \leq -4.$$

Since $\log v$ has a maximum at τ, we have

$$\begin{aligned}
0 \geq \Delta \log v(\tau) =&\Delta \log f^*\sigma(\tau) - \Delta \log \rho_r(\tau) \\
=& -\kappa_{f_*\sigma}(\tau) \cdot \left(f^*\sigma(\tau)\right)^2 \\
&+ \kappa_{\rho_r}(\tau) \cdot (\rho_r(\tau))^2 \\
\geq& 4\left(f^*\sigma(\tau)\right)^2 - 4\,(\rho_r(\tau))^2 .
\end{aligned}$$

This gives

$$\frac{f^*\sigma(\tau)}{\rho_r(\tau)} \leq 1$$

or

$$M \leq 1$$

as desired. □

Observe that the usual Schwarz lemma is a corollary of Ahlfors's version: We take U to be the disc with σ being the Poincaré metric. Let f be a holomorphic mapping of $\mathbb{D}$ to $U = \mathbb{D}$ such that $f(0) = 0$. Then σ satisfies the hypotheses of the theorem, and the conclusion is that

$$f^*\sigma(0) \leq \rho(0).$$

Unravelling the definition of $f^*\sigma$ yields

$$\left|f'(0)\right| \cdot \sigma(f(0)) \leq \rho(0).$$

But $\sigma = \rho$ and $f(0) = 0$, so this becomes

$$\left|f'(0)\right| \leq 1.$$

Exercise The property

$$d(f(P), f(Q)) \leq d(P, Q),$$

where $d = d_\rho$ is the Poincaré distance, is called the *distance nonincreasing property* of the Poincaré metric. Use this property to give a geometric proof of the other inequality in the classical Schwarz lemma.

With a little more notation we can obtain a more powerful version of the Ahlfors/Schwarz lemma. Let $\mathbb{D}(0, \alpha)$ be the open disc of radius α and center 0. For $A > 0$, define the metric ρ_α^A on $\mathbb{D}(0, \alpha)$ by

$$\rho_\alpha^A(z) = \frac{2\alpha}{\sqrt{A}\left(\alpha^2 - |z|^2\right)}.$$

This metric has constant curvature $-A$. Then we have the following theorem:

Theorem 12.2.7 *Let U be a planar domain that is equipped with a metric σ whose curvature is bounded above by a negative constant $-B$. Then every holomorphic function $f : \mathbb{D}(0, \alpha) \to U$ satisfies*

$$f^*\sigma(z) \leq \frac{\sqrt{A}}{\sqrt{B}}\rho_\alpha^A(z), \quad \forall z \in \mathbb{D}(0, \alpha).$$

It is a good exercise to construct a proof of this more general result, modeled of course on the proof of Theorem 12.2.6.

12.3 A Generalization by Ahlfors

The classical Schwarz-Pick lemma states that any holomorphic map of the unit disc into itself decreases the Poincaré metric. Later Ahlfors generalized this lemma to holomorphic mappings between two Riemann surfaces where curvatures of these Riemann surfaces were used in a very explicit way.

Let $\mathbb{D}(0, a)$ be the open disc of radius a in $\mathbb{C}$

$$\mathbb{D}(0, a) = \{z \in \mathbb{C}; |z| < a\}.$$

Then the metric

$$ds_a^2 = \frac{4a^2 dz d\bar{z}}{A\left(a^2 - |z|^2\right)^2} = 2g dz d\bar{z} \quad (A > 0)$$

on $\mathbb{D}(0, a)$ has Gaussian curvature $-A$.

A complex manifold M is called Hermitian if there is given on the fibers of its tangent bundle a Hermitian bilinear form $h\colon T(M) \times T(M) \to \mathbb{C}$ which is positive definite and twice smooth on vector fields of class $C^2(M)$.

In local coordinates we have in a neighborhood $U \subset M$

$$h_p(u, v) = \sum_{j,k=1}^{n} h_{jk}(p) dz_j(u) \overline{dz_k(u)}, \quad h_{jk} = h\left(\frac{\partial}{\partial z_j}, \frac{\partial}{\partial z_k}\right),$$

where the matrix $h = (h_{jk})$ is Hermitian symmetric $(h_{kj} = \bar{h}_{jk})$ and positive definite (all eigenvalues are positive) and $h_{jk} \in C^2(U)$. The form $\operatorname{Im} h$ is skew symmetric, and $g = \operatorname{Re} h$ is symmetric and defines a Riemannian metric on M; $h(u, u) = g(u, u)$.

On a Hermitian manifold one defines in a natural way the norm of a tangent vector: $\|u\|_h = \sqrt{h_p(u, u)}$, the length of a smooth path $\gamma : [0, 1] \to M$:

$$|\gamma| = \int_0^1 \sqrt{h_{\gamma(t)}(\gamma'(t), \gamma'(t))} dt,$$

and the distance $d_h(p, q)$ between two points $p, q \in M$ is the greatest lower bound of all paths connecting them; these are the same as in the Riemannian metric $g = \operatorname{Re} h$. The quantity $h_p(u, u) = H(u, u)$ gives a differential metric on M: It will likewise be denoted by $ds_h^2(u)$.

Let us fix on a Hermitian manifold a point p along with a vector $v \in T_p(M)$, and let us consider a holomorphic curve $f : \mathbb{D}_r \to M$ such that $f(0) = p$ and $f'(0) = v$. On the image $S = f(\mathbb{D}_r)$, we then have the induced metric

$$h \mid S = \sum_{j,k=1}^{n} h_{jk}(f(\zeta)) f_j'(\zeta) \overline{f_k'(\zeta)} d\zeta d\bar{\zeta} = H(\zeta) d\zeta d\bar{\zeta},$$

where $H > 0$ is also in $C^2(\mathbb{D}_r)$; we take r so small that S lies in a single coordinate neighborhood on M and $f' \neq 0$ on $\mathbb{D}_r$. Considering S as a real two-dimensional surface, we can compute its Gaussian curvature at the point p:

$$K_f(p) = -\left.\frac{1}{2H(0)} \frac{\partial^2 \ln H}{\partial \zeta \partial \bar{\zeta}}\right|_{\zeta=0}$$

(here $\operatorname{Re} \zeta$ and $\operatorname{Im} \zeta$ are thus isometric coordinates on S). It is not hard to see that the number K_f is invariant under conformal changes of the parameter ζ; it is called the holomorphic curvature of S. The least upper bound of $K_f(p)$ over all holomorphic curves f, $f(0) = p$, $f'(0) = v$, is called the holomorphic sectional curvature at p in the direction v.

Notice that the holomorphic sectional curvature can be defined for differential metrics that need not be Hermitian (example: the Carathéodory metric or the

Kobayashi metric), such that all restrictions to holomorphic curves $f : \mathbb{D}_r \to M$ have the form $hd\zeta d\bar{\zeta}$ with a positive function $h \in C^2(\mathbb{D}_r)$.

The following theorem of Ahlfors [10] generalizes the Schwarz lemma.

Theorem 12.3.1 *Let M be a one-dimensional Kähler manifold with metric ds_M^2 whose Gaussian curvature is bounded above by a negative constant $-B$. Then every holomorphic map $f : \mathbb{D}(0, a) \to M$ satisfies*

$$f^* ds_M^2 \leqq \frac{A}{B} ds_a^2.$$

Proof Let u be the nonnegative function on $\mathbb{D}(0, a)$ defined by

$$f^* ds_M^2 = u(z) ds_a^2 = 2h dz d\bar{z},$$

where $u \geq 0$ is a smooth function in $\mathbb{D}(0, a)$.

Clearly, we have

$$h = \frac{f^* ds_M^2}{2 ds_a^2}.$$

We want to prove that $u \leqq \frac{A}{B}$ on $\mathbb{D}(0, a)$. We fix $\zeta_0 \in \mathbb{D}(0, a)$, an arbitrary number r, $|\zeta_0| < r < a$, and set

$$ds_r^2 = \frac{4r^2 |dz|^2}{A\left(r^2 - |z|^2\right)^2}.$$

If we define $f^* ds_M^2 = u_r ds_r^2$, then direct computation shows that

$$u_r = u \frac{r^2 (a^2 - |z|^2)^2}{a^2 (r^2 - |z|^2)^2} = \frac{h}{g}.$$

Therefore $u/u_r \to 0$ as $|\zeta| \to r$, and hence u_r attains its maximum inside of $\mathbb{D}_r$. It suffices to prove the desired inequality for a maximum point; therefore we may assume that ζ_0 is a maximum point. Since $u_r(\zeta_0) \to u(\zeta_0)$ as $r \to a$, it suffices to prove the desired inequality for this maximum point.

If $u_r(\zeta_0) = 0$, then the inequality is trivial.

Assume that $u_r(\zeta_0) > 0$; then ζ_0 is not a critical point of the curve f.

The Gaussian curvature k of the metric $ds_M^2 = 2h dz d\bar{z}$ is given by

$$k = -\frac{1}{h} \frac{\partial^2 \log h}{\partial z \partial z}.$$

The Gaussian curvature of the metric $ds_a^2 = 2g dz d\bar{z}$ is given by

$$-\frac{1}{g}\frac{\partial^2 \log g}{\partial z \partial \bar{z}}$$

and equal to $-A$.

Since $k \leq -B$ by our assumption, we have

$$\frac{\partial^2 \log u_r}{\partial z \partial \bar{z}} = \frac{\partial^2 \log(h/g)}{\partial z \partial \bar{z}} = \frac{\partial^2 \log h}{\partial z \partial \ddot{z}} - \frac{\partial^2 \log g}{\partial z \partial \bar{z}} = -kh - Ag \geqq Bh - Ag.$$

Since $\log u_r$ attains its maximum at ζ_0, the left-hand side in the inequality above is nonpositive at ζ_0 and so is the right-hand side. Hence, $A/B \geqq h/g = u_r$ at ζ_0. Since u_r attains its maximum at ζ_0, it follows that $A/B \geq u_r$ everywhere. □

12.4 Compact Riemann Surfaces of Genus $\geqq 2$

In this section we prove the following statement:

Theorem 12.4.1 *Every compact Riemann surface M of genus $g \geqq 2$ admits a Kähler metric whose Gaussian curvature is bounded above by a negative constant.*

Proof Since the genus of M is g, there are g linearly independent holomorphic 1-forms $\omega_1, \ldots, \omega_g$ on M. We set

$$ds^2 = \omega_1 \bar{\omega}_1 + \omega_2 \bar{\omega}_2.$$

In terms of a local coordinate system z of M, we may write

$$ds^2 = \left(f_1 \bar{f}_1 + f_2 \bar{f}_2\right) dz d\bar{z}, \qquad \text{where } \omega_i = f_i dz.$$

If $p_1, \ldots, p_k$ is the set of common zeros of ω_1 and ω_2, then ds^2 is a Kähler metric $M \setminus \{p_1, \ldots, p_k\}$. By a simple calculation we see that the Gaussian curvature K is given by

$$K = -2\left|f_1' f_2 - f_1 f_2'\right|^2 / \left(f_1 \bar{f}_1 + f_2 \bar{f}_2\right)^3.$$

Since ω_1 and ω_2 are linearly independent, it follows that $f_2 \not\equiv 0$ and $f_1/f_2 \not\equiv$ constant, and consequently $K \not\equiv 0$. Hence there are at most finitely many points $p_{k+1}, \ldots, p_m$ in $M \setminus \{p_1, \ldots, p_k\}$ where the curvature K vanishes. For each p_i, choose a neighborhood V_i in such a way that $V_i \cap V_j = \emptyset$ for $i \neq j$. In each V_i we shall modify the metric so that the Gaussian curvature becomes negative everywhere on M. We fix i, $i = 1, \ldots, m$. Choosing a local coordinate system z in V_i such that $z(p_i) = 0$, let r be a positive number and

$$B = \{z : |z| < r\} \Subset V_i: \quad B' = \{z : |z| < r/2\}.$$

Let $a(z, \bar{z})$ be a C^∞ function on V_i such that

$$0 \leqq a(z, \bar{z}) \leqq 1, \quad a\,\big|B' \equiv 1, \quad a\big|\,(V_i \setminus B) \equiv 0.$$

Let c be a constant, $0 < c < 1$, and set

$$g = f + ch, \qquad \text{where } f = f_1\bar{f_1} + f_2\bar{f_2} \quad \text{and} \quad h = a \cdot (1 + z\bar{z}).$$

Then the metric $gdzd\bar{z}$ coincides with ds^2 on $V_i \setminus B$. Since the curvature of ds^2 is bounded above by a negative constant on the compact set $\bar{B} \setminus B'$, the metric $gdzd\bar{z}$ has a strictly negative curvature on $\bar{B} \setminus B'$ if c is sufficiently small. From the estimate on $K(g)$

$$\begin{aligned} K(g) &= K[c(f+h) + (1-c)f] \\ &\leqq \frac{c^2(f+h)^2 K[c(f+h)] + (1-c)^2 f^2 K[(1-c)f]}{(f+ch)^2} \\ &= \frac{c(f+h)^2 K(f+h) + (1-c) f^2 K(f)}{(f+ch)^2} \\ &\leqq \frac{c(f+h)^2 K(f+h) - (1-c) f^2 k'}{(f+ch)^2}, \end{aligned}$$

we see that the curvature of $gdzd\bar{z}$ is strictly negative in B' if c is sufficiently small. □

12.5 Holomorphic Mappings from an Annulus into an Annulus

Let A be the *annulus* in $\mathbb{C}$ defined by

$$A = \{z \in \mathbb{C} : 0 < r < |z| < R\}.$$

The number $M = \log(R/r)$ is called the *modulus of* A. Two annuli with the same modulus are clearly biholomorphic to each other under a suitable homothetic transformation. We may therefore assume that $rR = 1$.

Let A_1 and A_2 be two annuli with moduli M_1 and M_2, respectively:

$$A_k = \{z \in \mathbb{C} : 0 < r_k < |z| < 1/r_k\}, \quad k = 1, 2.$$

Let $f : A_1 \to A_2$ be a holomorphic mapping and $f_* : \pi_1(A_1) \to \pi_1(A_2)$ the induced homomorphism on the fundamental groups of A_1 and A_2. Let α_k be the generator of $\pi_1(A_k) = Z$, $k = 1, 2$. Then the degree of f, denoted by $\deg f$, is defined by

$$f_*(\alpha_1) = (\deg f)\alpha_2.$$

For each integer m such that $|m| \leqq M_2/M_1$, we define a holomorphic mapping $f_m : A_1 \to A_2$ with $\deg f = m$ as follows:

$$f_m(z) = z^m \quad \text{for } z \in A_1.$$

From topology we know that any mapping $f : A_1 \to A_2$ (holomorphic or not) of degree m with $|m| \leqq M_2/M_1$ is homotopic to f_m. In fact, any two mappings of A_1 into A_2 with the same degree are homotopic to each other. We know also that for any integer m there is a mapping (not necessarily holomorphic) of A_1 into A_2 with degree m. But we have the following theorem:

Theorem 12.5.1 *Let A_1 and A_2 be annuli with moduli M_1 and M_2 as above. If $f : A_1 \to A_2$ is a holomorphic mapping, then $|\deg f| \leqq M_2/M_1$. If M_2/M_1 is an integer, then a holomorphic mapping $f : A_1 \to A_2$ with degree $m = \pm M_2/M_1$ coincides with the mapping f_m up to a rotation.*

Proof We consider the band $B = \{z \in \mathbb{C} : -b < \operatorname{Im} z < b\}$ of width $2b$. We know that it is conformally equivalent to the open unit disc or equivalently to the upper half plane $H = \{w \in \mathbb{C} : \operatorname{Im} w > 0\}$ in $\mathbb{C}$. Indeed, the mapping $z \in B \to ie^{\pi z/2b} \in H$ is a biholomorphic mapping. The invariant metric ds_H^2 of curvature -1 on H is given by

$$ds_H^2 = \frac{dwd\bar{w}}{v^2}, \qquad \text{where } w = u + iv.$$

If we induce this metric to the band B by the holomorphic mapping given above, we obtain the following invariant metric ds_B^2 of curvature -1:

$$ds_B^2 = \frac{\pi^2 dzd\bar{z}}{4b^2\cos^2(\pi y/b)}, \qquad \text{where } z = x + iy.$$

We consider now a holomorphic mapping p from B onto the annulus $A = \{w \in \mathbb{C} : r < |w| < 1/r\}$, $r = e^{-2\pi b}$, defined by

$$p(z) = e^{2\pi iz}, \quad z \in B.$$

Then $p : B \to A$ is a covering projection. We denote by ds_A^2 the metric on A induced by ds_B^2. Then the rectangle $\{z \in B : 0 \leqq x \leqq 1\}$ is a fundamental domain for this covering space (B, A, p). The projection p maps the upper edge, the lower edge, and the two vertical edges of this rectangle onto the inner boundary, the outer boundary of the annulus A, and the segment $\{w = u + iv \in A : u > 0, v = 0\}$, respectively. It is also clear that p maps $\{z = x + iy \in B : 0 \leqq x \leqq 1, y = 0\}$ onto the unit circle $\{w \in A :: |w| = 1\}$, which is the generator of the fundamental group $\pi_1(A)$. Consider a curve $w(t)$, $0 \leqq t \leqq 1$, in A which represents the generator of $\pi_1(A)$. We may assume that $w(0) = w(1) > 0$. To compute its length with respect to ds_A^2, we consider a curve $z(t)$ in the rectangle $\{z \in B : 0 \leqq x \leqq 1\}$ such that $p[z(t)] = w(t)$, $\mathrm{Re}[z(0)] = 0$, and $\mathrm{Re}[z(1)] = 1$ and compute the length of $z(t)$ with respect to ds_B^2. From the expression of ds_B^2 given above, we see that the curve $w(t)$ has the least length when it is the circle $|w(t)| = 1$, i.e., when $z(t)$ is real for all t. This least length is given by

$$\int_0^1 \frac{\pi dx}{2b} = \frac{\pi}{2b}.$$

We have shown that the circle $w(t) = e^{2\pi it}$, $0 \leqq t \leqq 1$, represents the generator α of $\pi_1(A)$ and has arclength $\pi/2b$ with respect to ds_A^2 and that any closed curve in A representing α has arclength $\geqq \pi/2b$, where the equality holds only when the closed curve coincides with the unit circle up to a parametrization. Similarly, the curve $w(t) = e^{2m\pi it}$, $0 \leqq t \leqq 1$, representing $m\alpha \in \pi_1(A)$ has arclength $|m\pi/2b|$, and any closed curve representing $m\alpha$ has arclength $> |m\pi/2b|$ unless it coincides with $w(t) = e^{2m\pi it}$ up to a parametrization.

Given two annuli A_1 and A_2, we consider the bands B_1 and B_2 of widths b_1 and b_2 and covering projections $p_1 : B_1 \to A_1$ and $p_2 : B_2 \to A_2$ as above. Let $f : A_1 \to A_2$ be a holomorphic mapping. Since B_1 is holomorphically equivalent to the unit open disc, we may apply Schwarz-Pick-Ahlfors lemma (Theorem 12.3.1) to the map $f \circ p_1 : B_1 \to A_2$ and see that $f \circ p_1$ is distance-decreasing, i.e., $(f \circ p_1)^* ds_{A_2}^2 \leqq ds_{B_1}^2$. Since p_1 is a local isometry, we see that f itself is also distance-decreasing, i.e., $f^* ds_{A_2}^2 \leqq ds_{A_1}^2$. We consider the unit circle $w_1(t) = e^{2\pi it}, 0 \leqq t \leqq 1$, in A_1 and its image curve $w_2(t) = f[w_1(t)], 0 \leqq t \leqq 1$, in A_2. Since $w_1(t)$ has arclength $\pi/2b_1$ and f is distance-decreasing, $w_2(t)$ has arclength $\leqq \pi/2b_1$. If $\deg f = m$ so that $w_2(t)$ represents $m\alpha_2 \in \pi_1(A_2)$, where α_2 is the generator of $\pi_1(A_2)$, then $u_2(t)$ has arclength $\geqq |m\pi/2b_2|$. Hence, $|m\pi/2b_2| \leqq \pi/2b_1$, that is, $|m| \leqq b_2/b_1$. Since $r_k = e^{-2\pi b_k}$ and $M_k = \log\left(1/r_k^2\right) = -2\log r_k$ for $k = 1, 2$, we have $b_2/b_1 = M_2/M_1$. Hence, $|m| \leqq M_2/M_1$, which proves the first assertion of the theorem.

Assume that M_2/M_1 is an integer and $|m| = M_2/M_1$. Then $w_2(t)$ has arclength $|m\pi/2b_2| = \pi/2b_1$. Hence, the closed curve $w_2(t)$ coincides with $e^{2m\pi it}, 0 \leqq t \leqq 1$, up to a parametrization. Since f is distance-decreasing and $w_2(t)$ has arclength $\pi/2b_1$, f maps $w_1(t)$ onto $w_2(t)$ isometrically. This implies that $w_2(t)$ coincides actually with $e^{2m\pi it}$ up to a rotation. It follows that f and f_m coincide on the unit circle $w_1(t)$ of A_1 up to a rotation, and hence they coincide on A_1 up to a rotation.

□

Corollary 12.5.2 *Let f be a holomorphic mapping from an annulus $A = \{z \in \mathbb{C} : r < |z| < 1/r\}$ into itself. Then f is either homotopic to a constant map or of the form $f(z) = e^{\pm 2\pi i(z+a)}$, where a is a real number.*

Theorem 12.5.1 is due to Huber [250]. The proof presented here is perhaps a little more differential geometric. In connection with the result of this section, see also Schiffer [459], Jenkins [258], and Landau and Osserman [347, 348].

Chapter 13
Invariant Distances on Complex Manifolds

13.1 Poincaré Metric

One of the more useful tools in complex analysis is the hyperbolic metric of a planar domain. We discuss here the hyperbolic metric of the unit disc, sometimes referred to as the Poincaré plane or disc. Later we shall use this metric of $\mathbb{D}$, together with the uniformization theorem, to define the hyperbolic metric of an arbitrary planar domain (other than the full plane or the punctured plane). This is quite a technical and deep fact from complex analysis, in fact, one of the most important results and themes of nineteenth century mathematics, and we shall not offer proofs here. Mostly, we shall need only the hyperbolic metric on the disc and the triply punctured sphere.

For each $a \in \mathbb{D}$ and $\theta \in \mathbb{R}$, the linear fractional transformation

$$\phi_a(z) = e^{i\theta} \frac{z - a}{1 - \overline{a}z} \tag{13.1}$$

defines a holomorphic self-homeomorphism of the disc $\mathbb{D}$ to itself. We compute

$$\phi_a'(z) = e^{i\theta} \frac{1 - |a|^2}{(1 - \overline{a}z)^2}$$

from which we have the identity

$$\frac{|\phi_a'(z)|}{1 - |\phi_a(z)|^2} = \frac{1}{1 - |z|^2}. \tag{13.2}$$

Equation (13.2) expresses the fact that each $\phi_a : \mathbb{D} \to \mathbb{D}$ is an isometry of the Riemannian metric

P. V. Dovbush, S. G. Krantz, *The Geometric Theory of Complex Variables*,
https://doi.org/10.1007/978-3-031-77204-7_13

$$ds_{hyp}(z) = \frac{2|dz|}{1-|z|^2}, \quad z \in \mathbb{D}.$$

This corresponds to the choice of matrix

$$A = \begin{bmatrix} \frac{4}{(1-|z|^2)^2} & 0 \\ 0 & \frac{4}{(1-|z|^2)^2} \end{bmatrix}.$$

in (13.2). Integrating this metric provides the hyperbolic distance $\rho_{\mathbb{D}}$ on the unit disc $\mathbb{D}$,

$$\rho_{\mathbb{D}}(z, w) = \inf_{\gamma} \int_{\gamma} ds_{hyp}, \tag{13.3}$$

where the infimum is over all rectifiable curves γ joining z to w in $\mathbb{D}$. From the definition (13.3), the triangle inequality is clear, and so $\rho_{\mathbb{D}}(z, w)$ is a metric.

Next, symmetry considerations and an integration quickly reveal that

$$\rho_{\mathbb{D}}(0, z) = \log \frac{1+|z|}{1-|z|},$$

while using the transitivity of the group of linear fractional transformations of the disc gives us the more general formula

$$\rho_{\mathbb{D}}(z, w) = \log \frac{|1-\bar{z}w| + |z-w|}{|1-\bar{z}w| - |z-w|}.$$

As a consequence, we can make the useful observation that, for all $z \in \mathbb{D}$,

$$|z| = \tanh \frac{1}{2} \rho_{\mathbb{D}}(0, z). \tag{13.4}$$

Actually the group of all linear fractional transformations of $\mathbb{D}$, described by (13.1), is isomorphic to the group $\mathrm{PSL}(2, \mathbb{R})$, the projective group of 2×2 matrices with real entries and determinant 1. To see this, note that the map

$$\Phi : z \mapsto i\frac{1-z}{1+z}$$

is conformal from $\mathbb{D}$ onto the upper half-space $\mathbb{H} = \{z : \mathrm{Im}\, m(z) > 0\}$. In fact, Φ is an isometry from the hyperbolic metric of the disc to the metric

$$ds = \frac{|dz|}{\mathrm{Im}\, m(z)}, \quad z \in \mathbb{H},$$

giving us another model for the hyperbolic plane. The reflection principle quickly identifies the conformal (and hence isometric) transformations of $\mathbb{H}$ as the linear fractional transformations

$$z \mapsto \frac{az+b}{cz+d}, \quad a, b, c, d \in \mathbb{R}, \quad ad - bc \neq 0.$$

Since we may multiply the numerator and denominator of this fractional transformation by any nonzero constant without affecting the transformation, we may normalize so that $ad - bc = 1$. The reader may wish to verify that this establishes a topological isomorphism between the group of orientation-preserving isometries of $\mathbb{H}$ and $PSL(2, \mathbb{R})$.

Discrete subgroups of linear fractional transformations are called *Fuchsian groups*, and it is a fact that every Riemann surface, other than the sphere, torus, plane, and punctured plane, admits a complete hyperbolic metric. Further, each such surface can be identified with the orbit space (or quotient) of a Fuchsian group. Thus hyperbolic geometry plays a central role in the theory of surfaces. More of the basic facts concerning hyperbolic geometry, hyperbolic trigonometry, discrete groups, and Riemann surfaces can be found in the well-known texts of Beardon [45], Ahlfors-Sario [23], and Farkas-Kra [148], among many others.

Any Riemann surface admits a universal cover $\widetilde{\Sigma}$, a simply connected Riemann surface, for which the deck transformations (the fundamental group) will act as a discrete group Γ of conformal transformations without fixed points [189].

We have already identified the conformal transformations of $\mathbb{D}$ as the linear fractional transformations $z \mapsto e^{i\theta}(z-a)/(1-\overline{a}z)$, $|a| < 1$, $\theta \in \mathbb{R}$. These maps act as hyperbolic isometries, as discussed earlier, and so the quotient inherits a hyperbolic metric in a natural way from the covering projection $\mathbb{D} \to \mathbb{D}/\Gamma = \Sigma$.

Of importance to us later will be the case of the triply punctured sphere $\widehat{\mathbb{C}}\backslash\{0, 1, \infty\} = \mathbb{C}\backslash\{0, 1\}$. There, the universal covering projection $\mathbb{D} \to \mathbb{C}\backslash\{0, 1\}$ is the Weierstrass $\mathcal{P}$-function [18]. Another example is the planar domain $\mathbb{D}\backslash\{0\}$. There the universal covering projection is $\exp\left(\frac{z-1}{z+1}\right)$. If we recall the conformal equivalence between $\mathbb{D}$ and $\mathbb{H}$, one can easily identify the associated group of deck transformations acting on $\mathbb{H}$ as the group $\langle z \mapsto z + k : k \in \mathbb{Z}\rangle$.

In general, for a given domain $\Omega \subset \mathbb{C}$, with $\partial\Omega$ having more than one finite boundary point, there is a universal covering map $\varphi : \mathbb{D} \to \Omega$ given by the uniformization theorem (Theorem 9.3.4). The element of arclength for the hyperbolic metric of Ω is given at each $z \in \Omega$ by

$$ds_{hyp} = \frac{2\left|\psi'(z)\right| |dz|}{1 - |\psi(z)|^2}, \tag{13.5}$$

where ψ is a local inverse of φ defined near z. As all local inverses differ by composition with a linear fractional transformation of $\mathbb{D}$, the identity (13.2) shows that this metric is well defined. As an example, let the domain $\Omega = \mathbb{D}\setminus\{0\}$ and the covering map $\psi(z) = \exp\left(\frac{z-1}{z+1}\right)$. Using (13.5), a calculation shows that

$$ds_{\text{hyp}} = \frac{1}{|z| \log \frac{1}{|z|}}, \qquad z \in \mathbb{D}\setminus\{0\}.$$

There are many other ways to construct these hyperbolic metrics—for instance, through potential-theoretic methods or as the solutions to the PDE $\Delta \log \lambda = \lambda^2$, whence the metric $\lambda(z)|dz|$ has constant negative curvature; see [17].

The metric defined in (13.5) is complete and of constant curvature. If Ω is a planar domain with at least two finite points in its boundary, we call Ω a *hyperbolic domain.* Then we define the hyperbolic metric ρ_Ω by integrating the Riemannian metric in (13.5),

$$\rho_\Omega(z, w) = \inf_\gamma \int_\gamma ds_{hyp},$$

where the infimum is over all rectifiable curves γ joining z to w. This infimum is actually attained by a *hyperbolic geodesic.*

13.2 Bergman Metric

The Bergman Kernel

It is difficult to create an explicit integral formula, with holomorphic reproducing kernel, for holomorphic functions on an arbitrary domain[1] in $\mathbb{C}^n$. Classical studies that perform such constructions tend to concentrate on domains having a great deal of symmetry (see, for instance, [244]). We now examine one of several nonconstructive approaches to this problem. This circle of ideas, due to Bergman [75] and to Szegö [494] (some of the ideas presented here were anticipated by the thesis of Bochner [88]), will later be seen to have profound applications to the boundary regularity of holomorphic mappings.

Bungart [96] and Gleason [176] have shown that any bounded domain in $\mathbb{C}^n$ will have a reproducing kernel for holomorphic functions such that the kernel itself is holomorphic in the free variable. In other words, the formula has the form

$$f(z) = \int_\Omega f(\zeta) K(z, \zeta) dV(\zeta),$$

[1] Here a domain is a connected, open set.

and K is holomorphic in the z variable. Of course Bungart's and Gleason's proofs are highly nonconstructive, and one can say almost nothing about the actual form of the kernel K. The venerable Bochner-Martinelli kernel is easily constructed on any bounded domain with reasonable boundary (just as an application of Stokes's theorem) and the kernel is explicit—just like the Cauchy kernel in one complex variable. Also the kernel is the same for every domain. But the Bochner-Martinelli kernel is definitely *not* holomorphic in the free variable. On the other hand, Henkin [230], Kerzman [270], E. Ramirez [30], and Grauert-Lieb [185] have given very explicit constructions of reproducing kernels on strictly pseudoconvex domains (see the definition below). And their kernels *are* holomorphic in the z variable. This matter is treated in [331, Chap. 10]. In fact this last described result was considered to be quite a dramatic advance. For Henkin, Kerzman, Ramirez, and Grauert-Lieb provided us with a fairly explicit kernel, with an explicit and measurable singularity, that can not only reproduce but also create holomorphic functions. Such a kernel is very much like the Cauchy kernel in one complex variable. Thus, at least on strictly pseudoconvex domains, we can perform many of the activities to which we are accustomed from the function theory of one complex variable. We can get formulas for derivatives of holomorphic functions, we can analyze power series, we can consider an analog of the Cauchy transform, and (perhaps most importantly) we can write down solution operators for the $\overline{\partial}$ problem. People were optimistic that these new integral formulas would give a shot in the arm to the theory of function algebras—that they would now be able to study $H^\infty(\Omega)$ and $A(\Omega)$ on a variety of domains in $\mathbb{C}^n$ (see [165, Chap. II, IV] for the role model in $\mathbb{C}^1$). But this turned out to be too difficult.

The Bergman kernel is a canonical kernel that can be defined on any bounded domain. It has wonderful invariance properties and is a powerful tool for geometry and analysis. But it is difficult to calculate explicitly.

In this section we will see some of the invariance properties of the Bergman kernel. This will lead in later sections to the definition of the Bergman metric (in which all biholomorphic mappings become isometries) and to such other canonical constructions as representative coordinates. The Bergman kernel has certain extremal properties that make it a powerful tool in the theory of partial differential equations (see Bergman and Schiffer [74]). Also the form of the singularity of the Bergman kernel (calculable for some interesting classes of domains) explains many phenomena of the function theory of several complex variables.

Let $\Omega \subseteq \mathbb{C}^n$ be a bounded domain (it is possible, but often tricky, to treat unbounded domains as well). If the domain is smoothly bounded, then we may think of it as specified by a *defining function*:

$$\Omega = \left\{z \in \mathbb{C}^n : \rho(z) < 0\right\} .$$

It is customary to require that $\nabla \rho \neq 0$ on $\partial\Omega$. One can demonstrate the existence of a defining function by using the implicit function theorem. See [336] for the latter and [331] for a detailed consideration of defining functions. Given a domain Ω as described in the last paragraph and a point $P \in \partial\Omega$, we say that w is a *complex tangent vector* at P and write $w \in T_P(\partial\Omega)$ if

$$\sum_{j=1}^{n} \frac{\partial \rho}{\partial z_j}(P) w_j = 0 .$$

The point P is said to be *strongly pseudoconvex* if

$$\sum_{j,k=1}^{n} \frac{\partial^2 \rho}{\partial z_j \partial \overline{z}_k}(P) w_j \overline{w}_k > 0$$

for $0 \neq w \in T_P(\partial\Omega)$. In fact a little elementary analysis shows that we can write the defining property of strong pseudoconvexity as

$$\sum_{j,k=1}^{n} \frac{\partial^2 \rho}{\partial z_j \partial \overline{z}_k}(P) w_j \overline{w}_k \geq C|w|^2$$

and make the estimate uniform when P ranges over a compact, strongly pseudoconvex boundary neighborhood of Ω. Again, the book [331, Chap. 3] has extensive discussion of the notion of strong pseudoconvexity.

Now let us return to the Bergman theory. Let dV denote the Lebesgue volume measure on Ω. Define the *Bergman space*

$$A^2(\Omega) = \left\{ f \text{ holomorphic on } \Omega : \int_\Omega |f(z)|^2 dV(z)^{1/2} \equiv \|f\|_{A^2(\Omega)} < \infty \right\} .$$

Of course we equip the Bergman space with the inner product

$$\langle f, g \rangle = \int_\Omega f(z) \overline{g}(z) dV(z).$$

Lemma 13.2.1 *Let $K \subseteq \Omega \subseteq \mathbb{C}^n$ be compact. There is a constant $C_K > 0$, depending on K and on n, such that*

$$\sup_{z \in K} |f(z)| \leq C_K \|f\|_{A^2(\Omega)} \text{ all } f \in A^2(\Omega) .$$

Proof Since K is compact, there is an $r(K) = r > 0$ so that, for any $z \in K$, $\mathbb{B}(z, r) \subseteq \Omega$. Here $\mathbb{B}(z, r)$ is the usual Euclidean ball with center z and radius r. Therefore for each $z \in K$ and $f \in A^2(\Omega)$, the mean-value property for holomorphic functions implies that

$$\begin{aligned}
|f(z)| &= \left| \frac{1}{V(\mathbb{B}(z,r))} \int_{\mathbb{B}(z,r)} f(t)dV(t) \right| \\
&= \left| \frac{1}{V(\mathbb{B}(z,r))} \int f(t)\chi_{\mathbb{B}(z,r)}(t)dV(t) \right| \\
&\leq (V(\mathbb{B}(z,r)))^{-1/2} \|f\|_{L^2(\mathbb{B}(z,r))} \\
&\leq C(n)r^{-n} \|f\|_{A^2(\Omega)} \\
&\equiv C_K \|f\|_{A^2(\Omega)}.
\end{aligned}$$

□

Lemma 13.2.2 *The space $A^2(\Omega)$ is a Hilbert space with the inner product*

$$\langle f, g \rangle \equiv \int_\Omega f(z)\overline{g(z)}dV(z).$$

Proof Everything is clear except for completeness. Let $\{f_j\} \subseteq A^2$ be a sequence that is Cauchy in norm. Since L^2 is complete, there is an L^2 limit function f. We need to see that f is holomorphic. But Lemma 13.2.1 yields that norm convergence implies normal convergence (i.e., uniform convergence on compact sets). Certainly holomorphic functions are closed under normal limits (just use the Cauchy theory of one complex variable). Therefore f is holomorphic, and $A^2(\Omega)$ is complete. □

Lemma 13.2.3 *For each fixed $z \in \Omega$, the functional*

$$\phi_z : f \mapsto f(z), \quad f \in A^2(\Omega)$$

is a continuous linear functional on $A^2(\Omega)$.

Proof This is immediate from Lemma 13.2.1 if we take K to be the singleton $\{z\}$.

We may now apply the Riesz representation theorem to see that there is an element $K_z \in A^2(\Omega)$ such that the linear functional ϕ_z is represented by inner product with K_z: If $f \in A^2(\Omega)$, then, for all $z \in \Omega$, we have

$$f(z) = \phi_z(f) = \langle f, K_z \rangle .$$

□

Definition 13.2.4 The *Bergman kernel* is the function

$$K(z, \zeta) = K_\Omega(z, \zeta) \equiv \overline{K_z(\zeta)},$$

$z, \zeta \in \Omega$. It has the reproducing property

$$f(z) = \int_\Omega K(z,\zeta) f(\zeta) dV(\zeta), \quad \forall f \in A^2(\Omega).$$

Proposition 13.2.5 *The Bergman kernel* $K(z,\zeta)$ *is conjugate symmetric:* $K(z,\zeta) = \overline{K(\zeta,z)}$.

Proof By its very definition, $\overline{K(\zeta,\cdot)} \in A^2(\Omega)$ for each fixed ζ. Therefore the reproducing property of the Bergman kernel gives

$$\int_\Omega K(z,t)\overline{K(\zeta,t)} dV(t) = \overline{K(\zeta,z)}.$$

On the other hand,

$$\begin{aligned}\int_\Omega K(z,t)\overline{K(\zeta,t)} dV(t) &= \overline{\int K(\zeta,t)\overline{K(z,t)} dV(t)} \\ &= \overline{\overline{K(z,\zeta)}} = K(z,\zeta).\end{aligned}$$

□

Proposition 13.2.6 *The Bergman kernel is uniquely determined by the properties that it is an element of* $A^2(\Omega)$ *in z, is conjugate symmetric, and reproduces* $A^2(\Omega)$.

Proof Let $K'(z,\zeta)$ be another such kernel. Then

$$\begin{aligned}K(z,\zeta) = \overline{K(\zeta,z)} &= \int K'(z,t)\overline{K(\zeta,t)} dV(t) \\ &= \overline{\int K(\zeta,t)\overline{K'(z,t)} dV(t)} \\ &= \overline{\overline{K'(z,\zeta)}} = K'(z,\zeta).\end{aligned}$$

□

Since $L^2(\Omega)$ is a separable Hilbert space, then so is its subspace $A^2(\Omega)$. Thus there is a countable, complete orthonormal basis $\{\phi_j\}_{j=1}^\infty$ for $A^2(\Omega)$.

Proposition 13.2.7 *Let L be a compact subset of* Ω. *Then the series*

$$\sum_{j=1}^\infty \phi_j(z)\overline{\phi_j(\zeta)}$$

sums uniformly on $L \times L$ *to the Bergman kernel* $K(z,\zeta)$.

Proof By the Riesz-Fischer and Riesz representation theorems, we obtain

$$\begin{aligned}
\sup_{z\in L}\left(\sum_{j=1}^{\infty}|\phi_j(z)|^2\right)^{1/2} &= \sup_{z\in L}\left\|\{\phi_j(z)\}_{j=1}^{\infty}\right\|_{\ell^2} \\
&= \sup_{\substack{\|\{a_j\}\|_{\ell^2}=1\\ z\in L}}\left|\sum_{j=1}^{\infty}a_j\phi_j(z)\right| \\
&= \sup_{\substack{\|f\|_{A^2}=1\\ z\in L}}|f(z)| \\
&\leq C_L.
\end{aligned} \tag{13.6}$$

In the last inequality we have used Lemma 13.2.1. Therefore

$$\sum_{j=1}^{\infty}\left|\phi_j(z)\overline{\phi_j(\zeta)}\right| \leq \left(\sum_{j=1}^{\infty}|\phi_j(z)|^2\right)^{1/2}\left(\sum_{j=1}^{\infty}|\phi_j(\zeta)|^2\right)^{1/2},$$

and the convergence is uniform over $z, \zeta \in L$. For fixed $z \in \Omega$, (13.6) shows that $\{\phi_j(z)\}_{j=1}^{\infty} \in \ell^2$. Hence we have that $\sum\phi_j(z)\overline{\phi_j(\zeta)} \in \overline{A^2(\Omega)}$ as a function of ζ. Let the sum of the series be denoted by $K'(z,\zeta)$. Notice that K' is conjugate symmetric by its very definition. Also, for $f \in A^2(\Omega)$, we have

$$\int K'(\cdot,\zeta)f(\zeta)dV(\zeta) = \sum\hat{f}(j)\phi_j(\cdot) = f(\cdot)$$

where convergence is in the Hilbert space topology. (Here $\hat{f}(j)$ is the jth Fourier coefficient of f with respect to the basis $\{\phi_j\}$.) But Hilbert space convergence dominates pointwise convergence (Lemma 13.2.1), so

$$f(z) = \int K'(z,\zeta)f(\zeta)dV(\zeta), \text{ all } f \in A^2(\Omega).$$

Therefore K' is the Bergman kernel. □

Remark 13.2.8 It is worth noting explicitly that the proof of Proposition 13.2.7 shows that

$$\sum\phi_j(z)\overline{\phi_j(\zeta)}$$

equals the Bergman kernel $K(z,\zeta)$ no matter what the choice of complete orthonormal basis $\{\phi_j\}$ for $A^2(\Omega)$. This can be very useful information in practice.

Proposition 13.2.9 *If Ω is a bounded domain in $\mathbb{C}^n$, then the mapping*

$$P : f \mapsto \int_{\Omega} K(\cdot, \zeta) f(\zeta) dV(\zeta)$$

is the Hilbert space orthogonal projection of $L^2(\Omega, dV)$ onto $A^2(\Omega)$. We call P the Bergman projection.

Proof Notice that P is idempotent and self-adjoint and that $A^2(\Omega)$ is precisely the set of elements of L^2 that are fixed by P. □

Definition 13.2.10 Let $\Omega \subseteq \mathbb{C}^n$ be a domain, and let $f : \Omega \to \mathbb{C}^n$ be a holomorphic mapping, that is, $f(z) = (f_1(z), \ldots, f_n(z))$ with $f_1, \ldots, f_n$ holomorphic on Ω. Let $w_j = f_j(z)$, $j = 1, \ldots, n$. Then the holomorphic Jacobian matrix of f is the matrix

$$J_{\mathbb{C}} f = \frac{\partial (w_1, \ldots, w_n)}{\partial (z_1, \ldots, z_n)}.$$

Write $z_j = x_j + iy_j$, $w_k = \xi_k + i\eta_k$, $j, k = 1, \ldots, n$. Then the real Jacobian matrix of f is the matrix

$$J_{\mathbb{R}} f = \frac{\partial (\xi_1, \eta_1, \ldots, \xi_n, \eta_n)}{\partial (x_1, y_1, \ldots, x_n, y_n)}.$$

Proposition 13.2.11 *With the notation as in the definition, we have*

$$\det J_{\mathbb{R}} f = |\det J_{\mathbb{C}} f|^2 ,$$

whenever f is a holomorphic mapping.

Proof We exploit the functoriality of the Jacobian. Let $w = (w_1, \ldots, w_n) = f(z) = (f_1(z), \ldots, f_n(z))$. Write $z_j = x_j + iy_j$, $w_j = \xi_j + i\eta_j$, $j = 1, \ldots, n$. Then, using the fact that f is holomorphic,

$$d\xi_1 \wedge d\eta_1 \wedge \cdots \wedge d\xi_n \wedge d\eta_n = (\det J_{\mathbb{R}} f(x, y))\, dx_1 \wedge dy_1 \wedge \cdots \wedge dx_n \wedge dy_n. \tag{13.7}$$

On the other hand,

$$\begin{aligned} & d\xi_1 \wedge d\eta_1 \wedge \cdots \wedge d\xi_n \wedge d\eta_n \\ &= \frac{1}{(2i)^n} d\overline{w}_1 \wedge dw_1 \wedge \cdots \wedge d\overline{w}_n \wedge dw_n \\ &= \frac{1}{(2i)^n} \overline{(\det J_{\mathbb{C}} f(z))}\, (\det J_{\mathbb{C}} f(z))\, d\overline{z}_1 \wedge dz_1 \wedge \cdots \wedge d\overline{z}_n \wedge dz_n \\ &= |\det J_{\mathbb{C}} f(z)|^2\, dx_1 \wedge dy_1 \wedge \cdots \wedge dx_n \wedge dy_n. \end{aligned} \tag{13.8}$$

Equating (13.7) and (13.8) gives the result. □

Exercise for the Reader Prove Proposition 13.2.11 using only matrix theory (no differential forms). This will give rise to a great appreciation for the theory of differential forms (see Bers [79, Chap. 7] for help).

Now we can prove the holomorphic implicit function theorem:

Theorem 13.2.12 *Let $f_j(w, z), j = 1, \ldots, m$, be holomorphic functions of $(w, z) = ((w_1, \ldots, w_m), (z_1, \ldots, z_n))$ near a point $\left(w^0, z^0\right) \in \mathbb{C}^m \times \mathbb{C}^n$. Assume that*

$$f_j\left(w^0, z^0\right) = 0, \quad j = 1, \ldots, m$$

and that

$$\det\left(\frac{\partial f_j}{\partial w_k}\right)_{j,k=1}^{m} \neq 0 \text{ at } \left(w^0, z^0\right).$$

Then the system of equations

$$f_j(w, z) = 0, j = 1, \ldots, m$$

has a unique holomorphic solution $w(z)$ in a neighborhood of z^0 that satisfies $w\left(z^0\right) = w^0$.

Proof We rewrite the system of equations as

$$\operatorname{Re} f_j(w, z) = 0, \operatorname{Im} f_j(w, z) = 0$$

for the $2m$ real variables $\operatorname{Re} w_k, \operatorname{Im} w_k, k = 1, \ldots, m$. By Proposition 13.2.11, the determinant of the Jacobian over $\mathbb{R}$ of this new system is the modulus squared of the determinant of the Jacobian over $\mathbb{C}$ of the old system. By our hypothesis, this number is nonvanishing at the point $\left(w^0, z^0\right)$. Therefore the classical implicit function theorem (see Rudin [449] or [336]) implies that there exist C^1 functions $w_k(z), k = 1, \ldots, m$, with $w\left(z^0\right) = w^0$ and which solve the system. Our job is to show that these functions are in fact holomorphic. When properly viewed, this is purely a problem of geometric algebra:

Applying exterior differentiation to the equations

$$0 = f_j(w(z), z), \quad j = 1, \ldots, m$$

yields that

$$0 = df_j = \sum_{k=1}^{m} \frac{\partial f_j}{\partial w_k} dw_k + \sum_{k=1}^{m} \frac{\partial f_j}{\partial z_k} dz_k.$$

There are no $d\overline{z}_j$'s and no $d\overline{w}_k$'s because the f_j's are holomorphic.

The result now follows from linear algebra only: The hypothesis on the determinant of the matrix $\left(\partial f_j/\partial w_k\right)$ implies that we can solve for dw_k in terms of dz_j. Therefore w is a holomorphic function of z.

A holomorphic mapping $f : \Omega_1 \to \Omega_2$ of domains $\Omega_1 \subseteq \mathbb{C}^n$, $\Omega_2 \subseteq \mathbb{C}^n$ is said to be biholomorphic if it is one-to-one, onto, and $\det J_{\mathbb{C}} f(z) \neq 0$ for every $z \in \Omega_1$. □

Exercise for the Reader Use Theorem 13.2.12 to prove that a biholomorphic mapping has a holomorphic inverse (hence the name).

Remark 13.2.13 It is true, but not at all obvious, that the nonvanishing of the Jacobian determinant is a superfluous condition in the definition of "biholomorphic mapping"; that is, the nonvanishing of the Jacobian follows from the univalence of the mapping. A proof of this assertion is sketched in [331, Chap. 11, Exercise 37].

In what follows we shall frequently denote the Bergman kernel for a given domain Ω by K_Ω.

Proposition 13.2.14 *Let Ω_1, Ω_2 be domains in $\mathbb{C}^n$. Let $f : \Omega_1 \to \Omega_2$ be biholomorphic. Then*

$$\det J_{\mathbb{C}} f(z) K_{\Omega_2}(f(z), f(\zeta)) \det \overline{J_{\mathbb{C}} f(\zeta)} = K_{\Omega_1}(z, \zeta).$$

Proof Let $\phi \in A^2(\Omega_1)$. Then, by change of variable,

$$\begin{aligned}&\int_{\Omega_1} \det J_{\mathbb{C}} f(z) K_{\Omega_2}(f(z), f(\zeta)) \det \overline{J_{\mathbb{C}} f(\zeta)} \phi(\zeta) dV(\zeta)\\ &= \int_{\Omega_2} \det J_{\mathbb{C}} f(z) K_{\Omega_2}(f(z), \tilde{\zeta}) \det \overline{J_{\mathbb{C}} f\left(f^{-1}(\tilde{\zeta})\right)} \phi\left(f^{-1}(\tilde{\zeta})\right)\\ &\quad \times \det J_{\mathbb{R}} f^{-1}(\tilde{\zeta}) dV(\tilde{\zeta}).\end{aligned}$$

By Proposition 13.2.11 this simplifies to

$$\det J_{\mathbb{C}} f(z) \int_{\Omega_2} K_{\Omega_2}(f(z), \tilde{\zeta}) \left\{ \left(\det J_{\mathbb{C}} f\left(f^{-1}(\tilde{\zeta})\right)\right)^{-1} \phi\left(f^{-1}(\tilde{\zeta})\right) \right\} dV(\tilde{\zeta}).$$

By change of variables, the expression in braces { } is an element of $A^2(\Omega_2)$. So the reproducing property of K_{Ω_2} applies, and the last line equals

$$= \det J_{\mathbb{C}} f(z) \left(\det J_{\mathbb{C}} f(z)\right)^{-1} \phi\left(f^{-1}(f(z))\right) = \phi(z).$$

By the uniqueness of the Bergman kernel, the proposition follows. □

Proposition 13.2.15 *For $z \in \Omega \subset\subset \mathbb{C}^n$, it holds that $K_\Omega(z, z) > 0$.*

Proof Now

$$K_{\Omega}(z, z) = \sum_{j=1}^{\infty} |\phi_j(z)|^2 \geq 0.$$

If in fact $K(z, z) = 0$ for some z, then $\phi_j(z) = 0$ for all j; hence, $f(z) = 0$ for every $f \in A^2(\Omega)$. This is absurd. □

Definition 13.2.16 For any bounded domain $\Omega \subseteq \mathbb{C}^n$, we define a Hermitian metric on Ω by

$$g_{ij}(z) = \frac{\partial^2}{\partial z_i \partial \overline{z}_j} \log K(z, z), \quad z \in \Omega.$$

This means that the square of the length of a tangent vector $\xi = (\xi_1, \dots, \xi_n)$ at a point $z \in \Omega$ is given by

$$|\xi|^2_{\mathbb{B},z} = \sum_{i,j} g_{ij}(z)\xi_i\overline{\xi}_j.$$

The metric that we have defined is called the *Bergman metric*.

In a Hermitian metric $\{g_{ij}\}$, the length of a C^1 curve $\gamma : [0, 1] \to \Omega$ is given by

$$\ell(\gamma) = \int_0^1 \left(\sum_{i,j} g_{i,j}(\gamma(t))\gamma_i'(t)\overline{\gamma_j'(t)} \right)^{1/2} dt.$$

If P, Q are points of Ω, then their distance $d_{\Omega}(P, Q)$ in the metric is defined to be the infimum of the lengths of all piecewise C^1 curves connecting the two points.

Remark 13.2.17 It is not a priori obvious that the Bergman metric for a bounded domain Ω is given by a positive definite matrix at each point. We now outline a proof of this fact.

First we generate an orthonormal basis for the Bergman space. Fix $z_0 \in \Omega$. Let ϕ_0 be the (unique) element of A^2 with $\phi_0(z_0)$ real, $\|\phi_0\| = 1$, and $\phi_0(z_0)$ maximal. (Why does such a ϕ_0 exist?) Let ϕ_1 be the (unique) element of A^2 with $\phi_1(z_0) = 0$, $(\partial\phi_1/\partial z_1)(z_0)$ real, $\|\phi_1\| = 1$, and $(\partial\phi_1/\partial z_1)(z_0)$ maximal. (Why does such a ϕ_1 exist?) Now ϕ_1 is orthogonal to ϕ_0, else ϕ_1 has nonzero projection on ϕ_0, leading to a contradiction. Continue this process to create an orthogonal system on Ω. Use Taylor series to see that it is complete. This circle of ideas comes from the elegant paper Kobayashi [291].

Exercise for the Reader Now let $\Omega \subseteq \mathbb{C}^n$ be a bounded domain, and let (g_{ij}) be its Bergman metric. Use the ideas in the last paragraph to prove that the matrix $(g_{ij}(z))$ is positive definite, each $z \in \Omega$. [Hint: The crucial fact is that, for each $z \in \Omega$ and each j, there is an element $f \in A^2(\Omega)$ such that $\partial f/\partial z_j(z) \neq 0$.]

Proposition 13.2.18 *Let $\Omega_1, \Omega_2 \subseteq \mathbb{C}^n$ be domains, and let $f : \Omega_1 \to \Omega_2$ be a biholomorphic mapping. Then f induces an isometry of Bergman metrics:*

$$|\xi|_{\mathbb{B},z} = |(J_{\mathbb{C}}f)\xi|_{\mathbb{B},f(z)}$$

for all $z \in \Omega_1, \xi \in \mathbb{C}^n$. Equivalently, f induces an isometry of Bergman distances in the sense that

$$d_{\Omega_2}(f(P), f(Q)) = d_{\Omega_1}(P, Q).$$

Proof This is a formal exercise, but we include it for completeness: From the definitions, it suffices to check that

$$\sum g^{\Omega_2}_{i,j}(f(z))\,(J_{\mathbb{C}}f(z)w)_i\left(\overline{J_{\mathbb{C}}f(z)w}\right)_j = \sum_{i,j} g^{\Omega_1}_{ij}(z)w_i\overline{w}_j \tag{13.9}$$

for all $z \in \Omega$, $w = (w_1, \ldots, w_n) \in \mathbb{C}^n$. But, by Proposition 13.2.14,

$$\begin{aligned} g^{\Omega_1}_{ij}(z) &= \frac{\partial^2}{\partial z_i \overline{z}_j} \log K_{\Omega_1}(z, z) \\ &= \frac{\partial^2}{\partial z_i \overline{z}_j} \log\left\{|\det J_{\mathbb{C}}f(z)|^2 K_{\Omega_2}(f(z), f(z))\right\} \qquad (13.10) \\ &= \frac{\partial^2}{\partial z_i \overline{z}_j} \log K_{\Omega_2}(f(z), f(z)) \end{aligned}$$

since $\log|\det J_{\mathbb{C}}f(z)|^2$ is locally

$$\log(\det J_{\mathbb{C}}f) + \log\left(\overline{\det J_{\mathbb{C}}f}\right) + C$$

and hence is annihilated by the mixed second derivative. But line (13.10) is nothing other than

$$\sum_{\ell,m} g^{\Omega_2}_{\ell,m}(f(z))\frac{\partial f_\ell(z)}{\partial z_i}\frac{\partial \overline{f_m(z)}}{\partial \overline{z}_j},$$

and (13.9) follows. □

Proposition 13.2.19 *Let $\Omega \subset\subset \mathbb{C}^n$ be a domain. Let $z \in \Omega$. Then*

$$K(z, z) = \sup_{f \in A^2(\Omega)} \frac{|f(z)|^2}{\|f\|_{A^2}^2} = \sup_{\|f\|_{A^2(\Omega)}=1} |f(z)|^2.$$

Proof Now

$$\begin{aligned} K(z, z) &= \sum |\phi_j(z)|^2 \\ &= \left(\sup_{\|\{a_j\}\|_{\ell^2=1}} \left| \sum \phi_j(z) a_j \right| \right)^2 \\ &= \sup_{\|f\|_{A^2=1}} |f(z)|^2, \end{aligned}$$

by the Riesz-Fischer theorem,

$$= \sup_{f \in A^2} \frac{|f(z)|^2}{\|f\|_{A^2}^2}.$$

□

We shall use this proposition in a moment. Meanwhile, we should like to briefly mention some open problems connected with the Bergman kernel:

The Lu Qi-Keng Conjecture

We have already noticed that $K_\Omega(z, z) > 0$, all $z \in \Omega$, any bounded Ω. It is reasonable to ask whether $K_\Omega(z, \zeta)$ is ever equal to zero. In fact various geometric constructions connected with the Bergman metric and associated biholomorphic invariants (which involve division by K) make it particularly desirable that K be nonvanishing.

If $\Omega = \mathbb{D}$, the unit disc, then explicit calculation (which we perform below) shows that

$$K(z, \zeta) = \frac{1}{\pi} \frac{1}{(1 - z\bar{\zeta})^2};$$

hence, $K(z, \zeta)$ is nonvanishing on $\mathbb{D} \times \mathbb{D}$. Proposition 13.2.14 and the Riemann mapping theorem then show that the Bergman kernel for any proper simply connected subdomain of $\mathbb{C}$ is nonvanishing.

The Bergman kernel for the annulus was studied in Skwarczynski [476] and was seen to vanish at some points. It is shown in Suita and Yamada [493] that if $\Omega \subseteq \mathbb{C}$ is a multiply connected domain with smooth boundary, then K_Ω must vanish—this is proved by an analysis of differentials on the Riemann surface consisting of the double of Ω. By using the easy fact that the Bergman kernel for a product domain is the product of the Bergman kernels (exercise), we may conclude that any domain in

$\mathbb{C}^2$ of the form $A \times \Omega$, where A is multiply connected, has a Bergman kernel with zeroes. The Lu Qi-Keng conjecture can be formulated as follows:

Conjecture A topologically trivial domain in $\mathbb{C}^n$ has nonvanishing Bergman kernel.

It is known (Greene and Krantz [192, 194]) that a domain that is C^∞ sufficiently close to the ball in $\mathbb{C}^n$ has nonvanishing Bergman kernel. Also, if a domain Ω has Bergman kernel that is bounded from zero (and satisfies a modest geometric condition), then all "nearby" domains have Bergman kernel that is bounded from zero. Thus it came as a bit of a surprise when in Boas [84] and [85], it was shown that there exist topologically trivial domains—even ones with real analytic boundary and satisfying all reasonable additional geometric conditions—for which the Bergman kernel has zeroes. See also Wiegerinck [528], where interesting ideas contributing to the solution of this problem first arose. In Sect. 13.2.2 we treat the results of Boas.

Exercise for the Reader The set of smoothly bounded domains for which the Lu Qi-Keng conjecture is true is closed in the Hausdorff topology on domains.

13.2.1 Bergman Representative Coordinates

The theory of the Bergman kernel gives rise to many important geometric invariants. Among these are the not-very-well-known *Bergman representative coordinates*. This is a local coordinate system in which a biholomorphic mapping is realized as a linear mapping. Such a result, while initially quite startling, is in fact completely analogous to the result in Riemannian geometry regarding geodesic normal coordinates. But geodesic normal coordinates are almost never holomorphic—unless the Kähler metric is flat. By contrast, the Bergman representative coordinates are *always* holomorphic. Our presentation in this section is inspired by [190].

The Bergman representative coordinates are of considerable intrinsic and theoretical interest. But they are also useful in understanding the boundary behavior of biholomorphic mappings. And Greene-Krantz made good use of them in proving their noted semicontinuity theorem (see [333, Sect. 5.2]).

The Bergman representative coordinates are also essential in the proof of Lu Qi-Keng's theorem on bounded domains with the Bergman metrics of constant holomorphic sectional curvature. This result will be stated and proved in the present section.

Now let Ω be a bounded domain in $\mathbb{C}^n$, and let q be a point of Ω. The Bergman kernel $K_\Omega(q, q)$ on the diagonal is of course real and positive so that there is a neighborhood U of q such that, for all z, w in U, $K_\Omega(z, w) \neq 0$. Then for all z, w in U, we define

$$b_j(z) = b_{j,q}(z) = \frac{\partial}{\partial \overline{w}_j} \log \frac{K(z, w)}{K(w, w)}\bigg|_{w=q} .$$

Note that these coordinates are well defined, independent of the choice of logarithmic branch. Each representative coordinate $b_j(z)$ is clearly a holomorphic function of z. The mapping

$$z \longmapsto (b_1(z), \ldots, b_n(z)) \in \mathbb{C}^n$$

is defined and holomorphic in a neighborhood of the point q (a neighborhood on which the kernel does not vanish). Note also that $(b_1(q), \ldots, b_n(q)) = (0, \ldots, 0)$. We are hoping to use these functions as holomorphic local coordinates in a neighborhood of q. By the holomorphic inverse function theorem, these functions give local coordinates if the holomorphic Jacobian

$$\det \left(\frac{\partial b_j}{\partial z_k} \right)_{j,k=1,\ldots,n}$$

is nonzero at q. But in fact the nonvanishing of this determinant at q is an immediate consequence of the fact that the Bergman metric is positive definite. To see this relationship, observe that

$$\begin{aligned} \left. \frac{\partial b_j}{\partial z_k} \right|_{z=q} &= \left. \frac{\partial}{\partial z_k} \left(\frac{\partial}{\partial \overline{w}_j} \log K(z, w) \right) \right|_{z=w=q} \\ &= \left. \frac{\partial^2}{\partial z_k \partial \overline{z}_j} \log K(z, z) \right|_{z=q} . \end{aligned}$$

This last term is of course the Hermitian inner product $\left. \left\langle \frac{\partial}{\partial z_k}, \frac{\partial}{\partial z_j} \right\rangle \right|_q$ with respect to the Bergman metric. Thus the expression

$$\left. \det \left(\frac{\partial b_j}{\partial z_k} \right) \right|_q$$

is the determinant of the inner product matrix of a positive definite Hermitian inner product. Hence this determinant is positive.

The utility of the new coordinates in studying biholomorphic mappings comes from the following:

Lemma 13.2.20 *Let Ω_1 and Ω_2 be two bounded domains in $\mathbb{C}^n$ with $q^1 \in \Omega_1$ and $q^2 \in \Omega_2$ fixed points. Denote by $b_1^1, \ldots, b_n^1$ the Bergman coordinates as defined near q^1 in Ω_1 and $b_1^2, \ldots, b_n^2$ the Bergman coordinates defined in the same way near q^2 in Ω_2. Suppose that there is a biholomorphic mapping $F : \Omega_1 \to \Omega_2$ with $F(q^1) = q^2$. Then the function defined near $0 \in \mathbb{C}^n$ by b^1 coordinate $\alpha \longmapsto b^2$ coordinate of the image of the α-point under F is a $\mathbb{C}$-linear transformation.*

In short, we say that the biholomorphic mapping F is linear when expressed in the Bergman representative coordinates b^j. In point of fact, the linear mapping induced by the introduction of the Bergman representative coordinates is nothing other than the complex Jacobian of the mapping F at the point q^1.

Proof of the Lemma To avoid confusion, we write $(z_1, \ldots, z_n)$ and $(w_1, \ldots, w_n)$ for the $\mathbb{C}^n$-coordinates in Ω_1 and $(Z_1, \ldots, Z_n)$ and $(W_1, \ldots, W_n)$ for the $\mathbb{C}^n$-coordinates in Ω_2. Now observe that, for each $j = 1, \ldots, n$,

$$\frac{\partial}{\partial \overline{w}_j} \log \frac{K_{\Omega_2}(F(z), F(w))}{K_{\Omega_2}(F(w), F(w))} = \frac{\partial}{\partial \overline{w}_j} \log \frac{K_{\Omega_1}(z, w)}{K_{\Omega_1}(w, w)}.$$

The reason for this identity is

$$\frac{K_{\Omega_2}(F(z), F(w))}{K_{\Omega_2}(F(w), F(w))} = \frac{K_{\Omega_1}(z, w)}{K_{\Omega_1}(w, w)} \times (\text{ a holomorphic function of } z)$$
$$\times (\text{ a holomorphic function of } w).$$

This last expression follows from the transformation law. Thus we obtain (from the complex chain rule) that

$$\begin{aligned} b_j^1(z) &\stackrel{\text{def}}{=} \frac{\partial}{\partial \overline{w}_j} \log \frac{K_{\Omega_1}(z, w)}{K_{\Omega_1}(w, w)}\bigg|_{w=q_1} \\ &= \frac{\partial}{\partial \overline{w}_j} \left[\log K_{\Omega_2}(F(z), F(w)) - \log K_{\Omega_2}(F(w), F(w))\right]\bigg|_{w=q_1} \\ &= \sum_k \left[\frac{\partial \overline{F}^k}{\partial \overline{w}_j} \cdot \frac{\partial}{\partial \overline{W}_k} \log \frac{K_{\Omega_2}(F(z), W)}{K_{\Omega_2}(W, W)}\right]\bigg|_{W=F(q_1)}, \end{aligned}$$

where F^k is the kth coordinate of $F\ (w_1, \ldots, w_k)$. But this last expression is exactly

$$\sum_k \frac{\partial \overline{F}^k}{\partial \overline{w}_j}\bigg|_{w=q_1} \cdot b_k^2(F(z)).$$

Hence

$$b_j^1(z) = \sum_k \frac{\partial \overline{F}^k}{\partial \overline{w}_j}\bigg|_{w=q_1} \cdot b_k^2(F(z)).$$

Since the Jacobian matrix $\left(\partial F^k/\partial w_j\right)$ of F is invertible at q, it follows that the $b_k^2(F)$ are linear functions of the b_j^1 coordinates. □

The reader may wish to try calculating the Bergman representative coordinates on the unit ball. You may then use these to rediscover the Möbius transformations on the ball.

Note that the whole concept of representative coordinates extends essentially automatically to complex manifolds for which the Bergman metric construction for $(n, 0)$ forms already discussed above yields a positive definite metric. The construction can still be done locally, using general local holomorphic coordinates, and it remains true that the Bergman coordinates linearize holomorphic mappings.

13.2.2 The Lu Qi-Keng Conjecture

In the papers [84, 85], Harold Boas gave an example of a strictly pseudoconvex, topologically trivial domain on which the Bergman kernel has zeroes (off the diagonal, of course). This is a counterexample to an old question of Lu Qi-Keng. We provide the details here.

It should first be noted that, in one complex dimension, any topologically trivial domain (except for the entire complex plane itself) is conformally equivalent to the unit disc. Since the kernel for the disc certainly never vanishes, we may conclude that the kernel for the topologically trivial domain also does not vanish. Such an analysis is not valid in higher dimensions; in that context there is no Riemann mapping theorem.

We note that in [192] it is proved that, in the class of smoothly bounded strictly pseudoconvex domains, the collection of domains for which the Bergman kernel function is bounded away from zero is open. Also the collection of domains for which the Bergman kernel has no zeros is closed. Boas's example shows that the second collection of domains is strictly larger than the first.

It is clear that the Bergman kernel for the ball $\mathbb{B} \subseteq \mathbb{C}^n$,

$$K_{\mathbb{B}}(z, \zeta) = c_n \cdot \frac{1}{(1 - z \cdot \overline{\zeta})^{n+1}},$$

has no zeros. This is a useful feature, as many geometric constructions (see [179]) entail division by the Bergman kernel. For instance, construction of Bergman representative coordinates entails division by the Bergman kernel. Thus in 1966, Lu Qi-Keng conjectured that the Bergman kernel of any bounded domain does not vanish.

It was shown, by direct calculation, by Skwarczyński [476] and Rosenthal [442] that the Lu Qi-Keng conjecture fails on the annulus in the plane. By contrast, it is clear from conformal mapping, and the usual transformation formula for the Bergman kernel, that the Bergman kernel of any simply connected domain in the plane will be nonvanishing.

The operative form of the Lu Qi-King conjecture then became: The Bergman kernel of any topologically trivial domain in $\mathbb{C}^n$ has no zeros. Greene and Krantz [192,

194] proved that the Bergman kernel of a strictly pseudoconvex domain deforms stably under perturbation of the domain. Hence, if $\Omega_0 \subseteq \mathbb{C}^n$ is a domain on which the Lu Qi-Keng conjecture is true, then the conjecture will remain true on "nearby" domains, where the sense of nearby must be interpreted in a suitable topology on domains. Hurwitz's theorem also shows that the limit of a sequence of domains (again, in a suitable topology) on which the Lu Qi-Keng conjecture holds will also have the Lu Qi-Keng property.

Thus it was a bit of a surprise when Boas [84] proved that the Lu Qi-Keng conjecture is false. We provide the details here. Indeed, consider the unbounded, logarithmically convex, complete Reinhardt domain

$$\Omega = \left\{(z, w) \in \mathbb{C}^2 : |w| < (1 + |z|)^{-1}\right\}.$$

Of course Ω is a complete circular domain, so any holomorphic function on Ω will have a globally convergent power series. Which holomorphic monomials are square integrable on Ω? Consider the calculation

$$\begin{aligned}\iint_\Omega \left|z^j w^k\right|^2 \, dV(z, w) &= \int_z \int_0^{2\pi} \int_0^{(1+|z|)^{-1}} |z|^{2j} r^{2k} \cdot r \, dr \, d\theta dA(z) \\ &= 2\pi \int_z \frac{r^{2k}}{2k+2}\bigg|_0^{(1+|z|)^{-1}} \cdot |z|^{2j} \, dA(z) \\ &= 2\pi \int_z \frac{(1+|z|)^{-2k-2}}{2k+2} \cdot |z|^{2j} \, dA(z).\end{aligned}$$

Now it is plain that the integrand is integrable near the origin. As $|z| \to +\infty$, the size of the integrand is $|z|^{2j-2k-2}$. In order for this to be integrable at infinity, we must have $k > j$ (just use polar coordinates). But then we see that any holomorphic, square-integrable function on Ω will have power series expansion, in terms of monomials $z^j w^k$, with k at least 1. Hence, the function will vanish at the origin.

The domain that we have just constructed fails the Lu Qi-Keng property, but it is unbounded. To find a bounded example, let $\Omega_R = \Omega \cap \mathbb{B}(0, R)$. Ramadanov's theorem (see also [332]) now tells us that the Bergman kernel K_{Ω_R} for Ω_R converges uniformly on compact sets to the Bergman kernel K_Ω for Ω. By Hurwitz's theorem, we may conclude that when R is large enough, the kernels K_{Ω_R} will vanish.

The domain Ω_R is a bounded, topologically trivial domain on which the Bergman kernel has zeros. To obtain a smoothly bounded example, we simply exhaust Ω_R by smooth, strictly logarithmically convex complete Reinhardt domains. See [331, Chap. 1] for the details of this process. Then Hurwitz's theorem and Ramadanov's theorem give the result.

In the paper [85], Boas shows that the Lu Qi-Keng conjecture fails "generically" in the following precise sense. Let $\mathcal{H}$ denote the Hausdorff distance on domains. To describe this idea, first note that if $S \subseteq \mathbb{R}^N$ and $x \in \mathbb{R}^N$, then

$$\operatorname{dist}(x, S) = \inf_{s \in S} |s - x|.$$

Now, if Ω_1, Ω_2 are domains in $\mathbb{C}^n$, then set

$$\mathcal{H}(\Omega_1, \Omega_2) = \max \left\{ \sup_{z^1 \in \Omega_1} \operatorname{dist}\left(z^1, \Omega_2\right), \sup_{z^2 \in \Omega_2} \operatorname{dist}\left(z^2, \Omega_1\right) \right\}.$$

Following the notation and language of Boas, we now set

$$\rho_1(\Omega_1, \Omega_2) = \mathcal{H}\left(\overline{\Omega}_1, \overline{\Omega}_2\right) + \mathcal{H}(\partial \Omega_1, \partial \Omega).$$

The main result in [85] is as follows:

Theorem 13.2.21 *The domains in $\mathbb{C}^n$ having the Lu Qi-Keng property are nowhere dense in the collection of all domains with any of the following secondary properties: (a) bounded, pseudoconvex and (b) bounded, strictly pseudoconvex.*

The proof of this result would take us far afield, and we omit it. The details may be found in [85].

13.2.3 The Lu Qi-Keng Theorem

Next we would like to present an application of Bergman representative coordinates to the proof of a remarkable theorem of Lu Qi-Keng on domains with a Bergman metric of constant holomorphic sectional curvature. Our presentation here owes a debt to [190].

Theorem 13.2.22 (Lu Qi-Keng) *Let Ω be a bounded domain in $\mathbb{C}^n$. Assume that the Bergman metric is complete and has constant holomorphic sectional curvature. Then Ω is biholomorphic to the unit ball.*

Notice that this result is certainly specific to the Bergman metric. For example, the annulus $\{\zeta \in \mathbb{C} : 1 < |\zeta| < R\}$, $R > 1$, admits a complete metric of constant (holomorphic sectional) curvature. But it is not even homeomorphic to the unit disc, much less biholomorphic to it.

Proof of Theorem 13.2.22

Background Information
If the curvature c were positive, then Ω would be a complete Riemannian manifold with all sectional curvatures greater than or equal to $c/4 > 0$. [This is because of the formula for Riemannian sectional curvature in case the holomorphic sectional curvature is constant.] Hence, Ω would be compact by standard Riemannian

geometry. [Myers's theorem: A complete Riemannian manifold with sectional curvature everywhere $\geq \epsilon > 0$ has diameter $\leq \pi/\sqrt{\epsilon}$ and is hence compact [410].]

If instead the curvature c were zero, then the universal cover of Ω would be a complete, simply connected Kähler manifold of sectional curvature 0 and hence would be biholomorphically isometric to $\mathbb{C}^n$. But then, since Ω is bounded, the covering map into Ω would be constant by Liouville's theorem. This would contradict surjectivity of the covering map.

The only remaining case is that the curvature c is negative. If g_Ω is the Bergman metric of Ω (with constant negative holomorphic sectional curvature c), then the metric

$$g \equiv -\frac{c(n+1)}{4} g_\Omega$$

has constant (negative) holomorphic sectional curvature $-4/(n+1)$. [Here g_Ω is, as usual, the Bergman metric.] Thus the simply connected covering space $\widehat{\Omega}$ of Ω with the pullback $\widehat{g}$ of the metric g is a complete, simply connected Kähler manifold with constant holomorphic sectional curvature $-4/(n+1)$. By standard Kähler geometry (cf. [299]), $(\widehat{\Omega}, \widehat{g})$ is biholomorphically isometric to $\mathbb{B}^n$ with its Bergman metric. Thus we obtain a holomorphic covering map $F : \mathbb{B}^n \to \Omega$ which is locally isometric for the Bergman metric on $\mathbb{B}^n$ and g on Ω, respectively.

To prove the theorem, we need only show that F is in fact injective.

The Heart of the Proof

For this let $q = F(0)$. Since F is a covering map, it is locally invertible. Namely, there exist an open neighborhood U of q and a neighborhood V of 0 such that $F|_V : V \to U$ is a biholomorphism. Denote by H_0 the inverse of $F|_V$. With $z, w \in U$, let

$$K_0(z, w) \equiv K_{\mathbb{B}^n}(H_0(z), H_0(w)) .$$

Then

$$\frac{\partial^2}{\partial z_j \partial \overline{z}_k} \log K_0(z, z) = g_{j\overline{k}} = \lambda g_{\Omega j\overline{k}}$$

by the condition on F above, where $\lambda = -\frac{c(n+1)}{4}$. This implies that

$$\frac{\partial^2}{\partial z_j \partial \overline{z}_k} \log K_0(z, z) - \frac{\partial^2}{\partial z_j \partial \overline{z}_k} \lambda \log K_\Omega(z, z) = 0$$

for every $z \in U$ and furthermore that

$$\log K_0(z, w) - \lambda \log K_\Omega(z, w) = \varphi(z) + \overline{\varphi(w)}$$

for every $z, w \in U$, for some holomorphic function $\varphi : U \to \mathbb{C}$. For this one may need to replace U by a smaller, simply connected neighborhood. Consequently one obtains

$$\frac{\partial}{\partial \overline{w}_j} \log \frac{K_0(z, w)}{K_0(w, w)} - \frac{\partial}{\partial \overline{w}_j} \lambda \log \frac{K_\Omega(z, w)}{K_\Omega(w, w)} = 0$$

for every $z, w \in U$. This gives rise to a direct computation with the Bergman representative coordinate systems $b^1 : V \to \mathbb{C}^n$ and $b^2 : U \to \mathbb{C}^n$. One obtains that

$$H_0(\zeta) = (F|_V)^{-1} = \left(b^1\right)^{-1} \circ A \circ b^2(\zeta) \tag{13.11}$$

for every $\zeta \in U$. Here A is the linear map represented by the matrix with the (j, k)th entry

$$\lambda \overline{\left.\frac{\partial F_k}{\partial z_j}\right|_0}.$$

Now look at the expressions in (13.11). The map b^1 is in fact a constant multiple of the Euclidean coordinate system. Therefore it extends to all of $\mathbb{C}^n$ holomorphically. So does the linear map A. The map $\zeta \to b^2(\zeta)$ extends to a holomorphic mapping of $\Omega \backslash Z_q$, where

$$Z_q = \{\zeta \in \Omega \mid K_\Omega(\zeta, q) = 0\}.$$

Since $K_\Omega(\cdot, q)$ is a holomorphic function on Ω with $K_\Omega(q, q) \neq 0$, the set Z_q is an analytic variety whose complex codimension in Ω is 1. Hence, $\Omega \backslash Z_q$ is a connected, dense, and open subset of Ω. Therefore using the expression of H_0 in (13.11), the map H_0 extends to a holomorphic mapping of $\Omega \backslash Z_q$ into $\mathbb{C}^n$. Let H denote this extension. Now, let $X \equiv F^{-1}\left(Z_q\right)$. Then one immediately sees that

$$X = \left\{z \in \mathbb{B}^n \mid K_\Omega(F(z), q) = 0\right\}.$$

Since $K_\Omega(F(0), q) = K_\Omega(q, q) \neq 0$, we see that X is again a complex analytic subvariety of $\mathbb{B}^n$ with complex codimension 1. Thus $\mathbb{B}^n \backslash X$ is a connected, dense, and open subset of $\mathbb{B}^n$. Furthermore, $H \circ F : \mathbb{B}^n \backslash X \to \mathbb{C}^n$ is holomorphic with $H \circ F(z) = z$ for every $z \in V$, as $H = H_0$ on V. This means that $H \circ F(z) = z$ for every $z \in \mathbb{B}^n \backslash X$. Now, for every $\zeta \in \Omega \backslash Z_q$, choose $x \in \mathbb{B}^n$ such that $F(x) = \zeta$. Then

$$H(\zeta) = H(F(x)) = x.$$

This implies that $H\left(\Omega\backslash Z_q\right) \subset \mathbb{B}^n$. We see that H is holomorphic on $\Omega\backslash Z_q$. The removable singularity theorem for bounded holomorphic maps (the Riemann removable singularities theorem) yields that H extends to a holomorphic mapping of Ω into $\mathbb{C}^n$. Since H continues to play the role of left inverse of F, it follows easily that F must be injective. That completes the proof and

$$X_2 = \left\{(z, w) \in \mathbb{C}^2 : |z| < 1/(|w| \log |w|), |w| > e\right\}.$$

Set

$$\Omega = X_1 \cup X_2 \cup \left\{(z, w) \in \mathbb{C}^2 : |z| < 2e, |w| < 2e\right\}.$$

□

13.3 The Kobayashi and Carathéodory Metrics

Definition 13.3.1 Let M be a complex manifold, p a point in $M, andv \in T_p(M)$ a tangent vector. The *Carathéodory metric* is defined as

$$F_M^C(p, v) = \sup_f |f_*(v)|, \tag{13.12}$$

where the least upper bound extends over all holomorphic maps $f : M \to \mathbb{D}$ and $f_* = df$ is the differential of f at p.

This differential can be expressed in local coordinates; by the theorem on the invariance of the differential, $df(v)$ does not depend on the choice of these coordinates. If in local coordinates

$$v = \sum_{k=1}^{n} v_k \frac{\partial}{\partial z_k},$$

then

$$df(v) = \sum_{k=1}^{n} \frac{\partial f}{\partial z_k} v_k;$$

in other words, $df(v) = v(f)$ is the derivation in the direction of v. Let us further remark that in (13.12) we may restrict ourselves to maps such that $f(p) = 0$: If $f(p) = \alpha \neq 0$, then, upon replacing f by the composition $g \circ f$, where $g(\zeta) = (\zeta - \alpha)/(1 - \overline{\alpha}\zeta)$, $|f_*(v)|$ gets multiplied by $\left(1 - |\alpha|^2\right)^{-1} > 1$, that is, it increases.

In the same paper [105] Carathéodory proved that F_M^C satisfies the triangle inequality: For arbitrary $p \in M$ and $u, v \in T_p(M)$,

$$F_M^C(p, u + v) \leqq F_M^C(p, u) + F_M^C(p, v) .$$

It also has the contraction property: For each holomorphic map $f : M \to N$ and arbitrary $p \in M$, $v \in T_p(M)$,

$$F_N^C(f(p), f_*(v)) \leqq F_M^C(p, v) .$$

From this property it follows that F_M^C is invariant under biholomorphic maps. This may be viewed as a multivariate generalization of Schwarz's lemma in invariant formulation, stating that for each holomorphic map $f : \Delta \to \Delta$

$$\frac{|f'(z)|\,|dz|}{1 - |f(z)|^2} \leqq \frac{|dz|}{1 - |z|^2} .$$

Let us also remark that c_M (the integrated distance) is continuous on $M \times M$ and that F_M^C is continuous on the tangent bundle $T(M)$.

Although Carathéodory states in one of his papers that he is not sure how useful his metric will be in the future study of functions of several complex variables, it has found numerous applications both by him and by Cartan.

The Kobayashi/Royden pseudometric, to be described below, is one of many invariant metrics that can be defined on a complex manifold.

Definition 13.3.2 Let M be a complex manifold, p a point in M, and $v \in T_p(M)$ a tangent vector. The *Kobayashi seminorm* is defined as

$$F_M^K(p, v) = \inf_f \left\{ \frac{1}{r} : f(0) = p,\ f'(0) = v \right\},$$

where the greatest lower bound is taken over all holomorphic maps f of discs $D_r = \{\zeta \in \mathbb{C} : |\zeta| < r\}$ into M, with the normalization indicated.

Royden's definition has its classical source in the Schottky-Landau theorem, according to which for any holomorphic map f of the disc $\Delta_r = \{|\zeta| < r\}$ into the Riemann sphere $\widehat{\mathbb{C}}$ with three points a, b, c removed, with $f'(0) \neq 0$, and the radius r is bounded from below by a constant depending only on a, b, c and the values of $f(0)$, $f'(0)$.

In [444] Royden proved that F_M^K, considered on $T_p(M)$, has the contraction property under holomorphic maps: If $f : M \to N$ is a holomorphic map, then for each $v \in T_p(M)$

$$F_N^K(f(p), f_*(v)) \leqq F_M^K(p, v) .$$

A key property of the Carethéodory and Kobayashi/Royden pseudometrics, and one that we shall use decisively in what follows, is that holomorphic mappings are distance nonincreasing in the Carathéodory and Kobayashi/Royden pseudometrics.

See [331, Ch. 11] for details. We sometimes instead use the phrase "distance-decreasing."

The following proposition is an immediate consequence of the definition of F_M^K:

Proposition 13.3.3 *If $f : N \to M$ is a holomorphic map, then*

$$F_M^K\left(f(p), f^*v\right) \leqq F_N^K(p, v).$$

In particular, if $N \subset M$, then

$$F_M(x, \xi) \leqq F_N(x, \xi).$$

If $\langle p, q; m, n\rangle$ is an element of the tangent bundle of $M \times N$ with $\langle p, m\rangle$ and $\langle q, n\rangle$ elements of the tangent bundles of M and N, then

$$F_{M\times N}(p, q; m, n) = \max\left(F_M(p, m), F_N(q, n)\right).$$

Proposition 13.3.4 *The function F_M^K is nonnegative, and we have*

$$F_M(p, \alpha m) = |\alpha| F_M(p, m).$$

If K is a compact set contained in a coordinate polydisc, there is a constant C_K such that

$$F_M(p, m) \leqq C_K |m|$$

for all $p \in K$ where $\|n\| = \max |n_i|$.

Proof The first statement follows from the fact that if $\psi(z) = \phi(\alpha z)$, then $\psi'(0) = \alpha\phi'(0)$, and ψ maps the disc of radius $|\alpha|^{-1}R$ into M. The second statement follows from the fact that for the coordinate polydisc P, we have

$$F_M(p, m) \leqq F_P(p, m) = \max_i \frac{r_i\,|m_i|}{r_i^2 - |m_i|^2} \leqq C_K \|m\|$$

since $|m_i|$ is bounded away from r_i on K. □

We refer to [331, 444, 472] for the study of these metrics.

13.4 Kobayashi Distance

Let $\mathbb{D}$ denote the open unit disc in the complex plane $\mathbb{C}$, and let ϱ be the distance function on $\mathbb{D}$ defined by the Poincaré metric of $\mathbb{D}$.

Let M be a complex manifold. We define a pseudodistance k_M on M as follows. Given two points $p, q \in M$, we choose points $p = p_0, p_1, \ldots, p_{k-1}, p_k = q$ of M, points $a_1, \ldots, a_k, b_1, \ldots, b_k$ of $\mathbb{D}$, and holomorphic mappings $f_1, \ldots, f_k$ of $\mathbb{D}$ into M such that $f_i(a_i) = p_{i-1}$ and $f_i(b_i) = p_i$ for $i = 1, \ldots, k$. For each choice of points and mappings thus made, we consider the number

$$\varrho(a_1, b_1) + \cdots + \varrho(a_k, b_k) .$$

Let $k_M(p, q)$ be the infimum of the numbers obtained in this manner for all possible choices. This quantity is the *Kobayashi distance*.

The definition of k_M is analogous to the definition of the distance induced by a Riemannian metric. We can think of a chain from p to q as a piecewise analytic curve. The length of an analytic curve is determined by the Poincaré metric on $\mathbb{D}$, and the length of a chain is just the sum of the lengths of the analytic pieces. Then $k_M(p, q)$ is just the infimum of the lengths of curves joining p and q. The pseudodistance k_M gives a geometric description of the space $\mathcal{O}(\mathbb{D}, M)$ of holomorphic mappings of $\mathbb{D}$ into M. The larger the distance is, the smaller $\mathcal{O}(\mathbb{D}, M)$ is.

It is an easy matter to verify that $k_M : M \times M \to \mathbf{R}$ is continuous and satisfies the axioms for pseudodistance:

$$k_M(p, q) \geqq 0, \quad k_M(p, q) = k_M(q, p), \quad k_M(p, r) \leq k_M(p, q) + k_M(q, r) .$$

The most important property of k_M is that holomorphic mappings are distance-decreasing. This is an immediate consequence of the definition.

Proposition 13.4.1 *Let M and N be two complex manifolds, and let $f : M \to N$ be a holomorphic mapping. Then*

$$k_M(p, q) \geqq k_N(f(p), f(q)) \quad \textit{for } p, q \in M.$$

Corollary 13.4.2 *Every biholomorphic mapping $f : M \to N$ is an isometry, i.e.,*

$$k_M(p, q) = k_N(f(p), f(q)) \quad \textit{for } p : q \in M.$$

In order to establish regularity properties of F_M^K, we need the following lemma:

Lemma 13.4.3 *Let ϕ be an analytic map of the disc D_R into M with $\phi'(0) \neq 0$. Then for each $r < R$, there is a $\lambda > 0$ and a holomorphic map f of the polydisc $D_r \times D_\lambda \times \cdots \times D_\lambda$ into M such that f is a biholomorphic map on some neighborhood of the origin, and $f|_{D_r \times 0 \times \cdots \times 0} = \phi|_{D_r}$.*

The proof of this lemma in the general case is quite complicated, and Royden [444] considers only the case when $M \subset \mathbb{C}^n$. In this case, choose coordinates in $\mathbb{C}^n$ so that $\phi'(0) = \langle 1, 0, \ldots, 0 \rangle$, and define a map $g : D_R \times \mathbb{C}^{n-1} \to \mathbb{C}^n$ by setting

$$g^1(t_1, \ldots, t_n) = \phi^1(t_1) \text{ and}$$

$$g^k(t_1, \ldots, t_n) = \phi^k(t_1) + t_k, \quad 2 \leqq k \leqq n\,.$$

Then g is holomorphic on $D_R \times \mathbb{C}^{n-1}$ and biholomorphic on a neighborhood of the origin. Moreover, $g \mid D_R \times 0 \times \cdots \times 0 = \phi$. Let $S = g^{-1}[M]$, and set $P_\lambda = D_r \times D_\lambda \times \cdots \times D_\lambda$. Then $\{\overline{P}_\lambda\}$ is a decreasing family of compact sets whose intersection is contained in S. Hence $\overline{P}_\lambda \subset S$ for some λ. Take $f = g \mid P_\lambda$.

The following lemma is an immediate consequence of Lemma 13.4.3 and the definition of F_M^K:

Lemma 13.4.4 *Let $\varepsilon > 0$ and an element $\langle x, \xi \rangle$ of the tangent bundle of M be given. Let τ_1 be that element of the tangent space of $\mathbb{C}^n$ whose coordinates are $(1, 0, \ldots, 0)$. Then there is a holomorphic map f of a polydisc P into M which is biholomorphic on a neighborhood of 0 and for which $f(0) = x$, $f^*\tau_1 = \xi$, and*

$$F_{\mathbb{D}}^K(0, \tau_1) < F_M^K(x, \xi) + \varepsilon.$$

Proposition 13.4.5 *The function F_M^K is upper semicontinuous on the tangent bundle of M.*

Proof Choose $\varepsilon > 0$ and an element $\langle x, \xi \rangle$ of the tangent bundle of M. Let $f : \mathbb{D} \to M$ be the holomorphic mapping satisfying the conditions of Lemma 13.4.4. Since $F_{\mathbb{D}}^K$ is continuous, there is a neighborhood U of $< 0, \tau_1 >$ in $\mathbb{D}$ such that $F_{\mathbb{D}}(t, \tau) < F_{\mathbb{D}}^K(0, \tau_1) + \varepsilon$ for $\langle t, \tau \rangle \in \mathbb{D}$. Since f is biholomorphic on a neighborhood of 0, $\langle f, f^* \rangle$ maps U onto a set containing a neighborhood V of $\langle x, \xi \rangle$. If $\langle y, \eta \rangle \in V$, then $y = f(t), \quad \eta = f^*(\tau)$ for some $\langle t, \tau \rangle \in U$, and we have

$$\begin{aligned} F_M^K(y, \eta) &= F_M^K\left(f(t), f^*(\tau)\right) \\ &\leqq F_D^K(t, \tau) \\ &< F_D^K(0, \tau_I) + \varepsilon \\ &< F_M^K(x, \xi) + 2\varepsilon. \end{aligned}$$

Thus F_M is upper semicontinuous. □

Let γ be a differentiable curve in M, and let $x : [a, b] \to M$ be a parametrization of γ. Since $\langle x, \dot{x} \rangle$ is a continuous map into the tangent bundle of M and F_M^K is upper semicontinuous, the function $F_M^K(x, \dot{x})$ is upper semicontinuous on $[a, b]$. Hence it is bounded and measurable, and we may define the length of γ by

$$l(\gamma) = \int_\gamma F_M^K = \int_a^b F_M^K(x(t), \dot{x}(t))dt.$$

Since F_M^K is homogeneous with respect to ξ, this integral is independent of the parametrization chosen. The following theorem tells us that d_M is the integrated form of F_M^K:

Theorem 13.4.6 *The Kobayashi metric k_M is the integrated form of F_M^K; that is,*

$$k_M(p, q) = \inf \int_\gamma F_M^K,$$

where the infimum is taken over all differentiable curves γ joining p to q.

Proof Define $\overline{d}(p, q)$ to be $\inf \int_\gamma F_M^K$ over all differentiable curves γ joining p to q. From the upper bound on F_M^K given in Proposition 13.3.4, it follows that this is the same as the infimum taken over all piecewise differentiable curves joining p to q. Thus $\overline{d}$ is a pseudometric, and in order to show that $\overline{d} \leqq d_M$, it will suffice to show that $\overline{d}(p, q) \leqq d_M^*(p, q)$. Let ϕ be an analytic map of D_R into M with $\phi(0) = p$, $\phi(1) = q$, and

$$\frac{1}{2} \log \frac{R+1}{R-1} < d_M^*(p, q) + \varepsilon.$$

Let γ be the image of [0, 1] under ϕ. Then

$$\begin{aligned} \overline{d}(p, q) &\leqq \int_\gamma F_M^K \\ &\leqq \int_0^1 F_{D_R}^K = \frac{1}{2} \log \frac{R+1}{R-1} \\ &\leqq d_M^*(p, q) + \varepsilon. \end{aligned}$$

Thus $\overline{d} \leqq d_M$.

To prove the inequality in the opposite direction, let γ be a differentiable curve joining p to q such that $\int_\gamma F_M^K < \overline{d}(p, q) + \varepsilon$, and let $x(t)$ be a parametrization of γ with $0 \leqq t \leqq 1$. Since $F_M^K(x(t), \dot{x}(t))$ is upper semicontinuous on [0, 1], it is the limit of a decreasing sequence of continuous functions on [0, 1]. Thus by the Lebesgue convergence theorem, there is a continuous function h on [0, 1] such that $h(t) > F_M^K(x(t), \dot{x}(t))$ and

$$\int_0^1 h(t) dt < \overline{d}(p, q) + \varepsilon.$$

Since h is continuous, it is Riemann integrable, and there is a $\delta > 0$ such that if $0 = t_0 < t_1 < \cdots < t_k = 1$ is any subdivision of [0, 1] with $t_{i+1} - t_i < \delta$ and $s_1, \ldots, s_k$ are points of [0, 1] with $|s_i - t_i| < \delta$, then

$$\sum_{i=1}^{k} h(\varepsilon_i)(t_i - t_{i-1}) < \overline{d}(p, q) + \varepsilon.$$

For each $s \in [0, 1]$, there is a map f of $P = \Delta_R \times \Delta_\lambda \times \cdots \times \Delta_\lambda$ into M such that $\frac{1}{R} < h(s)$, $f(0) = x(s)$, and $f'(0) = \dot{x}(s)$ and such that f is biholomorphic in a neighborhood of the origin. Thus there is an open interval I_s containing s so that on I_s we have $x = f \circ y$, where y is a differentiable map of I_s into $\mathbb{D}$ with $y(s) = (0, 0, \ldots, 0)$ and $y'(s) = (1, 0, \ldots, 0)$.

Since $y(t) = (t - s, 0, \ldots, 0) + O\left(|t - s|^2\right)$, it follows from the explicit expression for d_P on a polydisc that

$$d_P(y(t), y(u)) \leqq (1 + \varepsilon)\frac{|t - u|}{R} < (1 + \varepsilon)h(s)|t - u|$$

for all t and u in some sufficiently small interval I_s' containing s. We may suppose that the length of each I_s' is less than δ. The Kobayashi metric decreases under holomorphic maps, and so

$$d_M(x(t), x(u)) \leqq (I + \varepsilon)h(\varepsilon)|t - u|$$

for $t, u \in I_s'$.

By the Lebesgue covering lemma [267, p. 154], there is an $\eta > 0$ such that if $t, u \in [0, 1]$ and $|t - u| < \eta$, then there is an s with $t, u \in I_s'$. Let $0 = t_0 < t_1 < \cdots < t_k = 1$ be a subdivision of $[0, 1]$ with $t_i - t_{i-1} < \eta$, and choose s_i so that $t_i, t_{i-1} \in I_{s_i'}'$. Then $|t_i - s_i| < \delta$ and $t_i - t_{i-1} < \delta$ since the length of I_{s_i}' is less than δ. Hence

$$\begin{aligned} d_M(p, q) = d_M(x(0), x(1)) &\leqq \sum_{i=1}^{k} d_M\left(x\left(t_i\right), x\left(t_{i-1}\right)\right) \\ &\leqq (1 + \varepsilon)\sum_{i=1}^{k} h\left(s_i\right)\left(t_i - t_{i-1}\right) \\ &< (1 + \varepsilon)(\overline{d}(p, q) + \varepsilon). \end{aligned}$$

Since ε was arbitrary, we have

$$d_M(p, q) \leqq \overline{d}_M(p, q).$$

□

The pseudodistance k_M may be considered as a generalization of the Poincaré-Bergman metric for a unit disc. We have the following proposition:

Proposition 13.4.7 *For the open unit disc $\mathbb{D}$ in $\mathbb{C}$, $k_{\mathbb{D}}$ coincides with the distance ϱ defined by the Poincaré-Bergman metric.*

Proof By the Schwarz lemma (cf. Theorems 12.1.1 and 12.1.2 of Chap. 12.1), every holomorphic mapping $f : \mathbb{D} \to \mathbb{D}$ is distance-decreasing with respect to ϱ. From the very definition of $k_{\mathbb{D}}$, we have

$$k_D(p, q) \geqq \varrho(p, q) \quad \text{for } p, q \in \mathbb{D}.$$

Considering the identity transformation of $\mathbb{D}$, we obtain the inequality $k_{\mathbb{D}}(p, q) \leqq \varrho(p, q)$. □

Example 13.4.8 We have that $k_{\mathbb{C}} \equiv 0$. To see this, let $f_n : \mathbb{D} \to \mathbb{C}$ be defined by $f_n(z) = nz$. For $|z'| < n$, we have

$$k_{\mathbb{C}}\left(z', 0\right) = k_{\mathbb{C}}\left(f_n\left(\frac{z'}{n}\right), f_n(0)\right) \leq k_{\mathbb{D}}\left(\frac{z'}{n}, 0\right).$$

Since $k_{\mathbb{D}}\left(\frac{z'}{n}, 0\right) \to 0$ as $n \to \infty$, it follows that $d_{\mathbb{C}}\left(z', 0\right) = 0$.

The following proposition says that k_M is the largest pseudodistance on M for which $f : \mathbb{D} \to M$ is distance-decreasing.

Proposition 13.4.9 *Let M be a complex manifold and d' any pseudodistance on M such that*

$$d'(f(a), f(b)) \leqq \varrho(a, b) \quad \textit{for } a, b \in \mathbb{D}$$

for every holomorphic mapping $f : \mathbb{D} \to M$. Then

$$k_M(p, q) \geqq d'(p, q) \quad \textit{for } p, q \in M.$$

Proof Let $p_0, \ldots, p_k, a_1, \ldots, a_k, b_1, \ldots, b_k, f_1, \ldots, f_k$ be as in the definition of k_M. Then

$$\begin{aligned} d'(p, q) &= \sum_{i=1}^{k} d'\left(p_{i-1}, p_i\right) = \sum_{i=1}^{k} d'\left(f_i\left(a_i\right), f_i\left(b_i\right)\right) \\ &\leqq \sum_{i=1}^{k} \varrho\left(a_i, b_i\right). \end{aligned}$$

Hence

$$d'(p, q) \leqq \inf \sum_{i=1}^{k} \varrho\left(a_i, b_i\right) = k_M(p, q).$$

□

Proposition 13.4.10 *Let M and M' be two complex manifolds. Then*

$$k_M(p,q)+k_{M'}\left(p',q'\right) \geqq k_{M\times M'}\left(\left(p,p'\right),\left(q,q'\right)\right)$$
$$\geqq \operatorname{Max}\left[k_M(p,q),k_{M'}\left(p',q'\right)\right]$$
$$\text{for } p,q\in M \quad \text{and} \quad p',q'\in M'.$$

Proof We have

$$k_M(p,q)+k_{M'}\left(p',q'\right) \geqq k_{M\times M'}\left(\left(p,p'\right),\left(q,p'\right)\right)+k_{M\times M'}\left(\left(q,p'\right),\left(q,q'\right)\right)$$
$$\geqq k_{M\times M'}\left(\left(p,p'\right),\left(q,q'\right)\right),$$

where the first inequality follows from the fact that the mappings $f\colon M\to M\times M'$ and $f':M'\to M\times M'$ defined by $f(x)=\left(x,p'\right)$ and $f'\left(x'\right)=\left(q,x'\right)$ are distance-decreasing, and the second inequality is a consequence of the triangle inequality. The inequality $k_{M\times M'}\left(\left(p,p'\right),\left(q,q'\right)\right)\geqq \operatorname{Max}\left[k_M(p,q),k_{M'}\left(p',q'\right)\right]$ follows from the fact that the projections $M\times M'\to M$ and $M\times M'\to M'$ are both distance-decreasing. □

Example 13.4.11 If $\mathbb{D}$ is the open unit disc in $\mathbb{C}$, then

$$k_{\mathbb{D}\times\mathbb{D}}\left(\left(p,p'\right):\left(q,q'\right)\right)=\operatorname{Max}\left[k_{\mathbb{D}}(p,q),k_{\mathbb{D}}\left(p',q'\right)\right] \quad \text{for } p,q,p',q'\in\mathbb{D}.$$

To prove this assertion we may assume (because of the homogeneity of $\mathbb{D}$) that $p=p'=0$ and $k_{\mathbb{D}}(0,q)\geqq k_{\mathbb{D}}\left(0,q'\right)$, i.e., $|q|\geqq\left|q'\right|$. Consider the holomorphic mapping $f:\mathbb{D}\to\mathbb{D}\times\mathbb{D}$ defined by $f(z)=\left(z,\left(q'/q\right)z\right)$. Since f is distance-decreasing, we have

$$k_{\mathbb{D}\times\mathbb{D}}\left((0,0),\left(q,q'\right)\right)=k_{\mathbb{D}\times\mathbb{D}}(f(0),f(q))\leqq d_{\mathbb{D}}(0,q).$$

More generally, if $\mathbb{D}^k$ denotes the k-dimensional polydisc $\mathbb{D}\times\cdots\times\mathbb{D}$, then

$$k_{\mathbb{D}^k}\left(\left(p_1,\ldots,p_k\right),\left(q_1,\ldots,q_k\right)\right)=\operatorname{Max}\left[k_{\mathbb{D}}\left(p_i,q_i\right);i=1,\ldots,k\right].$$

This shows that $k_{\mathbb{D}^k}$ does not coincide with the distance defined by the Bergman metric of $\mathbb{D}^m$ unless $m=1$.

As we see from the definition of the pseudodistance, k_M is defined in a manner similar to the distance function on a Riemannian manifold. It is therefore quite natural to expect the following result:

Proposition 13.4.12 *Let M be a complex manifold and $\widetilde{M}$ a covering manifold of M with covering projection $\pi:\widetilde{M}\to M$. Let $p,q\in M$ and $\widetilde{p},\widetilde{q}\in\widetilde{M}$ such that $\pi(\widetilde{p})=p$ and $\pi(\widetilde{q})=q$. Then*

$$k_M(p,q) = \inf_{\widetilde{q}} k_{\widetilde{M}}(\widetilde{p},\widetilde{q}),$$

where the infimum is taken over all $\widetilde{q} \in \widetilde{M}$ such that $q = \pi(\widetilde{q})$.

Proof Since $\pi : \widetilde{M} \to M$ is distance-decreasing, we have

$$k_M(p,q) \leqq \inf_{\widetilde{q}} k_{\widetilde{M}}(\widetilde{p},\widetilde{q}).$$

Assuming the strict inequality, let

$$k_M(p,q) + \varepsilon < \inf_{\widetilde{q}} k_{\widetilde{M}}(\widetilde{p},\widetilde{q}),$$

where ε is a positive number. By the very definition of k_M, there exist points $a_1, \ldots, a_k, b_1, \ldots, b_k$ of the unit disc $\mathbb{D}$ and holomorphic mappings $f_1, \ldots, f_k$ of $\mathbb{D}$ into M such that

$$p = f_1(a_1), \quad f_1(b_1) = f_2(a_2), \ldots, f_{k-1}(b_{k-1}) = f_k(a_k), \quad f_k(b_k) = q$$

and

$$k_M(p,q) + \varepsilon > \sum_{i=1}^{k} \varrho(a_i, b_i).$$

Then we can lift $f_1, \ldots, f_k$ to holomorphic mappings $\widetilde{f_1}, \ldots, \widetilde{f_k}$ of $\mathbb{D}$ into $\widetilde{M}$ in such a way that

$$\begin{aligned} \widetilde{p} &= \overline{f}_1(a_1) \\ \widetilde{f_i}(b_i) &= \widetilde{f}_{i+1}(a_{i+1}) \quad && \text{for } i = 1, \ldots, k-1, \\ \pi \circ \overline{f}_i &= f_i && \text{for } i = 1, \ldots, k. \end{aligned}$$

If we set $\widetilde{q} = \widetilde{f_k}(b_k)$, then $\pi(\widetilde{q}) = q$ and $d_{\widetilde{M}}(\widetilde{p},\widetilde{q}) \leqq \sum_{i=1}^{k} \varrho(a_i, b_i)$. Hence $d_{\widetilde{M}}(\overline{p},\overline{q}) < k_M(p,q) + \varepsilon$, which contradicts our assumption. □

Remark 13.4.13 It is not clear whether the following is true:

$$k_M(p,q) = \min_{\widetilde{q}} k_{\widetilde{M}}(\widetilde{p},\widetilde{q}),$$

i.e., whether $\inf_{\widetilde{q}} k_{\widetilde{M}}(\widetilde{p},\widetilde{q})$ is really attained by some $\widetilde{q}$.

Example 13.4.14 Let M be a Riemann surface whose universal covering is a unit disc $\mathbb{D}$, e.g., a compact Riemann surface of genus $\geqq 2$. From Propositions 13.4.7

and 13.4.12, we see that k_M is the distance defined by a Kähler metric of M with constant curvature <0.

Example 13.4.15 It can be seen easily from the definition of the pseudodistance k_M that for a complex Euclidean space $\mathbb{C}^n$ the pseudodistance $k_{\mathbb{C}^n}$ is trivial, i.e., $k_{\mathbb{C}^n}(p, q) = 0$ for all $p, q \in \mathbb{C}^n$. More generally, if M is a complex manifold on which a complex Lie group G acts transitively, then k_M is trivial. For each point $p \in M$, let U be a neighborhood of p such that every point $q \in U$ lies on the orbit of a complex one-parameter subgroup of G through p. Hence, for every point $q \in U$, there is a holomorphic mapping $f : \mathbb{C} \to M$ whose image contains both p and q. Since f is distance-decreasing and $k_{\mathbb{C}}$ is trivial, we have $k_M(p, q) = 0$. To prove that k_M is trivial for any pair of points on M, we connect the pair by a chain of small open sets U and apply the triangle inequality. In particular, for a compact homogeneous complex manifold M or for $M = \mathbb{C} \setminus \{0\}$, k_M is trivial. If M is a compact Riemann surface of genus 0 or 1, then k_M is trivial.

Example 13.4.16 Let $M = \{z \in \mathbb{C}^n : r < |z| < R\}$ and $\mathbb{B}(0, R) = \{z \in \mathbb{C}^n; |z| < R\}$. If $n = 1$, the universal covering space of M is the upper half plane or equivalently the unit disc.

From Proposition 13.4.12 it follows that the invariant distance k_M coincides with the distance defined by ds^2_M. For $n > 1$, we do not know what k_M looks like, although we know that k_M is not complete, since M is not holomorphically complete.

13.4.1 Completeness with Respect to the Carathéodory Distance

Let M be a complex manifold. Throughout this section we are interested in the case where the Carathéodory pseudodistance c_M is a distance, i.e., the case where the family of holomorphic mappings $f : M \to \mathbb{D}$ separates the points of M. Since every bounded holomorphic function on M, multiplied by a suitable constant, yields a holomorphic mapping of M into the unit disc $\mathbb{D}$, the necessary and sufficient condition that c_M must be a distance is the following: For any two distinct points p and q of M, there is a bounded holomorphic function f on M such that $f(p) \neq f(q)$. This is very close to assuming that M is a bounded domain in a Stein manifold. In particular, all compact complex manifolds are excluded.

In general, we say that a metric space M is complete if, for each point p of M and each positive number r, the closed ball of radius r around p is a compact subset of M. If M is complete in this sense, then every Cauchy sequence in M converges. The converse is true for a Riemannian metric, but not in general.

Proposition 13.4.17 *If M and M' are complex manifolds with complete Carathéodory distance, then so is $M \times M'$.*

Proof This is immediate from Proposition 13.5.6. □

Proposition 13.4.18 *A closed complex submanifold M' of a complex manifold M with complete Carathéodory distance is also complete with respect to its own Carathéodory distance.*

Proof This is immediate from the fact that the injection $M' \to M$ is distance-decreasing. □

Proposition 13.4.19 *Let M and M_i, $i \in I$, be complex submanifolds of a complex manifold N such that $M = \bigcap_i M_i$. If each M_i is complete with respect to its Carathéodory distance, so is M.*

Proof Since each injection $M \to M_i$ is distance-decreasing, the proposition follows from the following trivial lemma. □

Lemma 13.4.20 *Let M and M_i, $i \in I$, be subsets of a topological space N such that $M = \bigcap_i M_i$. Let $\mathbb{D}$ and d_i be distances on M and M_i such that $d(p, q) \geqq d_i(p, q)$ for $p, q \in M$. If each M_i is complete with respect to d_i, then M is complete with respect to $\mathbb{D}$.*

We shall now give a large class of bounded domains that are complete with respect to their Carathéodory distances. Let G be a domain in $\mathbb{C}^n$ and $f_1, \ldots, f_k$ holomorphic functions defined in G. Let P be a connected component of the open subset of G defined by

$$|f_1(z)| < 1, \ldots, |f_k(z)| < 1.$$

Assume that the closure of P is compact and is contained in G. Then P is called an *analytic polyhedron.*

Theorem 13.4.21 *An analytic polyhedron P is complete with respect to its Carathéodory distance c_P.*

Proof Let F be the set of holomorphic mappings of P into the unit disc $\mathbb{D}$. Let o be a point of P. Given a positive number a, choose a positive number b, $0 < b < 1$, such that

$$\left\{z \in \mathbb{D} : \varrho\left(f_i(o), z\right) \leqq a \text{ for } i = 1, \ldots, k\right\} \subset \{z \in \mathbb{D} : |z| \leqq b\}.$$

Then

$$\begin{aligned}\left\{p \in P : c_P(o, p) \leqq a\right\} &= \{p \in P : \varrho(f(o), f(p)) \leqq a \text{ for } f \in F\} \\ &\subset \left\{p \in P : \varrho\left(f_i(o), f_i(p)\right) \leqq a \text{ for } i = 1, \ldots, k\right\} \\ &\subset \left\{p \in P : |f_i(p)| \leqq b \text{ for } i = 1, \ldots, k\right\}.\end{aligned}$$

Since $\{p \in P : |f_i(p)| \leqq b\}$ for $i = \{1, \ldots, k\}$ is compact, it follows that $\{p \in P : c_P(o, p) \leqq a\}$ is also compact. □

In the definition of an analytic polyhedron, we used only finitely many holomorphic functions $f_1, \ldots, f_k$. If we take an arbitrary family of holomorphic functions $f_\alpha, \alpha \in A$, on G and assume that the set defined by $|f_\alpha(z)| < 1, \alpha \in A$, is open, then we call each connected component of this open subset of G a *generalized analytic polyhedron*. The proof of Theorem 13.4.21 implies the following:

Theorem 13.4.22 *A generalized analytic polyhedron P is complete with respect to its Carathéodory distance c_P.*

Let h_j be real analytic functions on G of the form

$$h_j = \sum_{m=1}^{\infty} |f_{jm}|^2, \quad j = 1, \ldots, k,$$

where each f_{jm} is holomorphic in G. Let S be the set of sequences $\{a_m\}$ of complex numbers such that $\sum_{m=1}^{\infty} |a_m|^2 = 1$. Then

$$\{z \in G; |h_j(z)| < 1\} = \bigcap_{\{a_m\} \in S} \left\{ z \in G; \left| \sum_{m=1}^{\infty} a_m f_{jm}(z) \right| < 1 \right\}.$$

This shows that a connected component of the open subset of G defined by

$$\{z \in G : |h_j(z)| < 1 \text{ for } j = 1, \ldots, k\}$$

is a generalized analytic polyhedron. To such a manifold, Theorem 13.4.22 is therefore applicable.

So far, we have given sufficient conditions for a manifold to be complete with respect to its Carathéodory distance. We shall give now some necessary conditions.

Horstmann [243] has shown that a domain in $\mathbb{C}^n$ which is complete with respect to its Carathéodory distance is a domain of holomorphy. We shall give a generalization of Horstmann's result. Given a family F of holomorphic functions on a complex manifold M and a subset K of M, we set

$$\widehat{K}_F = \{p \in M; |f(p)| \leqq \sup |f(K)| \text{ for all } f \in F\},$$

where $\sup |f(K)|$ denotes the supremum of $|f(q)|$ for $q \in K$. Then $\widehat{K}_F$ is a closed subset of M containing K and is called the *convex hull of K with respect to F*. The convex hull of $\widehat{K}_F$ with respect to F coincides with $\widehat{K}_F$ itself. If $\widehat{K}_F$ is compact for every compact subset K of M, then M is said to be *convex with respect to F*. If $F' \subset F$, then $\widehat{K}_{F'} \supset \widehat{K}_F$. Hence, if M is convex with respect to F' and if $F' \subset F$, then M is convex with respect to F. If M is convex with respect to the family of all holomorphic functions on M, then M is said to be *holomorphically convex*.

Theorem 13.4.23 *Let M be a complex manifold with Carathéodory distance. Fix a point o of M, and let F be the set of bounded holomorphic functions f on M such that $f(o) = 0$. If M is complete with respect to c_M, then M is convex with respect to F and hence is holomorphically convex.*

Proof Let a be a positive number and $\mathfrak{B}$ the closed ball of radius a around $o \in M$, i.e.,

$$\mathfrak{B} = \left\{p \in M; c_M(o, p) \leqq a\right\}.$$

Since M is complete, $\mathfrak{B}$ is compact. Since every compact subset K of M is contained in $\mathfrak{B}$ for a sufficiently large a, it suffices to prove that $\widehat{\mathfrak{B}}_F$ is compact. We shall actually show that $\widehat{\mathfrak{B}}_F = \mathfrak{B}$.

$$\begin{aligned}
\widehat{\mathfrak{B}}_F &= \{p \in M; |f(p)| \leqq \sup |f(\mathfrak{B})| \text{ for } f \in F\} \\
&= \left\{p \in M; \varrho(0, f(p)) \leqq \sup_{q \in \mathfrak{B}} \varrho(0, f(q)) \text{ for } f \in F\right\} \\
&\subset \left\{p \in M; \varrho(0, f(p)) \leqq \sup_{q \in \mathfrak{B}} c_M(o, q)\right\} \\
&= \{p \in M : \varrho(0, f(p)) \leqq a\} \\
&= \mathfrak{B}.
\end{aligned}$$

Since $\widehat{\mathfrak{B}}_F$ contains $\mathfrak{B}$, we conclude that $\widehat{\mathfrak{B}}_F = \mathfrak{B}$.

If f is a bounded holomorphic function on M, then, for a suitable positive constant c, cf is a holomorphic mapping of M into the open unit disc $\mathbb{D}$. This means that, in Theorem 13.4.23, we may choose F to be the family of holomorphic mappings $f : M \to \mathbb{D}$ such that $f(o) = 0$. □

It is not clear if the converse to Theorem 13.4.23 holds. As we shall see, M need not be complete with respect to c_M even if M is holomorphically convex. It seems, however, very reasonable to expect that M is complete with respect to c_M if it is convex with respect to the family of bounded holomorphic functions.

Let M be a complex manifold of complex dimension n and A an analytic subset of dimension $\leqq n - 1$. Since every bounded holomorphic function on $M \setminus A$ can be extended to a bounded holomorphic function on M by Riemann's extension theorem, it follows that $M \setminus A$ cannot be convex with respect to the family of bounded holomorphic functions if A is non-empty and that c_M coincides with $c_{M \setminus A}$ on $M \setminus A$. In this way we obtain many examples of holomorphically convex manifolds which are not convex with respect to the family of bounded holomorphic functions and not complete with respect to their Carathéodory distances. The punctured disc $\mathbb{D} \setminus \{0\}$ is the simplest example. The question of completeness with respect to the Carathéodory distance may be quite possibly related to that of completeness with respect to the Bergman metric (see Bremermann [94] and Kobayashi [291] on the latter question).

Since $\mathbb{D}$ is the completion of $\mathbb{D} \setminus \{0\}$ with respect to the Carathéodory distance $c_{\mathbb{D}\setminus\{0\}}$ and every holomorphic mapping $f : \mathbb{D} \setminus \{0\} \to M$ is distance-decreasing with respect to $c_{\mathbb{D}\setminus\{0\}}$ and c_M, we have the following proposition:

Proposition 13.4.24 *If M is a complex manifold with complete Carathéodory distance, then every holomorphic mapping f from the punctured disc $\mathbb{D} \setminus \{0\}$ into M can be extended to a holomorphic mapping of $\mathbb{D}$ into M.*

13.5 Carathéodory Distance

Let $\mathbb{D}$ be the open unit disc in $\mathbb{C}$ with the Poincaré metric ρ.

Let M be a complex manifold. The *Carathéodory pseudodistance c_M* of M is defined by

$$c_M(p, q) = \sup_f \underline{g}(f(p), f(q)) \quad \text{for } p, q \in M,$$

where the supremum is taken with respect to the family of holomorphic mappings $f : M \to \mathbb{D}$. We prove that the pseudodistance k_M defined in Sect. 13.4 is greater than or equal to c_M. That will in particular imply that $c_M(p, q)$ is finite.

It was shown by Reiffen [430] that if $\gamma : [0, 1] \to \Omega$ is a C^1 curve in Ω with Carathéodory length defined by

$$l_c(\gamma) := \sup_{0=t_0<t_1<\cdots<t_k=1} \sum_{i=1}^{k} c_\Omega(\gamma(t_{i-1}, \gamma(t_i))),$$

then

$$l_c(\gamma) = \int_0^1 F_\Omega^C(\gamma(t), \gamma'(t))dt.$$

This infinitesimal metric was first investigated in depth by Reiffen [430], who showed that it gives the same length for arcs as the Carathéodory metric, and is thus the inner metric corresponding to the Carathéodory metric. Set

$$c_\Omega^i(z, w) = \inf \int_0^1 F_\Omega^C(\gamma(t), \gamma'(t))dt,$$

where the infimum is taken over all smooth paths $\gamma : [0, 1] \to \Omega$ connecting z to w. This metric, which is properly called the *Carathéodory-Reiffen metric*, is, in its integrated form, equivalent to the Carathéodory metric. Since it is the inner metric derived from the Carathéodory metric, it is always at least as large as the Carathéodory metric, but it is actually larger in many cases. The Carathéodory-Reiffen metric is also nonincreasing under holomorphic maps, both in its integrated and in its differential form.

Proposition 13.5.1 *For any complex manifold M, we have*

$$k_M(p,q) \geqq c_M(p,q) \quad \textit{for } p,q \in M.$$

Proof As in the definition of $k_M(p,q)$, choose points $p = p_0, p_1, \dots p_{k-1}, p_k = q$ of M and points $a_1, \dots, a_k, b_1, \dots, b_k$ of $\mathbb{D}$ and also mappings $f_1, \dots, f_k$ of $\mathbb{D}$ into M such that $f_i(a_i) = p_{i-1}$ and $f_i(b_i) = p_i$. Let f be a holomorphic mapping of M into $\mathbb{D}$. Then

$$\begin{aligned}\sum_{i=1}^{k} \varrho(a_i, b_i) &\geqq \sum_{i=1}^{k} \varrho(f \circ f_i(a_i), f \circ f_i(b_i)) \\ &\geqq \varrho(f \circ f_1(a_1), f \circ f_k(b_k)) \\ &= \varrho(f(p), f(q)),\end{aligned}$$

where the first inequality follows from the Schwarz lemma and the second inequality is a consequence of the triangle inequality. Hence

$$k_M(p,q) = \inf \sum_{i=1}^{k} \varrho(a_i, b_i) \geqq \sup \varrho(f(p), f(q)) = c_M(p,q).$$

□

It is an easy matter to verify that $c_M : M \times M \to R$ is continuous and satisfies the axioms for a pseudodistance:

$$c_M(p,q) \geqq 0, \quad c_M(p,q) = c_M(q,p), \quad c_M(p,q) + c_M(q,r) \geqq c_M(p,r).$$

The Carathéodory pseudodistance c_M shares many properties with k_M.

Proposition 13.5.2 *Let M and N be two complex manifolds, and let $f : M \to N$ be a holomorphic mapping. Then*

$$c_M(p,q) \geqq c_N(f(p), f(q)) \quad \textit{for } p,q \subset M.$$

The proof is trivial.

Corollary 13.5.3 *Every biholomorphic mapping $f : M \to N$ is an isometry, i.e.,*

$$c_M(p,q) = c_N(f(p), f(q)) \quad \textit{for } p,q \in M.$$

The Carathéodory pseudodistance may also be considered as a generalization of the Poincaré-Bergman metric for $\mathbb{D}$.

Proposition 13.5.4 *For the open unit disc $\mathbb{D}$ in $\mathbb{C}$, $c_{\mathbb{D}}$ coincides with the distance ϱ defined by the Poincaré-Bergman metric.*

Proof Using the Schwarz lemma for a holomorphic mapping $f : \mathbb{D} \to \mathbb{D}$, we obtain

$$\varrho(p, q) \geqq c_{\mathbb{D}}(p, q) \quad \text{for } p, q \in \mathbb{D}$$

from the very definition of $c_{\mathbb{D}}$. Considering the identity transformation of $\mathbb{D}$, we obtain the inequality $\varrho(p, q) \leqq c_{\mathbb{D}}(p, q)$. □

The following proposition says that c_M is the smallest pseudodistance on M for which $f : M \to \mathbb{D}$ is distance-decreasing.

Proposition 13.5.5 *Let M be a complex manifold and c' any pseudodistance on M such that*

$$c'(p, q) \geqq \varrho(f(p), f(q)) \quad p, q \in M$$

for every holomorphic mapping $f : M \to \mathbb{D}$. Then

$$c_M(p, q) \leqq c'(p, q) \quad \text{for } p, q \in M.$$

The proof is trivial.

The following proposition can be proved in the same way as Proposition 13.4.10.

Proposition 13.5.6 *Let M and M' be two complex manifolds. Then*

$$\begin{aligned} c_M(p, q) + c_{M'}(p', q') &\geqq c_{M\times M'}\left((p, p'), (q, q')\right) \\ &\geqq \text{Max}\left[c_M(p, q), c_{M'}(p', q')\right] \\ &\quad \text{for } p, q \in M \quad \text{and} \quad p', q' \in M'. \end{aligned}$$

Example 13.5.7 If $\mathbb{D}^m$ denotes the m-dimensional polydisc $\mathbb{D} \times \cdots \times \mathbb{D}$, then

$$c_{\mathbb{D}^m}\left((p_1, \ldots, p_m), (q_1, \ldots, q_m)\right) = \text{Max}\left[c_D(p_i, q_i); i = 1, \ldots, m\right]$$

for $p_i, q_i \in \mathbb{D}$. The proof is similar to the one in Example 13.4.11 in Sect. 13.4. Hence

$$k_{\mathbb{D}^m} = c_{\mathbb{D}^m}.$$

Example 13.5.8 Let $M = G/K$ be a bounded symmetric domain of rank l in $\mathbb{C}^n$ (in its natural realization). For a suitable subspace $\mathbb{C}^l$ of $\mathbb{C}^n$, $M \cap C^l$ is a polydisc

$\mathbb{D}^l = \mathbb{D} \times \cdots \times \mathbb{D}$, and M can be written as a union of polydiscs $k\left(\mathbb{D}^l\right)$, $k \in K$, where k is considered as a unitary transformation of $\mathbb{C}^n$. In particular, we have

$$\mathbb{D}^l \subset M \subset \mathbb{D}^n.$$

Since the injections $\mathbb{D}^l \to M$ and $M \to \mathbb{D}^n$ are distance-decreasing, we have $c_{\mathbb{D}^l} \geqq c_M \geqq c_{\mathbb{D}^n}$ on $\mathbb{D}^l$. From Example 13.5.7, we see that $c_{\mathbb{D}^l} = c_{\mathbb{D}^n}$ on $\mathbb{D}^l$. Hence,

$$c_{\mathbb{D}^l} = c_M \text{ on } \mathbb{D}^l.$$

Using the result in Example 13.5.7 of Sect. 13.4, we obtain a similar result for $k_{\mathbb{D}^l}$ and k_M. Given any two points $p, q \in M$, there is an element of G which sends p and q into $\mathbb{D}^l$. We may conclude therefore that, for a bounded symmetric domain M, the two distances c_M and k_M coincide.

Example 13.5.9 Since $c_M \leqq k_M$ in general, we see from Example 13.5.9 of Sect. 13.4 that if M is a complex manifold on which a complex Lie group G acts transitively, then c_M is trivial.

Example 13.5.10 If M is a compact complex manifold, then c_M is trivial. This is clear since every holomorphic function on M must be a constant function. This shows that there is no analog of Proposition 13.4.12 for c_M.

Example 13.5.11 Let $M = \mathbb{C} \setminus A$, where A is a finite set of points, or more generally, let $M = \mathbb{C}^n \setminus A$, where A is an analytic subset of dimension $\leqq n-1$. Then c_M is trivial. A holomorphic mapping $f : M \to \mathbb{D}$ may be considered as a bounded holomorphic function on M. By the so-called Riemann extension theorem, f can be extended to a holomorphic mapping of $\mathbb{C}^n$ into $\mathbb{D}$. By Liouville's theorem (or by Example 13.5.9), f must be a constant function, and hence c_M is trivial.

Example 13.5.12 Let

$$M = \left\{z \in \mathbb{C}^n; r < |z| < R\right\}$$

and

$$\mathbb{B} = \left\{z \in \mathbb{C}^n; |z| < R\right\}.$$

If $n \geqq 2$, then $c_M = c_{\mathbb{B}}|_M$. This follows from the fact that the envelope of holomorphy of M is $\mathbb{B}$, so that every holomorphic mapping from M into the unit disc $\mathbb{D}$ can be extended to a holomorphic mapping from $\mathbb{B}$ into $\mathbb{D}$. On the other hand, we do not know what the Carathéodory distance c_M looks like if $n = 1$. This is in contrast to the situation in Example 13.5.10 of Sect. 13.4, where k_M is known for $n = 1$ but not for $n \geqq 2$.

The Carathéodory distance was introduced in Carathéodory [105, 106]. For results on the Carathéodory distance, see Reiffen [429, 430].

13.6 Sibony Metric

A variant of the Carathéodory pseudometric is suggested by Sibony [472]: These time bounded holomorphic functions are replaced by bounded plurisubharmonic ones. In that paper the connection of this notion with hyperbolicity in the sense of Carathéodory and Kobayashi and with Montel manifolds is investigated. A related circle of ideas is touched upon in Berg [71] and in Kerzman-Rosay [271].

The following proposition of Sibony [472] is a version of Schwarz lemma and will be used to define the Sibony pseudodistance. We shall denote by $\mathbb{D}$ the usual Laplacian in $\mathbb{C}$.

Proposition 13.6.1 *Let u be a function defined on $\mathbb{D}$ of class C^2 in a neighborhood of the origin. Suppose that $0 \leq u \leq 1$, $u(0) = 0$, and that $\log u$ is subharmonic in $\mathbb{D}$. Then:*

(a) $u(z) \leq |z|^2$ for $z \in \mathbb{D}$ with equality at some point different from 0 if and only if $u(z)$ is identically equal to $|z|^2$.
(b) $\mathbb{D}u(0) \leq 4$ with equality if and only if $u(z) = |z|^2$ for every $z \in \mathbb{D}$.

Proof

(a) Since $u(0) = 0$ and $u \geq 0$, $grad\, u(0) = 0$. Define in $\mathbb{D} \setminus \{0\}$ the function $v(z) = u(z)|z|^{-2}$. $\log v$ is subharmonic in $\mathbb{D} \setminus \{0\}$; thus v is subharmonic in $\mathbb{D} \setminus \{0\}$. Since v is bounded, there is a subharmonic extension $\dot{v}$ of v in $\mathbb{D}$. Observe that $\overline{\lim} v(z) \leq 1$, it follows that $z \to e^{i\theta}$, $\tilde{v} \leq 1$, and therefore $u(z) \leq |z|^2$. If at some point $z_0 \neq 0$ we have $u(z_0) = |z_0|^2$, then $\tilde{v}(z_0) = 1$, which implies that v is identically 1.

(b) Let $z = x + iy$. Denoting partial differentiation by subscripts, we have

$$|z|^2 \geq u(z) = \frac{1}{2}\left(u_{xx}(0)x^2 + 2u_{xy}(0)xy + u_{yy}(0)y^2\right) + o\left(|z|^2\right).$$

Replacing y by $-y$ and adding the resulting inequality to the above, we obtain

$$u_{xx}(0)x^2 + u_{yy}(0)y^2 + o\left(|z|^2\right) \leq 2|z|^2.$$

Hence,

$$u_{z\bar{z}}(0) = \frac{1}{4}\left(u_{xx}(0) + u_{yy}(0)\right) \leq 1.$$

If $u_{z\bar{z}}(0) = 1$, then $\tilde{v}(0) = u_{z\bar{z}}(0) = 1$. Hence, $\tilde{v} \equiv 1$.

□

Remark 13.6.2 We note that in Proposition 13.6.1 if u is of class C^2 in a neighborhood of 0, then $u(z)/|z|^2$ is automatically bounded in $\mathbb{D}$.

Remark 13.6.3 If u is in $\mathcal{C}^2(\mathbb{D})$, $0 \leq u < 1$, and $\log u$ is subharmonic, then the preceding proposition shows that $Du(z) \leq 4\left(1-|z|^2\right)^{-2}$ in a neighborhood of $u^{-1}(0)$. But this inequality does not hold in the whole disc.

For historical comments concerning Schwarz-Pick-Ahlfors lemma, see Royden [445].

Following Sibony [472], we shall now define also an infinitesimal pseudometric F_M^S on a complex manifold M. For this purpose we fix some notations and definitions.

If u is a function of class $\mathcal{C}^2$ in an open set of $\mathbb{C}^n$ and $v = (v_1, \ldots, v_n) \in \mathbb{C}^n$, we denote

$$L_p(u, w) = \sum_{i,j=1}^{n} \frac{\partial^2 u(p)}{\partial z_i \partial \overline{z}_j} v_i \overline{v}_j$$

the Levi form of u at p.

When u is a function on a neighborhood U of a point p in a complex manifold M and $v \in T_p M$, where $T_p M$ is the tangent space to M at p, we define $L_p(u, v)$ as follows. Let ϕ be a chart from a neighborhood of p to an open neighborhood of the origin 0 in $\mathbb{C}^n$ with $\phi(p) = 0$. Then

$$L_p(u, v) = L_0(u \circ \phi^{-1}, \phi'(p)v).$$

It is clear that the preceding definition is independent of the chart ϕ. In fact if ψ is another chart, then

$$L_0(u \circ \psi^{-1}, \psi'(p)v) = L_0(u \circ \phi^{-1} \circ \phi \circ \psi^{-1}, \psi'(p)v) =$$
$$L_0(u \circ \phi^{-1} \circ \phi \circ \psi^{-1}, (\phi \circ \psi^{-1})'(0)\psi'(p)v) = L_0(u \circ \phi^{-1}, \phi'(p)v).$$

Let $\mathcal{S}_p$ be the family of functions u defined on M satisfying $0 \leq u \leq 1$, $u(p) = 0$, u of class $\mathcal{C}^2$ in a neighborhood of p and $\log u$ p.s.h. on M.

We define the *Sibony pseudometric* F_M^S on the tangent bundle as follows:

$$F_M^S(p, v) = \sup_{u \in \mathcal{S}_p} (L_p(u, v))^{1/2},$$

where $\langle p, v\rangle$ is an element of the tangent bundle.

Proposition 13.6.4 *The pseudometric F_M^S is locally bounded on the tangent bundle T_M. Furthermore we have:*

(a) $F_M^S(p, \lambda v) = |\lambda| F_M^S(p, v)$ *for* $\lambda \in \mathbb{C}$, $\langle p, v\rangle \in T_M$.
(b) $F_M^S(p, v_1 + v_2) \leq F_M^S(p, v_1) + F_M^S(p, v_2)$.

If f is a holomorphic map from M_1 to M_2, then

$$F^S_{M_2}\left(f(p), f'(p)v\right) \le F^S_{M_1}(p, v).$$

Proof Let ϕ be a chart from a neighborhood of p to the ball $\mathbb{B}(0, r)$ in $\mathbb{C}^n$. For $u \in \mathcal{S}_p$, the function $\log\left(u \circ \phi^{-1}\right)$ is p.s.h. in $\mathbb{B}(0, r)$ and $\left(u \circ \phi^{-1}\right)(0) = 0$. Therefore by Proposition 13.6.1,

$$L_p(u, v) = L_0(u \circ \phi^{-1}, \phi'(p)v) \le \frac{1}{r^2}|v|^2 \quad \text{for} \quad v \in \mathbb{C}^n.$$

Consequently

$$P(p, v) \le \frac{1}{r}\left|\phi'(p)v\right| < \infty.$$

Property (a) is clear from the definitions. We now prove the property (b).

$$\begin{aligned}
\left(F^S_M(p, v_1 + v_2)\right)^2 &= \sup_{u \in \mathcal{S}_p} \partial\overline{\partial}u\,(v_1 + v_2, \overline{v}_1 + \overline{v}_2) \\
&= \sup_u \left[\partial\overline{\partial}u\,(v_1, \overline{v}_1) + \partial\overline{\partial}u\,(v_2, \overline{v}_2) + 2\,\mathrm{Re}\,\partial\overline{\partial}u\,(v_1, \overline{v}_2)\right] \\
&\le \sup_u \Big[\partial\overline{\partial}u\,(v_1, \overline{v}_1) + \partial\overline{\partial}u\,(v_2, \overline{v}_2) \\
&\quad + 2\left(\partial\overline{\partial}u\,(v_1, \overline{v}_1)\right)^{1/2}\left(\partial\overline{\partial}u\,(v_2, \overline{v}_2)\right)^{1/2}\Big] \\
&= \sup_u \left[\left(\partial\overline{\partial}u\,(v_1, \overline{v}_1)\right)^{1/2} + \left(\partial\overline{\partial}u\,(v_2, \overline{v}_2)\right)^{1/2}\right]^2 \\
&\le \left[\left(\sup_u \partial\overline{\partial}u\,(v_1, \overline{v}_1)\right)^{1/2} + \left(\sup_u \partial\overline{\partial}u\,(v_2, \overline{v}_2)\right)^{1/2}\right]^2 \\
&= \left(F^S_M(p, v_1) + F^S_M(p, v_2)\right)^2.
\end{aligned}$$

□

Suppose f is a holomorphic map from a complex manifold M_1 to a complex manifold M_2. It is clear that $f^*\left(\mathcal{S}_{f(p)}(M_2)\right) \subset \mathcal{S}_p(M_1)$ for $p \in M_1$. If $q = f(p)$ and $u \in \mathcal{S}_q(M_2)$, we have

$$L_q(u, f'(p)v) = L_p(u \circ f, v).$$

It follows then that

$$F^S_{M_2}\left(f(p), f'(p)v\right) \le F^S_{M_1}(p, v). \tag{13.13}$$

We recall here the definitions of the infinitesimal Carathéodory metric E and the infinitesimal Kobayashi metric F.

Let M and N be two complex manifolds. From Proposition 13.6.1, it follows that for the unit disc $\mathbb{D}$ we have $F^C_{\mathbb{D}}(p, v) = F^S_{\mathbb{D}}(p, v) = F^K_{\mathbb{D}}(p, v)$. Therefore, since P is distance-decreasing for holomorphic maps, it follows from the functorial properties of F^C_M and F^K_M [296, Th. 2.2] that, for every complex manifold M, we have

$$F^C_M(p, v) \le F^S_M(p, v) \le F^K_M(p, v)\,.$$

Remark 13.6.5

(a) If Ω is a bounded symmetric domain, then $F^S_\Omega = F^C_\Omega = F^K_\Omega$. But in general, even for strictly pseudoconvex domains with smooth boundary, the three pseudometrics are different.

(b) The *indicatrix* I_p of the metric F^S_M defined by

$$\mathrm{I}_p = \left\{ v \mid v \in T_pM,\, F^S_M(p, v) < 1 \right\}$$

is a convex disced domain, while the indicatrix for the Kobayashi metric could be any disced pseudoconvex domain.

Chapter 14
Hyperbolic Manifolds

About the strongest statement one can make about a complex manifold from the standpoint of holomorphic mappings is that it is hyperbolic or complete hyperbolic. These concepts were introduced by Kobayashi, and their implications are discussed in his book [297].

14.1 Hyperbolicity

This section will briefly recall the definitions and facts we will need.

Let M be a complex manifold and k_M the Kobayashi pseudodistance defined in Sect. 13.4. If k_M is a distance, i.e., $k_M(p, q) > 0$ for $p \neq q$, then M is called a *hyperbolic manifold*.

Proposition 14.1.1 *A complex submanifold Y of a hyperbolic manifold M is hyperbolic.*

Proof This is immediate from the inequality

$$k_Y(p, q) \geqq k_M(p, q),$$

which follows from Proposition 13.4.1. □

A simple application of distance-decreasing property of the Kobayashi pseudodistance is the following generalization of the little Picard theorem.

Proposition 14.1.2 *If M is hyperbolic, then any holomorphic map $f : \mathbb{C} \to M$ is a constant map.*

Proof For any, $z_1, z_2 \in \mathbb{C}$, we have

$$k_M(f(z_1), f(z_2)) \leq k_{\mathbb{C}}(z_1, z_2) = 0.$$

P. V. Dovbush, S. G. Krantz, *The Geometric Theory of Complex Variables*,
https://doi.org/10.1007/978-3-031-77204-7_14

Since M is hyperbolic, $f(z_1) = f(z_2)$. □

A hyperbolic manifold M is said to be *complete* if it is complete with respect to k_M. As in Sect. 13.4.1 M is complete with respect to k_M if for each point p of M and each positive number r, the closed ball of radius r around p is a compact subset of M. As we shall see later, for a hyperbolic manifold M, this definition is equivalent to the usual definition in terms of Cauchy sequences. We note here that the idea of "hyperbolic" is a natural geometric generalization of the notion of bounded domain. The proofs of the following three propositions are almost identical to those of Propositions 13.4.17, 13.4.18, and 13.4.19.

Proposition 14.1.3 *If M and M' are (complete) hyperbolic manifolds, so is $M \times M'$.*

Proposition 14.1.4 *A (closed) complex submanifold M' of a (complete) hyperbolic manifold M is a (complete) hyperbolic manifold.*

Proposition 14.1.5 *Let M and $M_i, i \in I$, be closed complex submanifolds of a complex manifolds N such that $M = \bigcap_i M_i$. If each M_i is a (complete) hyperbolic manifold, so is M.*

From Proposition 13.5.1 we obtain the following:

Proposition 14.1.6 *A complex manifold M is (complete) hyperbolic if its Carathéodory pseudodistance c_M is a (complete) distance.*

Corollary 14.1.7 *Every bounded domain in $\mathbb{C}^n$ is hyperbolic. Every generalized analytic polyhedron is complete hyperbolic.*

The second statement follows from Theorem 13.4.22. The following proposition is immediate from Proposition 13.4.9.

Proposition 14.1.8 *If a complex manifold M admits a (complete) distance d' for which every holornorphic mapping f of the open unit disc $\mathbb{D}$ into M is distance-decreasing, i.e., $d'(f(a), f(b)) \leqq \varrho(a, b)$ for $a, b \in \mathbb{D}$, then it is (complete) hyperbolic.*

The results so far have their counterparts for Carathéodory distance. We shall now give results that are particular to the invariant distance k_M.

A Riemann surface is of hyperbolic type in the classical sense if its universal covering space is the unit disc. A Riemann surface of hyperbolic type is a hyperbolic space in our sense as is immediate from the following:

Theorem 14.1.9 *Let M be a complex manifold and $\widetilde{M}$ a covering manifold of M. Then $\widetilde{M}$ is (complete) hyperbolic if and only if M is (complete) hyperbolic.*

Proof Assume that $\widetilde{M}$ is hyperbolic. Let $p, q \in M$ and $k_M(p, q) = 0$. Let $\widetilde{p}$ be a point of $\widetilde{M}$ such that $\pi(\widetilde{p}) = p$, where $\pi : \widetilde{M} \to M$ is the covering projection. By Proposition 13.4.12, there exists a sequence of points $\widetilde{q}_1, \dots, \widetilde{q}_i, \dots$ of $\widetilde{M}$ such that $\pi(\widetilde{q}_i) = q$ and $\lim_i d_{\widetilde{M}}(\widetilde{p}, \widetilde{q}_i) = 0$. Then the sequence $\{\widetilde{q}_i\}$ converges to $\widetilde{p}$. Hence

$\pi(\widetilde{q}_i)$ converges to p. Since $\pi(\widetilde{q}_i) = q$, we obtain $p = q$. This proves that M is hyperbolic.

Assume that $\widetilde{M}$ is complete hyperbolic. Let $\overline{\mathfrak{B}}_r$ be the closed ball of radius r around $\widetilde{p} \in \widetilde{M}$, i.e., $\widetilde{\mathfrak{B}}_r = \{\widetilde{q} \in \overline{M}; d_{\widetilde{M}}(\widetilde{p}, \widetilde{q}) \leqq r\}$. Similarly, let $\mathfrak{B}_r$ be the closed ball of radius r around $p = \pi(\widetilde{p}) \in M$. By Proposition 13.4.12, we have

$$\mathfrak{B}_r \subset \pi\left(\widetilde{\mathfrak{B}}_{r+\delta}\right) \qquad \text{for } \delta > 0.$$

Since $\widetilde{\mathfrak{B}}_{r+\delta}$ is compact by assumption and $\mathfrak{B}_r$ is closed, $\mathfrak{B}_r$ must be compact. Hence, M is complete.

Assume that M is hyperbolic. Let $\widetilde{p}, \widetilde{q} \in \widetilde{M}$ and $d_{\widetilde{M}}(\overline{p}, \widetilde{q}) = 0$. Since the projection π is distance-decreasing, we have $k_M(\pi(\widetilde{p}), \pi(\widetilde{q})) = 0$, which implies that $\pi(\widetilde{p}) = \pi(\widetilde{q})$. Let $\widetilde{U}$ be a neighborhood of $\widetilde{p}$ in $\widetilde{M}$ such that $\pi\colon \widetilde{U} \to \pi(\widetilde{U})$ is a diffeomorphism and $\pi(\overline{U})$ is an ε-neighborhood of $\pi(\widetilde{p})$ with respect to k_M. In particular, $\widetilde{U}$ does not contain $\widetilde{q}$ unless $\widetilde{p} = \widetilde{q}$. Since $d_{\widetilde{M}}(\widetilde{p}, \widetilde{q}) = 0$ by assumption, there exist points $a_1, \ldots, a_k, b_1, \ldots, b_k$ of $\mathbb{D}$ and holomorphic mappings $f_1, \ldots, f_k$ of $\mathbb{D}$ into $\widetilde{M}$ such that $\widetilde{p} = f_1(a_1)$, $f_i(b_i) = f_{i+1}(a_{i+1})$ for $i = 1, \ldots, k-1$ and $f_k(b_k) = \overline{q}$ and that $\sum_{i=1}^{k} \varrho(a_i, b_i) < \varepsilon$. Let $\widehat{a_i b_i}$ denote the geodesic arc from a_i to b_i in $\mathbb{D}$.

Joining curves $f(\widehat{a_1 b_1}), \ldots, f(\widehat{a_n b_n})$ in $\widetilde{M}$, we obtain a curve from $\widetilde{p}$ to $\widetilde{q}$ in $\widetilde{M}$, which will be denoted by $\widetilde{C}$. Note that $\pi \circ f_1, \ldots, \pi \circ f_k$ are distance-decreasing mappings of $\mathbb{D}$ into M and $\widehat{a_1 b_1}, \ldots, \widehat{a_k b_k}$ are geodesics in $\mathbb{D}$. Every point of the curve $\pi(\widetilde{C})$ remains in the ε-neighborhood $\pi(\widetilde{U})$ of $\pi(\widetilde{p})$. Hence, the endpoint $\overline{q}$ must coincide with $\widetilde{p}$.

Assume that M is complete hyperbolic. We shall prove here that $\overline{M}$ is complete in the sense that every Cauchy sequence with respect to $d_{\widetilde{M}}$ is convergent. In the following section, we shall prove that the completeness in this usual sense implies the completeness defined at the beginning of this section. Let $\{\overline{p}_i\}$ be a Cauchy sequence in $\widetilde{M}$. Since the projection π is distance-decreasing, $\{\pi(\widetilde{p}_i)\}$ is a Cauchy sequence in M and hence converges to a point $p \in M$. Let ε bc a positive number and U the 2ε-neighborhood of p in M. Taking ε small, we may assume that π induces a homeomorphism of each connected component of $\pi^{-1}(U)$ onto U. Let N be a large integer such that $\pi(p_i)$ is within the ε-neighborhood of p for $i > N$. Then every point outside U is at least ε away from $\pi(\widetilde{p}_i)$. Let $\widetilde{U}_i$ be the connected component of $\pi^{-1}(U)$ containing $\widetilde{p}_i$. We shall show that the ε-neighborhood of $\widetilde{p}_i$ lies in $\widetilde{U}_i$ for $i > N$. Let $\widetilde{q}$ be a point of $\widetilde{M}$ with $d_{\dot{M}}(\widetilde{p}_i, \widetilde{q}) < \varepsilon$. We choose points $a_1, \ldots, a_k, b_1, \ldots, b_k$ of $\mathbb{D}$ and holomorphic mappings $f_1, \ldots, f_k$ of $\mathbb{D}$ into $\overline{M}$ in the usual manner so that $\sum_{j=1}^{k} \varrho\left(a_j, b_j\right) < \varepsilon$. Denoting by $\widehat{a_j b_j}$ the geodesic arc from a to b_j, let $\widetilde{C}$ be the curve from $\widetilde{p}_i$ to $\widetilde{q}$ obtained by joining $f_1\left(\widehat{a_1 b_1}\right), \ldots, f_k\left(\widehat{a_k b_k}\right)$ in M. Let $C = \pi(\overline{C})$. From the construction of C, it is clear that C is contained in the ε-neighborhood of $\pi(p_i)$ and hence in U. It follows that $\widetilde{C}$ lies in $\widetilde{U}_i$. Let $\widetilde{p}$ be the point of $\widetilde{U}_i$ defined by $p = \pi(\widetilde{q})$. Then $\{\widetilde{p}_j\}$ converges to $\widetilde{p}$. □

Corollary 14.1.10 *A Riemann surface M is hyperbolic if and only if it is a Kobayashi hyperbolic manifold.*

A complex manifold $\widetilde{M}$ is called a *spread domain* (domaine etalé in French) over a complex manifold M with projection π if every point $\widetilde{p} \in \widetilde{M}$ has a neighborhood $\widetilde{U}$ such that π is a holomorphic diffeomorphism of $\widetilde{U}$ onto the open set $\pi(\widetilde{U})$ of M. This is a concept more general than that of covering manifold.

From the proof of Theorem 14.1.9 above, we obtain the following:

Proposition 14.1.11 *A spread domain $\widetilde{M}$ over a hyperbolic manifold M is hyperbolic.*

There is a strong relation between hyperbolicity and negative curvature which is not yet completely understood. We will use only the simplest aspect of this relationship, the Ahlfors-Schwarz lemma.

From the definition of the Kobayashi pseudodistance and the Ahlfors-Schwarz lemma (Theorem 12.3.1), we have the following proposition:

Proposition 14.1.12 *Let M be a complex manifold and d_Ω a pseudo-Hermitian metric (it may degenerate on a proper analytic subvariety) which has curvature $\le -k < 0$ when restricted to the holomorphic image of any disc. Then*

$$d_M \ge k d_\Omega,$$

where d_M is the Kobayashi pseudodistance and d_Ω is the pseudodistance associated with Ω.

Remark 14.1.13 The condition hypothesized for Ω is equivalent to Ω having all holomorphic sectional curvatures $\le -k$. In the interest of maintaining an elementary level, we will avoid this concept.

14.1.1 On Completeness of Kobayashi Distance

We saw in Theorem 14.1.9 a similarity between the pseudodistance k_M and a Riemannian metric on M. We shall point out here another important similarity. Given any subset A of a complex manifold M and a positive number r, let $U(A; r)$ be the open set defined by

$$U(A; r) = \{p \in M; k_M(p, a) < r \text{ for some point } a \in A\}.$$

With this notation, we have the following proposition:

Proposition 14.1.14 *Let o be a point of a complex manifold M, and let r and r' be positive numbers. Then*

$$U\left[U(o; r); r'\right] = U\left(o; r + r'\right).$$

Proof The inclusion $U\left[U(o;r);r'\right]\subset U\left(o;r+r'\right)$ is true for any pseudodistance and makes use of the triangle inequality only. In order to prove the inclusion in the opposite direction, let $p\in U\left(0;r+r'\right)$, and set $k_M(o,p)=r+r'-3\varepsilon$ for ϵ small. Then there are points $a_i,b_i\in\mathbb{D}$ and holomorphic mappings $f_i:\mathbb{D}\to M$, $i=1,\ldots,k$, such that

$$\begin{aligned} f_1\left(a_1\right) &=0 \\ f_i\left(b_i\right) &=f_{i+1}\left(a_{i+1}\right) \qquad \text{for } i=1,\ldots,k-1, \\ f_k\left(b_k\right) &=p \end{aligned}$$

$$\sum_{i=1}^{k}\varrho\left(a_i,b_i\right)<r+r'-2\varepsilon.$$

Let j be the largest integer, $1\leqq j\leqq k$, such that

$$\sum_{i=1}^{j-1}\varrho\left(a_i,b_i\right)<r-\varepsilon.$$

Let c_j be the point on the geodesic from a_j to b_j in $\mathbb{D}$ such that

$$\sum_{i=1}^{j-1}\varrho\left(a_i,b_i\right)+\varrho\left(a_j,c_j\right)=r-\varepsilon.$$

If we set

$$q=f_j\left(c_j\right),$$

then $k_M(o,q)<r$ and $k_M(q,p)<r'$ so that $p\in U\left(q;r'\right)\subset U(q';r)$, $q'\in U(o;r)$. □

In the proof of Theorem 14.1.9 we promised to show that if a hyperbolic manifold M is complete in the sense that every Cauchy sequence converges, then the closure $\overline{U}(o;r)$ of $U(o;r)$ is compact for all $o\in M$ and all positive numbers r. We shall show that this assertion is true for a larger class of metric spaces.

Theorem 14.1.15 *Let M be a locally compact metric space with distance function d satisfying the equality*

$$U\left[U(o;r);r'\right]=U\left(o;r+r'\right)$$

for all $o\in M$ and all positive numbers r and r'. Then M is complete in the sense that every Cauchy sequence converges if and only if the closure $\overline{U}(o;r)$ of $U(o;r)$ is compact for all $o\in M$ and all positive numbers r.

Proof We have only to prove that if M is (Cauchy-)complete, then $\overline{U}(o; r)$ is compact; the implication in the opposite direction is trivial. □

Lemma 14.1.16 *$\overline{U}(o; r)$ is compact if there exists a positive number b such that $\overline{U}(p; b)$ is compact for every $p \in U(o; r)$.*

Proof Since M is locally compact, there is a positive number $s < r$ such that $\overline{U}(o; s)$ is compact. It suffices to show that if $\overline{U}(o; s)$ is compact, so is $\overline{U}[o; s + (b/2)]$. Let $p_1, p_2, \ldots$ be points of $\overline{U}[o; s + (b/2)]$. Let $q_1, q_2, \ldots$ be points of $\overline{U}(o; s)$ such that $d(p_i, q_i) < \frac{3}{4}b$. Since $\overline{U}(o; s)$ is compact, we may assume (by choosing a subsequence if necessary) that $q_1, q_2, \ldots$ converges to some point, say q, of $\overline{U}(o; s)$. Then $\overline{U}(q; b)$ contains all p_i for large i. Since $\overline{U}(q; b)$ is compact by assumption, a suitable subsequence of $p_1, p_2, \ldots$ converges to a point p of $\overline{U}(q; b)$. Since $p_1, p_2, \ldots$ is a sequence in $\overline{U}[o; s + (b/2)]$ and $\overline{U}[o; s + (b/2)]$ is a closed set, the limit point p lies in $\overline{U}[o; s+(b/2)]$. This completes the proof of lemma. □

Proof The proof of Theorem 14.1.15 is now reduced to showing that there exists a positive number b such that $\overline{U}(p; b)$ is compact for all $p \in M$. Assume the contrary. Then there exists a point $p_1 \in M$ such that $\overline{U}\left(p_1; \frac{1}{2}\right)$ is noncompact. Applying the lemma to $\overline{U}\left(p_1; \frac{1}{2}\right)$, we see that there exists a point $p_2 \in \overline{U}\left(p_1; \frac{1}{2}\right)$ such that $\overline{U}\left[p_2; (1/2^2)\right]$ is noncompact. In this way we obtain a Cauchy sequence $p_1, p_2, p_3, \ldots$ such that $p_k \in \overline{U}\left[p_{k-1}; (1/2^{k-1})\right]$ and $\overline{U}\left[p_k; (1/2^k)\right]$ is noncompact. Let p be the limit point of the Cauchy sequence $p_1, p_2, \ldots$. Since M is locally compact, for a suitable positive number c, $\overline{U}(p; c)$ is compact. For a sufficiently large k, $\overline{U}\left[p_k; (1/2^k)\right]$ is a closed set contained in $\overline{U}(p; c)$ and hence must be compact. This is a contradiction. □

Remark 14.1.17 Let M be a hyperbolic manifold and M^* its completion with respect to the distance k_M. Unfortunately, M^* need not be locally compact, and hence it need not be complete in the strong sense that every closed ball of radius r in M^* is compact. An example of a hyperbolic manifold M such that M^* is not locally compact has been found by Kiernan [279]. A similar example of a Riemannian manifold has been found by D.B.A. Epstein (see Kobayashi [293]).

14.2 Kobayashi-Hurwitz Theorem

It has been known for a long time that the automorphism group of a compact Riemann surface of genus greater than 1 is a finite group, Klein (see Poincaré [417]), Hurwitz [251]. In this section, we shall generalize this classical result to higher dimensional compact complex manifolds. The first generalization is the following (Kobayashi [292]).

Theorem 14.2.1 *Let M be a compact hyperbolic manifold. Then the group* Aut(M) *is finite.*

This is a generalization of the following famous and very precise result of Hurwitz (1893) (a proof of which appears as XVI.21.22):

Theorem 14.2.2 *Let M be a compact Riemann surface of genus $g > 1$. Then the order of* $\mathrm{Aut}(M)$ *is at most* $84(g-1)$.

Remark 14.2.3 This result of Hurwitz cannot be improved, since the upper bound $84(g-1)$ is attained in some important cases. (That $\mathrm{Aut}(M)$ is finite for $g > 1$ was first proved by Schwarz.)

On the other hand, we know that for the following classes of Riemann surfaces $\mathrm{Aut}(M)$ is obviously a Lie group:

1. $\mathbb{CP}$, a Riemann sphere (a surface of genus zero)
2. $\mathbb{C}$
3. $\mathbb{C}^*$ (i.e., $\mathbb{CP}$ with two points removed)
4. $\mathbb{T}$ (tori, i.e., surfaces of genus 1)
5. $\mathbb{D}$
6. $\mathbb{D}^*$ (once-punctured disc)
7. Annuluses

We know that tori and annuluses contain an infinite class of distinct conformal types (cf. [374, Exercise 2, p. 556]); 5–7 above are hyperbolic surfaces. The following natural question, put first by Felix Klein, arises: Are these seven types the only surfaces with continuous automorphism groups? The answer is "yes."

Theorem 14.2.4 (Klein) *The group of automorphisms of a Riemann surface is always (properly) discontinuous, except in the seven cases enumerated above.*

Remark 14.2.5 The preceding theorem is due in essence to Klein, but the first precise proof was given by Poincaré in 1883 in the case of compact surfaces. But, as Weyl remarks, "Poincaré's proof, word for word, remains valid for open surfaces." We now leave this fascinating subject and advise the reader to consult Kobayashi's monographs: [294, 295].

14.3 Extension Theorems

We pass to a generalization of the big Picard theorem, which for functions of one variable can be stated as follows: Any holomorphic mapping of the punctured disc $\mathbb{D}_* = \{0 < |\zeta| < 1\}$ into the closed plane $\mathbb{CP}$ with three points omitted extends to a holomorphic mapping of the disc U into $\mathbb{CP}$. In contrast to the little Picard theorem, the big theorem does not extend to arbitrary hyperbolic manifolds.

The first extension theorem that used the intrinsic distance was proved in 1968 by Mrs. Kwack [345, Theorem 5, 18].

Theorem 14.3.1 (Kwack) *Let A be a closed complex subspace of a complex manifold X, and let M be a hyperbolic complex space. Then every holomorphic* *map* $F : X \setminus A \to M$ *extends to a holomorphic map* $F : X \to M$ *if one of the following conditions is satisfied:*

(i) M *is compact.*
(ii) M *is complete hyperbolic and* $codim\, A \geq 2$.

This shows that the concept of hyperbolic is useful for attacking our extension problem, but it does not explain the big picard theorem since $\mathbb{CP}^1 \setminus \{0, 1, \infty\}$ is not compact. It is natural to ask if condition (i) in Kwack's theorem can be replaced by the assumption that M is a hyperbolic, relatively compact, open set in a complex space Y. To see that the answer is no, consider the following example.

Example 14.3.2 The manifold

$$M = \left\{[z_0, z_1, z_2] \in \mathbb{P}^2 : z_0 = 1, 0 < |z_1| < 1, |z_2| < \left|e^{1/z_1}\right|\right\}$$

is hyperbolic, since the mapping $(1, z_1, z_2) \to \left(z_1, z_2 e^{-1/z_1}\right)$ transforms it biholomorphically into the hyperbolic manifold $\mathbb{D}_* \times \mathbb{D}$. However the mapping $f : \mathbb{D}_* \to M$ defined by the formula $f(\zeta) = \left[1, \zeta, \frac{1}{2}e^{1/\zeta}\right]$ cannot be extended holomorphically to $\mathbb{D}$.

However, the following result does hold.

Theorem 14.3.3 ([345, Theorem 4, 16]) *If* M *is a hyperbolic manifold, then either the holomorphic mapping* $f : \mathbb{D}_* \to M$ *extends holomorphically to* $\mathbb{D}$ *or its limit set at the point* $\zeta = 0$ *does not contain points of* M.

Proof Suppose the limit set of f at the point $\zeta = 0$ contains a point $p_0 \in M$, i.e., there exists a sequence $\zeta_\nu \to 0$ of points of $\mathbb{D}_*$ for which $f(\zeta_\nu) \to p_0$.

Let V denote a neighborhood of p_0 which in local coordinates on M with center p_0 is described as $\{p \in M : |z(p)| < \varepsilon\}$ and $W = \{p \in M : |z(p)| < \varepsilon/2\}$; let $r_\nu = |\zeta_\nu|$ and $\gamma_\nu = \{|\zeta| = r_\nu\}$.

The Kobayashi metric of the punctured disc is

$$ds^2_{U_*} = \frac{d\zeta d\bar{\zeta}}{|\zeta|^2 \ln^2 \frac{1}{|\zeta|}}.$$

It follows that in this metric the length of the circle $\{|\zeta| = r\}$ is equal to $2\pi / \ln \frac{1}{r}$ and tends to zero as $r \to 0$.

Since the Kobayashi diameter $f(\gamma_\nu) \to 0$, then $f(\gamma_\nu) \subset W$ for $\nu \geq \nu_0$, and without loss of generality we may assume that $\nu_0 = 1$ and that r_ν decreases to zero.

It suffices to prove the continuous extendability of f at the point $\zeta = 0$, i.e., that the image of $\{0 < |\zeta| < \delta\}$ for sufficiently small δ belongs to W. If only a finite number of images of the annuli $\{r_{\nu+1} < |\zeta| < r_\nu\}$ extends beyond the limits of W, then everything is proved; in the case when there are infinitely many such annuli, we will arrive at a contradiction. Passing to a subsequence, we may assume that the images of all the annuli extend beyond the limits of W.

We denote by $A_\nu = \{\alpha_\nu < |\zeta| < \beta_\nu\}$ the largest of the annuli for which $\alpha_\nu < r_\nu < \beta_\nu$ and $f(A_\nu) \subset W$ and by $\gamma'_\nu = \{|\zeta| = \beta_\nu\}, \gamma''_\nu = \{|\zeta| = \alpha_\nu\}$ the boundary circles of A_ν (their images are contained in $\bar{W}$, but not in W). Since the diameters of $f(\gamma'_\nu)$ and $f(\gamma''_\nu)$ tend to zero, then, passing to a subsequence if necessary, we may assume that $f(\gamma'_\nu) \to p', f(\gamma''_\nu) \to p''$. The points $p', p'' \in \partial W$, and hence, are different from p_0; without loss of generality, we assume that $z_1(p')$ and $z_1(p'')$ are different from $z_1(p_0) = 0$ ($z_1(p)$ is the first coordinate of $z(p)$).

On the set $f^{-1}(W)$, the mapping f is represented in local coordinates in the form $(f_1, \dots, f_n)$, and, in particular, for all ν a holomorphic mapping $f_1 : A_\nu \to \mathbb{C}$ is defined. For sufficiently large ν, the image $f_1(\gamma_\nu)$ lies in a neighborhood of the point $z_1 = 0$, not intersecting neighborhoods of the points $z_1(p')$ and $z_1(p'')$, which contain $f_1(\gamma'_\nu)$ and $f_1(\gamma''_\nu)$. The union of the latter forms $f_1(\partial A_\nu)$, and by the domain preservation principle, the mapping f_1 cannot take interior points of A_ν to $\partial f_1(A_\nu)$, i.e., $\partial f_1(A_\nu) \subset f_1(\partial A_\nu)$. On the other hand, $f_1(A_\nu)$ is a bounded domain containing $f_1(\gamma_\nu)$, and under our hypotheses there does not exist a bounded domain whose boundary belongs to the union $f_1(\gamma'_\nu) \cup f_1(\gamma''_\nu)$ and which contains $f_1(\gamma_\nu)$, which is a contradiction. □

In particular, for $M = \mathbb{CP} \setminus \{$ three points $\}$, we obtain from Theorem 14.3.3 the big Picard theorem for functions of one variable. Indeed, in this case the limit set, since it is connected, must reduce to one of the excluded points, and then f extends continuously to the point $\zeta = 0$. If M is a compact hyperbolic manifold, then any holomorphic mapping $f : \mathbb{D}_* \to M$ extends to a mapping from $\mathcal{O}(U, M)$, since here the second possibility of Theorem 14.3.3 cannot be realized.

In Example 14.3.2, we could construct a mapping that has no extension precisely because the boundary of M in $\mathbb{CP}^2$ is so badly behaved. For example, the boundary is not even an algebraic subvariety of $\mathbb{CP}^2$. The next definition gives sufficient restrictions on the boundary in order to imply a strong extension theorem.

Definition 14.3.4 Let Y be a complex space, and let M be a relatively compact, hyperbolic complex subspace of Y. We say that M is *hyperbolically imbedded* in Y if either of the following (equivalent) conditions are satisfied:

(i) If p and q are boundary points of M and if $\{p_n\}$ and $\{q_n\}$ are sequences in M such that $p_n \to p, q_n \to q$, and $d_M(p_n, q_n) \to 0$, then $p = q$.
(ii) If p is a boundary point of M and U is a neighborhood of p in Y, there exists a neighborhood V of p in Y such that $\bar{V} \subset U$ and the distance between $M \cap (Y \setminus U)$ and $M \cap V$ with respect to d_M is positive.

Intuitively, a relatively compact subspace M is hyperbolically imbedded in Y if M is hyperbolic and if the "distance" between any two distinct boundary points is greater than zero.

The following are some examples of hyperbolically imbedded spaces:

(1) Let M be a compact hyperbolic manifold. Then M is hyperbolically imbedded in $Y = M$.
(2) Let $Y = \mathbb{CP}$ and $M = \mathbb{CP} \setminus \{\text{three points}\}$. M is covered by the unit disc, and therefore it is complete hyperbolic. Since M has only isolated boundary points, M is hyperbolically imbedded in Y.
(3) If M_1 and M_2 are hyperbolically imbedded in Y_1 and Y_2, respectively, then $M_1 \times M_2$ is hyperbolically imbedded in $Y_1 \times Y_2$. To see this, let $\pi : M_1 \times M_2 \to M_j$ be the projection onto M_j, and let $p = (p', p''), q = (q', q'')$, $\{p_n = (p'_n, p''_n)\}$, and $\{q_n = (q'_n, q''_n)\}$ be as in (i) above. If $d_{M_1 \times M_2}(p_{n'} q_n) \to 0$, then $d_{M_1}(p'_n, q'_n) \to 0$ and $d_{M_2}(p''_n, q''_n) \to 0$ since π_1 and π_2 are distance-decreasing. Thus $p' = q'$ and $p'' = q''$, and therefore $M_1 \times M_2$ is hyperbolically imbedded in $Y_1 \times Y_2$.

The following theorem is the basis for extension theorems. The proof is essentially the same as Mrs. Kwack's proof of [345, Theorem 3], although Theorem 14.3.5 is slightly more general. The winding number argument used below is due to Grauert and Reckziegel.

Theorem 14.3.5 (Kiernan) *Let M be hyperbolically imbedded in Y, and let $f_k : \mathbb{D}^* \to M$ be a sequence of holomorphic maps where $\mathbb{D}^* = \{z \in \mathbb{C} : 0 < |z| < 1\}$. Let $\{z_k\}$ and $\{z'_k\}$ be sequences in $\mathbb{D}^*$ converging to 0 and such that $f_k(z'_k) \to q \in Y$. Then:*

(i) $f_k(z_k) \to q$.
(ii) Any bolomorphic map $f : \mathbb{D}^* \to M$ extends to a bolomorphic map $f : \mathbb{D} \to Y$.
(iii) $f_k(0) \to q$.

Proof

(i) Assume that $f_k(z_k) \to p \neq q$ and that $|z_k| \leq |z'_k|$. Choose a coordinate neighborhood U of p such that U is an analytic subset of

$$W = \left\{(w_1, \ldots, w_n) \in \mathbb{C}^n \,|\, |w_1|^2 + \cdots + |w_n|^2 < 2\right\},$$

$q \notin U$, and $p = (0, \ldots, 0)$. Let $\rho_k(t) = z_k e^{it}$ for $0 \leq t \leq 2\pi$. Then the diameters of the sets ρ_k with respect to $d_{\mathbb{D}^*}$ are converging to 0. Since M is hyperbolically imbedded in Y and since f_k is distance-decreasing, this implies $f_k(\rho_k)$ is converging to p. Thus for all sufficiently large k, there exists an annulus R'_k centered at 0 such that $\rho_k \subset R'_k$ and

$$f_k(R'_k) \subset \mathbb{B} = \left\{(w_1, \ldots, w_n) \subset \mathbb{C}^n : |w_1|^2 + \cdots + |w_n|^2 < 1\right\}.$$

Let R_k be the largest such annulus. We can assume either that R_k is not a punctured disc for any k or that R_k is a punctured disc for all k. We consider the former case first. Since $f_k(z'_k) \to q \notin U$, there exist a_k and b_k in $\mathbf{R}_k$

with $|a_k| < |b_k|$ and $|f_k(a_k)| = |f_k(b_k)| = 1$. By taking a subsequence and relabeling, we may assume that $f_k(a_k) \to q' \in \bar{B}1$ and $f_k(b_k) \to q'' \in \mathbf{B1}$. Let σ_k and τ_k be the curves defined by $\sigma_k(t) = a_k e^{it}$ and $\tau_k(t) = b_k e^{it}$ for $0 \le t \le 2\pi$. By the same reasoning used for ρ_k, we see that $f_b(\sigma_b) \to q'$ and $f_k(\tau_k) \to q''$.

Let $f_k = \left(f_k^1, \ldots, f_k^n\right)$ in $f_k^{-1}(U)$. By rotating the coordinate ball W if necessary, we see that the winding number of the curves $f_k^1(\sigma_k)$ and $f_k^1(\tau_k)$ around the point $f_k^1(z_k)$ is 0 for all k sufficiently large. By Cauchy's theorem, we have

$$\int_{\sigma_k} \frac{f_k^{1\prime}(z)}{f_k^1(z) - f_k^1(z_k)} dz = \int_{f_k^1(\sigma_k)} \frac{dw_1}{w_1 - f_k^1(z_k)} = 0$$

and

$$\int_{\tau_k} \frac{f_k^1(z)}{f_k^1(z) - f_k^1(z_k)} dz = \int_{f_k^1(\tau_k)} \frac{dw_1}{w_1 - f_k^1(z_k)} = 0.$$

Thus

$$0 = \int_{\sigma_k} \frac{f_k^{1\prime}(z)}{f_k^1(z) - f_k^1(z_k)} dz - \int_{\tau_k} \frac{f_k^{1\prime}(z)}{f_k^1(z) - f_k^1(z_k)} dz = 2\pi i(P - N),$$

where N and P are the number of zeros and poles of the function $f_k^1(z) - f_k^1(z_k)$ on the annulus R_k. This is a contradiction since $N > 0$ and $P = 0$.

If R_k is a punctured disc for all k, then the boundary of $\bar{R}_k$ is τ_k, where b_k and τ_k are as above. Since $f_k(R_k) \subset \mathbb{B}$, f_k extends holomorphically to $\bar{R}_k$ with $f_k(0)$ in $\mathbb{B}$. Setting $\sigma_k = \varnothing$, the argument used in the preceding paragraph leads to a contradiction. This proves (i).

(ii) Let $\{z_k'\}$ be any sequence in $\mathbb{D}^*$ such that $f(z_k') \to q \in Y$. Define $f(0) = q$. By (i), f is continuous. The Riemann extension theorem now implies that f is holomorphic at 0.

(iii) If $f_k(0)$ does not converge to q, then without loss of generality we may assume that $f_k(0) \to p \neq q$. Since each f_k is continuous, there exists a sequence $\{z_k\}$ in $\mathbb{D}^*$ such that $z_k \to 0$ and $f_k(z_k) \to p$. This contradicts (i).

□

The following theorem was proven by Kwack in the case when M is compact and by Kobayashi in the case when A is nonsingular.

Theorem 14.3.6 (Kiernan) *Let A be a closed complex subspace of a (nonsingular) complex manifold X. If the singularities of A are normal crossings and if M is byperbolically imbedded in Y, then any bolomorphic map $f : X \setminus A \to M$ extends to a bolomorphic map $f : X \to Y$.*

Proof Since the singularities of A are normal crossings, we can assume that

$$X = \mathbb{D} \times \cdots \times \mathbb{D} = \mathbb{D}^n \times \mathbb{D}^l = \left\{(w_1, \ldots, w_n, t_1, \ldots, t_l) \,||w_i\, |< 1 \text{ and } \left|t_j\right| < 1\right\}$$

and that

$$X \setminus A = \left(\mathbb{D}^*\right)^n \times \mathbb{D}^l = \left\{(w_1, \ldots, w_n, t_1, \ldots, t_l) \,|0 < |w_i\, |< 1 \text{ and } \left|t_j\right| < 1\right\}.$$

The proof is by induction on n.

Case 1. $X \setminus A = \mathbb{D}^*$ This is part (ii) of Theorem 14.3.5.

Case 2. Assume we can extend f when $X \setminus A = (\mathbb{D}^*)^n$. We show that this implies we can extend f if $X \setminus A = (\mathbb{D}^*)^n \times \mathbb{D}^l$.

Let $w = (w_1, \ldots, w_n) \in \mathbb{D}^n$ and $t = (t_1, \ldots, t_l) \in \mathbb{D}^l$. We denote the point $(w_1, \ldots, w_n, t_1, \ldots, t_l) \in \mathbb{D}^n \times \mathbb{D}^l$ by (w, t). Let $f : (\mathbb{D}^*)^n \times \mathbb{D}^l \to M$ be holomorphic, and define $f_t : (\mathbb{D}^*)^n \to M$ by $f_t(w) = f(w, t)$. Since each map f_t extends to $\mathbb{D}^n$, we can extend f to a map from $\mathbb{D}^n \times \mathbb{D}^l$ into Y by defining $f(w, t) = f_t(w)$. By the Riemann extension theorem, it suffices to show that the extended map f is continuous.

We assume that f is not continuous at some point, say $(w, 0)$ for simplicity, and then obtain a contradiction. If f is not continuous at $(w, 0)$, then there exists a sequence of points $\left(w^k, t^k\right)$ in $(\mathbb{D}^*)^n \times (\mathbb{D}^*)^l$ with $\left(w^k, t^k\right) \to (w, 0)$ and such that the sequence $f\left(w^k, t^k\right) \to q \neq f(w, 0)$. Define $f_k : \mathbb{D}^* \to M$ by $f_k(z) = f\left(w^k, zt^k/\left|t^k\right|\right)$. Let $z_k' = \left|t^k\right|$. Since $z_k' \to 0$ and $f_k\left(z_k'\right) = f\left(w^k, t^k\right) \to q$, part (iii) of Theorem 14.3.5 implies that $f_k(0) = f\left(w^k, 0\right) \to q$. But f_t is continuous for each t, and therefore $f\left(w^k, 0\right) \to f(w, 0) \neq q$. Thus the assumption that f is not continuous is false.

Case 3. Assume that f can be extended if $X \setminus A = (\mathbb{D}^*)^n \times \mathbb{D}^l$. We show that this implies that f can be extended if $X \setminus A = (\mathbb{D}^*)^{n+1}$. By the inductive hypothesis, we can extend f to $\mathbb{D}^{n+1} - \{(0, \ldots, 0)\}$. The map $g : \mathbb{D}^* \to M$ defined by $g(z) = f(z, \ldots, z)$ extends to $\mathbb{D}$. Define $f(0, \ldots, 0) = g(0)$. It suffices to show that f is continuous.

If f is not continuous, there exists a sequence $\left(w^k, t^k\right) = \left(w_1^k, \ldots, w_n^k t^k\right)$ in $(\mathbb{D}^*)^{n+1}$ such that $f\left(w^k, t^k\right) \to q \neq f(0, \ldots, 0)$. By applying Theorem 14.3.5 with $f_k(z) = f\left(zw^k/\left|w^k\right|, t^k\right)$ and $z_k' = \left|w^k\right|$, we see that $f\left(0, t^k\right) = f_k(0) \to q$. On the other hand, if we apply Theorem 14.3.5 with

$$f_k(z) = f\left(zt^k/\left|t^k\right|, \ldots, zt^k/\left|t^k\right|, t^k\right) \quad \text{and} \quad z_k' = \left|t^k\right|,$$

we see that $f\left(0, t^k\right) \to f(0, \ldots, 0) \neq q$. This is a contradiction, and therefore f extends holomorphically. □

Given a holomorphic map $F = (f_1, \ldots, f_m) : \Omega \to \mathbb{C}^m$ on a domain $\Omega \subset \mathbb{C}^n$, we denote by $\operatorname{rank}_p F$ the complex rank of F at a point $p \in \Omega$, that is, the rank of the complex $m \times n$ Jacobian matrix

$$F'(p) = \left(\frac{\partial f_j}{\partial z_k}(p) \right)$$

This matrix represents the differential $dF_p = \partial F_p : T_p\mathbb{C}^n \to T_{F(p)}\mathbb{C}^m$ in standard bases on the tangent spaces $T_p\mathbb{C}^n$, $T_{F(p)}\mathbb{C}^m$, respectively. (See Sect. 7.6.) Clearly $\operatorname{rank}_p F \le \min\{m, n\}$. The map F is an *immersion* at p if $\operatorname{rank}_p F = n$ and is a submersion at p if $\operatorname{rank}_p F = m$. These notions coincide when $n = m$, and in this case F is said to be locally biholomorphic at p. These notions, being local, extend to holomorphic maps between complex manifolds.

If M and N are complex manifolds, a holomorphic map $F : N \to M$ is called a *holomorphic immersion* if the corresponding map $F_* : T(N) \to T(M)$ of the tangent spaces has maximal rank.

Trivially, the composition of two immersions is an immersion.

Let X be a complex manifold of dimension n. A subset M of X is a complex submanifold of dimension $m \in \{0, 1, \dots, n\}$ (and codimension $d = n - m$) if every point $p \in M$ admits an open neighborhood $U \subset X$ and a holomorphic chart $\phi : U \to U' \subset \mathbb{C}^n$ such that $\phi(U \cap M) = U' \cap \left(\mathbb{C}^m \times \{0\}^{n-m}\right)$. Any such chart (U, ϕ) on X is said to be adapted to M. Let $\pi : \mathbb{C}^n \to \mathbb{C}^m$ denote the coordinate projection $\pi(z_1, \dots, z_m, \dots, z_n) = (z_1, \dots, z_m)$. For each adapted complex chart (U, ϕ) on X, we get a complex chart $(U \cap M, \pi \circ \phi|_{U \cap M})$ on M with values in $\mathbb{C}^m$. The collection of all such charts is a complex atlas on M, and the corresponding complex structure on M is the complex submanifold structure induced by the inclusion map $\iota : M \hookrightarrow X$. Considering M with this submanifold structure as a complex manifold in its own right, the inclusion ι is a holomorphic embedding of M in X, that is, an injective holomorphic immersion of M onto the complex submanifold $\iota(M)$ of X.

The image of an injective holomorphic immersion $f : M \to X$ need not be a submanifold of X, not even a topological one, due to possible accumulation of the image on itself.

Similar reasoning as in Proposition 14.1.11 yields also the following:

Proposition 14.3.7 *If a complex manifold M' is holomorphically immersed in a hyperbolic manifold M, then M' is also hyperbolic.*

But it is complete hyperbolic. In fact, the upper half plane $\mathbb{H}$ is the universal covering space of $\mathbb{D} \setminus \{0\}$, the projection being given by $z \to e^{iz}$. We may also consider the unit disc $\mathbb{D}$ as the universal covering space of $\mathbb{D} \setminus \{0\}$. Our assertion follows from Theorem 14.1.9. More generally, we have the following theorem:

Theorem 14.3.8 *Let M be a complete hyperbolic manifold and f a bounded holomorphic function on M. Then the open submanifold $M' = \{p \in M;\ f(p) \neq 0\}$ of M is also complete hyperbolic.*

Proof Multiplying f by a suitable constant, we may assume that f is a holomorphic mapping of M into the open unit disc $\mathbb{D}$. We denote by $\mathbb{D}^*$ the punctured disc $\mathbb{D} \setminus \{0\}$. Let o be a point of M', and let a and b be positive numbers. Since $\mathbb{D}^*$ is complete hyperbolic, for a given positive number a we can choose a small positive number b such that

$$\{z \in \mathbb{D};\, |z| \geqq b\} \supset \left\{z \in \mathbb{D}^*;\, d_{\mathbb{D}^*}(f(o), z) \leqq a\right\}.$$

We set

$$\begin{aligned} A &= \left\{p \in M;\, k_M(o, p) \leqq a\right\}, & A' &= \left\{p \in M';\, d_{M'}(o, p) \leqq a\right\} \\ B &= \{p \in M;\, |f(p)| \geqq b\}, & B' &= \left\{p \in M';\, |f(p)| \geqq b\right\}. \end{aligned}$$

Since $d_{M'}(o, p) \geqq k_M(o, p)$ by Proposition 13.4.1, we have

$$A \supset A'.$$

Since b is positive and $M' = \{p \in M;\, f(p) \neq 0\}$, we have

$$B = B'.$$

Since $f : M' \to \mathbb{D}^*$ is distance-decreasing, we have

$$A' \subset \left\{p \in M';\, d_D(f(o), f(p)) \leqq a\right\} \subset \left\{p \in M';\, |f(p)| \geqq b\right\} = B' = B.$$

Since A is a compact subset of M by the completeness of M and B is clused in M, the intersection $A \cap B$ is compact. Since $B = B'$, it follows that $A \cap B$ is in M'. Since $A \cap B$ is a compact subset of M' and A' is closed in M', the intersection $A' \cap (A \cap B)$ is a compact subset of M'. Since both A and B contain A', it follows that A' coincides with $A' \cap (A \cap B)$ and hence is a compact subset of M', thus proving that M' is complete hyperbolic. □

It should be observed that M' above can never be complete with respect to its Carathéodory distance $c_{M'}$ unless $M = M'$. In fact, $c_{M'} = c_M \mid_{M'}$ (see Sect. 13.4.1).

14.4 Geometric Methods

Given a complex manifold M, it is often a difficult problem to determine whether or not M is hyperbolic. In this section we want to give differential geometric criteria for hyperbolicity and measure hyperbolicity. They are based on generalizations of the Schwarz lemma.

Theorem 14.4.1 *A (complete) Hermitian manifold M whose holomorphic sectional curvature is bounded above by a negative constant is (complete) hyperbolic.*

Proof Let $ds^2_{\mathbb{D}}$ denote the Poincaré-Bergman metric on the open unit disc $\mathbb{D}$. Let ds^2_M be the Hermitian metric of M. If we multiply ds^2_M by a suitable positive constant, we have by Theorem 12.3.1 the following inequality for every holomorphic mapping $f : \mathbb{D} \to M$:

$$f^* \left(ds_M^2\right) \leqq ds_{\mathbb{D}}^2 .$$

If we denote by d' and ϱ the distance functions on M and $\mathbb{D}$ defined by ds_M^2 and $ds_{\mathbb{D}}^2$, respectively, then this inequality implies that f is distance-decreasing with respect to d' and ϱ, i.e., $d'(f(a), f(b)) \leqq \varrho(a, b)$ for $a, b \in \mathbb{D}$. Now the theorem follows from Proposition 14.1.8. □

Corollary 14.4.2 *The Gaussian plane minus two points $\mathbb{C} \setminus \{a, b\}$ is a complete hyperbolic manifold.*

This follows from the result in Sect. 13.4.1 and Theorem 14.4.1. Similarly, the following corollary follows from Theorem 12.4.1.

Corollary 14.4.3 *Every compact Riemann surface of genus $g \geqq 2$ is a compact hyperbolic manifold.*

As a consequence of and also as a generalization of Corollary 14.4.2, we have the following example (due to P.J. Kiernan).

Example 14.4.4 In $\mathbb{CP}^2$, let L_1, L_2, L_3, and L_4 be four (complex) lines in general position. Let $a = L_1 \cap L_2$ and $b = L_3 \cap L_4$. Let L_0 be the lines through a and b. Then $M = \mathbb{CP}^2 \setminus \bigcup_{i=0}^4 L_i$ is complete hyperbolic. In fact, M is biholomorphic to the direct product of two copies of $\mathbb{C} \setminus \{0, 1\}$. To see this, we consider L_0 as the line at infinity so that $\mathbb{CP}^2 \setminus L_0 = \mathbb{C}^2$. Then L_1 is parallel to L_2, since $L_1 \cap L_2 = a \in L_0$. Similarly, L_3 is parallel to L_4. But L_1 and L_3 are not parallel, since L_1, L_2, L_3, and L_4 are in general position in $\mathbb{CP}^2$. It is now clear that $M = \mathbb{C}^2 \setminus \bigcup_{i=1}^4 L_i$ is affinely equivalent to the direct product of two copies of $\mathbb{C} \setminus \{0, 1\}$. On the other hand, $\mathbb{CP}^2 \setminus \bigcup_{i=1}^4 L_i$ is not hyperbolic, since it contains $L_0 \setminus \{a, b\} = \mathbb{CP}^2 \setminus \{a, b\}$, which is not hyperbolic. Let p_j, $j = 1, \ldots, 4$, be four points in general position in $\mathbb{CP}^2$. Connecting every pair of these points, we obtain six lines L_j, $j = 0, 1, \ldots, 5$. Since $\mathbb{CP}^2 \setminus \bigcup_{i=0}^5 L_j$ is contained in $M = \mathbb{CP}^2 \setminus \bigcup_{i=0}^4 L_j$, it is also hyperbolic. It is not clear whether it is also complete. For higher dimensional analogs of this example, see Kiernan [277].

The following result is also due to Kiernan [279].

Theorem 14.4.5 *Let E be a holomorphic fiber bundle over M with fiber F and projection π. Then E is (complete) hyperbolic if M and F are (complete) hyperbolic.*

Proof Assume that M and F are hyperbolic. Let $p, q \in E$. If $\pi(p) \neq \pi(q)$, then $d_E(p, q) \geqq k_M(\pi(p), \pi(q)) > 0$. Assume $\pi(p) = \pi(q)$. Choose a neighborhood U of $\pi(p)$ in M such that $\pi^{-1}(U) = U \times F$. Let B_s be the ball in M centered at $\pi(p)$ and of radius s with respect to k_M. Choose $s > 0$ and $r > 0$ in such a way that $B_{2s} \subset U$ and $d_{\mathbb{D}}(z, 0) < s$ for $z \in \mathbb{D}(0, r)$. Thus, if $f : \mathbb{D} \to E$ is holomorphic and $f(0) \in \pi^{-1}(B_s)$, then $f(\mathbb{D}(0, r)) \subset U \times F$. Choose $c > 0$ such that $d_{\mathbb{D}}(0, a) \geqq c d_{\mathbb{D}(0,r)}(0, a)$ for all $a \in \mathbb{D}(0, r/2)$. Let $f_i : \mathbb{D} \to E$ be holomorphic mappings, and let a_i, b_i be points of $\mathbb{D}$ such that $p = f_1(a_1)$, $f_1(b_1) = f_2(a_2), \ldots, f_k(b_k) = q$.

By homogeneity of $\mathbb{D}$, we may assume that $a_i = 0$ for all i. By inserting extra terms in this chain if necessary, we may also assume that $b_i \in \mathbb{D}(0, r/2)$ for all i. We set $p_0 = p, p_1 = f_1(b_1), \ldots, p_k = f_k(b_k) = q$.

We have two cases to consider. Consider first the case where at least one of the p_i's is not contained in $\pi^{-1}(B_s)$. Then it is easy to see that

$$\sum_{i=1}^{k} d_{\mathbb{D}}(0, b_i) \geqq \sum_{i=1}^{k} d_E(f_i(0), f_i(b_i)) = \sum_{i=1}^{k} d_E(p_{i-1}, p_i)$$
$$\geqq \sum_{i=1}^{k} k_M(\pi(p_{i-1}), \pi(p_i)) \geqq s.$$

Consider next the case where all p_i's are in $\pi^{-1}(B_s)$. Then

$$\sum_{i=1}^{i} d_{\mathbb{D}}(0, b_i) \geqq c \sum_{i=1}^{k} d_{\mathbb{D}(0,r)}(0, b_i)$$
$$\geqq c \sum_{i=1}^{k} d_F(\varphi(p_{i-1}), \varphi(p_i)) \geqq c d_F(p, q),$$

where $\varphi : U \times F \to F$ is the projection. This shows that

$$d_E(p, q) \geqq \min[s, c d_F(p, q)] > 0.$$

Thus E is hyperbolic.

Assume that M and F are complete hyperbolic. Let $\{p_n\}$ be a Cauchy sequence in E. Then $\{\pi(p_n)\}$ is a Cauchy sequence in M and therefore $\pi(p_n) \to x_0$ for some $x_0 \in M$. Choose a neighborhood U of x_0 in M such that $\pi^{-1}(U) = U \times F$. Choose $s > 0$, $r > 0$, and $c > 0$ as above. Given $\varepsilon > 0$ with $2\varepsilon < s$, choose an integer N such that $p_n \in \pi^{-1}(B_s)$ and $d_E(p_n, p_m) < \varepsilon$ for $n, m > N$. We shall show that $d_F(\varphi(p_n), \varphi(p_m)) < \varepsilon/c$ for $n, m > N$. We fix $n, m > N$. We choose holomorphic mappings $f_i : \mathbb{D} \to E$ and points $b_i \in \mathbb{D}$ such that

$$p_n = f_1(0), \quad f_1(b_1) = f_2(0), \ldots, f_{k-1}(b_{k-1}) = f_k(0), \quad f_k(b_k) = p_m,$$

and $\sum_{i=1}^{k} d_{\mathbb{D}}(0, b_i) < \varepsilon$. We may again assume without loss of generality that $b_i \in \mathbb{D}(0, r/2)$ for all i. Since

$$\varepsilon > \sum_{i=1}^{k} d_{\mathbb{D}}(0, b_i) \geqq \sum_{i=1}^{k} d_E(f_i(0), f_i(b)) \geqq \sum_{i=1}^{k} k_M(\pi \circ f_i(0), \pi \circ f_i(b)),$$

it follows that $f_i(0) \in \pi^{-1}\left(B_{2_\varepsilon}\right) \subset \pi^{-1}(B_s)$. Hence

$$\begin{aligned}\varepsilon > \sum_{i=1}^{k} d_{\mathbb{D}}(0, b_i) &\geqq c \sum_{i=1}^{k} d_{\mathbb{D}(0,r/2)}(0, b_i) \\ &\geqq c \sum_{i=1}^{k} d_F(\varphi \circ f_i(0), \varphi \circ f_i(b_i)) \\ &\geqq c d_F(\varphi(p_n), \varphi(p_m)).\end{aligned}$$

This shows that $\{\varphi(p_n)\}$ is a Cauchy sequence in F, and therefore $\varphi(p_n) \to y_0$ for some $y_0 \in F$. Clearly, $p_n \to (x_0, y_0) \in U \times F$. Thus E is complete. □

Remark 14.4.6 Since F may be considered as a closed complex submanifold of E, it follows that if E is (complete) hyperbolic, then F is also (complete hyperbolic). On the other hand, M need not be hyperbolic even if both E and F are complete hyperbolic. Let $\mathbb{B}^* = \left\{(z, w) \in \mathbb{C}^2 : 0 < |z|^2 + |w|^2 < 1\right\}$. Then $\mathbb{B}^*$ is a holomorphic fiber bundle over $\mathbb{CP}$ with fiber $\mathbb{D}^* = \{z \in \mathbb{C} : 0 < |z| < 1\}$. The bundle space $\mathbb{B}^*$ is hyperbolic, and the fiber $\mathbb{D}^*$ is complete hyperbolic, but the base space $\mathbb{CP}$ is not hyperbolic. If we set $E = \left\{(z, w) \in \mathbb{C}^2 : z \neq 0 \text{ and } |z|^2 + |w|^2 < 1\right\}$, the E is a fiber bundle over $\mathbb{C}$ with fiber $\mathbb{D}^*$. This furnishes an example where E is even complete.

By a reasoning similar to the proof of Theorem 14.4.5, Kiernan [279] proves that if E is a Hermitian vector bundle over a hyperbolic manifold M, then the open unit ball bundle $\{X \in E : \|X\| < 1\}$ is hyperbolic.

Theorem 14.4.5 was generalized by M.A. Illarionov [252].

We fix some notations and definitions.

The notion of a *complex space* was first introduced in 1951 by H. Behnke and K. Stein [50] and H. Cartan [107]; their definitions correspond to what is now called a normal complex space (see [186]). The definition that is accepted as the standard one, which is also used in this book, was given by J.-P. Serre in his GAGA paper [467]: A *reduced complex space* is a pair $(X, \mathcal{O}_X)$, where X is a paracompact Hausdorff topological space and $\mathcal{O}_X$ is a sheaf of rings of continuous functions on X (a subsheaf of the sheaf $\mathcal{C}_X$ of germs of continuous functions) such that for every point $x \in X$ there is a neighborhood $U \subset X$ and a homeomorphism $\phi : U \to A \subset \mathbb{C}^n$ onto a locally closed complex subvariety of $\mathbb{C}^n$ so that the homomorphism $\phi^* : \mathcal{C}_A \to \mathcal{C}_X$, $f \mapsto f \circ \phi$, induces an isomorphism of $\mathcal{O}_A$ onto $\mathcal{O}_U = \mathcal{O}_X|_U$. Intuitively speaking, X is obtained by gluing pieces of subvarieties in Euclidean spaces using biholomorphic transition maps.

Definition 14.4.7 A holomorphic fiber bundle is a triple (Z, π, X), where X and Z are complex spaces and $\pi : Z \to X$ is a holomorphic map of Z onto X such that there exist a complex manifold Y, an open cover $\mathcal{U} = \{U_\alpha\}$ of X, and for every α a biholomorphic map

$$\theta_\alpha : Z|_{U_\alpha} = \pi^{-1}(U_\alpha) \to U_\alpha \times Y, \quad \theta_\alpha(z) = (\pi(z), \vartheta_\alpha(z)). \tag{14.1}$$

The manifold Z is the *total space*, X is the *base space*, and Y is the *fiber*.

The simplest example is a product bundle $\pi : Z = X \times Y \to X$, $(x, y) \mapsto x$. By definition, every fiber bundle is isomorphic to the product bundle over small open sets in X. The fiber $Z_x = \pi^{-1}(x)$ over any point $x \in X$ is biholomorphic to Y. A map θ_α (14.1) is a fiber bundle chart on Z, and the collection $\{(U_\alpha, \theta_\alpha)\}$ is a holomorphic fiber bundle atlas on Z. The transition maps $\theta_{\alpha,\beta} = \theta_\alpha \circ \theta_\beta^{-1} : U_{\alpha,\beta} \times Y \to U_{\alpha,\beta} \times Y$ are of the form

$$\theta_{\alpha,\beta}(x, y) = \big(x, \vartheta_{\alpha,\beta}(x, y)\big), \quad x \in U_{\alpha,\beta}, y \in Y, \tag{14.2}$$

and they satisfy the cocycle condition

$$\theta_{\alpha,\alpha} = \mathrm{Id}, \quad \theta_{\alpha,\beta} \circ \theta_{\beta,\gamma} \circ \theta_{\gamma,\alpha} = \mathrm{Id} \quad \text{on } U_{\alpha,\beta,\gamma} \times Y. \tag{14.3}$$

For every fixed $x \in U_{\alpha,\beta}$, we have $\vartheta_{\alpha,\beta}(x, \cdot) \in \mathit{Aut}Y$. Conversely, given an open cover $\mathcal{U} = \{U_\alpha\}$ of X and a collection of biholomorphic self-maps (14.2) satisfying the cocycle condition (14.3), we get a holomorphic fiber bundle $Z \to X$ with these transition maps by taking Z to be the disjoint union of all $U_\alpha \times Y$, modulo the identifications provided by the transition maps. A *section* of $\pi : Z \to X$ is a map $f : X \to Z$ such that $\pi \circ f$ is the identity on X; that is, $f(x) \in Z_x$ for every $x \in X$. Any section of the product bundle $X \times Y \to X$ is of the form $f(x) = (x, g(x))$, where $g : X \to Y$ is a map to the fiber. If $\{(U_\alpha, \theta_\alpha)\}$ is a holomorphic fiber bundle atlas on $Z \to X$ with the transition maps $\theta_{\alpha,\beta}$ (14.2), then a holomorphic section $f : X \to Z$ is given by a collection of holomorphic maps $f_\alpha : U_\alpha \to Y$ satisfying the compatibility conditions

$$f_\alpha(x) = \vartheta_{\alpha,\beta}\left(x, f_\beta(x)\right), \quad x \in U_{\alpha,\beta}. \tag{14.4}$$

Definition 14.4.8 A holomorphic isomorphism of holomorphic fiber bundles $\pi : Z \to X$, $\pi' : Z' \to X$ is a biholomorphic map $\Phi : Z \to Z'$ such that $\pi' \circ \Phi = \pi$; if such Φ exists, then the bundles are holomorphically isomorphic. A fiber bundle is trivial if it is isomorphic to the product bundle.

Isomorphisms of a fiber bundle onto itself are *fiber bundle automorphisms*. Holomorphic automorphisms of a product bundle $X \times Y \to X$ are biholomorphic self-maps of $X \times Y$ of the form $(x, y) \mapsto (x, \varphi(x, y))$, with $\varphi(x, \cdot) \in \mathit{Aut}Y$ for every $x \in X$. In general we choose an open cover $\mathcal{U} = \{U_\alpha\}$ of X and fiber bundle atlases $\{(U_\alpha, \theta_\alpha)\}$ for (Z, π, X) and $\big\{(U_\alpha, \theta'_\alpha)\big\}$ for $\big(Z', \pi', X\big)$. A fiber bundle isomorphism Φ of (Z, π, X) to $\big(Z', \pi', X\big)$ is then given by a collection of fiber-preserving biholomorphic self-maps $\phi_\alpha : U_\alpha \times Y \to U_\alpha \times Y$ of the form $\phi_\alpha(x, y) = (x, \varphi_\alpha(x, y))$ so that the following diagrams commute:

$$\begin{array}{ccc} U_{\alpha,\beta} \times Y & \stackrel{\theta_{\alpha,\beta}}{\longleftarrow} & U_{\alpha,\beta} \times Y \\ \downarrow \phi_\alpha & & \downarrow \phi_\beta \\ U_{\alpha,\beta} \times Y & \stackrel{\theta'_{\alpha,\beta}}{\longleftarrow} & U_{\alpha,\beta} \times Y. \end{array}$$

If there exists a fiber bundle atlas on $Z \to X$ (in the given isomorphism class) such that all transition maps belong to a certain subgroup G of $AutY$, then we say that G is the structure group of the bundle or that the structure group of the bundle has been reduced to G. If $\pi : Z \to X$ is a holomorphic fiber bundle and X' is a complex subvariety of X, then the restriction $\pi : Z' = Z|_{X'} \to X'$ is a holomorphic fiber bundle over X', called the restricted bundle. Given a holomorphic fiber bundle $\pi : Z \to X$ with fiber Y and a holomorphic map $f : W \to X$, the pullback bundle $\pi' : f^*Z \to W$ and the map $F : f^*Z \to Z$ are defined as follows:

$$f^*Z = \{(w, z) \in W \times Z : f(w) = \pi(z)\}, \quad \pi'(w, z) = w, \quad F(w, z) = z. \tag{14.5}$$

Let $\{(U_\alpha, \theta_\alpha)\}$ be a fiber bundle atlas for $\pi : Z \to X$ with

$$\theta_\alpha(z) = (\pi(z), \vartheta_\alpha(z)) \in U_\alpha \times Y.$$

Set $V_\alpha = f^{-1}(U_\alpha) \subset W$, and define a map $\theta'_\alpha : f^*Z|_{V_\alpha} \to V_\alpha \times Y$ by $\theta'_\alpha(w, z) = (w, \vartheta_\alpha(z))$. (The map f appears implicitly in the above definition by the condition $\pi(z) = f(w)$.) This is a holomorphic fiber bundle atlas on $\pi' : f^*Z \to W$ with transition maps $\vartheta'_{\alpha,\beta}(w, y) = \vartheta_{\alpha,\beta}(f(w), y)$; hence $f^*Z \to W$ is indeed a holomorphic fiber bundle with fiber Y.

Definition 14.4.9 *A holomorphic vector bundle of rank n over a complex space X is a holomorphic fiber bundle $E \to X$ (see Definition 14.4.7) with fiber $Y = \mathbb{C}^n$ and structure group $GL_n(\mathbb{C})$. A vector bundle of rank $n = 1$ is also called a (holomorphic) line bundle.*

This means that we have an open cover $\mathcal{U} = \{U_\alpha\}$ of X and vector bundle charts $\theta_\alpha : E|_{U_\alpha} \xrightarrow{\cong} U_\alpha \times \mathbb{C}^n$ with transition maps of the form

$$\theta_\alpha \circ \theta_\beta^{-1}(x, v) = \theta_{\alpha,\beta}(x, v) = \left(x, g_{\alpha,\beta}(x)v\right), \quad x \in U_{\alpha,\beta}, v \in \mathbb{C}^n, \tag{14.6}$$

where $g_{\alpha,\beta} : U_{\alpha,\beta} \to GL_n(\mathbb{C})$ is a holomorphic multiplicative 1-cocycle:

$$g_{\alpha,\alpha} = 1, \quad g_{\alpha,\beta}g_{\beta,\alpha} = 1, \quad g_{\alpha,\beta}g_{\beta,\gamma}g_{\gamma,\alpha} = 1. \tag{14.7}$$

Every fiber E_x is a complex vector space such that the fiber bundle charts $\theta_\alpha : Z|_{U_\alpha} \to U_\alpha \times \mathbb{C}^n$ are $\mathbb{C}$-vector space isomorphisms on each fiber. If X is a real manifold of class $\mathcal{C}^r$ and $E \to X$ is a complex vector bundle whose transition maps $g_{\alpha,\beta} : U_{\alpha,\beta} \to GL_n(\mathbb{C})$ are of class $\mathcal{C}^r$, then we have a complex vector bundle of class $\mathcal{C}^r$ over X. For $r = 0$, we have a topological complex vector bundle over X.

Replacing $\mathbb{C}$ by any field F, we get topological F-vector bundles over topological spaces or smooth F-vector bundles over smooth manifolds. For $F = \mathbb{R}$, we speak of real vector bundles. Every vector bundle has the zero section sending each point $x \in X$ to the origin $0_x \in E_x$. Given a holomorphic vector bundle atlas $\{(U_\alpha, \theta_\alpha)\}$ on E with transition maps $g_{\alpha,\beta}$ (14.6), a section $f : X \to E$ is determined by a collection of maps $f_\alpha : U_\alpha \to \mathbb{C}^n$ satisfying the compatibility conditions

$$f_\alpha = g_{\alpha,\beta} f_\beta \quad \text{on } U_{\alpha,\beta}. \tag{14.8}$$

A complex manifold G that is also a group with holomorphic group operations is called a *complex Lie group*. The main examples include the general linear group $GL_n(\mathbb{C})$ (the group of invertible complex $n \times n$ matrices) and its subgroups such as $SL_n(\mathbb{C}) = \{A \in GL_n(\mathbb{C}) : \det A = 1\}$, the symplectic subgroup $Sp_n(\mathbb{C}) \subset GL_{2n}(\mathbb{C})$, certain quotients such as the projective linear group $PGL_n(\mathbb{C}) = GL_{n+1}(\mathbb{C})/\mathbb{C}^* = \operatorname{Aut} \mathbb{P}^n$ (the holomorphic automorphism group of $\mathbb{P}^n$), universal coverings of Lie groups, etc. A complex manifold X is said to be G-homogeneous if there exists a transitive holomorphic action $G \times X \to X$ of G on X by holomorphic automorphisms. Let Z be the total space of a holomorphic bundle over the complex space X with the fiber Y and projection $\pi : Z \to X$ and the structure group G—the finite-dimensional complex Lie group.

Theorem 14.4.10 *If the holomorphic bundle (Z, π, X) satisfies the condition $k_Y = 0$, then for any points $x, y \in Z$ the equality $k_Z(x, y) = k_X(\pi(x), \pi(y))$ holds.*

Proof Let us fix $\varepsilon > 0$. By the definition of the Kobayashi pseudometric, there are points $\pi(x) = p_0, p_1, \ldots, p_k = \pi(y)$ from X, points $a_1, a_2, \ldots, a_k, b_1, b_2, \ldots, b_k$ from $\mathbb{D}$, and holomorphic maps $f_1, f_2, \ldots, f_k$ of the disc $\mathbb{D}$ into the space X, such that $f_i(a_i) = p_{i-1}$ and $f_i(b_i) = p_i$, $i = 1, 2, \ldots, k$, satisfying the following inequality:

$$\sum_{i=1}^{k} \rho(a_i, b_i) < k_X(\pi(x), \pi(y)) + \varepsilon.$$

We construct holomorphic mappings $\varphi_1, \varphi_2, \ldots, \varphi_k$ from $\mathbb{D}$ into Z such that $\pi \circ \varphi_i = f_i$ for $i = 1, 2, \ldots, k$; $\varphi_i(b_i) = \varphi_{i+1}(a_{i+1})$, $i = 1, 2, \ldots, k-1$, and $\varphi_1(a_1) = x$. Suppose that $\varphi_1, \varphi_2, \ldots, \varphi_{j-1}$ are constructed. Let $\left(f_j^*(Z), f_j^*(\pi), \mathbb{D}\right)$ be the inverse image of the bundle (Z, π, X) for mapping $f_j : \mathbb{D} \to X$. The existence of equivalence of $\mathbb{D} \times Y \to f_j^*(Z)$ bundles over $\mathbb{D}$ is a consequence of Grauert's theorem [184], formulated as follows: classification of Steenrod holomorphic bundles over Stein space, whose structure group is a finite-dimensional complex analytic Lie group G and coincides with the topological classification. There is a point $c \in Y$ such that

$$\Psi \circ \Phi\left(a_j, c\right) = \varphi_{j-1}\left(b_{j-1}\right),$$

where $\Psi : f_j^*(Z) \to Z$ is a bundle map over $f_j : \mathbb{D} \to X$. Let us put

$$\varphi_j(z) = \Psi \circ \Phi(z, c).$$

Using the contraction property of the Kobayashi pseudometric and the condition $k_Y = 0$, we obtain

$$k_X(p(x), p(y)) \leq k_Z(x, y) \leqslant k_Z(x, \varphi_k(b_k)) + k_Z(\varphi_k(b_k), y) \leq$$

$$\leq \sum_{i=1}^{k} \rho(a_i, b_i) + k_F(\varphi_k(b_k), y) \leq k_X(p(x), p(y)) + \varepsilon.$$

As ε is arbitrary, $k_Z(x, y) = k_X(p(x), p(y))$. □

The following results are also due to Peter Kiernan. Let M be a complex manifold, consider a point $p \in M$, let $z = (z_1, \ldots, z_n), z(p) = 0$, be local coordinates on M in a neighborhood of p, and let $\mathfrak{B}_r = \{p' \in M : |z(p')| < r\}$, $\mathfrak{B}_1 = \mathfrak{B}$.

Lemma 14.4.11 *If there exist numbers r and δ such that for all holomorphic mappings $f : \mathbb{D} \to M$ the condition $f(0) \in \mathfrak{B}_r$ implies that $f(\mathbb{D}(0, \delta)) \subset \mathfrak{B}$, then for all $q \in M \setminus \mathfrak{B}$ the Kobayashi distance $k_M(p, q) > 0$.*

Proof Let $q \in M \setminus \mathfrak{B}$, and let $\sigma = \left\{f^j, \zeta_j', \zeta_j''\right\}_{j=1}^m$ be a chain on M from p to q; carrying out auxiliary Lobachevskii motions in $\mathbb{D}$, we may assume that all the $\zeta_j' = 0$. We need to prove that for all such chains the lengths $|\sigma| = \sum_{j=1}^m \rho\left(0, \zeta_j''\right)$ are bounded from below by a positive constant.

If in the chain σ there is a point $\zeta_j'' \notin \mathbb{D}(0, \delta/2)$, then $|\sigma| \geq \rho(0, \delta/2) > 0$, and hence it suffices to consider chains for which all $\zeta_j'' \in \mathbb{D}(0, \delta/2)$. We write $p_j = f^j\left(\zeta_j''\right) = f^{j+1}(0)$; there is a number l, $1 \leq l \leq m$, such that $p_1, \ldots, p_{l-1} \in \mathfrak{B}_r$, and $p_l \notin \mathfrak{B}_r$. Since $f^l(0) = p_{l-1} \in \mathfrak{B}_r$ and $\zeta_l'' \in \mathbb{D}(0, \delta)$, then by the hypothesis of the lemma $f^l\left(\zeta_l'\right) = p_l \in \mathfrak{B}$, and hence, $p_l \in B \setminus \mathfrak{B}_r$. We choose a constant $\lambda > 0$ such that for all $\zeta \in \mathbb{D}(0, \delta/2)$ the Lobachevskii distance $\rho_{\mathbb{D}}(0, \zeta) \geq \lambda\rho_{\mathbb{D}(0,\delta)}(0, \zeta)$. Since f^j, $1 \leq j \leq l$, maps $\mathbb{D}(0, \delta)$ holomorphically into $\mathfrak{B}$, then by the contractibility property of the Kobayashi metric and the triangle inequality we have

$$\sum_{j=1}^{m} \rho\left(0, \zeta_j''\right) \geq \lambda \sum_{j=1}^{l} \rho_{\mathbb{D}(0,\delta)}\left(0, \zeta_j''\right) \geq \lambda \sum_{j=1}^{l} k_B\left(p_{j-1}, p_j\right) \geq \lambda k_B\left(p, p_l\right).$$

Since $k_{\mathfrak{B}}$ is a metric and $p_l \in \mathfrak{B} \setminus \mathfrak{B}_r$, the quantity on the right-hand side is bounded from below by a positive constant that does not depend on the chain. □

As before let $\mathcal{O}(\mathbb{D}, M)$ be the family of holomorphic mappings from the unit disc $\mathbb{D}$ into a complex manifold M.

Theorem 14.4.12 *If in some metric d that induces the topology on M the family $\mathcal{O}(\mathbb{D}, M)$ is equicontinuous, then M is a hyperbolic manifold. Conversely, if M is a hyperbolic manifold, then $\mathcal{O}(\mathbb{D}, M)$ is equicontinuous in the Kobayashi metric k_M.*

Proof Suppose that $\mathcal{O}(\mathbb{D}, M)$ is equicontinuous in the metric d, and let $\mathfrak{B}_\varepsilon^d(p) = \{p' \in M : d(p', p) < \varepsilon\}$. For any point $q \in M$, different from p, we take local coordinates z with center p, as in Lemma 14.4.11 (so that $q \notin \mathfrak{B}$), and we choose $\varepsilon > 0$ so that $\mathfrak{B}_{2\varepsilon}^d(p) \subset \mathfrak{B}$. By the equicontinuity there is a $\delta > 0$ such that for any $f \in \mathcal{O}(\mathbb{D}, M)$ for which $f(0) \in \mathfrak{B}_\varepsilon^d(p)$, we certainly have $f(D_\delta) \subset \mathfrak{B}_{2\varepsilon}^d(p) \subset B$. If we also take $r > 0$ such that $\mathfrak{B}_r \subset \mathfrak{B}_\varepsilon^d(p)$, then the hypothesis of Lemma 14.4.11 will be satisfied, and hence, $d_M(p, q) > 0$. We have proved that M is hyperbolic.

The second assertion of the theorem follows from the fact that holomorphic mappings do not increase the Kobayashi metric. □

In [293], Professor Kobayashi defined hyperbolic and complete hyperbolic manifolds. In [535], Professor Wu defined tight and taut complex manifolds.

The complex manifolds M for which the family $\mathcal{O}(\mathbb{D}, M)$ is equicontinuous in some metric that induces the topology on M are termed *tight*. We will say that a complex manifold M possesses the Montel property if the family $\mathcal{O}(\mathbb{D}, M)$ is normal, i.e., from any sequence $f^\nu \in \mathcal{O}(\mathbb{D}, M)$, we can extract a subsequence f^{ν_j} that either converges uniformly on compact subsets of $\mathbb{D}$ or diverges compactly (this means that, for any compact subsets $K \subset \mathbb{D}$ and $K' \subset M$, there is a j_0 such that $f^{\nu_j}(K) \cap K' = \varnothing$ for all $j \geq j_0$). In the literature manifolds possessing this property are termed *taut*.

Theorem 14.4.13 *If a manifold M possesses the Montel property, then it is hyperbolic.*

Proof If M is not hyperbolic, then there are distinct points $p, q \in M$ such that $k_M(p, q) = 0$. We use Lemma 14.4.11, in which we take $r = 1/2$ and $\delta = 1/\nu$; by this lemma for any natural number ν, there is a mapping $f^\nu \in \mathcal{O}(\mathbb{D}, M)$ such that $f^\nu(0) \in \mathfrak{B}_{1/2}$, but $f^\nu(\mathbb{D}(0, 1/\nu)) \not\subset \mathfrak{B}$. It is obviously impossible to extract from the sequence $\{f^\nu\}$ a subsequence that is either uniformly convergent on compact subsets of $\mathbb{D}$ or compactly divergent. Thus, M does not possess the Montel property. □

The converse is not true. This follows from another theorem, and in order to state it we need a definition.

Definition 14.4.14 Let $\overline{\mathbb{D}} = \{|z| \leq 1\}$ be the closed disc, and let $A_r = \{r \leq |z| \leq 1\}$ be an annulus in it. We consider an arbitrary sequence $\{f^\nu\}$ of holomorphic mappings of $\overline{\mathbb{D}}$ into a complex manifold M, and we denote by $f^\nu|_{A_r}$ the restriction of f^ν to the annulus A_r for some $r < 1$. We will say that M satisfies the *disc condition* if the fact that $f^\nu|_{A_r}$ converges uniformly in A_r to a mapping

$f \in \mathcal{O}(A_r, M)$ implies that f^ν converges uniformly in $\overline{\mathbb{D}}$ to some mapping $\tilde{f} \in \mathcal{O}(\overline{\mathbb{D}}, M)$.

The disc condition is one of the forms of complex convexity; it implies the pseudoconvexity of manifolds (this is a generalization of the continuity principle).

Theorem 14.4.15 *If a complex manifold M possesses the Montel property, then it satisfies the disc condition.*

Proof We take the unit disc $\overline{\mathbb{D}}$, an annulus A_r in it, and an arbitrary sequence $f^\nu \in \mathcal{O}(\overline{\mathbb{D}}, M)$ such that the restriction $f^\nu|_{A_r} \to f \in \mathcal{O}(A_r, M)$. Since $\bigcup_{\nu=1}^{\infty} f^\nu(A_r)$ is a compact subset of M, then it is not possible to extract a compactly divergent sequence from the sequence $\left\{f^\nu|_{A_r}\right\}$. But then by the Montel property, we can extract from it a subsequence that converges uniformly on $\overline{\mathbb{D}}$ to a mapping $\tilde{f} \in \mathcal{O}(\overline{\mathbb{D}}, M)$. Since $\tilde{f}\Big|_{A_r} = f$ (since on A_r the whole sequence converges to f), then by the uniqueness theorem $\tilde{f}^r$ does not depend on the choice of subsequence. But it follows from this that the whole sequence $f^\nu \to \tilde{f}$ converges in $\overline{\mathbb{D}}$, and the disc condition holds. □

Example 14.4.16 A bounded domain in $\mathbb{C}^n$ that is not a domain of holomorphy is a hyperbolic manifold, but the disc condition does not hold, and hence, it also does not possess the Montel property, and a converse theorem to Theorem 14.4.13 does not hold.

We say that a hyperbolic manifold M is *complete hyperbolic* if any set on it that is bounded in the Kobayashi metric k_M is compact. For such manifolds, a converse theorem to Theorem 14.4.13 does indeed hold.

Theorem 14.4.17 *A complete hyperbolic manifold M possesses the Montel property.*

Proof Since M is hyperbolic, then by Theorem 14.4.12 the family $\mathcal{O}(\mathbb{D}, M)$ is equicontinuous in the metric k_M. Suppose that it is not possible to isolate a compactly divergent sequence from the sequence $f^\nu \in \mathcal{O}(\mathbb{D}, M)$. Then there are sets $K_0 \subset\subset \mathbb{D}$, $K_0' \subset\subset M$ and a subsequence of $\{f^\nu\}$ (which we again denote by $\{f^\nu\}$) such that $f^\nu(K_0) \cap K_0' \neq \varnothing$ for all ν. Therefore there is a sequence $\zeta_\nu \in K_0$ such that $f^\nu(\zeta_\nu) \in K_0'$ for all ν.

Let K be an arbitrary compact subset of $\mathbb{D}$; we shall prove that there exists a bounded set K' in the Kobayashi metric that contains all the $f^\nu(K)$. Without loss of generality we may assume that $K \supset K_0$; by the contractibility property there is a number R_1 such that $k_M(f^\nu(\zeta), f^\nu(\zeta_0)) \leq R_1$ for all $\zeta \in K$, $\zeta_0 \in K_0$, and all ν. On the other hand, for any fixed point $p \in M$, there exists a number R_2 such that $k_M(p, p_0) \leq R_2$ for all $p_0 \in K_0'$. Then

$$k_M\left(f^\nu(\zeta), p\right) \leq k_M\left(f^\nu(\zeta), f^\nu(\zeta_\nu)\right) + k_M\left(f^\nu(\zeta_\nu), p\right) \leq R_1 + R_2$$

for all $\zeta \in K$ and all ν. Thus, we can take as K' the Kobayashi ball with center p and radius $R_1 + R_2$.

We have proved that the sequence $\{f^\nu\}$ is not only equicontinuous but also uniformly bounded in the metric k_M. Since by hypothesis any bounded set in this metric is compact, then, exactly as in the proof of Montel's theorem, we will conclude from this that we can extract from $\{f^\nu\}$ a subsequence that converges uniformly on compact subsets of $\mathbb{D}$. We have proved that the family $\mathcal{O}(\mathbb{D}, M)$ is normal, i.e., possesses the Montel property. □

Combining Theorems 14.4.15 and 14.4.17 we obtain the following theorem:

Theorem 14.4.18 *Any complete hyperbolic manifold satisfies the disc condition.*

In particular, any domain in $\mathbb{C}^n$ which is a complete hyperbolic manifold is also a domain of holomorphy.

We want to give an infinitesimal formulation of hyperbolicity. For this purpose we let M be a connected, complex manifold having a Hermitian metric $|v|$ in the complexified tangent bundle of M. (We will also write $|p, q|$ for the distance, with respect to this metric, between points $p, q \in M$.) For a C^∞ mapping $f : \mathbb{D} \to M$, we define

$$\left|f'(z_0)\right| = \left|f_*(\partial/\partial z)_{z=z_0}\right|.$$

We recall the definition of the Kobayashi pseudodistance on a connected complex manifold M. Denote by $\mathbb{D}(0, r) = \{z \in \mathbb{C} : |z| < r\}$ the disc of radius r and by ω_r the invariant metric $r^4 dzdz/\left(r^2 - |z|^2\right)^2$, normalized so that $\omega_r(0) = dzdz$ is the Euclidean metric. We agree to use $\omega_r(p, q)$ to denote the distance between points $p, q \in \mathbb{D}(0, r)$ with respect to the metric ω_r.

Lemma 14.4.19 (Brody) *M is hyperbolic if, and only if,* $\sup\left|f'(0)\right| < \infty$, *for all* $f \in \mathcal{O}(\mathbb{D}, M)$.

Proof Because $\mathbb{D}$ is homogeneous, $\sup\left|f'(0)\right| = \sup|f_*(v)|$, $v \in T(\mathbb{D})$, $\omega(v) = 1$. Now suppose that this supremum is equal to $c < \infty$; then $d_M(p, q) \geqslant c^{-1}|p, q| > 0$, $p \neq q \in M$; therefore M is hyperbolic.

Conversely, if this supremum is infinite, there is a sequence of mappings $f_j \in \mathcal{O}(\mathbb{D}, M)$ with $\left|f_j'(0)\right| \nearrow \infty$; by compactness we may assume $f_j(0) \to p$. Let U and $V \supset U$ be two coordinate neighborhoods of p. The Cauchy estimates tell us that for any positive integer m and sufficiently large j, $f_j\left(\mathbb{D}(0, 1/m)\right) \cap \partial U \neq \varnothing$. In particular, we may choose a sequence of points $\{x_m\}$, $x_m \in \partial U$, such that $x_m \in f_j\left(\mathbb{D}(0, 1/m)\right)$ for some large n, and therefore

$$d_M\left(f_j(0), x_m\right) \leqslant \omega(0, 1/m) \to 0 \quad \text{as } m \to 0.$$

By the obvious continuity of d_M and the compactness of ∂U, we obtain a point $x \in \partial U$ with $d_M(p, x) = 0$. □

Theorem 14.4.20 (Brody) *Let M be a compact complex manifold. Suppose that M is not Kobayashi hyperbolic. Then there exists a non constant holomorphic $f : \mathbb{C} \to M$.*

Proof Suppose that M is not Kobayashi hyperbolic. Then by Lemma 14.4.19 there exists a sequence of holomorphic maps $\{f_j\}_{j=1}^{\infty}$, $f_j : \mathbb{D} \to M$, $f_j(0) = p_j$, $\left|f_j'(0)\right| \nearrow \infty$.

We may assume that f_j is holomorphic in a slightly larger disc.

Define $H_j : \overline{\mathbb{D}} \to \mathbb{R}^+$, $H_j(z) = \left|f_j'(z)\right| \left(1 - \left|z^2\right|\right)$. Then there exists $z_j \in \mathbb{D}$, and $H_j\left(z_j\right) = \max_{\mathbb{D}} H_j$. Let $g_j : \mathbb{D} \to M$, $g_j = f_j \circ \phi_j(w)$, where

$$\phi_j(w) = \frac{w + z_j}{1 + wz_j}.$$

Then

$$\left|g_j'(w)\right| \left(1 - \left|w^2\right|\right) = \left|f_j'(z)\right| \left|\phi_j'(w)\right| \left(1 - \left|w^2\right|\right) = \left|f_j'(z)\right| \left(1 - \left|z^2\right|\right)$$

(the last equality follows by a short computation).

So now,

$$\left|g_j'(w)\right| \leq \frac{\left|g_j'(0)\right|}{1 - \left|w^2\right|}.$$

Rescale the disc: Let $R_j := \left|g_j'(0)\right|$ (note that $\left|g_j'(0)\right| \to \infty$, $j \to \infty$), and define $h_j : \mathbb{D}\left(0, R_j\right) \to M$, $h_j(z) = g_j\left(z/R_j\right)$. Then

$$\left|h_j'(z)\right| = \frac{\left|g_j'\left(z/R_j\right)\right|}{R_j} \leq \frac{\left|g_j'(0)\right|}{R_j} \frac{1}{1 - \left|z/R_j\right|^2} \leq 2,$$

say on $\mathbb{D}\left(0, R_j/2\right)$. Also $\left|h_j'(0)\right| = 1$, so using a normal families argument we can find a holomorphic map $f : \mathbb{C} \to M$ such that $\left|f'(0)\right| = 1$; hence f is nonconstant. □

14.5 Invariant Volume Forms

Definition and Simplest Properties

Let $\mathbb{B}$ be an n-dimensional unit ball with Bergman metric

$$|\xi|^2_{\mathbb{B},z} = \sum_{i,j} g_{ij}(z)\xi_i\overline{\xi}_j$$

(see Example 13.2.16). The volume element v of $\mathbb{B}$ is given by

$$v_M = i^n G dz^1 \wedge d\bar{z}^1 \wedge \cdots \wedge dz^n \wedge d\bar{z}^n,$$

where

$$G = \det\left(g_{ij}\right).$$

The volume of a set $E \subset \mathbb{B}^n$ computed in the Bergman metric of $\mathbb{B}$ is

$$vol(E) = \int_E v.$$

In [294] Kobayashi introduced a volume analog of the metrics c_M and k_M considered in Sects. 13.4 and 13.5.

Definition 14.5.1 Let M be a complex n-dimensional manifold and $Q \subset M$ any subset of M. The invariant C-volume (or Carathéodory volume) of Q is defined as

$$V^c_M(Q) = \sup \mathrm{vol}\, f(Q),$$

where the least upper bound is taken over all holomorphic maps $f : M \to \mathbb{B}^n$.

The invariant K-volume (or Kobayashi volume) of Q is defined as

$$V^k_M(Q) = \inf \sum_{j=1}^{m} \mathrm{vol}\, E_j,$$

where the greatest lower bound is taken over all possible families of Borel measurable sets $E_j \subset B^n$ and holomorphic maps $f^j : B^n \to M$ such that $\bigcup_{j=1}^n f^j\left(E_j\right) \supset Q$.

Instead of the unit ball $\mathbb{B}^n$, one can also take the unit polydisc Δ^n with the Bergman volume

$$\beta(\zeta) = \left(\frac{i}{2}\right)^n \prod_{k=1}^{n} \frac{d\zeta_k \wedge d\bar{\zeta}_k}{\left(1 - |\zeta_k|^2\right)^2}. \tag{14.9}$$

Eisenman [145] and Kobayashi [296] gave infinitesimal analogs of these invariant volumes.

Definition 14.5.2 Let M be a complex n-dimensional manifold; the invariant Carathéodory pseudo-volume form of M at the point $p \in M$ is defined as

$$\Omega^c_M(p) = \sup f^*\beta\big|_p,$$

where the least upper bound extends over all holomorphic maps $f : M \to \Delta^n$, $f(p) = 0$, while the Kobayashi pseudo-volume form is

$$\Omega^k_M(p) = \inf\left(f^{-1}\right)^* \beta\Big|_0,$$

where the greatest lower bound extends over all holomorphic maps $f : \Delta^n \to M$, $f(0) = p$, nondegenerate at the point $\zeta = 0$.

Let us make explicit these definitions. Let $z = (z_1, \dots, z_n)$ be local coordinates near the point p, and let $\Phi = (i/2)^n \prod_{k=1}^n dz_k \wedge d\bar{z}_k$ be the Euclidean volume form. In the first case f is locally of the form $\zeta_j = f_j(z)$, $j = 1, \dots, n$, and as $f(p) = 0$ then, taking account of (14.9), we get $f^*\beta|_p = |J_f(p)|^2 \Phi$, where $J_f(p) = \det f'(p)$ is the Jacobian of the map f. Therefore

$$\Omega^c_M(p) = \sup_f |J_f(p)|^2 \Phi = H^c(p)\Phi; \tag{14.10}$$

it is clear that this expression does not depend on the local coordinates.

In the second case we must consider, instead of f, the inverse f^{-1} in a neighborhood of $\zeta = 0$. As $\det\left(f^{-1}\right)' = \left(\det f'\right)^{-1}$, we get

$$\Omega^k_M(p) = \sup_f \frac{1}{|J_f(0)|^2}\Phi = H^k(p)\Phi. \tag{14.11}$$

The coefficients H^c and H^k in the local expressions for Ω^c_M and Ω^k_M are nonnegative but may vanish at some points of M (this is independent of the choice of local coordinates) so that in the general case Ω^c_M and Ω^k_M are only pseudo-volume forms.

The form Ω^c_M is continuous on M, Ω^k_M only semicontinuous. Both forms are invariant under biholomorphic maps, and under holomorphic maps $f : M \to N$ they do not grow in the sense that

$$f^*\Omega_N(p) \leqq \Omega_M(p) \text{ for all } p \in M$$

(the pullback $f^*\Omega_N$ and the form Ω_M are compared by their coefficients as forms of maximal degree).

We have also the following theorem:

Theorem 14.5.3

(a) The form Ω^k_M is the largest among the pseudo-volume forms on M which do not increase under holomorphic maps $\Delta^n \to M$, whereas Ω^c_M is the least among those which enjoy the same property for maps $M \to \Delta^n$.

(b) *On every complex manifold* $\Omega^c_M(p) \leqq \Omega^k_M(p)$.

The Kobayashi form agrees with passage to holomorphic coverings: If $\pi : \tilde{M} \to M$ is such a covering, then $\Omega^k_{\tilde{M}}(p) = \pi^* \Omega^k_M(p)$ at each point $p \in \tilde{M}$. The Carathéodory form does not have this property.

14.6 Measure Hyperbolicity

Let us call a subset E of a complex m-dimensional manifold M a set of zero measures if locally, in the neighborhood of each point, it has zero $2n$-dimensional Hausdorff measure in a coordinate neighborhood of the point.

Definition 14.6.1 The manifold M is called measure hyperbolic (respectively, C-measure hyperbolic) if $\Omega^k_M(p) \neq 0$ (respectively, $\Omega^c_M(p) \neq 0$) outside a set of zero measures.

It is interesting to compare measure hyperbolicity to hyperbolicity with respect to invariant metrics. It is not hard to see that if $C_M(p, v) \neq 0$ for all $v \neq 0$, then the coefficient $H^c(p)$ in the expression (14.10) for the form $\Omega^c_M(p)$ in local coordinates z near the point p satisfies $H^c(p) \leqq \pi^n / \operatorname{vol} B^c(p)$, where $\operatorname{vol} B^c(p)$ is the volume of the Carathéodory unit ball in $T_p(M)$ evaluated with the aid of the Euclidean volume form Φ. Analogously, if $K_M(p, v) \neq 0$, then in the same coordinates $H^k(p) \geqq \pi^n / \operatorname{vol} B^k(p)$. As always $B^k(p) \subset B^c(p)$, it follows that

$$H^c(p) \leqq \frac{\pi^n}{\operatorname{vol} B^c(p)} \leqq \frac{\pi^n}{\operatorname{vol} B^k(p)} \leqq H^k(p).$$

M.N. L'vovskii [367] proved that the class of C-volume hyperbolic manifolds coincides with the class of complex manifolds M for which the Carathéodory distance satisfies $c_M(p, q) > 0$ for all points $p \in M$ and all points $q \in M \backslash A$ disjoint from p, where A is a certain analytic subset of M. It is likewise easy to see that $C_M(p, v) = 0$ for some vector $v \neq 0$ iff $\Omega^c_M(p) \neq 0$. The analogous statement for the Kobayashi metric is not true.

Application: The monotonicity property of volumes and forms makes it possible to obtain generalizations of the classical theorems of Picard and Schottky-Landau. Here is one result of this nature.

Theorem 14.6.2 ([296]) *Let M be a complex manifold such that $\Omega^k_M(p) = 0$ on a nonempty open subset, and let N be measure hyperbolic. Then each holomorphic map $f : M \to N$ is degenerate at all points of M.*

14.7 Existence of Bounded p.s.h. Functions and Hyperbolicity

We first recall some results due to H. Royden.

A complex manifold M is hyperbolic at a point p if there is a neighborhood U of p and a positive constant c such that $F_M(q, \xi) \geq c\|\xi\|$ for all $q \in \mathrm{U}$, where $\|\cdot\|$ denotes the norm of the tangent vector ξ with respect to a fixed Hermitian metric on M.

The manifold M is said to be hyperbolic if the Kobayashi pseudodistance d_M is a distance; recall that d_M is the integrated form of M.

In [444] Royden proved that M is hyperbolic at every point if and only if d_M is a distance.

Theorem 14.7.1 *Let M be a complex manifold. Suppose there is a p.s.h. bounded function u on M which is strictly p.s.h. on a neighborhood of p; then M is hyperbolic at p.*

Proof For simplicity we shall suppose that u is of class $\mathcal{C}^2$ in a neighborhood of p. The modifications are needed when u is only supposed uppersemicontinuous and

$$L_z(u, v) \geq c\|v\|^2 d\sigma$$

as distributions are given at the end of the proof (here $d\sigma$ denotes Lebesgue measure and c is a positive constant).

Let Φ be a chart from a neighborhood U of p to the ball $B(0, 2r)$ such that $\Phi(p) = 0$. We can suppose that the function u is negative and that on $\mathrm{B}(0, 2r)$

$$L_z(u \circ \Phi^{-1}, w) \geq c|w|^2, \quad c > 0, \text{ for } \quad w \in \mathbb{C}^n.$$

Let θ be a smooth nondecreasing function defined on $\mathbb{R}^+$ such that $\theta(x) = x$ for $x \leq \frac{1}{2}$ and $\theta(x) = 1$ for $x \geq \frac{3}{4}$. For $\lambda > 0$, we define

$$\begin{aligned} \Psi_\lambda(z) &= \theta\left(\frac{|\Phi(z)|^2}{\mathbf{r}^2}\right) \exp(\lambda u(z)) \quad \text{for} \quad z \in U \\ \Psi_\lambda(z) &= \exp(\lambda u(z)) \qquad\qquad\qquad \text{for} \quad z \in M \setminus U. \end{aligned}$$

It is clear that on a neighborhood of $\overline{M \setminus \mathrm{U}}$ and on $\Phi^{-1}(B(0, r/2))$ the function $\log \Psi_\lambda$ is p.s.h. Consider in $B(0, 2r)$ the function

$$\tilde{\Psi}_\lambda(z) = \Psi_\lambda \circ \Phi^{-1}(z) = \theta\left(\frac{|z|^2}{r^2}\right) \exp\left(\lambda \left(u_o \Phi^{-1}\right)(z)\right).$$

Set $h = \log \theta$. An easy computation shows that there is a constant $A > 0$ such that for all $z \in B(0, 1)$ and $v \in \mathbb{C}^n$

$$L_{|z|^2}(u, v) \geq -A\|v\|^2.$$

But then

$$L_z(\Psi_\lambda, v) \geq \left(-\frac{A}{r^2} + c\lambda\right) \|v\|^2.$$

If we choose $\lambda = \frac{A}{r^2 c}$, we find that $\Psi_\lambda \in \mathcal{S}_p$ and

$$L_z(\Psi_\lambda, v) = \frac{|\Phi'(p)v|^2}{r^2} \exp(\lambda u(p)).$$

Therefore

$$P(p, v) \geq \frac{|\Phi'(p)v|}{r} \exp\left(\frac{\text{A}}{2r^2 c} u(p)\right).$$

Since $F(p, v) \geq P(p, v)$ and the majorization above is independent of $p \in \Phi^{-1}(B(0, r))$, we see that M is hyperbolic at p, and we have even a lower bound for P in terms of geometric objects.

When u is only upper semicontinuous (u.s.c.), we will show that the above estimate still holds for $F_M(p, v)$. The proof that $\log \Psi_\lambda$ is p.s.h. is the same. Let f be in $\mathcal{O}(\mathbb{D}, M)$ with $f(0) = p$ and $f'(0) = v/\alpha$. Then $\Phi(f(\zeta)) = \zeta \text{h}(\zeta)$ with $\text{h}(0) = \Phi'(p)v/\alpha$. Since $(\log \Psi_\lambda \circ f)$ is subharmonic on $\mathbb{D}$ and $\Psi_\lambda(f(\zeta)) \cdot |\zeta|^{-2}$ is bounded on $\mathbb{D}$, we find, as in Proposition 13.6.1, that

$$\frac{1}{r^2} \left|\Phi'(p)v/a\right|^2 \exp(\lambda u(p)) = \overline{\lim_{\zeta \to 0}} \frac{\Psi_\lambda(f(\zeta))}{|\zeta|^2} \leq 1.$$

So

$$a \geq \left|\frac{\Phi'(p)v}{r}\right| \exp\left(\frac{\lambda}{2} u(p)\right)$$

and

$$F(p, v) \geq \frac{|\Phi'(p)v|}{r} \exp\left(\frac{\text{A}}{2r^2 c} u(p)\right).$$

□

Corollary 14.7.2 *Let M be a complex manifold with a smooth strictly p.s.h. function V. Suppose there exists on M a bounded continuous p.s.h. function u. Let K be a compact connected component of the set $(u \leq \lambda)$ $\lambda \in \mathbf{R}$. Then M is hyperbolic at every point of K.*

Proof We can suppose that the function v, which is not necessarily bounded, is positive on M. Let V be a relatively compact neighborhood of K such that $\overline{V}$ is disjoint from the other components of $(u \leq \lambda)$. Let $\mu = \inf_{z \in \partial V} u(z)$. Since u is continuous, we have $\mu > \lambda$. Let $c = \mathrm{Max}_{z \in \partial V}(u+v)$, and let h be a convex function defined on $\mathbb{R}$ such that $h(x) = \mathbf{x}$ for $x < \lambda + \frac{\mu-\lambda}{2}$ and $h(\mu) > c$. We define the function Ψ by

$$\Psi(z) = \sup((u+v)(z), h(u(z))) \quad \text{for } z \in V$$
$$\Psi(z) = h(u(z)) \quad \text{for } z \in M \setminus V.$$

The function Ψ is p.s.h. bounded and strictly p.s.h. on a neighborhood of K. Thus by Theorem 14.7.1, M is hyperbolic at every point of K. □

The following corollary partially answers problem 2 of [223] raised by H. Wu. The question in [223] is the following. Suppose a complex manifold M possesses a C^∞ strictly p.s.h. bounded exhaustion function. Is it complete hyperbolic?

Corollary 14.7.3 *Let M be a complex manifold admitting a smooth strictly p.s.h. function. If there exists on M a bounded p.s.h. exhaustion function, then M is hyperbolic. In fact M is a taut manifold.*

Proof Let u be the bounded exhaustion function. We suppose that u has its values in $[0, 1]$; since u is an exhaustion function, for every $a < 1$ the set $u^{-1}[0, a]$ is relatively compact in M. We recall that a complex manifold M is taut if for every sequence $\{f_n\}$ in $H(\mathbb{D}, M)$ either there exists a convergent subsequence for the compact open topology or $\{f_n\}$ is compactly divergent. For results concerning taut manifolds, we refer to [294, 296]. □

The proof of Corollary 14.7.2 can be easily adapted to show that M is hyperbolic at every point. Hence by Royden's result [444] M is hyperbolic.

Consider now a sequence $\{f_n\} \in \mathcal{O}(\mathbb{D}, M)$. Let $\{u_n\}$ be the sequence of subharmonic functions in the unit disc defined by $u_n(z) = u\,(f_n(z))$.

Suppose there is a point $z_0 \in \mathbb{D}$ such that $\overline{\lim_{n\to\infty}} u_n\,(z_0) < 1$. Then there is a subsequence f_{n_j} such that $f_{n_j}\,(z_0)$ is in a compact set $L \subset M$. Since M is hyperbolic, the sequence f_n is equicontinuous. Recall that the Kobayashi metric $\mathbb{D}$ is distance-decreasing under holomorphic maps and that this metric induces the topology of M. Thus as u is an exhaustion function, there exist $r > 0$ and $\varepsilon > 0$ such that $\lim_j u_{n_j}(z) \leq 1 - \varepsilon$ for $|z - z_0| < r$.

Consider now the function v defined on $\mathbb{D}$ by

$$v(z) = \overline{\lim_j}\, u_{n_j}(z).$$

The upper semicontinuous regularization v^* of v is subharmonic on $\mathbb{D}$; moreover it is well known that $v = v^*$ almost everywhere in $\mathbb{D}$. Since $v \leq 1 - \varepsilon$ on a disc, it follows by the maximum principle that $v^* < 1$ everywhere. Consequently for every $r < 1$ there exists a constant $c(r) < 1$ such that $v^* \leq c(r)$ on the disc $\mathbb{D}(0, r)$.

But then by Hartog's lemma, for $r_1 < r$ and $\varepsilon > 0$, such that $c(r) + \varepsilon < 1$, one has $u_{n_j}(z) \leq c(r) + \varepsilon$ for $|z| < r_1$ and $j > j_0$. Therefore for every $z \in \mathbb{D} f_{n_j}(z)$ is relatively compact, so Ascoli's lemma implies that a subsequence of (f_{n_j}) is convergent in the compact open topology.

The second possibility is that for every $z \in \mathbb{D}$ $\liminf_{j\to\infty} u_j(z) = 1$. Since $u < 1$, this implies that for every $z \in \mathbb{D}$ $f_n(z)$ tends to infinity on M. Let K be compact in $\mathbb{D}$ and L a compact in M. Choose $c < 1$ such that $L \subset (u \leq c)$. Let $\delta > 0$ be such that $L_1 = \{z \mid z \in M \quad d(z, L) \leq \delta\}$ is compact. Then there is $c_1 < 1$ with $L_1 \subset (u(z) \leq c_1)$. Choose $\eta > 0$ so that if z, z' are in K and $|z - z'| \leq \eta$, then $d(f_n(z), f_n(z')) \leq \delta$ for every n. As K is compact, there is a finite covering of K by discs $\mathbb{D}(p_j, \eta)$, $1 \leq j \leq s$. Choose n_0 so that $u_n(p_j) > c_1$ for $n \geq n_0$ and $1 \leq j \leq s$.

This implies that for $n \geq n_0$ and $z \in K$, $f_n(z) \notin L$. In fact if $z \in K$, there is j such that $z \in \mathbb{D}(p_j, \eta)$ and $d(f_n(z), f_n(p_j)) \leq \delta$. Thus if $f_n(z) \in L$, then $f_n(p_j) \in L_1$. This contradicts the choice of n_0; consequently the sequence f_n is compactly divergent.

Remark 14.7.4 Corollary 14.7.3 and the pseudometric P give some information about Theorem F of R. Greene and H. Wu [203]. In particular their inequality (6.3), p. 100, can be read as a lower bound estimate for the Kobayashi metric.

14.8 Hyperbolicity of a Complex Manifold and Other Equivalent Properties

Hahn and Kim [226] have given hyperbolicity criteria for Hermitian manifolds that are connected with classical theorems by Schottky and Landau. To formulate them we need the following notions.

Let M be a connected complex manifold of dimension m with Hermitian metric h_M. Let ρ_M denote the distance function associated with h_M.

Definition 14.8.1 A complex manifold M is said to satisfy the *Schottky property* if for each $p \in M$ and for each relatively compact open set W in a coordinate neighborhood of p and an $r \in (0, 1)$, there exists a positive constant $S = S(W, r)$ such that every

$$\rho_M(p, f(z)) \leqslant S \text{ for } |z| < r. \tag{14.12}$$

A complex manifold M is said to satisfy the *Landau property* if for each $p \in M$ and each relatively compact open set W in a coordinate neighborhood of p, there exists a positive constant $R = R(W)$ such that

$$\sup\left\{h_M\left(f(0), f'(0)\right) : f \in \mathcal{O}(\mathbb{D}, M) \text{ with } f(0) \in W\right\} \leqslant R. \tag{14.13}$$

Then we can prove the following.

Theorem 14.8.2 *The following statements are equivalent on any Hermitian manifold M:*

(a) M is hyperbolic.
(b) M satisfies the Schottky property.
(c) M satisfies the Landau property.

Proof It will be proved through the following implications: (a) $\Leftrightarrow$ (b), (a) $\Rightarrow$ (c) $\Rightarrow$ (a).

To prove (a) $\Rightarrow$ (b), let $p \in M$ and U be its local coordinate neighborhood. Let W be a relatively compact neighborhood of p in U. Since k_M is a metric, it induces the standard topology in M, see Theorem 14.8.6. Therefore, we may define $W = \{q : k_M(p,q) < \rho_0\}$ for some $\rho_0 > 0$ and $U = \{q : k_M(p,q) < \rho_0 + \rho_1\}$, where $\rho_1 = \tan h^{-1} r$, for any given $r \in (0,1)$. If $f \in \mathcal{O}(\mathbb{D}, M)$ satisfies $f(0) \in W$, then $k_M(p, f(0)) < \rho_0$ and $k_M(f(0), f(z)) \leqslant k_D(0,z) \leqslant \rho_D(0,r) = \tan h^{-1} r = \rho_1$ for $|z| \leqslant r$. Therefore, it follows from the triangle inequality that $k_M(p, f(z)) \leqslant \rho_0 + \rho_1$ whenever $|z| \leqslant r$. Since k_M is a metric, $k_M \geqslant c\rho_M$ for some $c > 0$, which proves (b) with $S = (\rho_0 + \rho_1)/c$.

For the proof of (b) $\Rightarrow$ (a), we follow the method of [278]: Let p and q be two distinct points of M. We may take a coordinate neighborhood U of p which does not contain q. Let $\varphi : U \to \mathbb{B} = \{w \in \mathbb{C}^m : |w| < 1\}$ be a biholomorphic map with $\varphi(p) = 0, \varphi(U) = \mathbb{B}$. Without loss of generality, we may assume that $\varphi(W) = \{w \in \mathbb{C}^m : |w| < \rho\} = \mathbb{B}(0,\rho)$ for some $\rho \in (0,1)$ and $\varphi(V) \subset \mathbb{B}$, $V = \{m \in M : \rho_M(p, M) < S\}$. Let $r \in (0,1)$ be given. If $|z| < r/2$, then there exists a constant $c > 0$ such that $\tan h^{-1}|z| \geqslant c \tan h^{-1}|z|/r$ or, equivalently, $\rho_{\mathbb{D}}(0,z) \geqslant c\rho_{\mathbb{D}(0,r/2)}(0,z)$. Let $\alpha = \{p = p_0, p_1, \ldots, p_l = q; a_1, \ldots, a_l; f_1, \ldots, f_l\}$ be any chain connecting p and q, used in the construction of the Kobayashi metric $k_M(p,q)$. Since $\varphi(q)$ lies outside $\mathbb{B}$, there must be an index, say l_0, $0 \leqslant l_0 < l$, such that $\varphi(p_0), \varphi(p_1), \ldots, \varphi(p_{l_0 - 1}) \in \mathbb{B}(0,\rho)$ and $\varphi(p_{l_0}) \notin \mathbb{B}(0,\rho)$. By the homogeneity of $\mathbb{D}$, we may assume that $a_1, \ldots, a_{l_0}$ lie in $\mathbb{D}(0, r/2)$. Then

$$\begin{aligned} |\alpha| &\geqslant \sum_{i=1}^{l_0} \rho_{\mathbb{D}}(0, a_i) \geqslant c \sum_{l=1}^{l_0} \rho_{\mathbb{D}(0,r/2)}(0, a_i) \geqslant c \sum_{i=1}^{l_0} k_{\mathbb{B}}(\varphi(p_{i-1}), \varphi(p_l)) \\ &\geqslant c k_{\mathbb{B}}(0, \varphi(p_{l_0})) = c \tan h^{-1} \|\varphi(p_{l_0})\| \geqslant c \tan h^{-1} \rho \equiv c'. \end{aligned}$$

Thus, $k_M(p,q) \geqslant c' > 0$. Here we used the fact that on $\mathbb{B}$ the Kobayashi metric agrees with the standard Kähler metric. It is given by

$$k_{\mathbb{B}}(0,z) = \frac{1}{2} \log \frac{1 + \|z\|}{1 - \|z\|} = \tan h^{-1} \|z\| \quad \text{for } z \in \mathbb{B}.$$

(a) $\Rightarrow$ (c). Suppose that M fails to satisfy the Landau property. Then there exist a point p_0 in an open neighborhood W which is relatively compact in a local coordinate neighborhood U of p_0 and a sequence $f_k \in \mathcal{O}(\mathbb{D}, M)$ such that $f_k(0) \in W$ and $h_M\left(f_R(0), f_k'(0)\right) \to \infty$ as $k \to \infty$. We claim that if M is hyperbolic, then there exists an $r \in (0,1)$ such that $\{f_k\}$ contains a subsequence which converges uniformly to a holomorphic map $f : \mathbb{D}(0,r) \to M$. First we observe that $\bar{W} \cap (M\backslash U) = \varnothing$, where $\bar{W}$ is compact and $M\backslash U$ is closed.

Therefore, there exists a number $\rho' > 0$ such that

$$Q = \left\{p \in M : k_M(\bar{W}, p) < \rho'\right\} \subset\subset U.$$

Since $k_M\left(f_k(0), f_k(z)\right) \leqslant \rho_{\mathbb{D}}(0,z) = \tan h^{-1}|z|$ for all $f_k \in \mathcal{O}(\mathbb{D}, M)$ with $f_k(0) \in W$, $f_k(z) \in Q$ whenever $|z| \leqslant r' = \tanh \rho'$. Since Q is relatively compact, it is bounded in $\mathbb{C}^m$. Therefore, by Montel's theorem there is a subsequence of $\{f_k\}$ which converges uniformly on $\mathbb{D}(0,r)$ to a holomorphic map $f : \mathbb{D}(0,r) \to M$ for $r < r'$. Denote again this convergent subsequence by $\{f_k\}$. By Weierstrass' theorem, $f_k'(z)$ converges to $f'(z)$ uniformly on $\mathbb{D}(0,r)$. In particular, $f_k'(0)$ converges to $f'(0)$. Thus, $h_M\left(f_k(0), f_k'(0)\right) \to h_M\left(f(0), f'(0)\right) < \infty$ as $k \to \infty$, which is a contradiction.

(c) $\Rightarrow$ (a). Given $(q,\xi) \in T(W)$ and $v \in \mathbb{C}$, let $f \in \mathcal{O}(\mathbb{D}, M)$ satisfy $f(0) = q$ and $f'(0)v = \xi$. Then by (14.14)

$$R|v| \geqslant h_M\left(f(0), f'(0)\right)|v| = h_M\left(f(0), f'(0)v\right) = h_M(q,\xi)$$

or

$$|v| \geqslant \frac{1}{R} h_M(q,\xi), \tag{14.14}$$

where R is the upper bound given in (14.13). Thus,

$$K_M(q,\xi) \geqslant c h_M(q,\xi), \tag{14.15}$$

$c = 1/R$, as desired. □

Definition 14.8.3 A map $f \in \mathcal{O}(\mathbb{D}, M)$ is said to be Bloch map if

$$\sup\left\{Q_f(z) : z \in \mathbb{D}\right\} < \infty, \tag{14.16}$$

where

$$Q_f(z) = \left(1 - |z|^2\right) h_M\left(f(z), f'(z)\right). \tag{14.17}$$

We remark that the nonnegative function Q_f is invariant under the group Aut($\mathbb{D}$) of holomorphic automorphisms of $\mathbb{D}$ in the sense that for all $\varphi \in$ Aut($\mathbb{D}$)

$$Q_{f\circ\varphi}(z) = Q_f(\varphi(z)), \quad z \in \mathbb{D}. \tag{14.18}$$

It is an easy consequence of the definition of Kobayashi metric that if (14.16) holds for all $f \in \mathcal{O}(\mathbb{D}, M)$, then M is hyperbolic. The converse is, however, not true in general (see Lemma 14.8.5). If M is compact, we obtain the following characterization of the hyperbolicity of M.

Theorem 14.8.4 *Let M be a connected compact Hermitian manifold. The following statements are equivalent:*

(a) *M is hyperbolic.*
(b) $\sup\left\{h_M\left(f(0), f'(0)\right) : f \in \mathcal{O}(\mathbb{D}, M)\right\} < \infty$.
(c) $\sup\left\{Q_f(z) : z \in \mathbb{D}\right\} < \infty$ *for all* $f \in \mathcal{O}(\mathbb{D}, M)$, *i.e., every* $f \in \mathcal{O}(\mathbb{D}, M)$ *is Bloch.*
(d) *M has the* Picard property, *i.e., there is no nonconstant holomorphic map* $f : \mathbb{C} \to M$.

Unlike in the one complex-dimensional case [225], the notion of the Picard property is a weaker notion than the Schottky property, the Landau property, or the hyperbolicity for a noncompact manifold [294].

We note that the notion of the Picard property used in [225] and here, somewhat reluctantly, coincides with the term previously used by S. Kobayashi [294] and others.

Proof of Theorem 14.8.4 By a result of R. Brody 14.4.19, (a) $\Leftrightarrow$ (d) holds on a connected compact manifold. Therefore, it is enough to prove the implications (a) $\Rightarrow$ *(b)* $\Rightarrow$ (c) $\Rightarrow$ (a). First we prove the following:

Lemma 14.8.5 *A complex manifold M is hyperbolic if*

$$\sup\left\{Q_f(z) : z \in \mathbb{D}\right\} < \infty$$

for all $f \in \mathcal{O}(\mathbb{D}, M)$. *The converse is not true in general.*

Proof The above hypothesis implies that there exists a constant $Q > 0$ such that for all $f \in \mathcal{O}(\mathbb{D}, M)$

$$h_M\left(f(z), f'(z)v\right) \leqslant QK_D(z, v) \quad (z \in \mathbb{D}, v \in \mathbb{C}).$$

Let $\tilde{\rho}_M$ be the distance function associated with the differential metric $\tilde{h}_M = Q^{-1}h_M$. By integrating both sides of the above inequality along the geodesic curve between two points z and w in $\mathbb{D}$ and using the definition of integrated distance $\tilde{\rho}_M$, we have

$$\tilde{\rho}_M(f(z), f(w)) \leqslant k_D(z, w)$$

for all $f \in \mathcal{O}(\mathbb{D}, M)$. Since k_M is the largest among those pseudometrics which satisfy (2), $\tilde{\rho}_M \leqslant k_M$ (see [294]). Thus, M is hyperbolic.

To show the second half, let M be the right half plane of $\mathbb{C}$ and h_M the euclidean metric. Then M is clearly hyperbolic. Set

$$f(z) = (1+z)/(1-z) \text{ and } z_n = 1 - 1/n \in \mathbb{D}.$$

Then $f \in \mathcal{O}(\mathbb{D}, M)$ and

$$Q_f(z_n) = \left(1 - |z_n|^2\right)\left|f'(z_n)\right| = 2(2n-1) \to \infty$$

as $n \to \infty$. This completes the lemma. □

From this lemma, (c) ⇒ (a) trivially follows. To prove (a) ⇒ (b), let M be hyperbolic. By (c) of Theorem 14.8.2, for each $p \in M$ there exists a relatively compact open neighborhood W_p of p such that

$$\sup\left\{\left\|f'(0)\right\| : f \in \mathcal{O}(\mathbb{D}, M), f(0) \in W_p\right\} < \infty.$$

Since M is compact, it can be covered by a finite number of such W_p's. Thus, (b) follows.

(b) ⇒ (c). Suppose (c) fails to hold. Then there exist sequences $\{z_n\}$ in $\mathbb{D}$ and $\{f_n\}$ in $\mathcal{O}(\mathbb{D}, M)$ such that $Q_{f_n}(z_n) > n$ for all n. Since $\mathbb{D}$ is homogeneous, for each z_n there exists $\varphi_n \in \mathrm{Aut}(\mathbb{D})$ such that $\varphi_n(0) = z_n$. Therefore, by the invariant property of Q_f (see (14.18)),

$$Q_{f_n}(z_n) = Q_{f_n}(\varphi_n(0)) = Q_{g_n}(0) = h_M\left(g_n(0), g_n'(0)\right),$$

where $g_n = f_n \circ \varphi_n$. Since h_M is Hermitian, it follows from the compactness of M, condition (b), and (14.15) that $Q_{f_n}(z_n)$ is bounded for all n, which is a contradiction.

Theorem 14.8.6 *Let X be a connected complex space. Then the Kobayashi pseudodistance d_X is continuous. If d_X is an actual distance, it induces the standard topology on X.*

Proof Suppose that $d_Y : X \times X \to R$ is not continuous. Since d_S satisfies the axioms for a pseudodistance, there is a point p in X for which the function $d_X(\cdot, p)$ is not continuous at p. Because X is first countable, there exist a sequence $\{p_u\}$ converging to p and a positive number ε such that $d_X(p_n, p) \geqq \varepsilon$ for all n. Hironaka's resolution of singularities [238, 2, Main Theorem I', pp. 151-152] gives an open neighborhood U of p, a complex manifold M, and a proper holomorphic map f from M onto U. (A concise explanation of the relevant part of Hironaka's terminology can be found in §1 of [239].) We may assume that $\{p_n\}$ is a sequence in U. Since f is onto, there is a sequence $\{q_n\}$ in M with $f(q_n) = p_n$ for all n. Because f is proper, we may, by taking a suitable subsequence if necessary, assume that $\{q_n\}$ converges to a point q of M. By continuity, $f(q) = p$. Let Q be an open neighborhood of q in M which is biholomorphically equivalent to a polydisc.

Because $f|_Q: Q \to X$ is distance-decreasing and d_Q is continuous, $d_X(p_n, p) = d_X(f(q_n), f(q)) \leqq d_Q(q_n, q) \to 0$ as $n \to \infty$. Thus $\varepsilon \leqq d_X(p_n, p) \to 0$ as $n \to \infty$, a contradiction.

Let p be a point of X, and let r be a positive real number. We will prove that the open ball $\mathfrak{B} = \{q \in X \mid d_X(p, q) < r\}$ is pathwise connected in X. Let q be a point of $\mathfrak{B}$. By the definition of the Kobayashi pseudodistance, there are points $p = p_0, p_1, \ldots, p_{k-1}, p_k = q$ of X, points $a_1, \ldots, a_k, b_1, \ldots, b_k$ of $\mathbb{D}$, and holomorphic maps $f_1, \ldots, f_k$ from $\mathbb{D}$ into X such that $f_j(a_j) = p_{j-1}$ and $f_j(b_j) = p_j$ for $j = 1, \ldots, n$, and $\rho(a_1, b_1) + \cdots + \rho(a_k, b_k) < r$. Let c_j denote the geodesic arc from a_j to b_j with respect to the Poincaré metric on $\mathbb{D}$. Since each c_j is a geodesic arc and each f_j is distance-decreasing, the path c from p to q formed by linking $f_1(c_1), \ldots, f_k(c_k)$ lies entirely in $\mathfrak{B}$.

Now assume that d_X is an actual distance. To prove that d_X induces the standard topology, we need only show that every open set in X is open with respect to the distance d_X. Let V be an open subset of X, and let p be a point of V. Since X is locally compact, there is a relatively compact open neighborhood W of p in X with $W \subset V$. Let r be the minimum value of the positive continuous function $d_X(p, \cdot)$ on the compact set ∂W, and let $\mathfrak{B} = \{q \in X \mid d_X(p, q) < r\}$. Then $\mathfrak{B} \cap \partial W = \varnothing$; since $\mathfrak{B}$ is connected and $p \in \mathfrak{B} \cap W$, we have $\mathfrak{B} \subset W \subset V$. Thus V is open with respect to d_X. □

Chapter 15
The Fatou Theory and Related Matters

15.1 Review of the Classical Theory of H^p Spaces on the Disc

Throughout this section we let $\mathbb{D} \subseteq \mathbb{C}$ denote the unit disc.

Let $0 < p < \infty$. We define

$$H^p(\mathbb{D}) = \left\{ f \text{ holomorphic on } \mathbb{D} : \sup_{0<r<1} \int_0^{2\pi} \left| f\left(re^{i\theta}\right)\right|^p d\theta^{1/p} \equiv \|f\|_{H^p} < \infty \right\}.$$

Also define

$$H^\infty(\mathbb{D}) = \left\{ f \text{ holomorphic on } \mathbb{D} : \sup_D |f| \equiv \|f\|_{H^p} < \infty \right\}.$$

The fundamental result in the subject of H^p, or *Hardy*, spaces (and also the fundamental result of this section) is that if $f \in H^p(\mathbb{D})$, then the limit

$$\lim_{r \to 1^-} f\left(re^{i\theta}\right) \equiv \widetilde{f}\left(e^{i\theta}\right)$$

exists for almost every $\theta \in [0, 2\pi)$. For $1 \leq p \leq \infty$, the function f can be recovered from $\widetilde{f}$ by way of the Cauchy or Poisson integral formulas; for $p < 1$ this "recovery" process is more subtle and must proceed by way of distributions. Once this pointwise boundary limit result is established, then an enormous and rich mathematical structure unfolds (see Y. Katznelson [264], K. Hoffman [241], and J. Garnett [170]).

Recall that the *Poisson kernel* for the disc is

$$P_r\left(e^{i\theta}\right) = \frac{1}{2\pi} \frac{1 - r^2}{1 - 2r\cos\theta + r^2}.$$

P. V. Dovbush, S. G. Krantz, *The Geometric Theory of Complex Variables*,
https://doi.org/10.1007/978-3-031-77204-7_15

For harmonic functions, let

$$\mathbf{h}^p(\mathbb{D}) = \left\{ f \in \mathcal{O}(\mathbb{D}) : \sup_{0<r<1} \left[\frac{1}{2\pi} \int_0^{2\pi} \left| f\left(re^{i\theta}\right) \right|^p d\theta \right]^{1/p} \equiv \|f\|_{h^p} < \infty \right\}$$

and

$$\mathbf{h}^\infty(\mathbb{D}) = \left\{ f \text{ harmonic on } \mathbb{D} : \sup_{\mathbb{D}} |f| \equiv \|f\|_{H^p} < \infty \right\}.$$

Throughout this section, arithmetic and measure theory on $[0, 2\pi)$ (equivalently on $\partial\mathbb{D}$) is done by identifying $[0, 2\pi)$ with $\mathbb{R}/2\pi\mathbb{Z}$. See Y. Katznelson [264] for more on this identification procedure.

Proposition 15.1.1 *Let* $1 < p \leq \infty$ *and* $f \in \mathbf{h}^p(\mathbb{D})$. *Then there is an* $\widetilde{f} \in L^p(\partial\mathbb{D})$ *such that*

$$f\left(re^{i\theta}\right) = \int_0^{2\pi} \widetilde{f}\left(e^{i\psi}\right) P_r\left(e^{i(\theta-\psi)}\right) d\psi.$$

Proof Define $f_r\left(e^{i\theta}\right) = f\left(re^{i\theta}\right), 0 < r < 1$. Then $\{f_r\}_{0<r<1}$ is a bounded subset of $\left(L^{p'}(\partial\mathbb{D})\right)^*$, $p' = p/(p-1)$. By the Banach-Alaoglu theorem (see W. Rudin [448]), there is a subsequence f_{r_j} that converges weak-* to some $\widetilde{f}$ in $L^p(\partial\mathbb{D})$. For any $0 < r < 1$, let $r < r_j < 1$. Then

$$f\left(re^{i\theta}\right) = f_{r_j}\left(\left(r/r_j\right)e^{i\theta}\right) = \int_0^{2\pi} f_{r_j}\left(e^{i\psi}\right) P_{r/r_j}\left(e^{i(\theta-\psi)}\right) d\psi$$

because $f_{r_j} \in C(\overline{\mathbb{D}})$. Now $P_{r/r_j} \in C(\partial\mathbb{D}) \subseteq L^{p'}(\partial\mathbb{D})$. Thus the right-hand side of the last equation is

$$\int_0^{2\pi} f_{r_j}\left(e^{i\psi}\right) P_r\left(e^{i(\theta-\psi)}\right) d\psi$$
$$+ \int_0^{2\pi} f_{r_j}\left(e^{i\psi}\right)\left[P_{r/r_j}\left(e^{i(\theta-\psi)}\right) - P_r\left(e^{i(\theta-\psi)}\right)\right] d\psi.$$

As $j \to \infty$, the second integral vanishes (because the expression in brackets converges uniformly to 0), and the first integral tends to

$$\int_0^{2\pi} \widetilde{f}\left(e^{i\psi}\right) P_r\left(e^{i(\theta-\psi)}\right) d\psi.$$

This is the desired result. □

Remark 15.1.2 It is easy to see that the proof breaks down for $p = 1$ since L^1 is not the dual of any Banach space. This breakdown is not merely ostensible: The harmonic function

$$f\left(re^{i\theta}\right) = P_r\left(e^{i\theta}\right)$$

satisfies

$$\sup_{0<r<1} \int_0^{2\pi} \left|f\left(re^{i\theta}\right)\right| d\theta < \infty,$$

but the Dirac δ mass is the only measure of which f is the Poisson integral.

Exercise for the Reader If $f \in \mathbf{h}^1$, then there is a Borel measure μ_f on $\partial\mathbb{D}$ such that $f\left(re^{i\theta}\right) = P_r\left(\mu_f\right)\left(e^{i\theta}\right)$.

Proposition 15.1.3 *Let $f \in L^p(\partial\mathbb{D})$, $1 \leq p < \infty$. Then $\lim_{r\to 1^-} P_r f = f$ in the L^p norm.*

Remark 15.1.4 The result is false for $p = \infty$ if f is discontinuous. The correct analog in the uniform case is that if $f \in C(\partial\mathbb{D})$, then $P_r f \to f$ uniformly.

As an exercise, consider a Borel measure μ on $\partial\mathbb{D}$. Show that its Poisson integral converges in the weak-* topology to μ.

Proof of Proposition 15.1.3 If $f \in C(\partial\mathbb{D})$, then the result is clear by the solution of the Dirichlet problem. If $f \in L^p(\partial\mathbb{D})$ is arbitrary, let $\epsilon > 0$, and choose $g \in C(\partial\mathbb{D})$ such that $\|f - g\| < \epsilon$. Then

$$\begin{aligned} \|P_r f - f\|_{L^p} &\leq \|P_r(f-g)\|_{L^p} + \|P_r g - g\|_{L^p} + \|g - f\|_{L^p} \\ &\leq \|P_r\|_{L^1} \|f-g\|_{L^p} + \|P_r g - g\|_{L^p} + \epsilon \\ &\leq \epsilon + o(1) + \epsilon \end{aligned}$$

as $r \to 1^-$. □

Proposition 15.1.5 *Let $K \subseteq \partial\mathbb{D}$ be compact, and let $\{I_\alpha\}_{\alpha\in A}$ be a covering of K by open intervals. Then there is a subcovering $\left\{I_{\alpha_j}\right\}_{j=1}^M$ such that every point of K is contained in at least one but not more than two of the I'_{α_j}s. [We call such a subcover a cover of "valence two."]*

Proof Since K is compact, $\{I_\alpha\}_{\alpha\in A}$ has a finite subcollection $\left\{I_{\alpha_j}\right\}_{j=1}^J$ that still covers K. Now, in sequence, discard any I_{α_j} that is contained in the union of the other intervals. □

Definition 15.1.6 If $f \in L^1(\partial\mathbb{D})$, let

$$Mf(\theta) = \sup_{R>0} \frac{1}{2R} \int_{-R}^{R} \left|f\left(e^{i(\theta-\psi)}\right)\right| d\psi.$$

The function Mf is called the *Hardy-Littlewood maximal function* of f.

Definition 15.1.7 Let (X, μ) be a measure space and $f : X \to \mathbb{C}$ be measurable. We say that f is of *weak type* $p, 0 < p < \infty$, if $\mu\{x : |f(x)| > \lambda\} \leq C/\lambda^p$, all $0 < \lambda < \infty$. The space weak type ∞ is defined to be L^∞.

Lemma 15.1.8 (Chebycheff's Inequality) *If $f \in L^p(X, d\mu)$, then f is weak type p, $1 \leq p < \infty$.*

Proof Let $\lambda > 0$. Then

$$\mu\{x : |f(x)| > \lambda\} \leq \int_{\{x:|f(x)|>\lambda\}} |f(x)|^p/\lambda^p d\mu(x) \leq \lambda^{-p}\|f\|_{L^p}^p.$$

□

Exercise for the Reader There exist functions that are of weak type p but not in $L^p, 1 \leq p < \infty$.

Definition 15.1.9 An operator $T : L^p(X, d\mu) \to$ {measurable functions } is said to be of *weak type* $(p, p), 0 < p < \infty$, if

$$\mu\{x : |Tf(x)| > \lambda\} \leq C\|f\|_{L^p}^p/\lambda^p, \quad \text{all} \quad f \in L^p, \quad \lambda > 0.$$

Proposition 15.1.10 *The operator M is of weak type* $(1, 1)$.

Proof Let $\lambda > 0$. Set $S_\lambda = \{\theta : |Mf(e^{i\theta})| > \lambda\}$. Let $K \subseteq S_\lambda$ be a compact subset with $2m(K) \geq m(S_\lambda)$. For each $k \in K$, there is an interval $I_k \ni k$ with $|I_k|^{-1} \int_{I_k} |f(e^{i\psi})| d\psi > \lambda$. Then $\{I_k\}_{k\in K}$ is an open cover of K. By Proposition 15.1.5, there is a subcover $\{I_{k_j}\}_{j=1}^M$ of K of valence not exceeding 2. Then

$$\begin{aligned} m(S_\lambda) &\leq 2m(K) \leq 2m\left(\bigcup_{j=1}^M I_{k_j}\right) \leq 2\sum_{j=1}^M m(I_{k_j}) \\ &\leq \sum_{j=1}^M \frac{2}{\lambda}\int_{I_{k_j}} \left|f\left(e^{i\psi}\right)\right| d\psi \\ &\leq \frac{4}{\lambda}\|f\|_{L^1}. \end{aligned}$$

□

Definition 15.1.11 If $e^{i\theta} \in \partial\mathbb{D}$, $1 < \alpha < \infty$, then define the *Stolz region* (or *non-tangential approach region* or *cone*) with vertex $e^{i\theta}$ and aperture α to be

$$\Gamma_\alpha\left(e^{i\theta}\right) = \left\{z \in \mathbb{D} : \left|z - e^{i\theta}\right| < \alpha(1 - |z|)\right\}.$$

Proposition 15.1.12 *If $e^{i\theta} \in \partial\mathbb{D}$, $1 < \alpha < \infty$, then there is a constant $C_\alpha > 0$ such that if $f \in L^1(\partial\mathbb{D})$, then*

$$\sup_{re^{i\phi}\in\Gamma_\alpha(e^{i\theta})} \left|P_r f\left(e^{i\phi}\right)\right| \le C_\alpha M f\left(e^{i\theta}\right).$$

Proof For $re^{i\phi} \in \Gamma_\alpha\left(e^{i\theta}\right)$, we have

$$|\theta - \phi| \le 2\alpha(1-r).$$

Therefore, for $1/\alpha \le r < 1$, we obtain

$$\begin{aligned}
&\left|P_r f\left(e^{i\phi}\right)\right| \\
&= \left|\frac{1}{2\pi}\int_0^{2\pi} f\left(e^{i(\phi-\psi)}\right)\frac{1-r^2}{1-2r\cos\psi + r^2}d\psi\right| \\
&= \left|\frac{1}{2\pi}\int_0^{2\pi} f\left(e^{i(\phi-\psi)}\right)\frac{1-r^2}{(1-r)^2+2r(1-\cos\psi)}d\psi\right| \\
&\le \frac{4}{2\pi}\sum_{j=0}^{\log_2(\pi/\alpha(1-r))}\int_{S_j}\left|f\left(e^{i(\phi-\psi)}\right)\right|\frac{1-r^2}{(1-r)^2+2r\left(2^{j-1}\alpha(1-r)\right)^2}d\psi \\
&\quad + \frac{1}{2\pi}\int_{|\psi|<\alpha(1-r)}\left|f\left(e^{i(\phi-\psi)}\right)\right|\frac{1-r^2}{(1-r)^2}d\psi,
\end{aligned}$$

where $S_j = \left\{\psi : 2^j\alpha(1-r) \le |\psi| < 2^{j+1}\alpha(1-r)\right\}$. Now this is

$$\begin{aligned}
&\le \frac{4\alpha}{4\pi\alpha^2}\sum_{j=0}^{\infty}\frac{1}{2^{2j-2}(1-r)}\int_{|\psi|<(2+2^{j+1})\alpha(1-r)}\left|f\left(e^{i(\theta-\psi)}\right)\right|d\psi \\
&\quad + \frac{2}{2\pi}\frac{1}{1-r}\int_{|\psi|<3\alpha(1-r)}\left|f\left(e^{i(\theta-\psi)}\right)\right|d\psi \\
&\le \frac{32}{\pi}\sum_{j=0}^{\infty}2^{-j}\left[\frac{1}{2\alpha\left(2+2^{j+1}\right)(1-r)}\int_{|\psi|<(2+2^{j+1})\alpha(1-r)}\left|f\left(e^{i(\theta-\psi)}\right)\right|d\psi\right] \\
&\quad + \frac{6\alpha}{\pi}\frac{1}{2\cdot 3\alpha(1-r)}\int_{|\psi|<3\alpha(1-r)}\left|f\left(e^{i(\theta-\psi)}\right)\right|d\psi \\
&\le \frac{32}{\pi}\cdot\sum_{j=0}^{\infty}2^{-j}Mf(\theta) + \frac{6\alpha}{\pi}Mf(\theta) \\
&\le \frac{64}{\pi}Mf(\theta) + \frac{6\alpha}{\pi}Mf(\theta).
\end{aligned}$$

If $0 < r \leq 1/\alpha$, then

$$\begin{aligned}|P_r f(\phi)| &\leq \frac{1}{2\pi}\int_0^{2\pi}\left|f\left(e^{i(\phi-\psi)}\right)\right|(2\alpha/(\alpha-1))d\psi \\ &\leq \frac{2\alpha}{\alpha-1}Mf(\theta).\end{aligned}$$

□

Theorem 15.1.13 *Let* $f \in \mathbf{h}^p(\mathbb{D})$ *and* $1 < p \leq \infty$. *Let* $\widetilde{f}$ *be as in Proposition 15.1.1 and* $1 < \alpha < \infty$. *Then*

$$\lim_{\Gamma_\alpha(e^{i\theta})\ni z\to e^{i\theta}} f(z) = \widetilde{f}\left(e^{i\theta}\right), \text{ a.e. } e^{i\theta} \in \partial\mathbb{D}.$$

Proof It suffices to handle the case $p < \infty$ and f real-valued. If $\epsilon > 0$, then choose $g \in C(\partial\mathbb{D})$ real-valued so that $\|\widetilde{f} - g\|_{L^p(\partial\mathbb{D})} < \epsilon^2$. We know by the theory of the Dirichlet problem that

$$\lim_{\Gamma_\alpha(e^{i\theta})\ni z\to e^{i\theta}} g(z) = g\left(e^{i\theta}\right), \quad \text{all} \quad e^{i\theta} \in \partial\mathbb{D}, \tag{15.1}$$

where $g\left(re^{i\theta}\right) \equiv P_r g\left(e^{i\theta}\right)$. Therefore

$$\begin{aligned}&m\{e^{i\theta} : \limsup_{\Gamma_\alpha(e^{i\theta})\ni z\to e^{i\theta}} |f(z) - \widetilde{f}(e^{i\theta})| > \epsilon\} \\ &\leq m\{e^{i\theta} : \limsup_{\Gamma_\alpha(e^{i\theta})\ni z\to e^{i\theta}} |f(z) - g(z)| > \epsilon/3\} \\ &\quad + m\{e^{i\theta} : \limsup_{\Gamma_\alpha(e^{i\theta})\ni z\to e^{i\theta}} |g(z) - g(e^{i\theta})| > \epsilon/3\} \\ &\quad + m\{e^{i\theta} : |g(e^{i\theta}) - \widetilde{f}(e^{i\theta})| > \epsilon/3\} \\ &\leq \{e^{i\theta} : C_\alpha M(\widetilde{f} - g) > \epsilon/3\} + 0 + (3\|g - \widetilde{f}\|_{L^p}/\epsilon)^p.\end{aligned}$$

In the last estimate we used Lemma 15.1.8 and (15.1). Now the last line is majorized by

$$\begin{aligned}C'_\alpha\|\widetilde{f} - g\|_{L^1}/(\epsilon/3) + 3^p\epsilon^p &\leq C''_\alpha\|\widetilde{f} - g\|_{L^p}/(\epsilon/3) + 3^p\epsilon^p \\ &\leq C'''_\alpha\epsilon.\end{aligned}$$

It follows that

$$\lim_{\Gamma_\alpha(e^{i\theta})\ni z\to e^{i\theta}} f(z) = \widetilde{f}\left(e^{i\theta}\right), \quad \text{a.e.} \quad e^{i\theta} \in \partial\mathbb{D}.$$

The informal statement of Theorem 15.1.13 is that f has non-tangential boundary limits almost everywhere. □

Theorem 15.1.13 contains essentially all that can be said about the boundary behavior of harmonic functions. Why are holomorphic functions better? The classical way of answering this question (which will, in the long run, be seen to be misleading) is to use Blaschke factorization:

Definition 15.1.14 If $a \in \mathbb{C}$, $|a| < 1$, then the *Blaschke factor* at a is

$$B_a(z) = \frac{z - a}{1 - \bar{a}z}.$$

It is elementary to verify that B_a is holomorphic on a neighborhood of $\overline{\mathbb{D}}$ and that $\left|B_a\left(e^{i\theta}\right)\right| = 1$ for all θ.

Lemma 15.1.15 *If* $0 < r < 1$ *and* f *is holomorphic on a neighborhood of* $\overline{\mathbb{D}}(0, r)$, *let* $p_1, \ldots, p_k$ *be the zeroes of* f *(listed with multiplicity) in* $\mathbb{D}(0, r)$. *Assume that* $f(0) \neq 0$ *and that* $f\left(re^{it}\right) \neq 0$, *all* t. *Then*

$$\log|f(0)| + \log \prod_{j=1}^{k} r\left|p_j\right|^{-1} = \frac{1}{2\pi}\int_0^{2\pi} \log\left|f\left(re^{it}\right)\right| dt.$$

Proof The function

$$F(z) = \frac{f(z)}{\prod_{j=1}^{k} B_{p_j/r}(z/r)}$$

is holomorphic on a neighborhood of $\overline{\mathbb{D}}(0, r)$; hence $\log|F|$ is harmonic on a neighborhood of $\overline{\mathbb{D}}(0, r)$. Thus

$$\begin{aligned} \log|F(0)| &= \frac{1}{2\pi}\int_0^{2\pi} \log\left|F\left(re^{it}\right)\right| dt \\ &= \frac{1}{2\pi}\int_0^{2\pi} \log\left|f\left(re^{it}\right)\right| dt \end{aligned}$$

or

$$\log|f(0)| + \log \prod_{j=1}^{k} \frac{r}{\left|p_j\right|} = \frac{1}{2\pi}\int_0^{2\pi} \log\left|f\left(re^{it}\right)\right| dt.$$

□

Notice that, by the continuity of the integral, Lemma 15.1.15 holds even if f has zeroes on $\left\{re^{it}\right\}$.

Corollary 15.1.16 *If f is holomorphic in a neighborhood of $\overline{\mathbb{D}}(0, r)$, then*

$$\log |f(0)| \leq \frac{1}{2\pi} \int_0^{2\pi} \log \left| f\left(re^{it}\right)\right| dt.$$

Proof The term $\log \prod_{j=1}^{k} \left|r/p_j\right|$ in Lemma 15.1.15 is positive. □

Corollary 15.1.17 *If f is holomorphic on $\mathbb{D}$, $f(0) \neq 0$, and $\{p_1, p_2, \ldots\}$ are the zeroes of f counting multiplicities, then*

$$\log |f(0)| + \log \prod_{j=1}^{\infty} \frac{1}{\left|p_j\right|} \leq \sup_{0<r<1} \frac{1}{2\pi} \int_0^{2\pi} \log^+ \left| f\left(re^{it}\right)\right| dt.$$

Proof Apply Lemma 15.1.15, letting $r \to 1^-$. □

Corollary 15.1.18 *If $f \in H^p(\mathbb{D})$, $0 < p \leq \infty$, and $\{p_1, p_2, \ldots\}$ are the zeroes of f counting multiplicities, then $\sum_{j=1}^{\infty} \left(1 - \left|p_j\right|\right) < \infty$.*

Proof Since f vanishes to finite order k at 0, we may replace f by $f(z)/z^k$ and assume that $f(0) \neq 0$. It follows from Corollary 15.1.17 that

$$\log \prod_{j=1}^{\infty} \left\{ \frac{1}{\left|p_j\right|} \right\} < \infty$$

or $\prod \left(1/\left|p_j\right|\right)$ converges; hence $\prod \left|p_j\right|$ converges. So $\sum_j \left(1- \mid p_j\right) < \infty$. □

Proposition 15.1.19 *If $\{p_1, p_2, \ldots\} \subseteq \mathbb{D}$ satisfy $\sum_j \left(1 - \left|p_j\right|\right) < \infty$, $p_j \neq 0$ for all j, then*

$$\prod_{j=1}^{\infty} \frac{-\overline{p}_j}{\left|p_j\right|} B_{p_j}(z)$$

converges normally on $\mathbb{D}$.

Proof Restrict attention to $|z| \leq r < 1$. Then the assertion that the infinite product converges uniformly on this disc is equivalent with the assertion that

$$\sum_j \left| 1 + \frac{\overline{p}_j}{\left|p_j\right|} B_{p_j}(z) \right|$$

converges uniformly. But

$$
\begin{aligned}
\left|1+\frac{\overline{p}_j}{|p_j|}B_{p_j}(z)\right| &= \left|\frac{|p_j|-|p_j|\overline{p}_j z+\overline{p}_j z-|p_j|^2}{|p_j|(1-z\overline{p}_j)}\right| \\
&= \left|\frac{(|p_j|+z\overline{p}_j)(1-|p_j|)}{|p_j|(1-z\overline{p}_j)}\right| \\
&\le \frac{(1+r)(1-|p_j|)}{1-r},
\end{aligned}
$$

so the convergence is uniform. □

Definition 15.1.20 Let $0 < p \le \infty$ and $f \in H^p(\mathbb{D})$. Let $\{p_1, p_2, \ldots\}$ be the zeroes of f counted according to multiplicities. Let

$$
B(z) = \prod_{j=1}^{\infty} \frac{-\overline{p}_j}{|p_j|} B_{p_j}(z)
$$

(where each $p_j = 0$ is understood to give rise to a factor of z). Then B is a well-defined holomorphic function on $\mathbb{D}$ by Proposition 15.1.19. Let $F(z) = f(z)/B(z)$. By the Riemann removable singularities theorem, F is a well-defined, nonvanishing holomorphic function on $\mathbb{D}$. The representation $f = F \cdot B$ is called the *canonical factorization* of f.

Exercise for the Reader All the assertions of Definition 15.1.20 hold for $f \in N(\mathbb{D})$, the Nevanlinna class.

Proposition 15.1.21 *Let $f \in H^p(\mathbb{D})$, $0 < p \le \infty$, and let $f = F \cdot B$ be its canonical factorization. Then $F \in H^p(\mathbb{D})$ and $\|F\|_{H^p(\mathbb{D})} = \|f\|_{H^p(\mathbb{D})}$.*

Proof Trivially, $|F| = |f/B| \ge |f|$, so $\|F\|_{H^p} \ge \|f\|_{H^p}$. If $N = 1, 2, \ldots$, let

$$
B_N(z) = \prod_{j=1}^{N} \frac{-\overline{p}_j}{|p_j|} B_{p_j}(z)
$$

(where the factors corresponding to $p_j = 0$ are just z). Let $F_N = f/B_N$. Since $\left|B_N\left(e^{it}\right)\right| = 1$, all t, it holds that $\|F_N\|_{H^p} = \|f\|_{H^p}$ (use the fact that $B_N\left(re^{it}\right) \to B_N\left(e^{it}\right)$ uniformly in t as $r \to 1^-$.) If $0 < r < 1$, then

$$
\begin{aligned}
\int_0^{2\pi} \left|F\left(re^{it}\right)\right|^p dt^{1/p} &= \lim_{N\to\infty} \int_0^{2\pi} \left|F_N\left(re^{it}\right)\right|^p dt^{1/p} \\
&\le \lim_{N\to\infty} \|F_N\|_{H^p} = \|f\|_{H^p}.
\end{aligned}
$$

Therefore $\|F\|_{H^p} \le \|f\|_{H^p}$. □

Corollary 15.1.22 *If* $\{p_1, p_2, \ldots\}$ *is a sequence of points in* $\mathbb{D}$ *satisfying* $\sum_j \left(1 - |p_j|\right) < \infty$ *and if* $B(z) = \prod_j \left(-\overline{p}_j / |p_j|\right) B_{p_j}(z)$ *is the corresponding Blaschke product, then*

$$\lim_{r \to 1^-} B\left(re^{it}\right)$$

exists and has modulus 1 almost everywhere.

Proof The conclusion that the limit exists follows from Theorem 15.1.13 and the fact that $B \in H^\infty$. For the other assertion, note that the canonical factorization for B is $B = 1 \cdot B$. Therefore, by Proposition 15.1.21,

$$\int \left|\widetilde{B}\left(e^{it}\right)\right|^2 dt^{1/2} = \|B\|_{H^2} = \|1\|_{H^2} = 1;$$

hence $\left|\widetilde{B}\left(e^{it}\right)\right| = 1$ almost everywhere. □

Theorem 15.1.23 *If* $f \in H^p(\mathbb{D}), 0 < p \leq \infty$*, and* $1 < \alpha < \infty$*, then*

$$\lim_{\Gamma_\alpha(\mathbb{D}) \ni z \to e^{i\theta}} f(z)$$

exists for almost every $e^{i\theta} \in \partial\mathbb{D}$ *and equals* $\widetilde{f}\left(e^{i\theta}\right)$*. Also,* $\widetilde{f} \in L^p(\partial\mathbb{D})$ *and*

$$\|\widetilde{f}\|_{L^p} = \|f\|_{H^p} \equiv \sup_{0<r<1} \int_0^{2\pi} \left|f\left(re^{i\theta}\right)\right|^p d\theta^{1/p}.$$

Proof By Definition 15.1.20, write $f = B \cdot F$, where F has no zeroes and B is a Blaschke product. Then $F^{p/2}$ is a well-defined H^2 function and thus has the appropriate boundary values almost everywhere. A fortiori, F has non-tangential boundary limits almost everywhere. Since $B \in H^\infty$, B has non-tangential boundary limits almost everywhere. It follows that f does as well. The final assertion follows from the corresponding fact for H^2 functions (exercise). □

See also [133] and the references therein.

15.2 The Szego and Poisson-Szego Kernels

The basic theory of the Szegö kernel is similar to that for the Bergman kernel; they are both special cases of a general theory of "Hilbert spaces with reproducing kernel" (see N. Aronszajn [33], Chang and Krantz [109]). Thus we only outline the basic steps here, leaving details to the reader.

Let $\Omega \subseteq \mathbb{C}^n$ be a bounded domain with C^2 boundary. Let $A(\Omega)$ be those functions continuous on $\bar{\Omega}$ that are holomorphic on Ω. Let $H^2(\partial\Omega)$ be the space consisting of the closure in the $L^2(\partial\Omega, d\sigma)$ topology of the restrictions to $\partial\Omega$ of elements of $A(\Omega)$. Then $H^2(\partial\Omega)$ is a proper Hilbert subspace of $L^2(\partial\Omega)$. Each element $f \in H^2(\partial\Omega)$ has a natural holomorphic extension to Ω given by its Poisson integral Pf.

For each fixed $z \in \Omega$, the functional

$$\psi_z : H^2(\partial\Omega) \ni f \mapsto Pf(z)$$

is continuous. (Why?) Let $k_z(\zeta)$ be the Hilbert space representative for the functional ψ_z. Define the Szegö kernel $S(z, \zeta)$ by the formula

$$S(z, \zeta) = \overline{k_z(\zeta)} \quad z \in \Omega, \zeta \in \partial\Omega.$$

If $f \in H^2(\partial\Omega)$, then

$$\mathrm{Pf}(z) = \int_{\partial\Omega} S(z, \zeta) f(\zeta) d\sigma(\zeta)$$

for all $z \in \Omega$. We shall not explicitly formulate and verify the various uniqueness and extremal properties for the Szegö kernel. The reader is invited to consider these topics.

Let $\{\phi_j\}_{j=1}^{\infty}$ be an orthonormal basis for $H^2(\partial\Omega)$. Define

$$S'(z, \zeta) = \sum_{j=1}^{\infty} \phi_j(z)\overline{\phi_j(\zeta)}, \qquad z, \zeta \in \Omega.$$

For convenience we tacitly identify each function with its Poisson extension to the interior of the domain. Then, for $K \subseteq \Omega$ compact, the series defining S' converges uniformly on $K \times K$. By a Riesz-Fischer argument, $S'(\cdot, \zeta)$ is the Poisson integral of an element of $H^2(\partial\Omega)$ and $S'(z, \cdot)$ is the conjugate of the Poisson integral of an element of $H^2(\partial\Omega)$. So S' extends to $(\bar{\Omega} \times \Omega) \cup (\Omega \times \bar{\Omega})$, where it is understood that all functions on the boundary are defined only almost everywhere. The kernel S' is conjugate symmetric. Also, by Riesz-Fischer theory, S' reproduces $H^2(\partial\Omega)$. Since the Szegö kernel is unique, it follows that $S = S'$.

The Szegö kernel may be thought of as representing a map

$$S : f \mapsto \int_{\partial\Omega} f(\zeta) S(\cdot, \zeta) d\sigma(\zeta)$$

from $L^2(\partial\Omega)$ to $H^2(\partial\Omega)$. Since $S = S'$ is self-adjoint and idempotent, it is the Hilbert space projection of $L^2(\partial\Omega)$ to $H^2(\partial\Omega)$.

The Poisson-Szegö kernel is obtained by a formal procedure from the Szegö kernel: This procedure manufactures a positive reproducing kernel from one that is not necessarily positive. Note in passing that, just as we argued for the Bergman kernel in the last section, $S(z, z)$ is never 0 when $z \in \Omega$.

Proposition 15.2.1 *Define*

$$\mathcal{P}(z, \zeta) = \frac{|S(z, \zeta)|^2}{S(z, z)}, \quad z \in \Omega, \zeta \in \partial\Omega.$$

Then for any $f \in A(\Omega)$ and $z \in \Omega$, it holds that

$$f(z) = \int_{\partial\Omega} f(\zeta)\mathcal{P}(z, \zeta)d\sigma(\zeta).$$

Proof Fix $z \in \Omega$ and $f \in A(\Omega)$, and define

$$u(\zeta) = f(\zeta)\frac{\overline{S(z, \zeta)}}{S(z, z)}, \quad \zeta \in \partial\Omega.$$

Then $u \in H^2(\partial\Omega)$; hence

$$\begin{aligned} f(z) = u(z) &= \int_{\partial\Omega} S(z, \zeta)u(\zeta)d\sigma(\zeta) \\ &= \int_{\partial\Omega} \mathcal{P}(z, \zeta)f(\zeta)d\sigma(\zeta). \end{aligned}$$

This is the desired formula. □

Remark 15.2.2 In passing to the Poisson-Szegö kernel, we gain the advantage of positivity of the kernel (for more on this circle of ideas, see [333, Chapter 8] and also Y. Katznelson [265, Chapter 1]). However, we lose something in that $\mathcal{P}(z, \zeta)$ is no longer holomorphic in the z variable nor conjugate holomorphic in the ζ variable. The literature on this kernel is rather sparse, and there are many unresolved questions.

As an exercise, use the paradigm of Proposition 15.2.1 to construct a positive kernel from the Cauchy kernel on the disc (be sure to first change notation in the usual Cauchy formula so that it is written in terms of arc length measure on the boundary). What familiar kernel results?

Like the Bergman kernel, the Szegö and Poisson-Szegö kernels can almost never be explicitly computed. They can be calculated asymptotically in a number of important instances, however (see C. Fefferman [159] and L. Boutet de Monvel and J. Sjöstrand [382]). We will give explicit formulas for these kernels on the ball.

Lemma 15.2.3 *The functions $\{z^\alpha\}$, where α ranges over multi-indices, are pairwise orthogonal and span $H^2(\partial B)$.*

Proof The orthogonality follows from symmetry considerations. For the completeness, notice that it suffices to see that the span of $\{z^\alpha\}$ is dense in $A(B)$ in the uniform topology on the boundary. By the Stone-Weierstrass theorem, the closed algebra generated by $\{z^\alpha\}$ and $\{\bar{z}^\alpha\}$ is all of $C(\partial B)$. But the monomials $\bar{z}^\alpha, \alpha \neq 0$, are orthogonal to $A(B)$ (use the power series expansion about the origin to see this). The claimed density follows. □

Lemma 15.2.4 *Let* $\mathbf{1} = (1, 0, \ldots, 0)$. *Then*

$$S(z, \mathbf{1}) = \frac{(n-1)!}{2\pi^n} \frac{1}{(1-z_1)^n}.$$

Proof We have that

$$S(z, \mathbf{1}) = \sum_\alpha \frac{z^\alpha \cdot \mathbf{1}^\alpha}{\|z_1^\alpha\|^2_{L^2(\partial B)}} = \sum_{k=0}^\infty \frac{z_1^k}{\eta(k)} = \frac{1}{2\pi^n} \sum_{k=0}^\infty \frac{z_1^k (k+n-1)!}{k!}$$

$$= \frac{(n-1)!}{2\pi^n} \sum_{k=0}^\infty \binom{k+n-1}{n-1} z_1^k = \frac{(n-1)!}{2\pi^n} \frac{1}{(1-z_1)^n}.$$

□

Lemma 15.2.5 *Let ρ be a unitary rotation on $\mathbb{C}^n$. For any $z \in \bar{B}$, $\zeta \in \partial B$, we have that $S(z, \zeta) = S(\rho z, \rho \zeta)$.*

Proof This is a standard change of variables argument, and we omit it. □

Theorem 15.2.6 *The Szegö kernel for the ball is*

$$S(z, \zeta) = \frac{(n-1)!}{2\pi^n} \frac{1}{(1 - z \cdot \bar{\zeta})^n}.$$

Proof Let $z \in B$ be arbitrary. Let ρ be the unique unitary rotation such that ρz is a multiple of 1. Then, by Lemma 15.2.5,

$$S(z, \zeta) = S\left(\rho^{-1}\mathbf{1}, \zeta\right) = S(\mathbf{1}, \rho\zeta) = \overline{S(\rho\zeta, \mathbf{1})} = \frac{(n-1)!}{2\pi^n} \frac{1}{(1 - (\overline{\rho\zeta}) \cdot \mathbf{1})^n}$$

$$= \frac{(n-1)!}{2\pi^n} \frac{1}{\left(1 - \bar{\zeta} \cdot (\rho^{-1}\mathbf{1})\right)^n} = \frac{(n-1)!}{2\pi^n} \frac{1}{(1 - z \cdot \bar{\zeta})^n}.$$

□

Corollary 15.2.7 *The Poisson-Szegö kernel for the ball is*

$$\mathcal{P}(z, \zeta) = \frac{(n-1)!}{2\pi^n} \frac{\left(1 - |z|^2\right)^n}{|1 - z \cdot \bar{\zeta}|^{2n}}.$$

Exercise for the Reader
Calculate the Szegö and Poisson-Szegö kernel for the polydisc.

Afterword to Sect. 15.2
It follows from work of A. Gleason [176] and L. Bungart [96] that Cauchy-type mtegral formulas that are holomorphic in the z variable (this is crucial for constructing holomorphic functions) exist on virtually any domain. However these references tell us almost nothing about the form of the integral kernel-whether its singularity is only on the diagonal or at what rate it blows up. The Bergman and Szegö theories give other methods for producing integral formulas on a wide class of domains.

Thus we have several "canonical kernels" on domains in $\mathbb{C}^n$: the Poisson kernel, the Bergman kernel, and the Szegö kernel. Unfortunately, these kernels are virtually never computable (however see C. Fefferman [150], L. Boutet de Monvel and J. Sjöstrand [382], and L. Hua [244]). Later on we shall learn of a construction by G.M. Henkin (variants were also constructed independently by E. Ramirez [30] and H. Grauert and I. Lieb [185]) that yields rather explicitly computable kernels on a large class of domains in $\mathbb{C}^n$. For many practical applications, the Henkin kernel is just as useful as the Szegö kernel. And it turns out that the Szegö kernel may be expressed as an asymptotic expansion in terms of powers of the Henkin kernel (see N. Kerzman and E.M. Stein [273]).

The connection between canonical kernels and computable kernels has opened up new vistas and has made accessible many deep results in function theory and geometry (see N. Kerzman and E.M. Stein [272, 273], R.E. Greene and S.G. Krantz [191–201], D.H. Phong and E.M. Stein [483], C. Fefferman [150, 151], L. Boutet de Monvel and J. Sjöstrand [382], and S. Bell and Boas [66]). The subject of integral representations will be a fertile area of research for some time to come.

15.3 Three Propositions About the Poisson Kernel

The crux of our arguments in Sect. 15.1 was Proposition 15.1.12, in which the Poisson integral was majorized by the Hardy-Littlewood maximal function.

In this estimation, the explicit form of the Poisson kernel for the disc was exploited. If we wish to use a similar program to study the boundary behavior of harmonic and holomorphic functions on general domains in $\mathbb{R}^N$ and $\mathbb{C}^n$, then we must again estimate the Poisson integral by a maximal function.

However there is no hope of obtaining an explicit formula for the Poisson kernel of an arbitrary smoothly bounded domain. In this section we shall instead obtain some rather sharp estimates that will suffice for our purposes.

We shall present here a classical treatment that was developed by N. Kerzman [269]. Kerzman's methods have the virtue of being self-contained and giving precisely the estimates that we need. We present the proofs of Propositions 15.3.2 and 15.3.7 with the generous permission of Norberto Kerzman.

Now the Poisson kernel for a C^2 domain $\Omega \subseteq \mathbb{R}^N$ is given by $P(x, y) = -\nu_y G(x, y), x \in \Omega, y \in \partial\Omega$—see S.G. Krantz [331]. Here ν_y is the unit outward normal vector field to $\partial\Omega$ at y, and $G(x, y)$ is Green's function for Ω. Recall that, for $N > 2$, we have $G(x, y) = c_n|x-y|^{-N+2} - F_x(y)$, where F depends on a $C^{2-\epsilon}$ fashion on x and y jointly and F is harmonic in y. By Now G is $C^{2-\epsilon}$ on $\overline{\Omega} \times \overline{\Omega}\setminus\{$ diagonal $\}$ and $G(x, y) = G(y, x)$. It follows that $P(x, y)$ behaves qualitatively like $|x - y|^{-N+1}$. [These observations persist in $\mathbb{R}^2$ by a slightly different argument.] In this section we use the method of comparison domains to refine this qualitative estimate on the Poisson kernel.

We begin with a geometric fact:

Geometric Fact Let $\Omega \subset\subset \mathbb{R}^N$ have C^2 boundary. There are numbers $r, \widetilde{r} > 0$ such that for each $y \in \partial\Omega$ there are balls $B\left(c_y, r\right) \equiv B_y \subseteq \Omega$ and $B\left(\widetilde{c}_y, \widetilde{r}\right) \equiv \widetilde{B}_y \subseteq \cap^c\, \overline{\Omega}$ that satisfy:

(i) $\overline{B}\left(\widetilde{c}_y, \widetilde{r}\right) \cap \overline{\Omega} = \{y\}$.
(ii) $\overline{B}\left(c_y, r\right) \cap {}^c\overline{\Omega} = \{y\}$.

Let us indicate why these balls exist. Fix $P \in \partial\Omega$. Applying the implicit function theorem to the mapping

$$\partial\Omega \times (-1, 1) \to \mathbb{R}^N$$
$$(\zeta, t) \mapsto \zeta + t\nu_\zeta$$

at the point $(P, 0)$, we find a neighborhood U_P of the point P on which the mapping is one-to-one. By the compactness of the boundary, there is thus a neighborhood U of $\partial\Omega$ such that each $x \in U$ has a unique nearest point in $\partial\Omega$. It further follows that there is an $\epsilon > 0$ such that if ζ_1, ζ_2 are distinct points of $\partial\Omega$, then $I_1 = \left\{\zeta_1 + t\nu_{\zeta_1} : |t| < 2\epsilon\right\}$ and $I_2 = \left\{\zeta_2 + t\nu_{\zeta_2} : |t| < 2\epsilon\right\}$ are disjoint sets (i.e., the normal bundle is locally trivial in a natural way). From this it follows that if $y \in \partial\Omega$, then we may take $c_y = y - \epsilon\nu_y, \widetilde{c}_y = y + \epsilon\nu_y$, and $r = \widetilde{r} = \epsilon$.

We may assume in what follows that $r = \widetilde{r} < \operatorname{diam}\Omega/2$. See Fig. 15.1.

We now consider estimates for $P_\Omega(x, y)$. Results are stated for all dimensions $N \geq 2$ but, for convenience, are proved only for $N \geq 3$. The basic technique is to compare Green's function and Poisson kernel for the given domain Ω with the corresponding functions for the internally and externally tangent balls $B\left(c_y, r\right)$ and $B\left(\widetilde{c}_y, \widetilde{r}\right)$ (and also for their complements).

Proposition 15.3.1 *Let $\Omega \subseteq \mathbb{R}^N$ be a domain with C^2 boundary. Let $P = P_\Omega : \Omega \times \partial\Omega \to \mathbb{R}$ be its Poisson kernel. Then for each $x \in \Omega$, there is a positive constant C_x such that*

$$0 < C_x \leq P(x, y) \leq \frac{C}{|x-y|^{N-1}} \leq \frac{C}{\delta(x)^{N-1}}.$$

Here $\delta(x) = \operatorname{dist}(x, \partial\Omega)$.

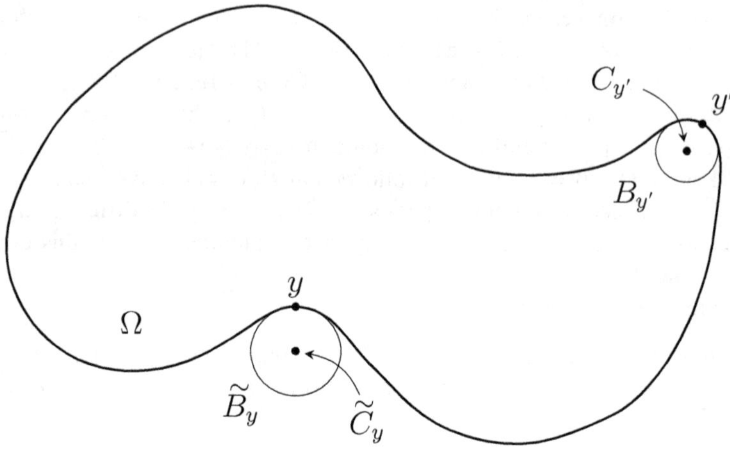

Fig. 15.1 Relation of the radii

Proof Recall that $G(x, y) = 0$ for $x \in \Omega$, $y \in \partial\Omega$. By the maximum principle applied to $-G(x, \cdot)$ on $\Omega\backslash B(x, \epsilon)$, we have that $G > 0$ on Ω. By the Hopf lemma, $-\nu_y G(x, y) > 0$ when $y \in \partial\Omega$. Since $\partial\Omega$ is compact, the bound from below on $P(x, y)$ is established.

Now for fixed $x \in \Omega$, $y \in \partial\Omega$, we know that $G(x, \cdot)$ equals 0 on $\partial\Omega$, whereas $G_{c\widetilde{B}_y}(x, \cdot) \geq 0$ on $\partial\Omega$. Since $G_{c\widetilde{B}_y}(x, \cdot) - G_\Omega(x, \cdot)$ is harmonic on Ω, it follows that

$$G_{c\widetilde{B}_y}(x, t) \geq G_\Omega(x, t), \quad t \in \overline{\Omega}$$
$$G_{cB_y}(x, y) = G_\Omega(x, y) = 0.$$

Therefore

$$P_{c\widetilde{B}_y}(x, y) = -\nu_y G_{c\widetilde{B}_y}(x, y) \geq -\nu_y G_\Omega(x, y) = P_\Omega(x, y). \tag{15.2}$$

The Poisson kernel for $\widetilde{B}_y$ is

$$P_{\widetilde{B}_y}(x, t) = \frac{1}{\omega_{N-1}\widetilde{r}} \cdot \frac{\widetilde{r}^2 - |x - \widetilde{c}_y|^2}{|x - t|^N}.$$

By Kelvin reflection, the Poisson kernel for ${}^c\widetilde{B}_y$ is

$$P_{c\widetilde{B}_y}(x, t) = \frac{1}{\omega_{N-1}\widetilde{r}} \cdot \frac{|x - \widetilde{c}_y|^2 - \widetilde{r}^2}{|x - t|^N}.$$

Thus (15.2) gives

$$
\begin{aligned}
P_\Omega(x,y) &\le \frac{1}{\omega_{N-1}\widetilde{r}} \cdot \frac{\left|x-\widetilde{c}_y\right|^2-\widetilde{r}^2}{|x-t|^N} \\
&= \frac{1}{\omega_{N-1}\widetilde{r}} \cdot \frac{\left(\left|x-\widetilde{c}_y\right|-\left|\widetilde{c}_y-y\right|\right)\left(\left|x-\widetilde{c}_y\right|+\widetilde{r}\right)}{|x-t|^N} \\
&\le \frac{2}{\omega_{N-1}\widetilde{r}} \frac{|x-y|(\operatorname{diam}\Omega+\widetilde{r})}{|x-y|^N} \\
&= C|x-y|^{-N+1 \le C\delta(x)^{-N+1}} .
\end{aligned}
$$

□

Exercises for the Reader Use the explicit formula for the Poisson kernel of $B(0, R) \subseteq \mathbb{R}^N$ to prove that if u is a nonnegative harmonic function on $B(0, R)$ and $\xi \in B(0, R)$, then:

(i) $u(\xi) \le 2(R/d(\xi))^{N-1} \cdot u(0)$.
(ii) $u(\xi) \ge [d(\xi)/(2^N R)] \cdot u(0)$, where $d(\xi) = \operatorname{dist}(\xi, {}^c B(0, R))$.
These first two inequalities are essentially the Harnack inequalities. Apply inequality (ii) to $G_\Omega(x, \cdot)$ on a ball B_y^* internally tangent to $\partial\Omega$ at $y \in \partial\Omega$ with $x \notin B_y^*$ to obtain:
(iii) $G_\Omega(x, t) \ge C_x[d(t)]G_\Omega(x, c_y)\,, \quad t \in B_y^*\,, \quad x \in \Omega$.
Conclude that:
(iv) $-\frac{\partial}{\partial \nu_y} G_\Omega(x, y) \ge C_x G_\Omega(x, c_y) > 0$,
so
(v) $P_\Omega(x, y) \ge C_x' > 0$.
If $x \in \Omega$, $\delta(x) < r$, let $\pi x \in \partial\Omega$ be the nearest point to x. Apply (ii) to $P_\Omega(\cdot, y)$ on $B_{\pi x}$ to obtain:
(vi) $P_\Omega(x, y) \ge C\delta(x) \cdot P_\Omega(c_{\pi x}, y)$ provided $\delta(x) < r$.
Conclude from (vi) and the fact that $\{c_{\pi x}\} \subset\subset \Omega$ that:
(vii) $P_\Omega(x, y) \ge C'\delta(x)$ for all $x \in \Omega$.
For the case $\delta(x) < r$ and $y = \pi x$, we may apply the argument in the proof of Proposition 15.3.1 to compare $P_\Omega(x, y)$ with $P_{B_y}(x, y)$ to obtain:
(viii) $P_\Omega(x, y) \ge c|x - y|^{-N+1}$ when $y = \pi x$.

With the results of the exercises in hand, it is now easy to prove the next proposition:

Proposition 15.3.2 *If $\Omega \subset\subset \mathbb{R}^n$ is a domain with C^2 boundary, then*

$$
P_\Omega(x, y) \ge C \cdot \frac{\delta(x)}{|x - y|^N}.
$$

Proof Choose $0 < \eta < r/2$ so small that the following is true: Let $x \in \Omega$, $y \in \partial\Omega$, $|x - y| < \eta$, and let $\theta = \theta(x, y) = \pi x - 2|x - y| \cdot \nu_{\pi x}$. Then there is an $\widetilde{x} \in B(\theta, 3|x - y|/2)$ such that $|\widetilde{x} - y| = \delta(\widetilde{x})$ (see Fig. 15.2).

Consider $y \in \partial\Omega$ to be fixed. To prove the result, we need only consider x such that $|x - y| < \eta$ (otherwise (vii) above applies). Notice that

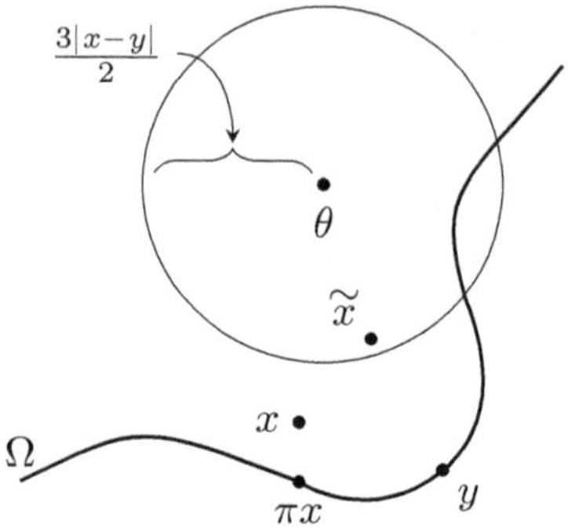

Fig. 15.2 Existence of $\widetilde{x}$

$$\begin{aligned}|\widetilde{x} - y| &\leq |\widetilde{x} - \theta| + |\theta - x| + |x - y| \\ &\leq 3\frac{|x - y|}{2} + 2|x - y| + |x - y| = 5|x - y|.\end{aligned} \tag{15.3}$$

Applying (ii) to $P_\Omega(\cdot, y)$ on $B(\theta, 2|x-y|)$ and then (i) to $P_\Omega(\cdot, y)$ on $B(\theta, 2|x-y|)$ gives

$$\begin{aligned}P_\Omega(x, y) &\geq C\frac{\delta(x)}{2|x - y|} \cdot P_\Omega(\theta, y) \\ &\geq C'\frac{\delta(x)}{|x - y|}\left(\frac{|x - y|/2}{2|x - y|}\right)^{N-1} P_\Omega(\widetilde{x}, y) \\ &= C''\delta(x)\frac{P_\Omega(\widetilde{x}, y)}{|x - y|}.\end{aligned}$$

But (viii), together with (15.3), shows that this majorizes $C'''\delta(x)/|x - y|^N$. □

The next estimate for P_Ω (Proposition 15.3.7) is the most useful and the deepest. In order to prove it, we must first consider an auxiliary function to which we will compare it.

Lemma 15.3.3 *Let* $B \subseteq \mathbb{R}^N$ *be the unit ball. Let* $\widetilde{B} = B((0, \ldots, 0, 1), 1)$ *and* $T = B \cap {}^c\widetilde{B}$. *Let* $\tau = \partial B \backslash \overline{\widetilde{B}}$. *There is a nonnegative harmonic function* H *on a neighborhood of* $\overline{T}$ *and a* $c > 0$ *such that:*

15.3.3 a $H(0) = 0.$
15.3.3 b $H(x) > 0$ *if* $x \in \overline{T} \setminus \{0\}.$
15.3.3 c $H \geq 1$ *on* $\tau.$
15.3.3 d $H(0, \ldots, 0, -t) \leq ct,$ *all* $0 \leq t \leq 1.$

Proof Let H be a large positive multiple of the Kelvin transform in $\widetilde{B}1$ of the function $x \mapsto x_N$ (see Fig. 15.3).

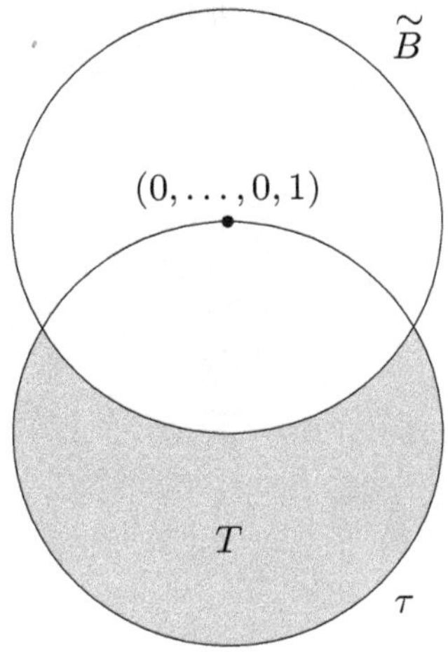

Fig. 15.3 Kelvin transform of the function $x \mapsto x_N$

A moment's reflection shows that this H works (do not calculate—think geometrically). □

Remark 15.3.4 If $0 < \alpha$, let T_α be the region obtained from T by dilation of order α (i.e., T_α is obtained from balls of radius α). Then $H_\alpha(x) = H(x/\alpha)$ has the same properties relative to T_α as H does relative to T.

Lemma 15.3.5 *There is a constant* $C = C(\Omega, N)$ *such that* $G_\Omega(x, y) \leq C \mid x- y|^{-N+2}$.

Proof Let $\epsilon > 0$. Both $G_\Omega(x, y) = -c_N|x - y|^{-N+2} - F_x(y)$ and $-2c_N \mid x- y|^{-N+2}$ are harmonic on $\Omega \backslash B(x, \epsilon)$, and the latter majorizes the former on $\partial(\Omega \backslash B(x, \epsilon))$ if ϵ is small. (Remember how F_x was constructed!) By the maximum principle, the majorization persists on $\Omega \backslash B(x, \epsilon)$ and hence on Ω. □

Lemma 15.3.6 *We have*

$$G_\Omega(x, y) \leq C(\Omega, N) \cdot \frac{\delta(y)}{|x - y|^{N-1}}, \tag{15.4}$$

$$G_\Omega(x, y) \leq C(\Omega, N) \cdot \frac{\delta(x)}{|x - y|^{N-1}}. \tag{15.5}$$

Proof We first prove (15.4). Recall that $\widetilde{r} < \operatorname{diam} \Omega/2$. If $\delta(y) \geq \widetilde{r} \mid x- y \mid /(8 \operatorname{diam} \Omega)$, then there is nothing to prove by Lemma 15.3.5. If $\delta(y) < \widetilde{r}|x - y|/(8 \operatorname{diam} \Omega)$, let $\alpha = \widetilde{r}|x - y|/(4 \operatorname{diam} \Omega) > 2\delta(y)$. Rotate and translate T_α so that 0 is sent to πy (the projection of y to $\partial\Omega$) and the resulting region T_α^* is tangent to $\partial\Omega$ at πy (see Fig. 15.4).

Observe in the figure that Ω and T_α^* have the same inward normal at πy. Let $t \in \partial T_\alpha^* \cap \Omega$. Then

Fig. 15.4 The Figure T_α^*

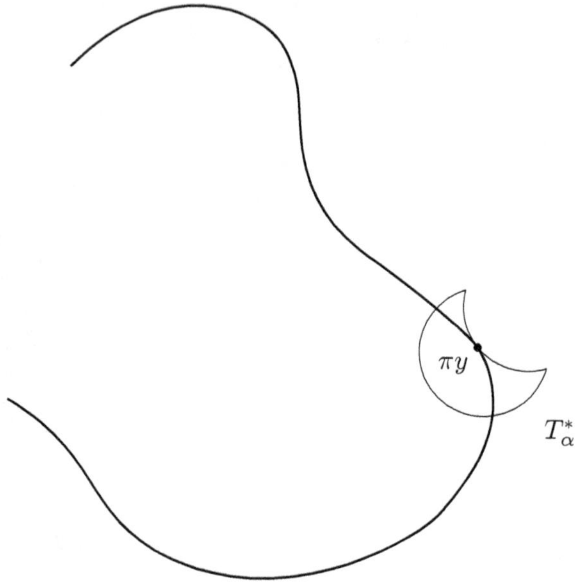

$$|x-t| \geq |x-y| - |y-t| \geq |x-y| - 2\alpha \geq \frac{|x-y|}{2}.$$

Therefore

$$G_\Omega(x,t) \leq C|x-t|^{-N+2} \leq C|x-y|^{-N+2}. \tag{15.6}$$

Let H_α^* be the harmonic function on T_α^* whose existence is guaranteed by Lemma 15.3.3. By (15.3.3 c) and (15.6),

$$H_\alpha^*(t)C|x-y|^{-N+2} \geq G_\Omega(x,t), \quad t \in \partial T_\alpha^* \cap \Omega. \tag{15.7}$$

But, trivially,

$$H_\alpha^*(t)C|x-y|^{-N+2} \geq 0 = G_\Omega(x,t), \quad t \in T_\alpha^* \cap \partial\Omega. \tag{15.8}$$

By the maximum principle,

$$H_\alpha^*(t)C|x-y|^{-N+2} \geq G_\Omega(x,t), \quad \text{all} \quad t \in T_\alpha^* \cap \Omega. \tag{15.9}$$

Therefore

$$H_\alpha^*(y)C|x-y|^{-N+2} \geq G_\Omega(x,y).$$

Now (15.3.3 d) yields that

$$C''(\delta(y)/\alpha)|x-y|^{-N+2} \geq G_\Omega(x,y); \tag{15.10}$$

hence the choice of α implies

$$C''\delta(y)|x-y|^{-N+1} \geq G_\Omega(x,y).$$

Now (15.5) follows from (15.4) by symmetry. □

The final estimate on $P_\Omega(x,y)$ is the following:

Proposition 15.3.7 *If $\Omega \subset\subset \mathbb{R}^N$ is a domain with C^2 boundary, then*

$$P_\Omega(x,y) \leq C\frac{\delta(x)}{|x-y|^N}.$$

Proof We exploit Lemma 15.3.6. We prove that

$$G_\Omega(x,y) \leq C\frac{\delta(x)\delta(y)}{|x-y|^N}. \tag{15.11}$$

For the case $\delta(x) \geq \widetilde{r}|x-y|/(8\,\mathrm{diam}\,\Omega)$, there is nothing to prove. So we may assume that $\delta(x) < \widetilde{r}|x-y|/(8\,\mathrm{diam}\,\Omega)$.

Use the notation of the proof of Lemma 15.3.6. We do a comparison on T^*_α. The argument used to prove (15.10) now shows that (15.11) holds because, with (15.5) replacing (15.6), we obtain

$$G_\Omega(x,t) \leq C\cdot\frac{\delta(x)H^*_\alpha(t)}{|x-t|^N}, \quad t \in \partial\left(T^*_\alpha \cap \Omega\right)$$

instead of (15.9).

By the maximum principle, (15.11) follows. Using the fact that

$$G_\Omega(x,\cdot)|_{\partial\Omega} = 0$$

and differentiating, the proposition follows. □

15.4 Subharmonicity, Harmonic Majorants, and Boundary Values

Let $\Omega \subseteq \mathbb{C}^N$ be a domain and $f : \Omega \to \mathbb{R}$ a function. The function f is said to have a harmonic majorant if there is a (necessarily) nonnegative harmonic u on Ω with $|f| \leq u$. We are interested in harmonic majorants for subharmonic functions.

Proposition 15.4.1 *If* $f : \mathbb{B} \to \mathbb{R}^+$*is subharmonic, then* f *has a* harmonic majorant *if and only if*

$$\sup_{0<r<1} \int_{\partial\mathbb{B}} f(r\zeta)d\sigma(\zeta) < \infty.$$

Proof Let u be a harmonic majorant for f. Then

$$\int_{\partial\mathbb{B}} f(r\zeta)d\sigma(\zeta) \le \int_{\partial\mathbb{B}} u(r\zeta)d\sigma(\zeta) = \omega_{N-1} \cdot u(0) \equiv C < \infty$$

as claimed.

Conversely, if f satisfies

$$\sup_r \int_{\partial\mathbb{B}} f(r\zeta)d\sigma(\zeta) < \infty,$$

then the functions $f_r : \partial B \to \mathbb{C}$ given by $f_r(\zeta) = f(r\zeta)$ form a bounded subset of $L^1(\partial\mathbb{B}) \subseteq \mathcal{M}(\partial\mathbb{B})$. Let $\widetilde{f} \in \mathcal{M}(\partial\mathbb{B})$ be a weak-* accumulation point of the functions f_r. Then $F(r\zeta) \equiv P\widetilde{f}(r\zeta)$ is harmonic on $\mathbb{B}$, and for any $x \in B$ and $1 > r > |x|$, we have

$$\begin{aligned} 0 \le f(x) &\le \int P(x/r, \zeta) f_r(\zeta)d\sigma(\zeta) \\ &\to \int P(x, \zeta)d\widetilde{f}(\zeta) = F(x) \quad \text{as} \quad r \to 1^-. \end{aligned}$$

A consequence of Proposition 15.4.1 is that not all subharmonic functions have harmonic majorants. Harmonic majorants play a significant role in the theory of boundary behavior of harmonic and holomorphic functions. Proposition 15.4.1 suggests why growth conditions may, therefore, play a role. The fact that f harmonic implies $|f|^p$ subharmonic only for $p \ge 1$, whereas f holomorphic implies $|f|^p$ subharmonic for $p > 0$ suggests that we may expect different behavior for harmonic and holomorphic functions.

By way of putting these remarks in perspective and generalizing Proposition 15.4.1, we consider $\mathbf{h}^p(\Omega)$ (resp., $H^p(\Omega)$), with Ω any smoothly bounded domain in $\mathbb{R}^N$ (resp., $\mathbb{C}^n$). First we require some preliminary groundwork.

Let $\Omega \subset\subset \mathbb{R}^N$ be a domain with C^2 boundary. Let $\phi : \mathbb{R} \to [0, 1]$ be a C^∞ function supported in $[-2, 2]$ with $\phi \equiv 1$ on $[-1, 1]$. Then, with $\delta_\Omega(x) \equiv \text{dist}(x, \partial\Omega)$ and $\epsilon_0 > 0$ sufficiently small, we see that

$$\rho(x) = \begin{cases} -\phi(|x|/\epsilon_0)\,\delta_\Omega(|x|) - (1 - \phi(|x|/\epsilon_0)) & \text{if } x \in \overline{\Omega} \\ \phi(|x|/\epsilon_0)\,\delta_\Omega(|x|) + (1 - \phi(|x|/\epsilon_0)) & \text{if } x \notin \overline{\Omega} \end{cases}$$

is a C^2 defining function for Ω. The implicit function theorem implies that if $0 < \epsilon < \epsilon_0$, then $\partial\Omega_\epsilon \equiv \{x \in \Omega : \rho(x) = -\epsilon\}$ is a C^2 manifold that bounds $\Omega_\epsilon \equiv \{x \in \Omega : \rho_\epsilon(x) \equiv \rho(x) + \epsilon < 0\}$. Now let $d\sigma_\epsilon$ denote area measure on $\partial\Omega_\epsilon$. Then it is natural to let

$$\begin{aligned}\mathbf{h}^p(\Omega) = \Bigg\{ f \text{ harmonic on } \Omega : \sup_{0<\epsilon<\epsilon_0} \int_{\partial\Omega_\epsilon} |f(\zeta)|^p d\sigma_\epsilon(\zeta)^{1/p} \\ \equiv \|f\|_{h^p(\Omega)} < \infty \Bigg\}, \quad 0 < p < \infty; \\ \mathbf{h}^\infty(\Omega) = \Bigg\{ f \text{ harmonic on } \Omega : \sup_{x\in\Omega} |f(x)| \equiv \|f\|_{\mathbf{h}^\infty} < \infty \Bigg\}.\end{aligned}$$

In case Ω is a subdomain of complex space, we define

$$H^p(\Omega) = \mathbf{h}^p(\Omega) \cap \{ \text{ holomorphic functions } \}, \quad 0 < p \leq \infty.$$

□

The next lemma serves to free the definitions from their somewhat artificial dependence on δ_Ω and ρ.

Lemma 15.4.2 (Stein) *Let ρ_1, ρ_2 be two C^2 defining functions for a domain $\Omega \subseteq \mathbb{R}^N$. For $\epsilon > 0$ small and $i = 1, 2$, let*

$$\begin{aligned}\Omega^i_\epsilon &= \{x \in \Omega : \rho_i(x) < -\epsilon\} \\ \partial\Omega^i_\epsilon &= \{x \in \Omega : \rho_i(x) = -\epsilon\}.\end{aligned}$$

Let σ^i_ϵ be area measure on $\partial\Omega^i_\epsilon$. Then, for f harmonic on Ω, we have

$$\sup_{\epsilon>0} \int_{\partial\Omega^1_\epsilon} |f(\zeta)|^p d\sigma^1_\epsilon(\zeta) < \infty$$

if and only if

$$\sup_{\epsilon>0} \int_{\partial\Omega^2_\epsilon} |f(\zeta)|^p d\sigma^2_\epsilon(\zeta) < \infty.$$

*(**Note:** Since f is bounded on compact sets, equivalently the supremum is only of interest as $\epsilon \to 0$, there is no ambiguity in this assertion.)*

Proof By definition of defining function, grad $\rho_i \neq 0$ on $\partial\Omega$. Since $\partial\Omega$ is compact, we may choose $\epsilon_0 > 0$ so small that there is a constant λ, $0 < \lambda < 1$, with $0 < \lambda \leq |\text{grad}\, \rho_i(x)| < 1/\lambda$ whenever $x \in \Omega$, $\delta_\Omega(x) < \epsilon_0$. If $0 < \epsilon < \epsilon_0$, then notice that for $x \in \partial\Omega^2_\epsilon$ we have

$$\mathbb{B}(x, \lambda\epsilon/2) \subseteq \Omega$$

and, what is stronger,

$$\mathbb{B}(x, \lambda\epsilon/2) \subseteq \left\{t : -3\epsilon/\lambda^2 < \rho_1(t) < -\lambda^2 \cdot \epsilon/3\right\} \equiv S(\epsilon). \tag{15.12}$$

Therefore

$$|f(x)|^p \leq \frac{1}{V(B(x, \lambda\epsilon/2))} \int_{\mathbb{B}(x,\lambda\epsilon/2)} |f(t)|^p dV(t).$$

As a result,

$$\begin{aligned}
\int_{\partial\Omega_\epsilon^2} |f(x)|^p d\sigma_\epsilon^2 &\leq C\epsilon^{-N} \int_{\partial\Omega_\epsilon^2} \int_{\mathbb{B}(x,\lambda\epsilon/2)} |f(t)|^p dV(t) d\sigma_\epsilon^2(x) \\
&= C\epsilon^{-N} \int_{\mathbb{R}^N} \int_{\partial\Omega_\epsilon^2} \chi_{\mathbb{B}(x,\lambda\epsilon/2)}(t) |f(t)|^p d\sigma_\epsilon^2(x) dV(t) \\
&\leq C\epsilon^{-N} \int_{S(\epsilon)} |f(t)|^p \int_{\partial\Omega_\epsilon^2 \cap \mathbb{B}(t,\lambda\epsilon/2)} d\sigma_\epsilon^2(x) dV(t) \\
&\leq C'\epsilon^{-N}\epsilon^{N-1} \int_{S(\epsilon)} |f(t)|^p dV(t) \\
&\leq C'' \sup_\epsilon \int_{\partial\Omega_\epsilon^1} |f(t)|^p d\sigma_\epsilon^1(t).
\end{aligned}$$

Of course the reverse inequality follows by symmetry.

One technical difficulty that we face on an arbitrary Ω is that the device of considering the dilated functions $f_r(\zeta) = f(r\zeta)$ as harmonic functions on $\overline{\Omega}$ is no longer available. However this notion is an unnecessary crutch, and it is well to be rid of it. As a substitute, we cover Ω by finitely many domains $\Omega_1, \ldots, \Omega_k$ with the following properties:

(i) $\Omega = \cup_j \Omega_j$.
(ii) For each j, the set $\partial\Omega \cap \partial\Omega_j$ is an $(N-1)$-dimensional manifold with boundary.
(iii) There is an $\epsilon_0 > 0$ and a vector ν_j transversal to $\partial\Omega \cap \partial\Omega_j$ and pointing out of Ω such that $\Omega_j - \epsilon\nu_j \equiv \{z - \epsilon\nu_j : z \in \Omega_j\} \subset\subset \Omega$, all $0 < \epsilon < \epsilon_0$.

We leave the detailed verification of the existence of the sets Ω_j satisfying (i)–(iii) as an exercise. See Fig. 15.5 for an illustration of these ideas. □

For a general C^2 bounded domain, the substitute for dilation will be to fix $j \in \{1, \ldots, k\}$ and consider the translated functions $f_\epsilon(x) = f\left(x - \epsilon\nu_j\right)$, $f_\epsilon : \Omega_j \to \mathbb{C}$, as $\epsilon \to 0^+$.

Theorem 15.4.3 *Let $\Omega \subseteq \mathbb{R}^N$ be a domain and f harmonic on Ω. Let $1 \leq p < \infty$. The following are equivalent:*

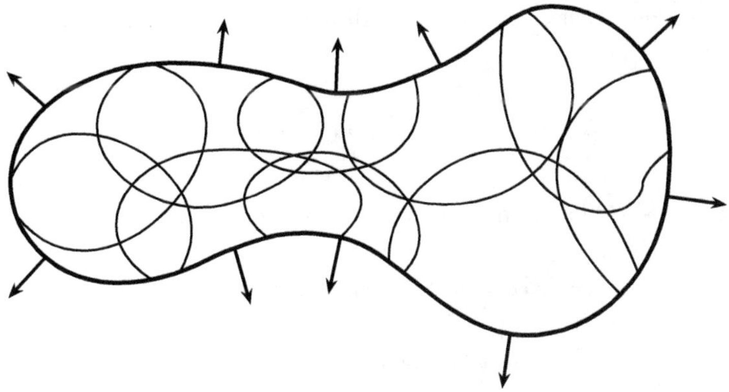

Fig. 15.5 Existence of the figures Ω_j

15.4.3 a $f \in \mathbf{h}^p(\Omega)$.

15.4.3 b *If* $p > 1$ *then there is an* $\widetilde{f} \in L^p(\partial\Omega)$ *such that* $f(x) = \int_{\partial\Omega} P(x, y)\widetilde{f}(y)d\sigma(y)$.
If $p = 1$ *then there is a* $\mu \in \mathcal{M}(\partial\Omega)$ *such that* $f(x) = \int_{\partial\Omega} P(x, y)d\mu(y)$.
Moreover, $\|f\|_{\mathbf{h}^p} \cong \|\widetilde{f}\|_{L^p}$.

15.4.3 c $|f|^p$ *has a harmonic majorant on* Ω.

Proof (15.4.3 b) $\Rightarrow$ (15.4.3 c). If $p > 1$, let

$$h(x) = \int_{\partial\Omega} P(x, y)|\widetilde{f}(y)|^p d\sigma(y).$$

Then, treating $P(x, \cdot)d\sigma$ as a positive measure of total mass 1, we have

$$\begin{aligned} |f(x)|^p &= \left| \int_{\partial\Omega} \widetilde{f}(y)P(x, y)d\sigma(y) \right|^p \\ &\overset{\text{(Jensen)}}{\leq} \int_{\partial\Omega} |\widetilde{f}(y)|^p P(x, y)d\sigma(y) \equiv h(x). \end{aligned}$$

The proof for $p = 1$ is similar.

(15.4.3 c) $\Rightarrow$ (15.4.3 a). If $\epsilon > 0$ is small, $x_0 \in \Omega$ is fixed, and G_Ω is Green's function for Ω; then $G_\Omega(x_0, \cdot)$ has nonvanishing gradient near $\partial\Omega$ (use Hopf's lemma). Therefore

$$\widetilde{\Omega}_\epsilon \equiv \{x \in \Omega : -G_\Omega(x, \cdot) < -\epsilon\}$$

are well-defined domains for ϵ small. Moreover (check the proof), the Poisson kernel for $\widetilde{\Omega}_\epsilon$ is $P_\epsilon(x, y) = -\nu_y^\epsilon G_\Omega(x, y)$. Here ν_y^ϵ is the normal to $\partial\widetilde{\Omega}_\epsilon$ at

$y \in \partial\widetilde{\Omega}_\epsilon$. Assume that $\epsilon > 0$ is so small that $x_0 \in \widetilde{\Omega}_\epsilon$. So if h is the harmonic majorant for $|f|^p$, then

$$h(x_0) = \int_{\partial\widetilde{\Omega}_\epsilon} -\nu_y^\epsilon G_\Omega(x_0, y)\, h(y) d\sigma(y). \tag{15.13}$$

Let $\pi_\epsilon : \partial\widetilde{\Omega}_\epsilon \to \partial\Omega$ be normal projection for ϵ small. Then

$$-\nu_y^\epsilon G_\Omega\left(x_0, \pi_\epsilon^{-1}(\cdot)\right) \to -\nu_y G_\Omega(x_0, \cdot)$$

uniformly on $\partial\Omega$ as $\epsilon \to 0^+$. We have that $-\nu_y G_\Omega(x_0, \cdot) \geq c_{x_0} > 0$ for some constant c_{x_0}. Thus $-\nu_y^\epsilon G_\Omega\left(x_0, \pi_\epsilon^{-1}(\cdot)\right)$ are all bounded below by $c_{x_0}/2$ if ϵ is small enough. As a result, (15.13) yields

$$\int_{\partial\widetilde{\Omega}_\epsilon} h(y) d\sigma(y) \leq 2h(x_0)/c_{x_0}$$

for $\epsilon > 0$ small. In conclusion,

$$\int_{\partial\widetilde{\Omega}_\epsilon} |f(y)|^p d\sigma(y) \leq 2h(x_0)/c_{x_0}.$$

(15.4.3 a) $\Rightarrow$ (15.4.3 b). Let Ω_j be as in ((i)) through ((iii)). Fix j. Define on Ω_j the functions $f_\epsilon(x) = f\left(x - \epsilon\nu_j\right), 0 < \epsilon < \epsilon_0$. Then the hypothesis and (a small modification of) Lemma 15.4.2 show that $\{f_\epsilon\}$ forms a bounded subset of $L^p\left(\partial\Omega_j\right)$. If $p > 1$, let $\widetilde{f}_j \in L^p\left(\partial\Omega_j\right)$ be a weak-* accumulation point (for the case $p = 1$, replace $\widetilde{f}_j$ by a Borel measure $\widetilde{\mu}_j$). The crucial observation at this point is that f is the Poisson integral of $\widetilde{f}_j$ on Ω_j. Therefore f on Ω_j is completely determined by $\widetilde{f}_j$ and conversely (see also the exercises at the end of the section). A moment's reflection now shows that $\widetilde{f}j = \widetilde{f}_k$ almost everywhere $[d\sigma]$ in $\partial\Omega_j \cap \partial\Omega_k \cap \partial\Omega$ so that $\widetilde{f} \equiv \widetilde{f}_j$ on $\partial\Omega_j \cap \partial\Omega$ is well defined. By appealing to a partition of unity on $\partial\Omega$ that is subordinate to the open cover induced by the (relative) interiors of the sets $\partial\Omega_j \cap \partial\Omega$, we see that $f_\epsilon = f \circ \pi_\epsilon^{-1}$ converges weak-* to $\widetilde{f}$ on $\partial\Omega$ when $p > 1$ (resp., $f_\epsilon \to \widetilde{\mu}$ weak-* when $p = 1$).

Referring to the proof of (15.4.3 c) $\Rightarrow$ (15.4.3 a), we write, for $x_0 \in \Omega$ fixed,

$$\begin{aligned} f(x_0) &= \int_{\partial\Omega_\epsilon} -\nu_y^\epsilon G_\Omega(x_0, y)\, f(y) d\sigma_\epsilon(y) \\ &= \int_{\partial\Omega} -\nu_y^\epsilon G_\Omega\left(x_0, \pi_\epsilon^{-1}(y)\right) f\left(\pi_\epsilon^{-1}(y)\right) \mathcal{J}^\epsilon(y) d\sigma(y), \end{aligned}$$

where $\mathcal{J}^\epsilon$ is the Jacobian of the mapping $\pi_\epsilon^{-1} : \partial\Omega \to \partial\Omega_\epsilon$. The fact that $\partial\Omega$ is C^2 combined with previous observations implies that the last line tends to

$$\int_{\partial\Omega} -\nu_y G_\Omega(x_0, y)\, \widetilde{f}(y) d\sigma(y) = \int_{\partial\Omega} P_\Omega(x_0, y)\, \widetilde{f}(y) d\sigma(y)$$

$$\left(\text{resp.} \int_{\partial\Omega} -\nu_y G_\Omega(x_0, y)\, d\widetilde{\mu}(y) = \int_{\partial\Omega} P_\Omega(x_0, y)\, d\widetilde{\mu}(y)\right)$$

as $\epsilon \to 0^+$. □

Exercise for the Reader

1. Prove the last statement in Theorem 15.4.3.

2. Imitate the proof of Theorem 15.4.3 to show that if μ is continuous and subharmonic on Ω and if

$$\sup_\epsilon \int_{\partial\Omega_\epsilon} |u(\zeta)|^p d\sigma(\zeta) < \infty, \quad p \geq 1,$$

then u has a harmonic majorant h. If $p > 1$, then h is the Poisson integral of an L^p function $\widetilde{h}$ on $\partial\Omega$. If $p = 1$, then h is the Poisson integral of a Borel measure $\widetilde{\mu}$ on $\partial\Omega$.

3. Let $\Omega \subseteq \mathbb{R}^N$ be a domain with C^2 boundary, and let ρ be a C^2 defining function for Ω. Define $\Omega_\epsilon = \{x \in \Omega : \rho(x) < -\epsilon\}, 0 < \epsilon < \epsilon_0$. Let $\partial\Omega_\epsilon$ and $d\sigma_\epsilon$ be as usual. Let $\pi_\epsilon : \partial\Omega_\epsilon \to \partial\Omega$ be orthogonal projection. Let $f \in L^p(\partial\Omega), 1 \leq p < \infty$. Define

$$F(x) = \int_{\partial\Omega} P_\Omega(x, y) f(y) d\sigma(y).$$

a. Prove that $\int_{\partial\Omega} P_\Omega(x, y) d\sigma(y) = 1$, any $x \in \Omega$.

b. There is a $C > 0$ such that, for any $y \in \partial\Omega$,

$$\int_{\partial\Omega_\epsilon} P_\Omega(x, y) d\sigma_\epsilon(x) \leq C, \quad \text{any} \quad 0 < \epsilon < \epsilon_0.$$

c. There is a $C' > 0$ such that

$$\int_{\partial\Omega_\epsilon} |F(x)|^p d\sigma_\epsilon(x) \leq C', \quad \text{any} \quad 0 < \epsilon < \epsilon_0.$$

d. If $\phi \in C(\Omega)$ satisfies $\|\phi - f\|_{L^p(\partial\Omega)} < \eta$ and

$$G(x) \equiv \int_{\partial\Omega} P_\Omega(x, y)(\phi(y) - f(y)) d\sigma(y),$$

then

$$\int_{\partial\Omega_\epsilon} |G(x)|^p d\sigma_\epsilon(x) \leq C'\eta, \quad \text{any} \quad 0 < \epsilon < \epsilon_0.$$

e. With ϕ as in part **d.** and $\Phi(x) = \int_{\partial\Omega} P_\Omega(x, y)\phi(y)d\sigma(y)$, then $\left(\Phi|_{\partial\Omega_\epsilon}\right) \circ \pi_\epsilon^{-1} \to \phi$ uniformly on $\partial\Omega$.

f. Imitate the proof of Proposition 15.1.3 to see that $F \circ \pi_\epsilon^{-1} \to f$ in the $L^p(\partial\Omega)$ norm.

15.5 Pointwise Convergence for Harmonic Functions on Domains in $\mathbb{R}^N$

Let $\Omega \subseteq \mathbb{R}^N$ be a domain. For $P \in \partial\Omega, \alpha > 1$, we define

$$\Gamma_\alpha(P) = \{x \in \Omega : |x - P| < \alpha\delta_\Omega(x)\}.$$

This is the N-dimensional analog of the Stolz region considered in Sect. 15.1. Our theorem is as follows:

Theorem 15.5.1 *Let $\Omega \subset\subset \mathbb{R}^N$ be a domain with C^2 boundary. Let $\alpha > 1$. If $1 < p \le \infty$ and $f \in \mathbf{h}^p$, then*

$$\lim_{\Gamma_\alpha(P)\ni x\to P} f(x) \equiv \widetilde{f}(P) \quad \textit{exists for almost every } P \in \partial\Omega.$$

Moreover,

$$\|\widetilde{f}\|_{L^p(\partial\Omega)} \cong \|f\|_{\mathbf{h}^p(\Omega)}.$$

Proof We may as well assume that $p < \infty$. We already know from (15.4.3 b) that there exists an $\widetilde{f} \in L^p(\partial\Omega)$, $\|\widetilde{f}\|_{L^p} \cong \|f\|_{\mathbf{h}^p}$, such that $f = P\widetilde{f}$. It remains to show that $\widetilde{f}$ satisfies the conclusions of the present theorem. This will follow just as in the proof of Theorem 15.1.13 as soon as we prove two things: First,

$$\sup_{x\in\Gamma_\alpha(P)} |P\widetilde{f}(x)| \le C_\alpha M_1\widetilde{f}(P), \tag{15.14}$$

where

$$M_1\widetilde{f}(P) \equiv \sup_{R>0} \frac{1}{\sigma(\mathbb{B}(P, R) \cap \partial\Omega)} \int_{\mathbb{B}(P,R)\cap\partial\Omega} |f(t)|d\sigma(t).$$

Second,

$$\sigma\left\{y \in \partial\Omega : M_1\widetilde{f}(y) > \lambda\right\} \le C\frac{\|\widetilde{f}\|_{L^1(\partial\Omega)}}{\lambda}, \quad \text{all} \quad \lambda > 0. \tag{15.15}$$

Now (15.14) is proved just as in Proposition 15.1.12. It is necessary to use the estimate given in Proposition 15.3.7. On the other hand, (15.15) is not obvious; we supply a proof in the paragraphs that follow. □

Lemma 15.5.2 *(Wiener) Let $K \subseteq \mathbb{R}^N$ be a compact set that is covered by the open balls $\{\mathbb{B}_\alpha\}_{\alpha\in A}$, $\mathbb{B}_\alpha = \mathbb{B}(c_\alpha, r_\alpha)$. There is a subcover $\mathbb{B}_{\alpha_1}, \mathbb{B}_{\alpha_2}, \ldots, \mathbb{B}_{\alpha_m}$, consisting of pairwise disjoint balls, such that*

$$\bigcup_{j=1}^{m} \mathbb{B}\left(c_{\alpha_j}, 3r_{\alpha_j}\right) \supseteq K.$$

Proof Since K is compact, we may immediately assume that there are only finitely many B_α. Let $\mathbb{B}_{\alpha_1}$ be the ball in this collection that has the greatest radius (this ball may not be unique). Let $\mathbb{B}_{\alpha_2}$ be the ball that has the greatest radius and also disjoint from $\mathbb{B}_{\alpha_1}$. At the jth step choose the (not necessarily unique) ball of greatest radius that is disjoint from $\mathbb{B}_{\alpha_1}, \ldots, \mathbb{B}_{\alpha_{j-1}}$ and continue. The process ends in finitely many steps. We claim that the $\mathbb{B}_{\alpha_j}$ chosen in this fashion do the job.

It is enough to show that $\mathbb{B}_\alpha \subseteq \cup_j \mathbb{B}\left(c_{\alpha_j}, 3r_{\alpha_j}\right)$ for every α. Fix an α. If $\alpha = \alpha_j$ for some j, then we are done. If $\alpha \notin \{\alpha_j\}$, let j_0 be the first index with $\mathbb{B}_{\alpha_j} \cap \mathbb{B}_\alpha \neq \emptyset$ (there must be one; otherwise the process would not have stopped). Then $r_{\alpha_{j_0}} \geq r_\alpha$; otherwise we selected $\mathbb{B}_{\alpha_{j_0}}$ incorrectly. But then clearly $\mathbb{B}\left(c_{\alpha_{j_0}}, 3r_{\alpha_{j_0}}\right) \supseteq \mathbb{B}(c_\alpha, r_\alpha)$ as desired. □

Corollary 15.5.3 *Let $K \subseteq \partial\Omega$ be compact, and let $\{\mathbb{B}_\alpha \cap \partial\Omega\}_{\alpha\in A}$, $\mathbb{B}_\alpha = \mathbb{B}(c_\alpha, r_\alpha)$, be an open covering of K by balls with centers in $\partial\Omega$. Then there is a pairwise disjoint subcover $\mathbb{B}_{\alpha_1}, B_{\alpha_2}, \ldots, \mathbb{B}_{a_m}$ such that $\cup_j \{\mathbb{B}(c_\alpha, 3r_\alpha) \cap \partial\Omega\} \supseteq K$.*

Proof The set K is a compact subset of $\mathbb{R}^N$ that is covered by $\{\mathbb{B}_\alpha\}$. Apply the preceding Lemma 15.5.2 and restrict to $\partial\Omega$. □

Lemma 15.5.4 *If $f \in L^1(\partial\Omega)$, then*

$$\sigma\{x \in \partial\Omega : M_1 f(x) > \lambda\} \leq C\frac{\|f\|_{L^1}}{\lambda},$$

all $\lambda > 0$.

Proof Let $S_\lambda = \{x \in \partial\Omega : M_1 f(x) > \lambda\}$. Let K be a compact subset of S_λ. It suffices to estimate $\sigma(K)$. Now for each $x \in K$, there is a ball $\mathbb{B}_x$ centered at x such that

$$\frac{1}{\sigma(\mathbb{B}_x \cap \partial\Omega)} \int_{\mathbb{B}_x \cap \partial\Omega} |f(t)| d\sigma(t) > \lambda. \tag{15.16}$$

The balls $\{\partial\Omega \cap \mathbb{B}_x\}_{x\in K}$ cover K. Choose, by Corollary 15.5.3, disjoint balls $\mathbb{B}_{x_1}, \mathbb{B}_{x_2}, \ldots, \mathbb{B}_{x_m}$ such that $\left\{\partial\Omega \cap 3\mathbb{B}_{x_j}\right\}$ cover K, where $3\mathbb{B}_{x_j}$ represents the threefold dilate of $\mathbb{B}_{x_j}$ (with the same center). Then

$$\sigma(K) \leq \sum_{j=1}^{m} \sigma\left(3\mathbb{B}_{x_j} \cap \partial\Omega\right)$$

$$\leq C(N, \Omega) \sum_{j=1}^{m} \sigma\left(\mathbb{B}_{x_j} \cap \partial\Omega\right),$$

where the constant C will depend on the curvature of $\partial\Omega$. But (15.16) implies that the last line is majorized by

$$C(N, \partial\Omega) \sum_{j} \frac{\int_{\mathbb{B}_{x_j} \cap \partial\Omega} |f(t)| d\sigma(t)}{\lambda} \leq C(N, \partial\Omega) \frac{\|f\|_{L^1}}{\lambda}.$$

This completes the proof of the theorem. □

15.6 Boundary Values of Holomorphic Functions in $\mathbb{C}^n$

Everything in Sect. 15.5 applies a fortiori to domains $\Omega \subseteq \mathbb{C}^n$. However, on the basis of our experience in the classical case, we expect $H^p(\Omega)$ functions to also have pointwise boundary values for $0 < p \leq 1$. That this is indeed the case is established in this section by two different arguments.

First, if $\Omega \subset\subset \mathbb{C}^n$ is a C^2 domain and $f \in H^p(\Omega)$, we shall prove through an application of Fubini's theorem (adapted from a paper of Lempert [356]) that f has pointwise boundary limits in a rather special sense at σ-almost every $\zeta \in \partial\Omega$. This argument is self-contained. After that, we derive some more powerful results using a much broader perspective. We shall not further develop the second methodology in this book, but we present an introduction to it for background purposes. Further details may be found in E.M. Stein [485] and C. Fefferman and E.M. Stein [152]. The logical progession of ideas in this chapter will proceed from the first approach based on Lempert [356].

Proposition 15.6.1 *Let $\Omega \subset\subset \mathbb{C}^n$ have C^2 boundary. Let $0 < p < \infty$ and $f \in H^p(\Omega)$. Write $\Omega = \cup_{j=1}^{k} \Omega_j$ as in (i) and (iii), and let $\nu_1, \ldots, \nu_k$ be the associated normal vectors. Then, for each $j \in \{1, \ldots, k\}$, it holds that*

$$\lim_{\epsilon \to 0^+} f\left(\zeta - \epsilon\nu_j\right) \equiv \widetilde{f}(\zeta)$$

exists for σ-almost every $\zeta \in \partial\Omega_j \cap \partial\Omega$.

Proof We may suppose that $p < \infty$. Fix $1 \leq j \leq k$. Assume for convenience that $\nu_j = \nu_P = (1 + i0, 0, \ldots, 0)$, $P \in \partial\Omega_j$, and that $P = 0$. If $z \in \mathbb{C}^n$, write $z = (z_1, \ldots, z_n) = \left(z_1, z'\right)$. We may assume that $\Omega_j = \cup_{|z'|<1} \left\{\left(z_1, z'\right) : z_1 \in \mathcal{D}_{z'}\right\}$, where $\mathcal{D}_{z'} \subseteq \mathbb{C}$ is a diffeomorph of $\mathbb{D} \subseteq \mathbb{C}$ with C^2 boundary. For each $|z'| < 1, k \in \mathbb{N}, 0 < 1/k < \epsilon_0$, let $b^k_{z'} = \left(\mathcal{D}_{z'} \times \{z'\}\right) \cap \{z \in \Omega_j : \text{dist}\left(z, \partial\Omega_j\right) = 1/k\}$. Define $B^k = \cup_{|z'|<1} b^k_{z'}$. Now a simple variant of Lemma 15.4.2 implies that

$$\sup_k \int_{B^k} |f(\zeta)|^p d\sigma_k \leq C_0 < \infty, \tag{15.17}$$

where σ_k is the surface measure on B^k. Formula (15.17) may be rewritten as

$$\sup_k \int_{|\zeta'|<1} \int_{b^k_{\zeta'}} \left|f\left(\zeta_1, \zeta'\right)\right|^p d\widetilde{\sigma}_k\left(\zeta_1\right) dV_{2n-2}\left(\zeta'\right) \leq C_0, \tag{15.18}$$

where $\widetilde{\sigma}_k$ is the surface (=linear) measure on $b^k_{\zeta'}$. If $M > 0, k \geq k_0 > 1/\epsilon_0$, we define

$$S^M_k = \left\{\zeta' : \left|\zeta'\right| < 1, \int_{b^k_{\zeta'}} \left|f\left(\zeta_1, \zeta'\right)\right|^p d\widetilde{\sigma}_k\left(\zeta_1\right) > M\right\}. \tag{15.19}$$

Then (15.18), (15.19), and Chebycheff's inequality together yield

$$V_{2n-1}\left(S^M_k\right) \leq \frac{C_0}{M}, \quad \text{all} \quad k\,.$$

Now let

$$\begin{aligned} S^M &= \Big\{\zeta' : |\zeta'| < 1, \int_{b^k_{\zeta'}} |f(\zeta_1, \zeta')|^p d\widetilde{\sigma}_k(\zeta_1) \leq M \\ &\qquad \text{for only finitely many } k\} \\ &= \bigcup_{\ell=k_0}^{\infty} \bigcap_{k=\ell}^{\infty} S^M_k\,. \end{aligned}$$

Then $V_{2n-1}\left(S^M\right) \leq C_0/M$. Since M may be made arbitrarily large, we conclude that for V_{2n-2} almost every $\zeta' \in \mathbb{D}^{n-1}(0, 1)$, there exist $k_1 < k_2 < \cdots$ such that

$$\int_{b^{k_m}_{\zeta'}} \left|f\left(\zeta_1, \zeta'\right)\right|^p d\widetilde{\sigma}_{k_m}\left(\zeta_1\right) = O(1) \quad \text{as} \quad m \to \infty.$$

It follows that the functions $f\left(\cdot, \zeta'\right) \in H^p\left(\mathcal{D}_{\zeta'}\right)$ for V_{2n-2} almost every $\zeta' \in \mathbb{D}^{n-1}(0, 1)$. Now Theorem 15.1.23 yields the desired result. □

At this point we could prove that f tends non-tangentially to the function $\widetilde{f}$ constructed in Proposition 15.6.1. However, we do not do so because a much stronger result is proved in the next section. The remainder of the present section consists of a digression to introduce the reader to the second point of view mentioned in the introduction. This second point of view involves far-reaching ideas arising from the "real variable" school of complex analysis. This methodology provides a more natural, and much more profound, approach to the study of boundary behavior.

The results that we present are due primarily to E.M. Stein [485]. To present Stein's ideas, we first need an auxiliary result of A.P. Calderón (see Stein [485]) and K.O. Widman [527]. Although it is well within the scope of this book to prove this auxiliary result on a half-space, a complete proof on smoothly bounded domains would entail a number of tedious ancillary ideas (such as the maximum principle for second order elliptic operators). Hence we only state the needed result and refer the reader to the literature for details.

Let $\Omega \subset\subset \mathbb{R}^N$ be a domain with C^2 boundary, and let $u : \Omega \to \mathbb{C}$ be harmonic. If $P \in \partial\Omega$, then we say that f is *non-tangentially bounded* at P if for each $\alpha > 1$ there is a $C_\alpha < \infty$ such that

$$\sup_{x \in \Gamma_\alpha(P)} |f(x)| < C_\alpha .$$

Now the result is as follows:

Theorem 15.6.2 *Let $\Omega \subset\subset \mathbb{R}^N$ have C^2 boundary. Let $u : \Omega \to \mathbb{C}$ be harmonic. Let $E \subseteq \partial\Omega$ be a set of positive σ-measure. Suppose that u is non-tangentially bounded at σ-almost every $P \in E$. Then u has non-tangential limits at σ-almost every $P \in E$.*

Remarks

(i) Obviously if u has a non-tangential limit at $P \in \partial\Omega$, then u is non-tangentially bounded at P.

(ii) Theorem 15.6.2 has no analog for radial boundedness and radial limits, even on the disc. Let $E \subseteq \partial\mathbb{D}$ be a F_σ set of first category and measure 2π. Let $f : \mathbb{D} \to \mathbb{C}$ be given by $f(z) = \sin\left(\frac{1}{1-|z|}\right)$. Then f is continuous, bounded by 1 on $\mathbb{D}$, and does not possess a radial limit at any point of $\partial\mathbb{D}$. Of course f is not holomorphic. But by a theorem of Bagemihl and Seidel [38], there is a holomorphic (!) function u on $\mathbb{D}$ with

$$\lim_{r \to 1^-} \left| u\left(re^{i\theta}\right) - f\left(re^{i\theta}\right) \right| = 0$$

for every $e^{i\theta} \in E$.

Theorem 15.6.3 *Let $\Omega \subset\subset \mathbb{C}^n$ have C^2 boundary. Let $f \in H^p(\Omega), 0 < p \le \infty$. Then f has non-tangential boundary limit at almost every $P \in \partial\Omega$.*

Proof We may assume that $p < \infty$. The function $|f|^{p/2}$ is subharmonic and uniformly square integrable over $\partial\Omega_\epsilon, 0 < \epsilon < \epsilon_0$. So $|f|^{p/2}$ has a harmonic majorant $h \in \mathbf{h}^2(\Omega)$. Since $h \in \mathbf{h}^2$, it follows that h has almost everywhere non-tangential boundary limits. Therefore h is non-tangentially bounded at almost every point of $\partial\Omega$. As a result, $|f|^{p/2}$, and therefore $|f|$ itself is non-tangentially bounded at almost every point of $\partial\Omega$. So, by Theorem 15.6.2, f has a non-tangential boundary limit $\widetilde{f}$ defined at almost every point of $\partial\Omega$. □

We conclude this section with a recasting of the ideas in the proof of Theorem 15.6.3 to make more explicit the role of the maximal function. We first need two lemmas.

Lemma 15.6.4 *Let $\{X, \mu\}$ be a measure space with $\mu \geq 0$. Let $f \geq 0$ on X be measurable and $0 < p < \infty$. Then*

$$\int_X f(x)^p d\mu(x) = \int_0^\infty ps^{p-1}\mu_f(s)ds = -\int_0^\infty s^p d\mu_f(s),$$

where $\mu_f(s) \equiv \mu\{x : f(x) \geq s\}, 0 \leq s < \infty$.

Proof We have

$$\begin{aligned}
\int_X f(x)^p d\mu(x) &= \int_X \int_0^{f(x)^p} dt d\mu(x) \\
&\overset{\text{(Fubini)}}{=} \int_0^\infty \int_{\{x:f(x)^p \geq t\}} d\mu(x)dt \\
&\overset{(t=s^p)}{=} \int_0^\infty ps^{p-1}\mu\{x : f(x) \geq s\}ds \\
&\overset{\text{(parts)}}{=} -\int_0^\infty s^p d\mu_f(x).
\end{aligned}$$

□

Lemma 15.6.5 *The maximal operator M_1 is bounded from $L^2(\partial\Omega)$ to $L^2(\partial\Omega)$.*

Proof We know that M_1 maps L^∞ to L^∞ (trivially) and is weakly bounded on L^1. After normalizing by a constant, we may suppose that

$$\|M_1 f\|_{L^\infty} \leq \frac{1}{3}\|f\|_{L^\infty}, \text{ all } \quad f \in L^\infty(\partial\Omega). \tag{15.20}$$

Also we may assume that

$$\sigma\{\zeta \in \partial\Omega : |M_1 f(\zeta)| > \lambda\} \leq \frac{\|f\|_{L^1}}{\lambda}, \text{ all } \quad \lambda > 0, \quad f \in L^1(\partial\Omega). \tag{15.21}$$

If $f \in L^2(\partial\Omega)$, $\|f\|_{L^2} = 1$, $0 < t < \infty$, we write

$$f(\zeta) = f(\zeta) \cdot \chi_{\{\zeta : |f(\zeta)| < t\}} + f(\zeta) \cdot \chi_{\{\zeta : |f(\zeta)| \geq t\}} \equiv f_1^t(\zeta) + f_2^t(\zeta).$$

Then

$$\begin{aligned} \|M_1 f\|_{L^2}^2 &= 2 \int_0^\infty \sigma_{M_1 f}(\lambda) \lambda d\lambda \\ &\leq \int_0^\infty \sigma_{M_1 f_2^\lambda}(\lambda/2) \lambda d\lambda + \int_0^\infty \sigma_{M_1 f_1^\lambda}(\lambda/2) \lambda d\lambda \\ &\equiv I + II. \end{aligned}$$

Now

$$\begin{aligned} I &\leq 2 \int_0^\infty \left(\| f_2^\lambda \|_{L^1} / \lambda \right) \lambda d\lambda \\ &= 2 \int_0^\infty \int_\lambda^\infty \sigma_f(t) dt d\lambda \\ &= 2 \int_0^\infty \sigma_f(t) \int_0^t d\lambda dt \\ &= 2 \int_0^\infty \sigma_f(t) \cdot t dt \\ &= \|f\|_{L^2}^2. \end{aligned}$$

Furthermore, notice that since $\left|f_1^\lambda(\zeta)\right| < \lambda$ for all ζ, it follows that $\left|M_1 f_1^\lambda(\zeta)\right| \leq \lambda/3$. Hence $\sigma_{M_1 f_1^\lambda}(\lambda/2) \equiv 0$. Therefore $II = 0$. That finishes the proof. □

Let us now examine more closely the interplay between harmonic majorants and maximal functions to obtain a quantitative version of the nontangential boundedness. Let the notation be as in Theorem 15.6.3. Let h be the least harmonic majorant for $|f|^{p/2}$ (assuming that $p < \infty$). Notice that the dominated convergence theorem (or Fatou's lemma) implies that $\widetilde{f} \in L^p(\partial\Omega)$. (*Exercise*: For a rigorous proof, you must either consider the Jacobian of π_ϵ^{-1} or restrict attention to the Ω_j's.) Now, for $\alpha > 1$ fixed and $P \in \partial\Omega$, we have

$$f_1^{*,\alpha}(P)^{p/2} \equiv \sup_{x \in \Gamma_\alpha(P)} |f(x)|^{p/2} \leq \sup_{x \in \Gamma_\alpha(P)} |h(x)| \leq C_\alpha M_1 \widetilde{h}(P).$$

Therefore

$$
\begin{aligned}
\left\| f_1^{*,\alpha} \right\|_{L^p(\partial\Omega)}^p &\le C_\alpha \left\| M_1 \widetilde{h} \right\|_{L^2(\partial\Omega)}^2 \\
&\le C \cdot C_\alpha \|\widetilde{h}\|_{L^2(\partial\Omega)}^2 \\
&\le C'_\alpha \sup_\epsilon \int_{\partial\Omega_\epsilon} |f(\zeta)|^p d\sigma(\zeta) \qquad (15.22) \\
&\le C''_\alpha \int_{\partial\Omega} |\widetilde{f}(\zeta)|^p d\sigma(\zeta) \\
&= C''_\alpha \|\widetilde{f}\|_{L^p(\partial\Omega)}^p .
\end{aligned}
$$

Inequality (15.22) is valid for $0 < p < \infty$. It is central to the so-called real variable theory of H^p spaces. For instance, one has the following theorem:

Theorem 15.6.6 *Let u be a real harmonic function on $\mathbb{D} \subseteq \mathbb{C}$. Let $0 < p < \infty$. Then u is the real part of an $f \in H^p(\mathbb{D})$ if and only if, for some $\alpha > 1$,*

$$
\left\| u_1^{*,\alpha} \right\|_{L^p(\partial\mathbb{D})} < \infty .
$$

Under these circumstances, $\left\| u_1^{*,\alpha} \right\|_{L^p(\partial\mathbb{D})} \cong \|f\|_{H^p(\mathbb{D})}$.

Theorem15.6.6 was originally proved by methods of Brownian motion (D. Burkholder, R. Gundy, and M. Silverstein [98]). C. Fefferman and E.M. Stein [152] gave a real-variable proof and extended the result in an appropriate sense to $\mathbb{R}^N$. The situation in several complex variables is rather more complicated (see J. Garnett and R. Latter [171] and S.G. Krantz and D. Ma [335]).

This ends our digression about the real-variable aspects of boundary behavior of harmonic and holomorphic functions. In the next section, we pick up the thread of Proposition 15.6.1 and prove a theorem that is strictly stronger. This requires a new notion of convergence.

15.7 Admissible Convergence

Let $\mathbb{B} \subseteq \mathbb{C}^n$ be the unit ball. The *Poisson kernel for the ball* has the form

$$
P(z, \zeta) = c_n \frac{1 - |z|^2}{|z - \zeta|^{2n}},
$$

whereas the *Poisson-Szegö kernel* has the form

$$
P(z, \zeta) = c_n \frac{\left(1 - |z|^2\right)^n}{|1 - z \cdot \overline{\zeta}|^{2n}}.
$$

As we know, an analysis of the convergence properties of these kernels entails dominating them by appropriate maximal functions. The maximal function involves the use of certain balls, and the shape of the ball should be compatible with the singularity of the kernel. That is why, when we study the real analysis of the Poisson kernel, we consider balls of the form

$$\beta_1(\zeta, r) = \{\xi \in \partial\mathbb{B} : |\xi - \zeta| < r\}, \quad \zeta \in \partial\mathbb{B}, \quad r > 0.$$

In studying the complex analysis of the Poisson-Szegö kernel (equivalently, the Szegö kernel), it is appropriate to use the balls

$$\beta_2(\zeta, r) = \{\xi \in \partial\mathbb{B} : |1 - \xi \cdot \overline{\zeta}| < r\}, \quad \zeta \in \partial\mathbb{B}, \quad r > 0.$$

These new non-isotropic balls are fundamentally different from the classical (or isotropic) balls β_1, as we shall see now. Assume without loss of generality that $\zeta = \mathbf{1} = (1, 0, \ldots, 0)$. Write $z' = (z_2, \ldots, z_n)$. Then

$$\beta_2(\zeta, r) = \{\xi \in \partial\mathbb{B} : |1 - \xi_1| < r\}.$$

Notice that, for $\xi \in \partial\mathbb{B}$,

$$\begin{aligned} |\xi'|^2 &= 1 - |\xi_1|^2 \\ &= (1 - |\xi_1|)\,(1 + |\xi_1|) \\ &\leq 2\,|1 - \xi_1|\,; \end{aligned}$$

hence

$$\beta_2(\zeta, r) \subseteq \left\{\xi \in \partial\mathbb{B} : |1 - \xi_1| < r, |\xi'| < \sqrt{2r}\right\}.$$

A similar computation shows that

$$\begin{aligned} \beta_2(\zeta, r) &= \left\{\xi \in \partial\mathbb{B} : |1 - \xi_1| < r, |\xi'| = \sqrt{1 - |\xi_1|^2}\right\} \\ &\supseteq \partial\mathbb{B} \cap \{\xi : |\mathrm{Im}\,\xi_1| < r/2, 1 - r/2 < \mathrm{Re}\,\xi_1 < 1, |\xi'| < \sqrt{r}\}. \end{aligned}$$

In short, the balls we are considering have dimension $\sim r$ in the complex space containing ν_1 and dimension $\sim \sqrt{r}$ in the orthogonal complement (see Fig. 15.6). Here, ν_1 is the unit outward normal to $\partial\mathbb{B}$ at the point $\mathbf{1} = (1, 0, \ldots, 0)$. The word "non-isotropic" means that we have different geometric behaviors in different directions.

In the classical setup, we considered *cones* modeled on the balls β_1:

$$\Gamma_\alpha(P) = \{z \in \mathbb{B} : |z - P| < \alpha(1 - |z|)\}, \quad P \in \partial\mathbb{B}, \quad \alpha > 1.$$

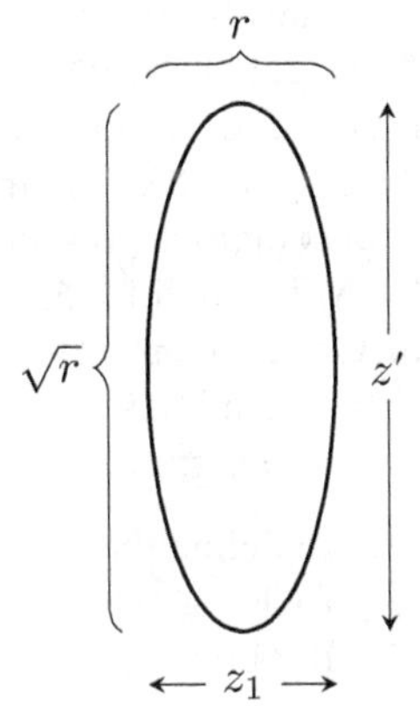

Fig. 15.6 Dimensions of the complex balls

In the new situation we consider *admissible regions* modeled on the balls β_2:

$$\mathcal{A}_\alpha(P) = \{z \in \mathbb{B} : |1 - z \cdot \overline{P}| < \alpha(1 - |z|)\}.$$

Our new theorem about boundary limits of H^p functions is as follows:

Theorem 15.7.1 *Let $f \in H^p(\mathbb{B})$, $0 < p \leq \infty$. Let $\alpha > 1$. Then the limit*

$$\lim_{\mathcal{A}_\alpha(P) \ni z \to P} f(P) \equiv \widetilde{f}(P)$$

exists for σ-almost every $P \in \partial\mathbb{B}$.

Since the Poisson-Szegö kernel is known explicitly on the ball, then for $p \geq 1$ the proof is deceptively straightforward: One defines, for $P \in \partial\mathbb{B}$ and $f \in L^1(\partial B)$,

$$M_2 f(P) = \sup_{r>0} \frac{1}{\sigma\,(\beta_2(P, r))} \int_{\beta_2(P,r)} |f(\zeta)| d\sigma(\zeta).$$

Also set $f(z) = \int_{\partial B} \mathcal{P}(z, \zeta) f(\zeta) d\sigma(\zeta)$ for $z \in \mathbb{B}$. Then, by explicit computation similar to the proof of Proposition 15.1.12,

$$f_2^{*,\alpha}(P) \equiv \sup_{z \in \mathcal{A}_\alpha(P)} |f(z)| \leq C_\alpha M_2 f(P), \quad \text{all} \quad f \in L^1(\partial\mathbb{B}).$$

This crucial fact, together with appropriate estimates on the operator M_2, enables one to complete the proof along classical lines for $p \geq 1$. For $p < 1$, matters are more subtle.

We forego the details of the preceding argument on B and instead develop the machinery for proving an analog of Theorem 15.7.1 on an arbitrary C^2 bounded domain in $\mathbb{C}^n$. In this generality, there is no hope of obtaining an explicit formula for the Poisson-Szegö kernel; indeed, there are no known techniques for obtaining estimates for this kernel on arbitrary domains (however see C. Fefferman [150] and A. Nagel, J.P. Rosay, E.M. Stein, and S. Wainger [391] for estimates on strongly pseudoconvex domains and on domains of finite type in $\mathbb{C}^2$). Therefore we must develop more geometric methods that do not rely on information about kernels. The results that we present were proved on the ball and on bounded symmetric domains by A. Koranyi [315, 316]. Many of these ideas were also developed independently in Gong Sheng [180, 181]. All of the principal ideas for arbitrary Ω are due to E.M. Stein [485].

Our tasks, then, are as follows:

(1) To define the balls β_2 on the boundary of an arbitrary Ω
(2) To define admissible convergence regions $\mathcal{A}_\alpha$
(3) To obtain appropriate estimates for the corresponding maximal function
(4) To couple the maximal estimates, together with the fact that "radial" boundary values are already known to exist (see Proposition 15.6.1) to obtain the admissible convergence result

If z, w are vectors in $\mathbb{C}^n$, we continue to write $z \cdot \overline{w}$ to denote $\sum_j z\overline{w}_j$. (Warning: It is also common in the literature to use the notation $z \cdot w = \sum_j z_j \overline{w}_j$.) Also, for $\Omega \subseteq \mathbb{C}^n$ a domain with C^2 boundary, $P \in \partial\Omega$, we let ν_P be the *unit outward normal* at P. Let $\mathbb{C}\nu_P$ denote the complex line generated by ν_P: $\mathbb{C}\nu_P = \{\zeta\nu_P : \zeta \in \mathbb{C}\}$.

By dimension considerations, if $T_P(\partial\Omega)$ is the $(2n-1)$-dimensional real tangent space to $\partial\Omega$ at P, then $\ell = \mathbb{C}\nu_P \cap T_P(\partial\Omega)$ is a (one-dimensional) real line. Let

$$\begin{aligned}\mathcal{T}_P(\partial\Omega) &= \left\{z \in \mathbb{C}^n : z \cdot \overline{\nu}_P = 0\right\} \\ &= \left\{z \in \mathbb{C}^n : z \cdot \overline{w} = 0 \text{ for all } w \in \mathbb{C}\nu_P\right\}.\end{aligned}$$

A fortiori, $\mathcal{T}_P(\partial\Omega) \subseteq T_P(\partial\Omega)$. If $z \in \mathcal{T}_P(\partial\Omega)$, then $iz \in \mathcal{T}_P(\partial\Omega)$. Therefore $\mathcal{T}_P(\partial\Omega)$ may be thought of as an $(n-1)$-dimensional complex subspace of $T_P(\partial\Omega)$. Clearly, $\mathcal{T}_P(\partial\Omega)$ is the complex subspace of $T_P(\partial\Omega)$ of maximal dimension. It contains all complex subspaces of $T_P(\partial\Omega)$. [The reader should check that $\mathcal{T}_P(\partial\Omega)$ is the same complex tangent space that was introduced when we first studied the Levi form.] We may think of $T_P(\partial\Omega)$ as the real orthogonal complement in $T_P(\partial\Omega)$ of ℓ.

Now let us examine the matter from another point of view. The complex structure is nothing other than a linear operator J on $\mathbb{R}^{2n}$ that assigns to $(x_1, x_2, \ldots, x_{2n-1}, x_{2n})$ the vector $(-x_2, x_1, -x_4, x_3, \ldots, -x_{2n}, x_{2n-1})$ (think of multiplication by i). With this in mind, we have that $J : \mathcal{T}_P(\partial\Omega) \to \mathcal{T}_P(\partial\Omega)$ both injectively and surjectively. Notice that $J\nu_P \in T_P(\partial\Omega)$, while $J(J\nu_P) = -\nu_P \notin T_P(\partial\Omega)$. We call $\mathbb{C}\nu_P$ the complex normal space to $\partial\Omega$ at P and $\mathcal{T}_P(\partial\Omega)$ the complex tangent space to $\partial\Omega$ at P. Let $\mathcal{N}_P = \mathbb{C}\nu_P$. Then we have $\mathcal{N}_P \perp \mathcal{T}_P$ and

$$\begin{aligned}\mathbb{C}^n &= \mathcal{N}_P \oplus_{\mathbb{C}} \mathcal{T}_P \\ T_P &= \mathbb{R}J\nu_P \oplus_{\mathbb{R}} \mathcal{T}_P.\end{aligned}$$

Example 15.7.2 Let $\Omega = \mathbb{B} \subseteq \mathbb{C}^n$ be the unit ball and $P = \mathbf{1} \in \partial\Omega$. Then $\mathbb{C}\nu_P = \{(z_1, 0, \ldots, 0) : z_1 \in \mathbb{C}\}$ and $\mathcal{T}_P = \{(0, z') : z' \in \mathbb{C}^{n-1}\}$.

Exercise for the Reader

1. Let $\Omega \subseteq \mathbb{C}^n$ be a domain. Let J be the real linear operator on $\mathbb{R}^{2n}$ that gives the complex structure. Let $P \in \partial\Omega$. Let $z = (z_1, \ldots, z_n) = (x_1 + iy_1, \ldots, x_n + iy_n) = (x_1, y_1, \ldots, x_n, y_n)$ be an element of $\mathbb{C}^n \cong \mathbb{R}^{2n}$. The following are equivalent:

   (i) $w \in \mathcal{T}_P(\partial\Omega)$.
   (ii) $Jw \in \mathcal{T}_P(\partial\Omega)$.
   (iii) $Jw \perp \nu_P$ and $w \perp \nu_P$.

2. With notation as in the previous exercise, let $A = \sum_j a_j(z)\partial/\partial z_j$, $B = \sum_j b_j(z)\partial/\partial z_j$ satisfy $A\rho|_{\partial\Omega} = 0$, $B\rho|_{\partial\Omega} = 0$, where ρ is any defining function for Ω. Then the vector field $[A, B]$ has the same property. (However note that $[A, \overline{B}]$ does not annihilate ρ on $\partial\Omega$ if Ω is the ball, for instance.) Therefore the holomorphic part of $\mathcal{T}_P$ is integrable (see G.B. Folland and J.J. Kohn [154]).
3. If $\Omega = \mathbb{B} \subseteq \mathbb{C}^2$, $P = (x_1 + iy_1, x_2 + iy_2) \approx (x_1, y_1, x_2, y_2) \in \partial\mathbb{B}$, then $\nu_P = (x_1, y_1, x_2, y_2)$ and $J\nu_P = (-y_1, x_1, -y_2, x_2)$. Also $\mathcal{T}_P$ is spanned over $\mathbb{R}$ by $(y_2, x_2, -y_1, -x_1)$ and $(-x_2, y_2, x_1, -y_1)$.

   The next definition is best understood in light of the foregoing discussion and the definition of $\beta_2(P, r)$ in the boundary of the unit ball $\mathbb{B}$. Let $\Omega \subset\subset \mathbb{C}^n$ have C^2 boundary. For $P \in \partial\Omega$, let $\pi_P : \mathbb{C}^n \to N_P$ be (real or complex) orthogonal projection.

Definition 15.7.3 If $P \in \partial\Omega$, let

$$\beta_1(P, r) = \{\zeta \in \partial\Omega : |\zeta - P| < r\}$$

$$\beta_2(P, r) = \left\{\zeta \in \partial\Omega : |\pi_P(\zeta - P)| < r, |\zeta - P| < r^{1/2}\right\}.$$

Exercise for the Reader The ball $\beta_2(P, r)$ has diameter $\sim \sqrt{r}$ in the $(2n - 2)$ complex tangential directions and diameter $\sim r$ in the one (normal) direction. Therefore $\sigma\,(\beta_2(P, r)) \approx (\sqrt{r})^{2n-2} \cdot r \approx Cr^n$.

If $z \in \Omega$, $P \in \partial\Omega$, we let

$$\delta_P(z) = \min\{\text{dist}(z, \partial\Omega), \text{dist}\,(z, T_P(\Omega))\}.$$

Notice that if Ω is convex, then $\delta_P(z) = \delta_\Omega(z)$.

Definition 15.7.4 If $P \in \partial\Omega, \alpha > 1$, let

$$\mathcal{A}_\alpha = \left\{z \in \Omega : |(z - P) \cdot \overline{\nu}_P| < \alpha\delta_P(z), |z - P|^2 < \alpha\delta_P(z)\right\}.$$

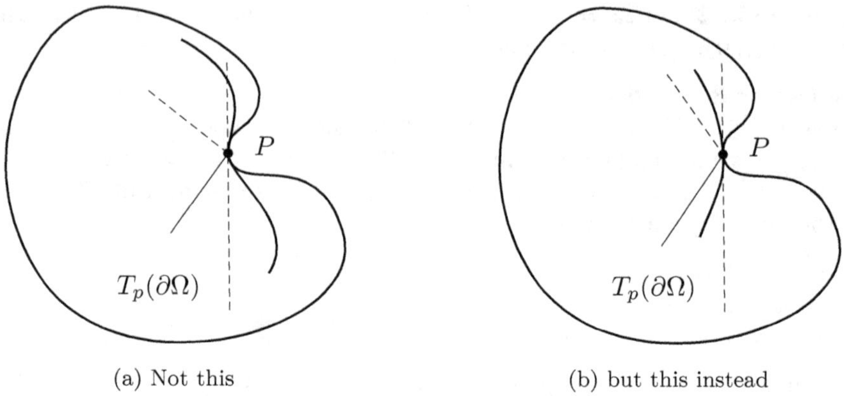

Fig. 15.7 The use of δ_P

Notice that δ_P is used because near non-convex boundary points we still want $\mathcal{A}_\alpha$ to have the fundamental geometric shape of (paraboloid × cone) as shown in Fig. 15.7.

Definition 15.7.5 If $f \in L^1(\partial\Omega)$ and $P \in \partial\Omega$, then we define

$$M_j f(P) = \sup_{r>0} \sigma\left(\beta_j(P,r)\right)^{-1} \int_{\beta_j(P,r)} |f(\zeta)| d\sigma(\zeta), \quad j = 1, 2.$$

Definition 15.7.6 If $f \in C(\Omega)$, $P \in \partial\Omega$, then we define

$$f_2^{*,\alpha}(P) = \sup_{z \in \mathcal{A}_\alpha(P)} |f(z)|.$$

The first step of our program is to prove an estimate for M_2. This will require a covering lemma (indeed, it is known that weak type estimates for operators like M_j are logically equivalent to covering lemmas—see A. Cordoba and R. Fefferman [119]). We exploit a rather general paradigm due to K.T. Smith [478]:

Definition 15.7.7 Let X be a topological space equipped with a positive Borel measure m, and suppose that for each $x \in X$ and each $r > 0$, there is a "ball" $\mathfrak{B}(x, r)$. The "K.T. Smith axioms" for this setting are:

15.7.7 a Each $\mathfrak{B}(x, r)$ is an open set of finite measure which contains x.
15.7.7 b If $r_1 \leq r_2$, then $\mathfrak{B}(x, r_1) \subseteq \mathfrak{B}(x, r_2)$.
15.7.7 c There is a constant $c_0 > 0$ such that if $\mathfrak{B}(x, r) \cap \mathfrak{B}(y, s) \neq \emptyset$ and $r \geq s$, then $\mathfrak{B}(x, c_0 r) \supseteq \mathfrak{B}(y, s)$.
15.7.7 d There is a constant K such that $m(\mathfrak{B}(x_0, c_0 r)) \leq K m(\mathfrak{B}(x_0, r))$ for all r.

Now we have the following theorem:

Theorem 15.7.8 *Let the topological space X, measure m, and balls $\mathfrak{B}(x, r)$ be as in the definition. Let K be a compact subset of X and $\{\mathfrak{B}(x_\alpha, r_\alpha)\}_{\alpha \in A}$ a covering of K by balls.*

Then, there is a finite pairwise disjoint subcollection:

$$\mathfrak{B}\left(x_{\alpha_1}, r_{\alpha_1}\right), \ldots, \mathfrak{B}\left(x_{\alpha_m}, r_{\alpha_m}\right)$$

such that $K \subseteq \cup_{j=1}^k \mathfrak{B}\left(x_{\alpha_j}, c_0 r_{\alpha_j}\right)$.

It follows that if we define

$$Mf = \sup_{r>0}(\mathfrak{B}(x, r))^{-1} \int_{B(x,r)} |f(t)| dm(t), \quad f \in L^1(X, dm),$$

then

$$m\{x : Mf(x) > \lambda\} \leq C \frac{\|f\|_{L^1}}{\lambda}.$$

Proof Exercise for the Reader. Imitate the proofs of Lemmas 15.5.2 and 15.5.4. □

Thus we need to see that the $\beta_2(P, r)$ on $X = \partial\Omega$ with $m = \sigma$ satisfy (15.7.7 a)–(15.7.7 d). Now (15.7.7 a) and (15.7.7 b) are trivial. Also (15.7.7 d) is easy if one uses the fact that $\partial\Omega$ is C^2 and compact (use the Exercise for the reader following Example 15.7.2). Thus it remains to check (15.7.7 c) (in many applications, this is the most difficult property to check).

Suppose that $\beta_2(x, r) \cap \beta_2(y, s) \neq \emptyset$. Thus there is a point $a \in \beta_2(x, r) \cap \beta_2(y, s)$. We may assume that $r = s$ by (15.7.7 b). We thus have $|x - a| \leq r^{1/2}$, $|y - a| \leq r^{1/2}$, and hence $|x - y| \leq 2r^{1/2}$. Let the constant $M \geq 2$ be chosen so that $\|\pi_x - \pi_y\| \leq M|x - y|$. (We must use here the fact that the boundary is C^2.) We claim that $\beta_2(x, (3 + 4M)r) \supseteq \beta_2(y, r)$. To see this, let $v \in \beta_2(y, r)$.

The easy half of the estimate is

$$|x - v| \leq |x - y| + |y - v| \leq 2r^{1/2} + r^{1/2} = 3r^{1/2}.$$

Also

$$\pi_x(x - v) = \pi_x(x - a) + \pi_y(a - v) + \left\{\pi_x - \pi_y\right\}(a - v).$$

Therefore

$$\begin{aligned}
|\pi_x(x - v)| &\leq r + 2r + \|\pi_x - \pi_y\||a - v| \\
&\leq 3r + M|x - y| \cdot |a - v| \\
&\leq 3r + M2r^{1/2}(|a - y| + |y - v|) \\
&\leq (3 + 4M)r.
\end{aligned}$$

This proves (15.7.7 c). Thus we have the following:

Corollary 15.7.9 *If $f \in L^1(\partial\Omega)$, then*

$$\sigma\{\zeta \in \omega : M_2 f(\zeta) > \lambda\} \leq C \frac{\|f\|_{L^1(\partial\Omega)}}{\lambda}, \qquad \textit{all } \lambda > 0.$$

Proof Apply the theorem. □

Corollary 15.7.10 *The operator M_2 maps $L^2(\partial\Omega)$ to $L^2(\partial\Omega)$ boundedly.*

Proof Exercise. Use the technique of Lemma 15.6.5. □

Remark 15.7.11 In fact there is a general principal at work here. If a linear operator T is bounded from L^∞ to L^∞ and is also weak type $(1, 1)$, then it is bounded on L^p, $1 < p < \infty$. Of course M_2 is not linear. Instead it is sublinear: $M_2(f + g)(x) \leq M_2 f(x) + M_2 g(x)$. This is sufficient for the result stated.

The results mentioned in the last paragraph are instances of "interpolation" theorems for operators. For a more thorough discussion of this topic, see E.M. Stein and G. Weiss [486] or J. Bergh and J. Löfström [73].

The next lemma is the heart of the matter: It is the technical device that allows us to estimate the behavior of a holomorphic function in the interior (in particular, on an admissible approach region) in terms of a maximal function on the boundary. The argument comes from Stein [485] and Barker [41].

Lemma 15.7.12 *Let $u \in \mathcal{C}(\overline{\Omega})$ be nonnegative and plurisubharmonic on Ω. Define $f = u|_{\partial\Omega}$. Then*

$$u_2^{*,\alpha}(P) \leq C_\alpha M_2 (M_1 f)(P)$$

for all $P \in \partial\Omega$ and any $\alpha > 1$.

Proof After rotating and translating coordinates, we may suppose that $P = 0$ and $\nu_P = (1 + i0, 0, \ldots, 0)$. Let $\alpha' > \alpha$. Then there is a small positive constant k such that if $z = (x_1 + iy_1, z_2, \ldots, z_n) \in \mathcal{A}_\alpha(P)$, then $\mathcal{D}(z) = \mathbb{D}(z_1, -kx_1) \times \mathbb{D}^{n-1}\left((z_2, \ldots, z_n), \sqrt{-kx_1}\right) \subseteq \mathcal{A}_{\alpha'}(P)$ (see Fig. 15.8).

We restrict attention to $z \in \Omega$ so close to $P = 0$ that the projection along ν_P given by

$$z = (x_1 + iy_1, \ldots, x_n + iy_n) \to (\widetilde{x}_1 + i\widetilde{y}_1, x_2 + iy_2, \ldots, x_n + iy_n) \equiv \widetilde{z} \in \partial\Omega$$

makes sense. [Observe that points z that are far from $P = 0$ are trivial to control using our estimates on the Poisson kernel.]

The projection of $\mathcal{D}(z)$ along ν_P into the boundary lies in a ball of the form $\beta_2(\widetilde{z}, Kx_1)$—this observation is crucial.

Notice that the subharmonicity of u implies that $u(z) \leq \mathrm{Pf}(z)$. Also there is a $\beta > 1$ such that $z \in \mathcal{A}_{\alpha'}(0) \Rightarrow z \in \Gamma_\beta(\widetilde{z})$. Therefore the standard argument leading up to (15.16) yields that

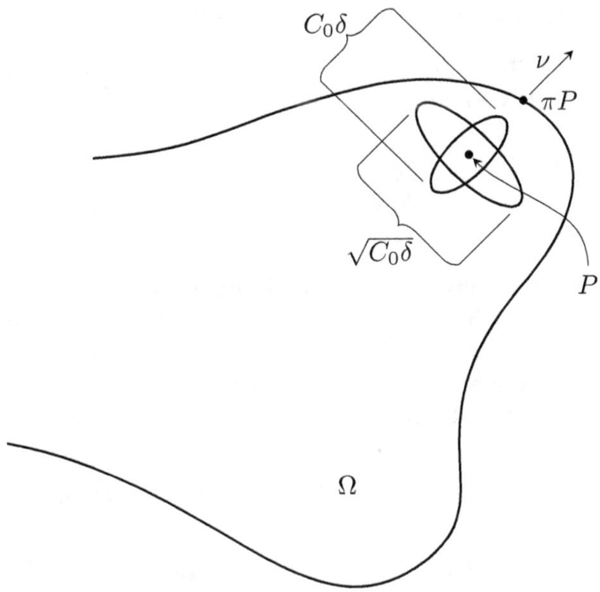

Fig. 15.8 Existence of the constant k

$$|u(z)| \leq |\operatorname{Pf}(z)| \leq C_\alpha M_1 f(\widetilde{z}). \tag{15.23}$$

Now we bring the complex analysis into play. For that we may exploit the plurisubharmonicity of $|u|$ on $\mathcal{D}(z)$ by invoking the subaveraging property in each dimension in succession. Thus

$$\begin{aligned}|u(z)| &\leq \left(\pi\, |kx_1|^2\right)^{-1} \cdot \left(\pi \left(\sqrt{-kx_1}\right)^2\right)^{-(n-1)} \int_{\mathcal{D}(z)} |u(\zeta)| dV(\zeta) \\ &= C x_1^{-n-1} \int_{\mathcal{D}(z)} |u(\zeta)| dV(\zeta).\end{aligned}$$

Notice that if $z \in \mathcal{A}_\alpha(P)$, then each ζ in the last integrand is in $\mathcal{A}_{\alpha'}(P)$. Thus the last line is

$$\begin{aligned}&\leq C' x_1^{-n-1} \int_{\mathcal{D}(z)} M_1 f(\widetilde{\zeta}) dV(\zeta) \\ &\leq C''' x_1^{-n-1} \cdot x_1 \int_{\beta_2(\widetilde{z}, K x_1)} M_1 f(t) d\sigma(t) \\ &\leq C''' x_1^{-n} \int_{\beta_2(0, K' x_1)} M_1 f(t) d\sigma(t) \\ &\leq C''''' \left(\sigma\left(\beta_2\left(0, K' x_1\right)\right)\right)^{-1} \int_{\beta_2(0, K' x_1)} M_1 f(t) d\sigma(t) \\ &\leq C'''' M_2\left(M_1 f\right)(0).\end{aligned}$$

□

Now we may prove our main result:

Theorem 15.7.13 *Let* $0 < p \leq \infty$. *Let* $\alpha > 1$. *If* $\Omega \subset\subset \mathbb{C}^n$ *has* C^2 *boundary and* $f \in H^p(\Omega)$, *then for* σ*-almost every* $P \in \partial\Omega$ *we have that*

$$\lim_{\mathcal{A}_\alpha(P) \ni z \to P} f(z)$$

exists.

Proof We already know that the limit exists almost everywhere in the special sense of Proposition 15.6.1. Call the limit function $\widetilde{f}$. We need only consider the case $p < \infty$. Let $\Omega = \cup_{j=1}^k \Omega_j$ as usual. It suffices to concentrate on Ω_1. Let $\nu = \nu_1$ be the outward normal to $\partial\Omega_1$ at the point ξ. Then, by (15.6.1), the Lebesgue dominated convergence theorem implies that, for $\partial\widetilde{\Omega}_1 \equiv \partial\Omega \cap \partial\Omega_1$,

$$\lim_{\epsilon \to 0} \int_{\partial\widetilde{\Omega}_1} |f(\zeta - \epsilon\nu) - \widetilde{f}(\zeta)|^p d\sigma(\zeta) = 0. \tag{15.24}$$

For each $j, k \in \mathbb{N}$, consider the function $f_{j,k} : \Omega_1 \to \mathbb{C}$ given by

$$f_{j,k}(z) = |f(z - \nu/j) - f(z - \nu/k)|^{p/2}.$$

Then $f_{j,k} \in C\left(\overline{\Omega}_1\right)$ and is plurisubharmonic on Ω_1. Therefore a trivial variant of Lemma 15.7.12 yields

$$\begin{aligned}
\int_{\partial\widetilde{\Omega}_1} \left|(f_{j,k})_2^{*,\alpha}(\zeta)\right|^2 d\sigma(\zeta) &\leq C_\alpha \int_{\partial\widetilde{\Omega}_1} \left|M_2\left(M_1 f_{j,k}(\zeta)\right)\right|^2 d\sigma(\zeta) \\
&\leq C'_\alpha \int_{\partial\widetilde{\Omega}_1} \left|M_1 f_{j,k}(\zeta)\right|^2 d\sigma(\zeta) \\
&\leq C''_\alpha \int_{\partial\widetilde{\Omega}_1} \left|f_{j,k}(\zeta)\right|^2 d\sigma(\zeta),
\end{aligned}$$

where we have used Corollary 15.7.9, Lemma 15.6.5, and the proof of Lemma 15.6.5. Now let $j \to \infty$, and apply (15.24) to obtain

$$\int_{\partial\widetilde{\Omega}_1} \sup_{z \in \mathcal{A}_\alpha(\zeta)} |f(z) - f(z - \nu/k)|^p d\sigma(\zeta) \leq C''_\alpha \int_{\partial\widetilde{\Omega}_1} |\widetilde{f}(\zeta) - f(\zeta - \nu/k)|^p d\sigma(\zeta). \tag{15.25}$$

Let $\epsilon > 0$. Then

$$\sigma\{\zeta \in \partial\widetilde{\Omega}_1 : \limsup_{\mathcal{A}_\alpha(\zeta)\ni z\to\zeta} |f(z) - \widetilde{f}(\zeta)| > \epsilon\Bigg\}$$

$$\le \sigma\left\{\zeta \in \partial\widetilde{\Omega}_1 : \limsup_{\mathcal{A}_\alpha(\zeta)\ni z\to\zeta} |f(z) - f(z - \nu/k)| > \epsilon/3\right\}$$

$$+ \sigma\left\{\zeta \in \partial\widetilde{\Omega}_1 : \limsup_{\mathcal{A}_\alpha(\zeta)\ni z\to\zeta} |f(z - \nu/k) - f(\zeta - \nu/k)| > \epsilon/3\right\}$$

$$+ \sigma\left\{\zeta \in \partial\widetilde{\Omega}_1 : \limsup_{\mathcal{A}_\alpha(\zeta)\ni z\to\zeta} |f(\zeta - \nu/k) - \widetilde{f}(\zeta)| > \epsilon/3\right\}$$

$$\le C\int_{\partial\widetilde{\Omega}_1} \sup_{z\in\mathcal{A}_\alpha(\zeta)} |f(z) - f(z - \nu/k)|^p d\sigma(\zeta)/\epsilon^p$$

$$+ 0 + C\int_{\partial\widetilde{\Omega}_1} |\widetilde{f}(\zeta) - f(\zeta - \nu/k)|^p d\sigma(\zeta)/\epsilon^p,$$

where we have used (the proof of) Chebycheff's inequality. By (15.25), the last line does not exceed

$$C'\int_{\partial\widetilde{\Omega}_1} |\widetilde{f}(\zeta) - f(\zeta - \nu/k)|^p d\sigma(\zeta)/\epsilon^p.$$

Now (15.24) implies that, as $k \to \infty$, this last quantity tends to 0. Since $\epsilon > 0$ was arbitrary, we conclude that

$$\limsup_{\mathcal{A}_\alpha(\zeta)\ni z\to\zeta} |f(z) - \widetilde{f}(\zeta)| = 0$$

almost everywhere. □

The theorem says that f has "admissible limits" at almost every boundary point of Ω. The considerations in the next section (indeed, an inspection of the arguments in the present section) suggest that Theorem 15.7.13 is best possible only for strongly pseudoconvex domains. At the boundary point $(1, 0)$ of the domain $\{(z_1, z_2) : |z_1|^2 + |z_2|^{2m} < 1\}$, the natural interior polydiscs to study are of the form

$$\left\{(1 - \delta + \xi_1, \xi_2) : |\xi_1| < c \cdot \delta, |\xi_2| < c \cdot \delta^{1/2m}\right\}.$$

This observation, together with an examination of the proof of Lemma 15.7.12, suggests that the aperture in complex tangential directions of the approach regions should vary from boundary point to boundary point, and this aperture should depend on the Levi geometry of the point. A theory of boundary behavior for H^p functions taking these observations into account, for a special class of domains in

$\mathbb{C}^2$, is enunciated in A. Nagel, E.M. Stein, and S. Wainger [392]. A more general paradigm for theories of boundary behavior of holomorphic functions is developed in S.G. Krantz [327]. Related ideas also appear in S.G. Krantz [326]. The key tool in the last two references is the Kobayashi metric, a geometric device that we studied in Chap. 13.

15.8 The Lindelöf Principle

In this section we formulate a version of the Lindelöf principle for domains in C^1 and $\mathbb{C}^n$. It is not optimal but serves to illustrate the main ideas. For further details, see E. Hille [235], J. Cima and S.G. Krantz [115], E. Chirka [116], and O. Lehto and K.I. Virtanen [354].

The classical Lindelöf Principle is as follows:

Theorem 15.8.1 *Let $\mathbb{D} \subseteq \mathbb{C}$ be the unit disc. Let $f \in H^\infty(\mathbb{D})$. Suppose that $\lim_{r\to 1^-} f\left(re^{i\theta}\right) = \ell \in \mathbb{C}$ exists. Then f has non-tangential limit ℓ at $e^{i\theta}$.*

Proof Fix $\alpha > 1$. We may assume that $\ell = 0$ and $\theta = 0$. Define, for $j \in \mathbb{N}$,

$$\Omega_j = \left\{z \in \mathbb{C} : 1 - 2^{-j+1}/(2\alpha) \le \operatorname{Re} z \le 1 - 2^{-j-1}/(2\alpha), |\operatorname{Im} z| < 2^{-j+1}\right\}.$$

For $j_0(\alpha)$ sufficiently large,

$$\mathbb{D} \supseteq \bigcup_{j=j_0}^{\infty} \Omega_j \supseteq \Gamma_\alpha(\mathbf{1}) \cap \left\{z_1 \in \mathbb{D} : \operatorname{Re} z \ge 1 - 2^{-j_0+1}/(2\alpha)\right\}.$$

Also, the map $\phi_j : \Omega_j \to \Omega_{j_0}$ given by $\phi_j(z) = 2^{j-j_0}(z-1)+1$ is biholomorphic. There is a K compact in Ω_{j_0} such that

$$\bigcup_{j\ge j_0} \phi_j^{-1}(K) \supseteq \Gamma_\alpha(\mathbf{1}) \cap \left\{z \in \mathbb{D} : \operatorname{Re} z \ge 1 - 2^{-j_0-1}/(2\alpha)\right\}.$$

Now the functions $\left\{f \circ \phi_j^{-1}\right\}$ form a normal family on Ω_{j_0}. Choose a subsequence $f \circ \phi_{j_k}^{-1}$ that converges uniformly on K. Call the limit function f_0. By hypothesis, $f_0(x + i0) = 0$, all $x + i0 \in K$. Thus $f_0 \equiv 0$. Therefore $f \circ \phi_j^{-1} \to 0$ uniformly on K. Hence

$$\lim_{\Gamma_\alpha(1)\ni z\to 1} f(z) = 0.$$

□

It is natural to conjecture that on the ball in $\mathbb{C}^n$ there is an analogous theorem with non-tangential convergence replaced by admissible convergence. This is unfortunately false.

Example 15.8.2 Let $\mathbb{B} \subseteq \mathbb{C}^2$ be the unit ball. Let $f(z_1, z_2) = z_2^2/(1 - z_1)$ for $(z_1, z_2) \in \mathbb{B}$. Then

$$\begin{aligned}|f(z_1, z_2)| &\le \frac{|z_2|^2}{|1 - z_1|} \\ &\le \frac{1 - |z_1|^2}{1 - |z_1|} \\ &\le 1 + |z_1| \\ &\le 2.\end{aligned}$$

Hence $f \in H^\infty(\mathbb{B})$. Also $\lim_{r \to 1^-} f(r\mathbf{1}) = 0$, where as usual $\mathbf{1} = (1, 0) \in \partial\mathbb{B}$. But the sequence

$$z_j \equiv \left(1 - \frac{1}{j}, \frac{1}{\sqrt{2j}}\right)$$

approaches $\mathbf{1}$ admissibly (i.e., through an admissible approach region) and

$$\lim_{j \to \infty} f\left(z_j\right) = \lim_{j \to \infty} \frac{1/(2j)}{1/j} = \frac{1}{2} \neq 0.$$

Thus f has no admissible limit at $\mathbf{1}$. We conclude that there is no Lindelöf principle for admissible convergence.

The correct substitute for admissible convergence is "hypoadmissible convergence":

Definition 15.8.3 Let $\mathbb{B} \subseteq \mathbb{C}^n$ be the unit ball and $P \in \partial\mathbb{B}$. Assume that $\left\{z^j\right\} \subseteq \mathbb{B}$ satisfies $\lim_{j \to \infty} z^j = P$. We say that $\left\{z^j\right\}$ converges hypoadmissibly to P if there is an $\alpha > 1$ such that

$$\frac{\left|1 - z^j \cdot \overline{P}\right|}{1 - \left|z^j\right|} < \alpha \quad \text{for all} \quad j$$

and

$$\limsup_{j \to \infty} \frac{\left|z^j - \left(z^j \cdot \overline{P}\right) P\right|^2}{\left|1 - z^j \cdot \overline{P}\right|} = 0.$$

In other words, $z^j \to P$ hypoadmissibly means that the smallest approach region containing $\{z^j\}$ is non-tangential in the complex normal direction and is asymptotically smaller than any admissible approach region in the complex tangential directions. Now we have the following theorem:

Theorem 15.8.4 *Let $\mathbb{B} \subseteq \mathbb{C}^n$ be the unit ball. Let $f \in H^\infty(\mathbb{B})$, and fix $P \in \partial\mathbb{B}$. If $\lim_{r\to 1^-} f(rP) = \ell \in \mathbb{C}$ exists, then, for any sequence $\{z^j\}_{j=1}^\infty \subseteq \mathbb{B}$ that approaches P hypoadmissibly, we have*

$$\lim_{j\to\infty} f\left(z^j\right) = \ell.$$

Proof Assume without loss of generality that $P = 1$. For each j, write $z^j = \left(z_1^j, \tau^j\right)$, where $\tau^j = \left(z_2^j, \ldots, z_n^j\right)$. Let $n^j = \left(z_1^j, 0, \ldots, 0\right)$. Fix $\alpha > 1$ as in Definition 15.8.3.

Pick $\epsilon > 0$. Let $\omega = \mathbb{B} \cap \{z_2 = \cdots = z_n = 0\}$ and $\hat{f} = f|_\omega$. Then $\hat{f}$ is a bounded holomorphic function on the disc ω. Also, the points $\omega \ni n^j \to 1$ non-tangentially. Since $\hat{f}$ has radial limit ℓ at 1, Theorem 15.8.1 implies that there is a J so large that $j \geq J$ implies that $\left|f\left(n^j\right) - \ell\right| < \epsilon/2$.

Let $\delta^j = 1 - \left|n^j\right|$. Then the $(n-1)$-dimensional polydisc

$$\mathbf{d}^j = \left\{\left(z_1^j, z_2, \ldots, z_n\right) : |z_i| < \sqrt{\alpha\delta^j/(2n)}, i = 2, \ldots, n\right\}$$

lies in $\mathbb{B}$. Let $M = \sup_{\mathbb{B}} |f|$. Then, applying the Cauchy estimates to f on $\mathbf{d}^j$, we have (since $\left|z^j - n^j\right| < \sqrt{\alpha\delta^j/(8n)}$ for j sufficiently large) that

$$\begin{aligned}
\left|f\left(z^j\right) - f\left(n^j\right)\right| &\leq \left|z^j - n^j\right| \cdot \sup_{0\leq t\leq 1} \left|\frac{d}{dt} f\left((1-t)z^j + tn^j\right)\right| \\
&\leq \left|z^j - n^j\right| \cdot \frac{M}{\sqrt{\alpha\delta^j/(8n)}} \\
&\leq C \cdot \frac{\left|z^j - \left(z^j \cdot \overline{P}\right) P\right|}{\sqrt{\left|1 - z^j \cdot \overline{P}\right|}} \to 0
\end{aligned}$$

as $j \to \infty$. Therefore there is a K so large that $j \geq K$ implies

$$\left|f\left(z^j\right) - f\left(n^j\right)\right| < \frac{\epsilon}{2}.$$

Thus for $j \geq \max(J, K)$, we have

$$\begin{aligned} \left|f\left(z^j\right) - \ell\right| &\leq \left|f\left(z^j\right) - f\left(n^j\right)\right| + \left|f\left(n^j\right) - \ell\right| \\ &< \frac{\epsilon}{2} + \frac{\epsilon}{2} = \epsilon. \end{aligned}$$

□

Exercise for the Reader

Let $\Omega_k \subseteq \mathbb{C}^2$ be given by $\Omega_k = \left\{z \in \mathbb{C}^2 : |z_1|^2 + |z_2|^{2k} < 1\right\}$.

If $P = \left(e^{i\theta}, 0\right) \in \partial\Omega_k$, then show that there is a Lindelöf principle for P in which the approach region can be taken to be (asymptotically smaller than) size $|1 - z \cdot \overline{P}|^{1/(2k)}$ in the complex tangential directions.

Results of the type in Theorem 15.8.4 were first considered by E. Chirka [116]. Generalizations to arbitrary domains are treated in J. Cima and S.G. Krantz [115]. In the latter work, as with the work described in S.G. Krantz [326, 327], the key is to use the language of the Kobayashi metric.

15.9 Additional Tangential Phenomena: Lipschitz Spaces

In this section we prove a special case of a remarkable result due to E.M. Stein [485] (see also Greiner and Stein [484]) that reflects the same aspects of the complex structure as does the existence of admissible limits. Explicit connections between the results of this section and those in Sects. 15.7 and 15.8 are explored in S.G. Krantz [319, 327] using the language of the Kobayashi metric.

Recall that if $U \subseteq \mathbb{R}^n$ is any open set, $0 < \alpha < 1$, then

$$\begin{aligned} \Lambda_\alpha(U) = \Bigg\{ f \in C(U) : \sup_{x, x+h \in U} & |f(x+h) - f(x)\,\|\,h|^\alpha + \|\,f\|_{L^\infty(U)} \\ & \equiv \|f\|_{\Lambda_\alpha(U)} < \infty \Bigg\}. \end{aligned}$$

If $k \in \mathbb{N}$, $0 < \alpha < 1$, then

$$\begin{aligned} \Lambda_{\alpha+k}(U) = \Bigg\{ f \in C^k(U) : \sum_{j=1}^{N} \left\| \frac{\partial f}{\partial x_j} \right\|_{\Lambda_{\alpha+k-1}(U)} & + \|f\|_{L^\infty(U)} \\ & \equiv \|f\|_{\Lambda_{\alpha+k}(U)} < \infty \Bigg\}. \end{aligned}$$

The definition of Λ_α when $\alpha \in \mathbb{N}$ is more technical:

$$\begin{aligned}\Lambda_1(U) = \Bigg\{ f \in \mathcal{C}(U) : \sup_{x, x+h, x-h \in U} \frac{|f(x+h) + f(x-h) - 2f(x)|}{|h|} \\ + \|f\|_{L^\infty(U)} \Big\} \\ \equiv \|f\|_{\Lambda_1(U)} < \infty \Big\} .\end{aligned}$$

This definition is ostensibly different from the more classical

$$\begin{aligned}\mathrm{Lip}_1(U) = \Bigg\{ f \in \mathcal{C}(U) : \sup_{x, x+h \in U} \frac{|f(x+h) - f(x)|}{|h|} + \|f\|_{L^\infty(U)} \\ \equiv \|f\|_{\mathrm{Lip}_1(U)} < \infty \Big\} .\end{aligned}$$

In fact $\mathrm{Lip}_1(U) \subsetneq \Lambda_1(U)$. While the space Lip_1 is rather natural in geometric applications (see H. Federer [149]), it is the space Λ_1 that is the right space for harmonic analysis and operator theory. A detailed study of Lipschitz spaces appears in S.G. Krantz [319].

We conclude this introductory material by recording the definition of Λ_k, $k \in \mathbb{N}$:

$$\Lambda_k = \left\{ f \in \mathcal{C}^{k-1}(U) : \sum_{j=1}^{N} \left\| \frac{\partial f}{\partial x_j} \right\|_{\Lambda_{k-1}(U)} + \|f\|_{L^\infty(U)} \equiv \|f\|_{\Lambda_k} < \infty \right\} .$$

Now we turn to the principal topic of this section: non-isotropic Lipschitz spaces on domains in $\mathbb{C}^n$. The spirit of the theorem, due to E.M. Stein [485], is that a holomorphic Λ_α function on a C^2 bounded domain in $\mathbb{C}^n$ is in fact automatically twice as smooth in complex tangential directions near the boundary. In order to give this result a precise formulation, we need some notation and terminology.

Let $\Omega \subset\subset \mathbb{C}^n$ be a domain with C^2 boundary. Define $\mathcal{C}^k(\Omega) = \{\gamma : [0,1] \to \Omega\}$ such that $\gamma \in \mathcal{C}^k$ and $|\gamma'(t)| \leq 1, \ldots, |\gamma^{(k)}(t)| \leq 1\}$. Let $U \supseteq \partial\Omega$ be a tubular neighborhood of $\partial\Omega$: Each $z \in U$ has a unique nearest point in $\partial\Omega$ (see J. Munkres [387], M. Hirsch [240], or Sect. 15.3 for a consideration of these matters). Let $\pi : U \cap \Omega \to \partial\Omega$ be normal projection. Write $\mathbb{C}^n = N_P \oplus T_P$ at each $P \in \partial\Omega$. Extend this decomposition to $P \in U \cap \Omega$ using $\pi^{-1} : N_P \equiv N_{\pi(P)}, T_P \equiv T_{\pi(P)}$. Now define

$$\mathcal{C}_1^k(U \cap \Omega) = \left\{ \gamma \in \mathcal{C}^k : \gamma'(t) \in T_{\gamma(t)}, 0 \leq t \leq 1 \right\} .$$

The elements of $\mathcal{C}_1^k$ are called *normalized complex tangential curves.*

Definition 15.9.1 Let $\Omega \subset\subset \mathbb{C}^n$ have C^2 boundary. Fix a neighborhood $U \supseteq \partial\Omega$ as above. Let $0 < \alpha < \beta < \infty$. We say that $f \in \Gamma_{\alpha,\beta}(\Omega)$ if $f \in \Lambda_\alpha(\Omega)$ and

$$\|f\|_{\Lambda_\alpha(\Omega)} + \sup_{\gamma \in \mathcal{C}^{[\beta]+1}(U\cap\Omega)} \|f \circ \gamma\|_{\Lambda_\beta([0,1])} \equiv \|f\|_{\Gamma_{\alpha,\beta}(\Omega)} < \infty.$$

Remark 15.9.2 Roughly speaking, $f \in \Gamma_{\alpha,\beta}(\Omega)$ means that $f \in \Lambda_\alpha$ in every direction and f is Λ_β in complex tangential directions. Although the definition of $\Gamma_{\alpha,\beta}$ depends on U and on the use of the Euclidean projection (as a method of retraction to the boundary), the choice of a different U or a different method of retraction results in the same space (exercise).

The theorem of Stein is as follows:

Theorem 15.9.3 *Let $\Omega \subset\subset \mathbb{C}^n$ have C^2 boundary. Let f be holomorphic on Ω. Then*

$$f \in \Lambda_\alpha(\Omega) \Leftrightarrow f \in \Gamma_{\alpha,2\alpha}(\Omega).$$

Proof We treat the case $0 < \alpha < 1/2$. Details for larger α may be found in Greiner and Stein [204] and S.G. Krantz [319]. Only the direction $\Rightarrow$ is nontrivial.

Let U be fixed as in the definition of $\Gamma_{\alpha,2\alpha}$. Let $\delta_0 > 0$ be so small that $\{z \in \Omega : \delta_\Omega(z) < 5\delta_0\} \subseteq U$. Let $z \in \Omega \cap U$, and let $\gamma \in \mathcal{C}^1_1(\Omega \cap U)$ satisfy $\gamma(0) = z$. It suffices to estimate

$$|f(\gamma(h)) - f(\gamma(0))| \tag{15.26}$$

when $0 < h < \delta_0$. If $\delta_\Omega(z) \geq 3\delta_0$ and

$$M = \sup_{\delta_\Omega(z)\geq\delta_0} |\nabla f(z)| < \infty, \tag{15.27}$$

then (15.26) does not exceed Mh. Therefore we may restrict attention to the case $\delta_\Omega(z) < 3\delta_0$. Let $\pi : U \to \partial\Omega$ be normal projection. Define

$$\begin{aligned} z' &= \gamma(0) - h^2 \nu_{\pi(z)} \\ z'' &= \gamma(h) - h^2 \nu_{\pi(\gamma(h))}. \end{aligned}$$

See Fig. 15.9.

Then (15.26) is majorized by

$$\begin{aligned} &\left|f(\gamma(h)) - f\left(z''\right)\right| + \left|f\left(z''\right) - f\left(z'\right)\right| + \left|f\left(z'\right) - f(\gamma(0))\right| \\ &\qquad = I + II + III. \end{aligned}$$

In order to analyze I, we let $P \in \Omega, \delta = \delta_\Omega(P) \leq 3\delta_0$. Let $\nu = \nu_{\pi(P)}$, and let τ be a unit element of $T_{\pi P}$. Elementary geometry shows that there is a $C_0 > 0$, independent of P and τ, such that the bidisc

Fig. 15.9 Existence of z' and z''

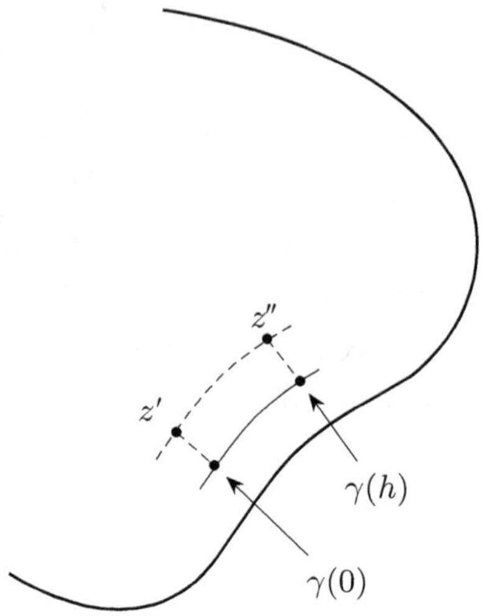

$$\left\{P - \lambda_1 \nu + \lambda_2 \tau : |\lambda_1| \le C_0 \delta, |\lambda_2| \le \sqrt{C_0 \delta}\right\} \subset\subset \Omega.$$

See Fig. 15.8.
Define

$$g(\lambda_1, \lambda_2) = f(P - \lambda_1 \nu + \lambda_2 \tau).$$

Then g is holomorphic, and we may use the Cauchy integral formula to write

$$g(\lambda_1, \lambda_2) = \frac{1}{2\pi} \oint_{\partial \mathbb{D}(0, C_0 \delta)} \frac{g(\zeta, \lambda_2)}{\zeta - \lambda_1} d\zeta.$$

Then

$$\left|\frac{\partial}{\partial \lambda_1} g(\lambda_1, \lambda_2)\right| = \left|\frac{1}{2\pi i} \oint_{\partial \mathbb{D}(0, C_0 \delta)} \frac{g(\zeta, \lambda_2)}{(\zeta - \lambda_1)^2} d\zeta\right|.$$

But the integrand $1/(\zeta - \lambda_1)^2$ integrates to zero. So we may subtract off a constant inside the integrand and write

$$\left|\frac{\partial}{\partial \lambda_1} g(\lambda_1, \lambda_2)\right| = \left|\frac{1}{2\pi i} \oint_{\partial \mathbb{D}(0, C_0 \delta)} \frac{g(\zeta, \lambda_2) - g(\lambda_1, \lambda_2)}{(\zeta - \lambda_1)^2} d\zeta\right|.$$

Now the Lipschitz hypothesis on f and elementary estimates show that

$$\left|\frac{\partial}{\partial \lambda_1} g(\lambda_1, \lambda_2)\right|, \le C \delta^{\alpha - 1} \tag{15.28}$$

all $|\lambda_1| \le C_0\delta/2$. We apply this derivative estimate as follows:

$$\begin{aligned} |I| &= \int_0^{h^2} \frac{d}{dt} f\left(\gamma(h) - t\nu_{\pi(\gamma(h))}\right) dt \\ &\le \int_0^{h^2} C \cdot t^{\alpha-1} dt \\ &\le C \cdot h^{2\alpha}. \end{aligned}$$

Thus I, and by symmetry III, is estimated. We turn our attention to II. First,

$$II = |f(\widetilde{\gamma}(h)) - f(\widetilde{\gamma}(0))|,$$

where $\widetilde{\gamma}(t) \equiv \gamma(t) - h^2\nu_{\pi(\gamma(t))}$. The expression II does not exceed

$$h \cdot \sup_{0\le t\le h} \left| \frac{d}{dt} f \circ \widetilde{\gamma}(t) \right|. \tag{15.29}$$

We claim that

$$\left| \frac{d}{dt} f \circ \mu(t) \right| \le C\delta^{\alpha-1/2}, \quad 0 \le t \le 1, \tag{15.30}$$

where μ is any element of $\mathcal{C}_1^1$ and $\delta \le 3\delta_0$ is its distance from $\partial\Omega$. This will complete the proof, for then (15.29) is majorized by $hC\left(h^2\right)^{\alpha-1/2} = Ch^{2\alpha}$. So it suffices to prove the claim. To do so, we apply the Cauchy estimates to the estimate (15.28) on the disc $\mathbb{D}\left(0, \sqrt{C_0\delta/2}\right)$ in the λ_2 variable to obtain

$$\left| \frac{\partial}{\partial\lambda_2} \frac{\partial}{\partial\lambda_1} g\left(\lambda_1, \lambda_2\right) \right| \le C\delta^{\alpha-1}(\sqrt{\delta})^{-1}$$

or

$$\left| \frac{\partial}{\partial\xi_1} \left(\frac{\partial}{\partial\lambda_2} g\left(\lambda_1, \lambda_2\right) \right) \right| \le C\delta^{\alpha-3/2}$$

all $|\lambda_1| \le C_0\delta/8$, $|\lambda_2| \le \sqrt{C_0\delta/8}$, $\xi_1 = \operatorname{Re}\lambda_1$. In particular, setting $\lambda = 0$ yields

$$\left| \frac{\partial}{\partial\nu} \left(\left(\frac{\partial}{\partial\tau} \right) f \right) \right|_P | \le C\delta^{\alpha-3/2}. \tag{15.31}$$

Now if $Q \in \Omega$ satisfies $\delta_\Omega(Q) \le 3\delta_0$, we write

$$\left|\left(\frac{\partial}{\partial\tau}\right) f(Q)\right| = \left|\int_{-2\delta_0}^{0} \frac{d}{ds}\left\{\left(\frac{\partial}{\partial\tau}\right) f(Q+s\nu)\right\} ds + \left(\frac{\partial}{\partial\tau}\right) f\left(Q - 2\delta_0\nu\right)\right|.$$

Estimate the first term with (15.31), letting $P = Q + s\nu$, and estimate the second term with (15.27). The result is

$$\left|\left(\frac{\partial}{\partial\tau}\right) f(Q)\right| \leq C\int_{-2\delta_0}^{0} (\delta_\Omega(Q)+s)^{\alpha-3/2}\, dx + M \leq C\delta_\Omega(Q)^{\alpha-1/2}. \tag{15.32}$$

This proves our claim. □

The following result was proved by W. Rudin [450] on the ball and generalized to arbitrary domains in S.G. Krantz [319]:

Theorem 15.9.4 *Let $\Omega \subset\subset \mathbb{C}^n$ be a domain with C^2 boundary. Suppose that $0 < \alpha < \infty$. Let $f : \Omega \to \mathbb{C}$ be holomorphic, and suppose that for each $P \in \partial\Omega$ the function*

$$t \mapsto f\left(P - t\nu_P\right)$$

is in $\Lambda_\alpha\,(0, \delta_0)$ with uniform Lipschitz bound C. Then $f \in \Gamma_{\alpha,2\alpha}(\Omega)$.

So it suffices to check smoothness of f along real normals.

In S.G. Krantz [319] these ideas were refined even further:

Theorem 15.9.5 *Let $\Omega \subset\subset \mathbb{C}^n$ have C^k boundary, $k \geq 2$. Let $0 < \alpha < k$. Let $\gamma : [0, 1] \to \Omega$ be called special complex normal if $\gamma : [0, 1] \to \partial\Omega$ satisfies $\gamma'(t) \in N_{\gamma(t)}$, all t. Let $f \in C(\overline{\Omega})$ be holomorphic on Ω. Suppose that for each special complex normal curve $\gamma : [0, 1] \to \partial\Omega$ with $\|\gamma\|_{C^k([0,1])} \leq 1$ we have $\|f\|_{\partial\Omega} \circ \gamma\|_{\Lambda_\alpha([0,1])} \leq C$. Then $f \in \Gamma_{\alpha,2\alpha}(\overline{\Omega})$.*

Remark 15.9.6 The complex normal curves locally foliate the boundary in a nice way. So the theorem says that it is enough to check $f|_{\partial\Omega}$ in just one set of directions. The hypothesis that f be continuous on $\overline{\Omega}$ may be weakened.

Theorem 15.9.3 is optimal only on strongly pseudoconvex domains. For instance, near a strongly pseudoconcave point, all holomorphic functions extend C^∞ (indeed holomorphically) past the boundary. And at weakly pseudoconvex points the degree of increased smoothing in the tangential direction will always be greater than 2. The paper S.G. Krantz [327] presents a natural way to study theorems like Theorem 15.9.3 and to derive the optimal estimates. The key to this work is the Kobayashi metric.

Given a bounded domain Ω in $\mathbb{C}^n$, it is well known that there exist a lot of Cauchy type integrals on bounded domains in the integral representation theory of several complex variables, and the definition of a Cauchy type integral depends on the geometric property of the whole boundary $\partial\Omega$ as the Leray integral.

Chapter 16
The Theorem of Bun Wong and Rosay

This section presents a remarkable characterization of the unit ball in terms of its automorphism group.

We begin our study by considering a self-mapping of a bounded domain Ω that possesses a fixed point.

Proposition 16.0.1 *Let $\Omega \subseteq \mathbb{C}^n$ be a bounded domain. Let $P \in \Omega$ and suppose that $\phi : \Omega \to \Omega$ satisfies $\phi(P) = P$. If $\text{Jac}_{\mathbb{C}}\, \phi(P) = \text{id}$, then ϕ is the identity.*

Proof We may assume that $P = 0$. Expanding ϕ in a power series about $P = 0$ (and remembering that ϕ is vector-valued hence so is the expansion), we have

$$\phi(z) = z + P_k(z) + O\left(|z|^{k+1}\right)$$

where P_k is the first homogeneous polynomial of order exceeding 1 in the Taylor expansion. Defining $\phi^j(z) = \phi \circ \cdots \circ \phi$ (j times), we have

$$\begin{aligned}
\phi^2(z) &= z + 2P_k(z) + O\left(|z|^{k+1}\right)\\
\phi^3(z) &= z + 3P_k(z) + O\left(|z|^{k+1}\right)\\
&\vdots\\
\phi^j(z) &= z + jP_k(z) + O\left(|z|^{k+1}\right)
\end{aligned}$$

Choose polydiscs $D^n(0,a) \subseteq \Omega \subseteq D^n(0,b)$. Then for $0 \le j \in \mathbb{Z}$, we know that $D^n(0,a) \subseteq \text{dom}\, \phi^j \subseteq D^n(0,b)$. Therefore, the Cauchy estimates imply that for any multi-index α with $|\alpha| = k$, we have

P. V. Dovbush, S. G. Krantz, *The Geometric Theory of Complex Variables*,
https://doi.org/10.1007/978-3-031-77204-7_16

$$j\left|D^\alpha\phi(0)\right| = \left|D^\alpha\phi^j(0)\right| \le n\frac{b\cdot\alpha!}{a^k}$$

Letting $j \to \infty$ yields that $D^\alpha\phi(0) = 0$.
We conclude that $P_k = 0$; this contradicts the choice of P_k unless $\phi(z) \equiv z$. □

Remark 16.0.2 Notice that this proposition is a generalization of the uniqueness part of the classical Schwarz lemma on the disc. In fact, a great deal of work has been devoted to generalizations of the Schwarz lemma to more general settings. We refer the reader to H.H. Wu [535], S.T. Yau [537], S.G. Krantz [324, 325], and Burns and Krantz [100] for more on this topic.

We begin by considering a rather sharp Schwarz lemma that generalizes Proposition 16.0.1. The exercise preceding it records a technical fact that will be needed in the proof.

Exercise for the Reader
Let $U_1 \subseteq \mathbb{C}^m, U_2 \subseteq \mathbb{C}^n$ be open and connected. Let $d_1, and d_2$ be metrics on $U_1, and U_2$, respectively, both of which induce the Euclidean topology. Let $\mathcal{F}$ be the family of functions $f : U_1 \to U_2$ such that

$$d_2(f(z), f(w)) \le d_1(z, w)$$

for all $z, w \in U_1$. If $p \in U_1, K \subset\subset U_2$, let $\mathcal{F}(p, K) \equiv \{f \in \mathcal{F} : f(p) \in K\}$. Then $\mathcal{F}(p, K)$ is compact in the compact-open topology. (Hint: Imitate the proof of the Ascoli/Arzela theorem. Use the fact that a metric ball of radius r is mapped into a metric ball of radius r. Do not fall into the trap of thinking that U_2 must be complete!) See Wu [535] for more on these matters.

We begin by considering a rather sharp Schwarz lemma that generalizes Proposition 16.0.1

Theorem 16.0.3 *(Carathéodory-Cartan-Kaup-Wu) Let $\Omega \subseteq \mathbb{C}^n$ be open, bounded, and connected. Let $f \in \Omega(\Omega)$ and $P \in \Omega$. Assume that $f(P) = P$. Let $f' = J_{\mathbb{C}} f(P)$ be the differential of f at P. Then:*

$$\text{The eigenvalues of } f' \text{ all have modulus not exceeding } 1. \tag{16.1}$$

$$\left|\det f'\right| \le 1; \tag{16.2}$$

$$\text{If } \left|\det f'\right| = 1, \text{ then } f \in \operatorname{Aut}(\Omega); \tag{16.3}$$

$$\text{If } f' = id, \text{ then } f = id. \tag{16.4}$$

Remark 16.0.4 This is a sort of Schwarz lemma for holomorphic mappings. For even more general results, see S.T. Yau [537].

Notice that (16.3) tells us that $f' = \mathrm{id}$ implies that f is biholomorphic. Thus, f is an isometry in the Bergman metric. But an isometry is completely determined by its first-order behavior (its value and differential at a point).

Proof of Theorem 16.0.3 We use the Carathéodory metric on Ω. All distances and balls are determined in that metric. Choose $r > 0$ small such that the metric ball $\mathbb{B}(P, r)$ is relatively compact in Ω. Let $\mathcal{F}_r$ be the family of distance nonincreasing maps from $\mathbb{B}(P, r)$ to $\mathbb{B}(P, r)$ that fix P (for small r, such balls are connected). By the exercise, $\mathcal{F}_r$ is compact. If f is as in the hypotheses of the theorem, λ is an eigenvalue of f', and $|\lambda| > 1$, then we obtain a contradiction as follows: The sequence $f_k = (f)^k \equiv f \circ \cdots \circ f \in \mathcal{F}_r$ (composition of f with itself k times) satisfies $(f_k)' = \left(f'\right)^k$. This mapping has λ^k as an eigenvalue. Since $\mathcal{F}_r$ is compact, the fact that $\lambda^k \to \infty$ is a contradiction. That proves (16.1). Of course, (16.2) follows immediately.

For (16.4), let D be some fixed differential monomial at P. Assume that $Df \neq 0$ and that D is of order at least 2. Then $D\left((f)^k\right) = k \cdot Df$ (the expression simplifies because $f' = \mathrm{id}$). This is a contradiction to the compactness of $\mathcal{F}_r$ as $k \to \infty$.

Finally, we address (16.3). This involves more work. We assume that $\left|\det f'\right| = 1$. Then the eigenvalues of f', by (16.1), each have modulus equal to 1. Put f' in Jordan canonical form. Then we claim that the resulting matrix is diagonal. If it is not, then there is a block of the form

$$\begin{pmatrix} \lambda & 1 & & \\ & \lambda & 1 & \\ & & \ddots & 1 \\ & & & \lambda \end{pmatrix}$$

where λ is one of the eigenvalues. Then $\left((f)^k\right)'$ has a corresponding block of the form

$$\begin{pmatrix} \lambda^k & k\lambda^{k-1} & & & \\ & \ddots & & & 0 \\ & \lambda^k & \ddots & & \\ & & \ddots & \ddots & \\ & & & \ddots & k\lambda^{k-1} \\ & 0 & & & \lambda^k \end{pmatrix}$$

As $k \to \infty$, the compactness of $\mathcal{F}_r$ is contradicted.

So f', in Jordan canonical form, is diagonal. By the compactness of $\mathcal{F}_r$, there is a sequence f^{k_j} such that $\left(f^{k_j}\right)' \to \mathrm{id}$ (treat each coordinate in succession) and

$f^{k_j} \to \tilde{f}$, some $\tilde{f} \in \mathcal{F}_r$. Since the convergence is normal, $\tilde{f}$ is holomorphic on $\mathbb{B}(P,r)$ and $\tilde{f}' = \text{id}$. Hence, (16.4) implies that $\tilde{f}\Big|_{\mathbb{B}(P,r)} = \text{id}$.

We somehow need to spread the last line out to all Ω. Let W be the largest open subset of Ω on which $\{f^k\}_{k=1}^{\infty}$ has a normally convergent subsequence. Changing notation, we may suppose that the f^{k_j} found in the preceding paragraph converge normally to the identity on W. Now let $z \in \bar{W}$. We claim that $z \in W$, so that W is closed. Since W is also open, it follows then that $W = \Omega$. Assuming the claim for the moment and making the temporary additional assumption that Ω is complete in the Carathéodory metric, let us complete the proof.

We have $\{f^{k_j}\}$ converging normally to id on Ω. Let $\mathcal{G}$ be the family of all distance nonincreasing maps of Ω into Ω that fix P. Then $\mathcal{G}$ is compact. Thus, there is a subsequence of $\{f^{k_j-1}\}$ that converges normally to some g on Ω. Call this subsequence f^{k_i-1}. Of course, g is then holomorphic. Finally,

$$f \circ g = f \circ \left(\lim_{i\to\infty} \left(f^{k_i-1} \right) \right) = \lim_{i\to\infty} f^{k_i} = id$$

and

$$g \circ f = \left(\lim_{i\to\infty} \left(f^{k_i-1} \right) \right) \circ f = \lim_{i\to\infty} f^{k_i} = id.$$

It follows that f is invertible and that $f^{-1} = g$.

To prove the claim, notice that, by continuity, f must fix z. Since f is distance nonincreasing, there are connected neighborhoods U, V of z such that $f^{k_3}(U) \subseteq V \subset\subset \Omega$ for all sufficiently large j. Let $\mathcal{F}_{V,U}$ be the set of all distance nonincreasing maps from U into V that fix z. Then $\mathcal{F}_{V,U}$ is compact. Once again, extract a normally convergent subsequence. Since this subsequence must converge to the identity on $U \cap W$, it, in fact, converges to the identity on all of U. Thus, $z \in W$ as claimed. We are finished in case Ω is complete in the Carathéodory metric.

In case Ω is not complete, we choose $r > 0$ so small that the metric ball $B(z,r)$ is relatively compact in Ω. Then the family of holomorphic mappings that send $\mathbb{B}(z,r)$ to $\mathbb{B}(z,r)$ and that fix z is compact. As before, we may argue that a subsequence $\{f^{k_j}\}$ of the maps converges to the identity on $\mathbb{B}(z,r)$, and therefore on $W \cup \mathbb{B}(z,r)$. This device, of not simply extending the set W, but of extending instead the functional element (f, W), enables us to show that $W = \Omega$. As an exercise, the reader should now complete the proof in case Ω is not complete. Complete details may be found in Kobayashi [291, pp. 75–77]. □

We now introduce two new notions that are related to the Carathéodory and Kobayashi metrics. These are holomorphic invariants that correspond loosely to volume forms. The exact nature of this correspondence is explored in D. Eisenman [145].

Definition 16.0.5 Let $\Omega \subseteq \mathbb{C}^n$ be open. Define the *Carathéodory volume form* on Ω by

$$M_\Omega^C(z) = \sup\left\{\left|\det f'(z)\right| : f \in \mathbb{B}(\Omega),\, f(z) = 0\right\}.$$

Define the *Kobayashi-Eisenman volume form* on Ω by

$$M_\Omega^K(z) = \inf\left\{1/\left|\det f'(0)\right| : f \in \Omega(\mathbb{B}),\, f(0) = z\right\}.$$

Remark 16.0.6 The quantities M^C and M^K transform in a natural manner under automorphisms of Ω. If Φ is such an automorphism, then

$$M_\Omega^C(\Phi(z)) = M_\Omega^C(z)\left|\det J_{\mathbb{C}}\Phi(z)\right|^{-1}$$

and

$$M_\Omega^K(\Phi(z)) = M_\Omega^K(z)\left|\det J_{\mathbb{C}}\Phi(z)\right|^{-1}.$$

Thus, M_Ω^K/M_Ω^C is a biholomorphic invariant. This fact has been exploited in Greene and Krantz [319, 320, 322, 323].

Proposition 16.0.7 (Bun Wong) *Let $\Omega \subseteq \mathbb{C}^n$ be a bounded, connected open set. Suppose that there is a point $P \in \Omega$ for which*

$$M_\Omega^C(P) = M_\Omega^K(P).$$

Then Ω is biholomorphic to the ball in $\mathbb{C}^n$.

Proof There is a holomorphic map $g \in \mathbb{B}(\Omega)$ with $M_\Omega^C(P) = \left|\det g'(P)\right|$. We may not, in general, choose a corresponding map h for $M_\Omega^K(P)$, but we may instead choose a sequence $h_j \in \Omega(\mathbb{B})$ with $\left|\det h_j'(P)\right|^{-1} \to M_\Omega^K(P)$. Consider the sequence $G_j = g \circ h_j \in \mathbb{B}(\mathbb{B})$. By a previously noted argument, there is a normally convergent subsequence $G_{j_k} \to G \in \mathbb{B}(\mathbb{B})$. By hypothesis, it must be that $\left|\det G'(0)\right| = 1$. Hence, by (16.3), $G \in \mathrm{Aut}(\mathbb{B})$. In particular, we may conclude that G is surjective. Therefore, g is surjective.

Now consider $H_j = h_j \circ g \in \Omega(\Omega)$. We would like, once again, to extract a normally convergent subsequence. However, we cannot immediately do this. What we can do is to restrict attention to a ball $\mathbb{B}(P, r)$ in the Carathéodory metric. Applying the exercise at the beginning of this section, we may extract a subsequence H_{j_k} that converges normally to a mapping H on $\mathbb{B}(P, r)$. Then $H_{j_k}'(P)$ converges to the identity so that $H' \equiv \mathrm{id}$ on $\mathbb{B}(P, r)$. Applying the connectedness argument of the proof of (16.3), we see that a subsequence $\left\{H_{j_{k_\ell}}\right\}$ of $\{H_j\}$ may be extracted that converges normally to some $\tilde{H} \in \Omega(\Omega)$ with $\tilde{H}'(P) = id$. Thus $\tilde{H} \equiv id$ by (16.4). As a result, $h_{j_{k_\ell}} \circ g \to \mathrm{id}$ normally on Ω. It follows that g must be injective.

We have established that $g \in \mathbb{B}(\Omega)$ is both surjective and injective. Thus, it is a biholomorphism of Ω to $\mathbb{B}$. □

We now turn to the principal result of this section (see B. Wong [533], J.P. Rosay [441]). It should be compared with the Riemann mapping theorem. We also refer the reader to S.G. Krantz [322], in which the crux of the proof is boiled down for application to one complex variable.

In what follows, when we say that $P \in \partial\Omega$ is a point of strong pseudoconvexity, then it is understood that $\partial\Omega$ is C^2 in a relative neighborhood of P (hence, $\partial\Omega$ is strongly pseudoconvex at points near P).

Theorem 16.0.8 *Let $\Omega \subseteq \mathbb{C}^n$ be any bounded, connected open set with a boundary point $P \in \Omega$ that is strongly pseudoconvex. Suppose further that there exist $K \subset\subset \Omega$, $z^k \in K$, and $f_k \in \mathrm{Aut}(\Omega)$ with $f_k\left(z^k\right) \to P$. Then Ω is biholomorphic to the unit ball in $\mathbb{C}^n$.*

Corollary 16.0.9 *If $\Omega \subseteq \mathbb{C}^n$ has C^2 boundary and has a transitive automorphism group, then Ω is biholomorphic to B.*

Proof Let $P \in \partial\Omega$ be the point in $\partial\Omega$ that is furthest from 0. Then P is a point of strong convexity, hence of strong pseudoconvexity. (To see this, let $R = \mathrm{dist}(0, P)$ and consider the Euclidean ball $\mathbb{B}(0, R)$.) Let $K = \left\{z^0\right\}$, where z^0 is some fixed interior point of Ω. Let $\Omega \ni w^j \to P$. By hypothesis, there are automorphisms f_j such that $f_j\left(z^0\right) = w^j \to P$. The result now follows from the theorem. □

Corollary 16.0.10 *Let $\Omega \subseteq \mathbb{C}^n$ be any strongly pseudoconvex set with a noncompact automorphism group. Then Ω is biholomorphic to $\mathbb{B} \subseteq \mathbb{C}^n$.*

Proof If $K, L \subset\subset \Omega$, then the set $\mathcal{F}(K, L) = \{f \in \mathrm{Aut}(\Omega) : f(K) \subseteq L\}$ is compact (exercise). Therefore, Aut Ω noncompact implies that there is a $z^0 \in \Omega$ and $f_j \in \mathrm{Aut}\,\Omega$ such that $f_j\left(z^0\right) \to P \in \partial\Omega$. Now apply the theorem. □

The remainder of the section is devoted to a proof of the theorem. The proof proceeds in stages by way of a sequence of preliminary results. Some of these have independent interest.

Lemma 16.0.11 *Let $\Omega \subseteq \mathbb{C}^n$ be open and bounded. Let $P \in \partial\Omega$ be a point of strong pseudoconvexity. Fix $A > 0$. For each $0 < \eta < 1$, there is a $\delta > 0$ such that if $z \in \Omega$, $|z - P| < \delta$, and $f \in \Omega(\mathbb{B})$ satisfies $f(0) = z$, then $|f(w) - P| < A$ for all w such that $|w| < 1 - \eta$.*

Proof Let L_P be the Levi polynomial at P. Then $\exp\left(L_P\right) \equiv g$ satisfies $g(P) = 1$ and $|g(z)| < 1$ for all $|z - P| < \epsilon$, $z \in \bar{\Omega}\setminus\{P\}$, ϵ sufficiently small.

If the conclusion of the lemma were not true, then there would be an $\eta > 0$ such that for each $j \geq 1/\epsilon$, there is a z_j with $\left|z_j - P\right| < 1/j$ and an $f_j \in \Omega(\mathbb{B})$ with $f_j(0) = z_j$ but $\left|f_j\left(w_j\right) - P\right| \geq A$ for some w_j with $\left|w_j\right| < 1 - \eta$. Consider the sequence $G_j = g \circ f_j$. Then $\left\{G_j\right\}$ is a normal family. Let G be the normal limit of a subsequence $\left\{G_{j_k}\right\}$. Then $G(0) = \lim_{k\to\infty} g \circ f_{j_k}(0) = \lim g\left(z_{j_k}\right) = 1$. But

$\{f_j\}$ is an equicontinuous family on $\left\{z : |z| \le \frac{3}{4}\right\}$. Thus $|G_j|$, and hence $|G|$ are bounded by 1 on some $\{z : |z| \le \mu\}$; some $\mu > 0$. Hence, $G \equiv 1$ on $\{z : |z| \le \mu\}$. But then $G \equiv 1$ on all of $\mathbb{B}$. Therefore, $G_{j_k} \to 1$ uniformly on $\{w : |w| \le 1 - \eta\}$. We may also suppose that $w_{j_k} \to w, |w| \le 1 - \eta$. But then $\left|f_{j_k}(w) - P\right| \ge A/2$ for all large k by equicontinuity. We conclude that $G(w)$ is bounded from 1. That is a contradiction. □

Lemma 16.0.12 *Let $\Omega \subseteq \mathbb{C}^n$ be open and bounded. Let $P \in \partial\Omega$ be a point of strong pseudoconvexity. If $\{z^j\} \subseteq \Omega$ is any sequence such that $z^j \to P$ and if $A > 0$, then*

$$\frac{M_\Omega^K(z^j)}{M_{\Omega \cap \mathbb{B}(P,A)}^K(z^j)} \to 1 \quad as \quad j \to \infty.$$

Proof Let $0 < \eta < 1$. Choose $\delta > 0$ according to the last lemma. Thus, if $\left|z^j - P\right| < \delta$ and $f \in \Omega(\mathbb{B})$, $f(0) = z^j$, then the map $f_\eta : w \mapsto f((1-\eta)w)$ is in $(\mathbb{B}(P, A) \cap \Omega)(\mathbb{B})$ with $f_\eta(0) = z^j$. Thus

$$M_{\Omega \cap \mathbb{B}(P,A)}^K\left(z^j\right) \le \frac{1}{\left|\det f_\eta'(0)\right|} = \frac{(1-\eta)^{-n}}{|\det f'(0)|}.$$

Taking the infimum over all such f gives

$$M_{\Omega \cap \mathbb{B}(P,A)}^K\left(z^j\right) \le (1-\eta)^{-n} M_\Omega^K\left(z^j\right).$$

Also

$$M_{\Omega \cap \mathbb{B}(P,A)}^K\left(z^j\right) \ge M_\Omega^K\left(z^j\right)$$

by the distance-decreasing property of the Kobayashi metric. This proves the result. □

Lemma 16.0.13 *Let $\Omega \subseteq \mathbb{C}^n$ be open and bounded. Let $P \in \partial\Omega$ be a point of strong pseudoconvexity. Fix $A > 0, z^0 \in \Omega$, and let $\phi_j \in \mathrm{Aut}(\Omega)$ be such that $\phi_j\left(z^0\right) \to P$. Then*

$$\frac{M_\Omega^C\left(\phi_j\left(z^0\right)\right)}{M_{\Omega \cap \mathbb{B}(P,A)}^C\left(\phi_j\left(z^0\right)\right)} \to 1 \quad as\ j \to \infty$$

Proof Let $\Omega_1 \subset\subset \Omega_2 \subset\subset \cdots \subset \Omega$ satisfy $\cup\Omega_j = \Omega$ and $z^0 \in \Omega_j$ for all j. Choose $f_j \in \mathbb{B}\left(\Omega_j\right)$ such that $M_{\Omega_j}^C = \left|\det f_j'\left(z^0\right)\right|$. Now there is an $f \in \mathbb{B}(\Omega)$

and a subsequence f_{j_k} such that $f_{j_k} \to f$ normally on Ω. Of course, $f(z_0) = 0$. It follows that $\left|\det f'_{j_k}(z_0)\right| \to \left|\det f'(z^0)\right|$.

Let $\eta > 0$. By the preceding paragraph, there is a k so large that $M^C_\Omega(z^0) \le M^C_{\Omega_{\Omega_k}}(z^0) \le (1+\eta)M^C_\Omega(z^0)$. By the proof of Lemma 16.0.11, there is an ℓ so large that $\phi_\ell(\Omega_{Jk}) \subseteq \mathbb{B}(P, A)$. Therefore

$$\begin{aligned}
1 &\le \frac{M^C_{\Omega \cap \mathbb{B}(P,A)}(\phi_\ell(z^0))}{M^C_\Omega(\phi_\ell(z^0))} \\
&\le \frac{M^C_{\phi_\ell(\Omega_{Jk})}(\phi_\ell(z^0))}{M^C_\Omega(\phi_\ell(z^0))} \\
&= \frac{M^C_{\Omega_{J_k}}(z^0)}{M^C_\Omega(z^0)} \le 1+\eta.
\end{aligned}$$

Since $\eta > 0$ was arbitrary, the result follows. □

Lemma 16.0.14 *Let Q be a positive definite quadratic form on $\mathbb{C}^n \times \mathbb{C}^n$. Define $\mathcal{D}_Q = \{w \in \mathbb{C}^n : 2\operatorname{Re} w_1 < -Q(w, w)\}$. Then $\mathcal{D}_Q$ is biholomorphic to $\mathbb{B} \subseteq \mathbb{C}^n$.*

Proof Write

$$Q(w, w) = \sum_{i,j} a_{ij} w_i \bar{w}_j$$

Since Q is positive definite, we can, by a linear change of coordinates, diagonalize $\sum_{i,j=2}^n a_{ij} w_i \bar{w}_j$. Therefore, $\mathcal{D}_Q$ is biholomorphic to a region of the form

$$2\operatorname{Re} w_1 < -\sum_{i=2}^n |\alpha_i w_i|^2 - 2\operatorname{Re} \sum_{i=2}^n a_{i1} w_i \bar{w}_1 - a_{11}|w_1|^2$$

A change of variable of the form

$$\begin{aligned}
z'_1 &= w_1 \\
z'_i &= w_i + \beta_i w_1, \quad i = 2, \ldots, n
\end{aligned}$$

eliminates the cross terms, and we arrive at a region of the form

$$\left|\alpha_0\left(z'_1 - \beta_1\right)\right|^2 + \sum_{i=2}^n |\alpha_i z'_i|^2 < \gamma$$

Finally, translation and dilation of z_1', together with dilations of $z_2', \ldots, z_n'$, give

$$\sum_{i=1}^{n} |z_i'|^2 < 1.$$

□

Corollary 16.0.15 *If $\mathcal{D}_Q$ is as in the preceding lemma and $P \in \mathcal{D}_Q$, then $M^C_{\mathcal{D}_Q}(P) = M^K_{\mathcal{D}_Q}(P)$.*

Proof It suffices to check the assertion at the origin in the ball. This is trivial. □

Finally, we may give a proof of our main result.

Proof of Theorem 16.0.8 By an easy compactness argument, we may find an $a \in \Omega$ and $f_k \in \operatorname{Aut} \Omega$ such that $f_k(a) \to P \in \partial\Omega$. We may choose a small neighborhood U of P in $\mathbb{C}^n$ such that every point $\zeta \in \partial\Omega \cap U$ is strongly pseudoconvex. If U is small enough, k large, and $f_k(a) \equiv w^k \in U$, then we may select $\zeta^k \in \partial\Omega$ with the following property: when Narasimhan's lemma is applied at the point ζ^k, then, in the new coordinates at ζ^k, the boundary has the defining function

$$\tilde{\rho}_k(\zeta) = 2\operatorname{Re}\zeta_1 + \mathcal{L}_{\zeta^k}(\zeta) + o\left(|\zeta|^2\right)$$

(where $\mathcal{L}_{\zeta^k}$ is the Levi form at ζ^k) and

$$w^k = (a_k, 0, \ldots, 0) \quad , \quad a_k < 0. \tag{16.5}$$

For this k and $\epsilon > 0$, we then have associated ellipsoids $E_\epsilon^+\left(\zeta^k\right)$ and $E_\epsilon^-\left(\zeta^k\right)$ given by

$$E_\epsilon^+\left(\zeta^k\right) = \left\{2\operatorname{Re} z_1 < -\mathcal{L}_{\zeta^k}(z) + \epsilon|z|^2\right\}$$
$$E_\epsilon^-\left(\zeta^k\right) = \left\{2\operatorname{Re} z_1 < -\mathcal{L}_{\zeta^k}(z) - \epsilon|z|^2\right\}$$

For $\epsilon > 0$ fixed, there exist $A > r > 0$ such that if k is large enough (w^k is near enough to P), then

$$E_\epsilon^-\left(\zeta^k\right) \cap \mathbb{B}\left(\zeta^k, r\right) \subseteq \Omega \cap \mathbb{B}\left(\zeta^k, A\right) \subseteq E_\epsilon^+\left(\zeta^k\right)$$

Refer to Fig. 16.1.

Notice that A, r depend on ϵ, but by the uniform continuity on U of the second derivatives of any defining function for Ω, they do not depend on ζ^k.

We need to calculate

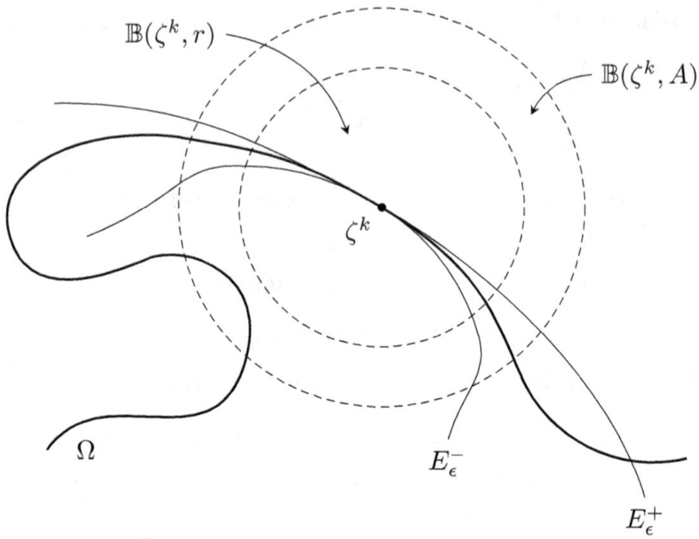

Fig. 16.1 Geometric results if k is large enough

$$\frac{M_\Omega^K(a)}{M_\Omega^C(a)} = \frac{M_\Omega^K\left(w^k\right)}{M_\Omega^C\left(w^k\right)} \tag{16.6}$$

Trivially, the expression is at least 1.

For an inequality in the opposite direction, we apply Lemmas 16.0.12 and 16.0.13 (and the *proof* of 16.0.12) to see that it is enough to estimate

$$\begin{aligned} \limsup_{k\to\infty} \frac{M^K_{\Omega\cap\mathbb{B}(\zeta^k,r)}\left(w^k\right)}{M^C_{\Omega\cap\mathbb{B}(\zeta^k,A)}\left(w^k\right)} &\le \limsup_{k\to\infty} \frac{M^K_{E^-(\zeta^k)\cap\mathbb{B}(P,r/2)}\left(w^k\right)}{M^C_{E^+(\zeta^k)}\left(w^k\right)} \\ &= \limsup_{k\to\infty} \frac{M^K_{E^-(\zeta^k)}\left(w^k\right)}{M^C_{E^+(\zeta^k)}\left(w^k\right)} \end{aligned} \tag{16.7}$$

Here the distance-decreasing property of the Kobayashi metric has been applied to obtain the last inequality. We have also used repeatedly the fact that letting $\tilde{r} = r/2$, it holds that $\mathbb{B}\left(\zeta^k, \tilde{r}\right) \subseteq \mathbb{B}(P, 2\tilde{r}) \subseteq \mathbb{B}\left(\zeta^k, 3\tilde{r}\right)$ for k large.

Now let

$$\Phi^-_{\epsilon,k} : E^-_\epsilon\left(\zeta^k\right) \to \mathbb{B}$$

$$\Phi^+_{\epsilon,k} : E^+_\epsilon\left(\zeta^k\right) \to \mathbb{B}$$

be the biholomorphisms constructed in Lemma 16.0.14. Then $\det J_{\mathbb{C}}\Phi^{-}_{\epsilon,k}\left(w^k\right)$ and $\det J_{\mathbb{C}}\Phi^{+}_{\epsilon,k}\left(w^k\right)$ both depend continuously on ϵ as $\epsilon \to 0^+$. Also their quotient may be made arbitrarily close to 1 for ϵ small, uniformly in k. Finally, notice that since $w^k = (a_k, 0, \ldots, 0)$, it holds that

$$\Phi^{-}_{\epsilon,k}\left(w^k\right) = (a_k\,(a_{11} + \epsilon) + 1, 0, \ldots, 0)$$

$$\Phi^{+}_{\epsilon,k}\left(w^k\right) = (a_k\,(a_{11} - \epsilon) + 1, 0, \ldots, 0)\,.$$

It follows that (16.7) equals

$$\limsup_{k\to\infty} \frac{M^K_B\,(a_k\,(a_{11} + \epsilon) + 1, 0, \ldots, 0)\det J_{\mathbb{C}}\Phi^{-}_{\epsilon,k}\left(w^k\right)}{M^C_B\,(a_k\,(a_{11} - \epsilon) + 1, 0, \ldots, 0)\det J_{\mathbb{C}}\Phi^{+}_{\epsilon,k}\left(w^k\right)}$$

$$= \limsup_{k\to\infty} \frac{M^K_B\,(a_k + 1, 0, \ldots, 0)\cdot \mathcal{J}^{+}\cdot \det J_{\mathbb{C}}\Phi^{-}_{\epsilon,k}\left(w^k\right)}{M^C_B\,(a_k + 1, 0, \ldots, 0)\cdot \mathcal{J}^{-}\cdot \det J_{\mathbb{C}}\Phi^{+}_{\epsilon,k}\left(w^k\right)}.$$

Here $\mathcal{J}^{\pm}$are the complex Jacobian determinants of the standard automorphism of $\mathbb{B}$ taking $(a_k\,(a_{11} \pm \epsilon) + 1, 0, \ldots, 0)$ to $(a_k + 1, 0, \ldots, 0)$. The last line is as near to 1 as we please if ϵ is small enough.

In conclusion, $M^K_\Omega(a) = M^C_\Omega(a)$. By Proposition 16.0.7, we conclude that Ω is biholomorphic to $\mathbb{B}$. □

Exercise for the Reader

Compute $\Phi^{+}_{\epsilon,k}$and $\Phi^{-}_{\epsilon,k}$explicitly and determine their Jacobian determinants exactly. Compute also $\mathcal{J}^{+}$and $\mathcal{J}^{-}$. Use these to render precise the last part of the proof of Theorem 16.0.8.

Chapter 17
Smoothness to the Boundary of Biholomorphic Mappings

Poincaré's theorem that the ball and polydisc are biholomorphically inequivalent shows that there is no Riemann mapping theorem (at least in the traditional sense) in several complex variables. More recent results of D. Burns, S. Shnider, and R. Wells [99] and of R.E. Greene and S.G. Krantz [192, 194] confirm how truly dismal the situation is. First, we need a definition.

Definition 17.0.1 Let ρ_0 be a $\mathcal{C}^k$ defining function for a domain $\Omega_0 \subset\subset \mathbb{R}^N$, $k \geq 2$. We define a neighborhood basis for Ω_0 in the $\mathcal{C}^k$ topology as follows: Let $\epsilon > 0$ be so small that if $\|\rho - \rho_0\|_{\mathcal{C}^k} < \epsilon$, then ρ has nonvanishing gradient on $\{x : \rho(x) = 0\}$. For any such ρ, let $\Omega_\rho = \{x \in \mathbb{R}^N : \rho(x) < 0\}$. Define

$$\mathcal{U}_\epsilon^k(\Omega_0) = \left\{\Omega_\rho \subseteq \mathbb{R}^N : \|\rho - \rho_0\|_{\mathcal{C}^k} < \epsilon\right\}.$$

Observe that $\mathcal{U}_\epsilon^k(\Omega_0)$ is a collection of domains. Then the sets $\mathcal{U}_\epsilon^k$ are called *neighborhoods of* Ω_0 in the $\mathcal{C}^k$ topology. Of course, neighborhoods in the $\mathcal{C}^\infty$ topology are defined similarly.

Theorem 17.0.2 *(Burns/Shnider/Wells) Let $k \in \mathbb{N}$. Let $\mathcal{U}_\epsilon^k(\mathbb{B})$ be any neighborhood of the ball $\mathbb{B} \subseteq \mathbb{C}^n$ in the $\mathcal{C}^k$ topology as defined above. If Ω_1, Ω_2 are domains in $\mathbb{C}^n$, then we say that $\Omega_1 \sim \Omega_2$ if Ω_1 is biholomorphic to Ω_2. If $n \geq 2$, then $\mathcal{U}_\epsilon^k/\sim$ is uncountable, no matter how small $\epsilon > 0$ is or how large k is (even $k = \infty$ or $k = \omega$).*

Theorem 17.0.2 is in striking contrast to the situation in $\mathbb{C}^1$, where $\mathcal{U}_\epsilon^k/\sim$ has only one element as soon as $k = 1$ and ϵ is, say, $\frac{1}{5}$. Greene and Krantz [192, 194] have refined the theorem to show that in fact each of the equivalence classes is closed and nowhere dense.

We now give a brief accounting of some of the differences between $n = 1$ and $n > 1$. A more detailed discussion appears in R.E. Greene and S.G. Krantz [198]. This subject begins with the following breakthrough of C. Fefferman [150].

P. V. Dovbush, S. G. Krantz, *The Geometric Theory of Complex Variables*,
https://doi.org/10.1007/978-3-031-77204-7_17

Theorem 17.0.3 *Let $\Omega_1, \Omega_2 \subseteq \mathbb{C}^n$ be strongly pseudoconvex domains with C^∞ boundary. If $\phi : \Omega_1 \to \Omega_2$ is biholomorphic, then ϕ extends to a C^∞ diffeomorphism of $\overline{\Omega}_1$ onto $\overline{\Omega}_2$.*

Fefferman's theorem enables one to see that if Ω_1 and Ω_2 are biholomorphic under ϕ, then there are certain differential invariants of $\partial\Omega_1, \partial\Omega_2$ that must be preserved under ϕ. More precisely, if k is large, then the kth order Taylor expansion of the defining function ρ_1 for Ω_1 (respectively of the defining function ρ_2 for Ω_2) has more coefficients than the k order Taylor expansion for ϕ (the disparity in the number grows rapidly with k). Since ρ_1 is mapped to ρ_2 under composition with ϕ^{-1}, it follows that some of these coefficients, or combinations thereof, must be invariant under biholomorphic mappings. N. Tanaka [495] and S.S. Chern and J. Moser [111] have made these remarks precise and have shown how to calculate these invariants. A more leisurely discussion of these matters appears in R.E. Greene and S.G. Krantz [198].

Now it is easy to see intuitively that two domains Ω and Ω_0 can be close in the C^k topology, any k, and have entirely different Chern/Moser/Tanaka invariants. This notion is made precise, for instance, in D. Burns, S. Shnider, and R. Wells [99], by using a transversality argument. (Note that everything we are saying is vacuous in $\mathbb{C}^1$ because the invariants must live in the complex tangent space to the boundary—which is empty in dimension one.) It is essentially a foregone conclusion that things will go badly in higher dimensions.

If one seeks positive results in the spirit of the Riemann mapping theorem in dimension $n \geq 2$, then one must find statements of a different nature. Fridman [161] has constructed a "universal domain" Ω^* which can be used to biomorphically exhaust any other. He has obtained a number of variants of this idea, using elementary but clever arguments. S. Semmes [466] has yet another approach to the Riemann mapping theorem that is more in the spirit of the work of Lempert. We present, mainly for background, a substitute for the Riemann mapping theorem whose statement and proof is more in the spirit of Fefferman's theorem.

Theorem 17.0.4 (Greene/Krantz [194]) *Let $\mathbb{B} \subseteq \mathbb{C}^n$ be the unit ball. Let $\rho_0(z) = |z|^2 - 1$ be the usual defining function for $\mathbb{B}$. If $\epsilon > 0$ is sufficiently small, $k = k(n)$ is sufficiently large, and $\Omega \in \mathcal{U}^k_\epsilon(\mathbb{B})$ then either*

$$\Omega \sim \mathbb{B} \text{ or} \tag{17.1}$$

$$\Omega \text{ is not biholomorphic to the ball and} \tag{17.2}$$

1. *Aut Ω is compact.*
2. *Aut Ω has a fixed point.*
3. *If $K \subset\subset \mathbb{B}$, $\epsilon > 0$ is sufficiently small (depending on K), and $\Omega \in \mathcal{U}^k_\epsilon(\mathbb{B})$ has the property that its fixed point set lies in K, then there is a biholomorphic mapping $\Phi : \Omega \to \Phi(\Omega) \equiv \Omega' \subseteq \mathbb{C}^n$ such that Aut (Ω') is the restriction to Ω' of a subgroup of the group of unitary matrices.*

The collection of domains to which (17.2) applies is both dense and open.

Theorem 17.0.4 shows, in a weak sense, that domains near the ball that have any automorphisms other than the identity are (biholomorphic to) domains with only Euclidean automorphisms. It should be noted that 17.2 (1) is already contained in the theorem of Bun Wong and Rosay and that the denseness of the domains to which (17.2) applies is contained in the work of Burns-ShniderWells. The proof of Theorem 17.0.4 involves a detailed analysis of Fefferman's asymptotic expansion for the Bergman kernel and of the $\overline{\partial}$-Neumann problem and would double the length of this book if we were to treat it in any detail.

The purpose of this lengthy introduction has been to establish the importance of Theorem 17.0.3 and to set the stage for what follows. It may be noted that the proof of the analogous result in $\mathbb{C}^1$, that a biholomorphic mapping of smooth domains extends smoothly to the boundary, was proved in P. Painlevé [406]. The result in one complex dimension has been highly refined, beginning with work of O. Kellogg [268] and of S. Warschawski [517–519], Rodin and Warschawski [439], and others. This classical work uses harmonic estimation and the Jordan curve theorem, devices that have no analog in higher dimensions. A short, self-contained, proof of the one-variable result—using ideas closely related to those presented here—appears in S. Bell and S.G. Krantz [70].

We conclude this section by presenting a short and elegant proof of Fefferman's theorem (Theorem 17.0.3). The techniques are due to S. Bell [54] and S. Bell and E. Ligocka [53]. The proof uses an important and nontrivial fact (known as "Condition R" of Bell and Ligocka) about the $\overline{\partial}$-Neumann problem. We will have to take Condition R on a strongly pseudoconvex domain for granted. (However, Condition R, and more generally the solution of the $\overline{\partial}$-Neumann problem, is considered in detail in S.G. Krantz [329].)

Let $\Omega \subset\subset \mathbb{C}^n$ be a domain with C^∞ boundary. We define Condition R as follows:

Condition R (S. Bell [54]): Define an operator on $L^2(\Omega)$ by

$$Pf(z) = \int_\Omega K(z,\zeta) f(\zeta) dV(\zeta),$$

where $K(z,\zeta)$ is the Bergman kernel for Ω. Then for each $j > 0$, there is an $m = m(j) > 0$ such that P satisfies the estimates

$$\|Pf\|_{W^j(\Omega)} \leq C_j \|f\|_{W^m(\Omega)}$$

for all testing functions f.

The deep fact, which we shall use without proof (but see the aforementioned reference S.G. Krantz [329]), is that Condition R holds on any strongly pseudoconvex domain.

We remark in passing that, in general, it does not matter whether $m(j)$ is much larger than j or whether $m(j)$ depends polynomially on j or exponentially on j. It so happens that for a strongly pseudoconvex domain, we may take $m = j$. This assertion is proved in S.G. Krantz [329] in detail. On the other hand, Barrett [44] has shown that on the Diederich-Fornaess worm domain, we must take $m(j) > j$. F.M. Christ [113] has shown that, on the worm domain, nothing works.

Now we build a sequence of lemmas leading to Fefferman's theorem. First we record some notation.

If $\Omega \subset\subset \mathbb{C}^n$ is any smoothly bounded domain and if $j \in \mathbb{N}$, we let
$WH^j(\Omega) = W^j(\Omega) \cap \{$ holomorphic functions on $\Omega\}$,
$WH^\infty(\Omega) = \bigcap_{j=1}^\infty WH^j(\Omega) = C^\infty(\overline{\Omega}) \cap \{$ holomorphic functions on $\Omega\}$.

Here W^j is the standard Sobolev space on a domain. Let $W_0^j(\Omega)$ be the W^j closure of $C_c^\infty(\Omega)$. (Exercise: If j is sufficiently large, then the Sobolev imbedding theorem implies trivially that $W_0^j(\Omega)$ is a proper subset of $W^j(\Omega)$.)

Let us say that $u, v \in C^\infty(\overline{\Omega})$ agree up to order k on $\partial\Omega$ if

$$\left(\frac{\partial}{\partial z}\right)^\alpha \left(\frac{\partial}{\partial \overline{z}}\right)^\beta (u - v)\bigg|_{\partial\Omega} = 0 \quad \forall \alpha, \beta \text{ with } |\alpha| + |\beta| \le k.$$

Lemma 17.0.5 *Let $\Omega \subset\subset \mathbb{C}^n$ be smoothly bounded and strongly pseudoconvex. Let $w \in \Omega$ be fixed. Let K denote the Bergman kernel. There is a constant $C_w > 0$ such that*

$$\|K(w, \cdot)\|_{\sup} \le C_w$$

Proof The function $K(z, \cdot)$ is harmonic. Let $\phi : \Omega \to \mathbb{R}$ be a radial, C_c^∞ function centered at w. Assume that $\phi \ge 0$ and $\int \phi(\zeta) dV(\zeta) = 1$. Then the mean value property implies that

$$K(z, w) = \int_\Omega K(z, \zeta)\phi(\zeta) dV(\zeta).$$

But the last expression equals $P\phi(z)$. Therefore,

$$\begin{aligned}\|K(w, \cdot)\|_{\sup} &= \sup_{z\in\Omega} |K(w, z)| \\ &= \sup_{z\in\Omega} |K(z, w)| \\ &= \sup_{z\in\Omega} |P\phi(z)|.\end{aligned}$$

By Sobolev's theorem, this is

$$\le C(\Omega) \cdot \|P\phi\|_{WH^{2n+1}}.$$

By Condition R, this is

$$\leq C(\Omega) \cdot \|\phi\|_{W^{m(2n+1)}} \equiv C_w .$$

□

Lemma 17.0.6 *Let $u \in C^\infty(\overline{\Omega})$ be arbitrary. Let $s \in \{0, 1, 2, \ldots\}$. Then there is a $v \in C^\infty(\overline{\Omega})$ such that $Pv = 0$, and the functions u and v agree to order s on $\partial\Omega$.*

Proof After a partition of unity, it suffices to prove the assertion in a small neighborhood U of $z_0 \in \partial\Omega$. After a rotation, we may suppose that $\partial\rho/\partial z_1 \neq 0$ on $U \cap \overline{\Omega}$, where ρ is a defining function for Ω. Define the differential operator

$$\nu = \frac{\mathrm{Re}\left\{\sum_{j=1}^{n} \frac{\partial\rho}{\partial z_j}\frac{\partial}{\partial \overline{z}_j}\right\}}{\sum_{j=1}^{n}\left|\frac{\partial\rho}{\partial z_j}\right|^2}$$

Notice that $\nu\rho = 1$. Now we define v by induction on s.

For the case $s = 0$, let

$$w_1 = \frac{\rho u}{\partial\rho/\partial\zeta_1}.$$

Define

$$v_1 = \frac{\partial}{\partial\zeta_1} w_1 = u + O(\rho).$$

Then u and v_1 agree to order 0 on $\partial\Omega$. Also,

$$Pv_1(z) = \int K(z,\zeta)\frac{\partial}{\partial\zeta_1} w_1(\zeta) dV(\zeta).$$

This equals, by integration by parts,

$$-\int \frac{\partial}{\partial\zeta_1} K(z,\zeta) w_1(\zeta) dV(\zeta).$$

Notice that the integration by parts is valid by Lemma 17.0.5 and because $w_1|_{\partial\Omega} = 0$. Also, the integrand in this last line is zero because $K(z,\cdot)$ is conjugate holomorphic. Thus, $Pv_1 \equiv 0$ and the case $s = 0$ is complete.

Suppose inductively that $w_{s-1} = w_{s-2} + \theta_{s-1}\rho^s$ and $v_{s-1} = (\partial/\partial z_1)(w_{s-1})$ have been constructed. We show that there is a w_s of the form

$$w_s = w_{s-1} + \theta_s \cdot \rho^{s+1}$$

such that $v_s = (\partial/\partial z_1)(w_s)$ agrees to order $(s-1)$ with u on $\partial\Omega$. By the inductive hypothesis,

$$\begin{aligned} v_s = \frac{\partial}{\partial z_1} w_s &= \frac{\partial w_{s-1}}{\partial z_1} + \frac{\partial}{\partial z_1} \left[\theta_s \cdot \rho^{s+1} \right] \\ &= v_{s-1} + \rho^s \left[(s+1)\theta_s \frac{\partial \rho}{\partial z_1} + \rho \cdot \frac{\partial \theta_s}{\partial z_1} \right] \end{aligned}$$

agrees to order $s - 1$ with u on $\partial\Omega$ so long as θ_s is smooth. So we need to examine $D(u - v_s)$, where D is an s-order differential operator. But if $\mathbb{D}$ involves a tangential derivative D_0, then write $D = D_0 \cdot D_1$. It follows that $D(u - v_s) = D_0(\alpha)$, where α vanishes on $\partial\Omega$ so that $D_0\alpha = 0$ on $\partial\Omega$. So we need only check $D = \nu^s$.

We have seen that θ_s must be chosen so that

$$\nu^s (u - v_s) = 0 \quad \text{on } \partial\Omega$$

Equivalently,

$$\nu^s (u - v_{s-1}) - \nu^s \left(\frac{\partial}{\partial z_1} \right) \left(\theta_s \rho^{s+1} \right) = 0 \quad \text{on } \partial\Omega$$

or

$$\nu^s (u - v_{s-1}) - \theta_s \left(\nu^s \frac{\partial}{\partial z_1} \rho^{s+1} \right) = 0 \quad \text{on } \partial\Omega$$

or

$$\nu^s (u - v_{s-1}) - \theta_s \cdot (s+1)! \frac{\partial \rho}{\partial z_1} = 0 \quad \text{on } \partial\Omega$$

It follows that we must choose

$$\theta_s = \frac{\nu^s (u - v_{s-1})}{(s+1)! \frac{\partial \rho}{\partial z_1}}$$

which is indeed smooth on U. As in the case $s = 0$, it holds that $P v_s = 0$. This completes the induction and the proof. □

Remark 17.0.7 A retrospection of the proof reveals that we have constructed v by subtracting from u a Taylor-type expansion in powers of ρ.

Lemma 17.0.8 *For each $s \in \mathbb{N}$, we have $WH^\infty(\Omega) \subseteq P\left(W_0^s(\Omega)\right)$.*

Proof Let $u \in C^\infty(\overline{\Omega})$. Choose v according to Lemma 17.0.6. Then $u - v \in W_0^s$ and $Pu = P(u - v)$. Therefore,

$$P\left(W_0^s\right) \supseteq P\left(C^\infty(\overline{\Omega})\right) \supseteq P\left(WH^\infty(\Omega)\right) = WH^\infty(\Omega).$$

□

Henceforth, let Ω_1, Ω_2 be fixed $\mathcal{C}^\infty$ strongly pseudoconvex domains in $\mathbb{C}^n$, with K_1, K_2 their Bergman kernels and P_1, P_2 the corresponding Bergman projections. Let $\phi : \Omega_1 \to \Omega_2$ be a biholomorphic mapping, and let $u = \det \mathrm{Jac}_{\mathbb{C}} \phi$. For $j = 1, 2$, let $\delta_j(z) = \delta_{\Omega_j}(z) = \operatorname{dist}\left(z, {}^c\Omega_j\right)$.

Lemma 17.0.9 *For any $g \in L^2(\Omega_2)$, we have*

$$P_1(u \cdot (g \circ \phi)) = u \cdot ((P_2(g)) \circ \phi) .$$

Proof Notice that $u \cdot (g \circ \phi) \in L^2(\Omega_1)$ by change of variables. Therefore,

$$\begin{aligned} P_1(u \cdot (g \circ \phi))(z) &= \int_{\Omega_1} K_1(z, \zeta) u(\zeta) g(\phi(\zeta)) dV(\zeta) \\ &= \int_{\Omega_1} u(z) K_2(\phi(z), \phi(\zeta)) \overline{u(\zeta)} u(\zeta) g(\phi(\zeta)) dV(\zeta). \end{aligned}$$

Change of variable now yields

$$\begin{aligned} P_1(u \cdot (g \circ \phi))(z) &= u(z) \int_{\Omega_2} K_2(\phi(z), \xi) g(\xi) dV(\xi) \\ &= u(z) \cdot [(P_2(g)) \circ \phi](z). \end{aligned}$$

□

Lemma 17.0.10 *Let $\psi : \Omega_1 \to \Omega_2$ be a C^j diffeomorphism that satisfies*

$$\left| \frac{\partial^\alpha \psi}{\partial z^\alpha}(z) \right| \leq C \cdot (\delta_1(z))^{-|\alpha|} \tag{17.3}$$

for all multi-indices α with $|\alpha| \leq j \in \mathbb{N}$ and

$$\left| \nabla \psi^{-1}(w) \right| \leq C \left(\delta_2(w)\right)^{-1} . \tag{17.4}$$

Suppose also that

$$\delta_2(\psi(z)) \leq C \delta_1(z) . \tag{17.5}$$

Then there is a number $J = J(j)$ such that whenever $g \in W_0^{j+J}(\Omega_2)$, then $g \circ \psi \in W_0^J(\Omega_1)$.

Proof The subscript 0 causes no trouble by the definition of W_0^j. Therefore, it suffices to prove an estimate of the form

$$\|g \circ \psi\|_{W_0^J} \leq C \|g\|_{W_0^{J+J}}, \qquad \text{all } g \in C_c^\infty(\Omega)$$

By the chain rule and Leibniz's rule, if α is a multi-index of modulus not exceeding j, then

$$\left(\frac{\partial}{\partial z}\right)^{\alpha}(g\circ\psi)=\sum\left[(D^{\beta}g)\circ\psi\right]\cdot D^{\gamma_1}\psi\cdots D^{\gamma_\ell}\psi,$$

where $|\beta|\le|\alpha|$, $\sum|\gamma_i|\le|\alpha|$, and the number of terms in the sum depends only on α (a classical formula of Faà de Bruno—see S. Roman [440]—actually gives this sum quite explicitly, but we do not require such detail). Note here that $D^{\gamma_2}\psi$ is used to denote a derivative of some component of ψ. By hypothesis, it follows that

$$\left|\left(\frac{\partial}{\partial z}\right)^{\alpha}(g\circ\psi)\right|\le C\sum\left|(D^{\beta}g)\circ\psi\right|\cdot(\delta_1(z))^{-j}.$$

Therefore,

$$\begin{aligned}\int_{\Omega_1}\left|\left(\frac{\partial}{\partial z}\right)^{\alpha}(g\circ\psi)\right|^2 dV \le& C\sum\int_{\Omega_1}\left|(D^{\beta}g)\circ\psi\right|^2(\delta_1(z))^{-2j}\,dV(z)\\ =&C\sum\int_{\Omega_2}\left|D^{\beta}g(w)\right|^2\delta_1\left(\psi^{-1}(w)\right)^{-2j}\\ &\times\left|\det J_{\mathbb{C}}\psi^{-1}\right|^2 dV(w).\end{aligned}$$

But (17.4) and (17.5) imply that the last line is majorized by

$$C\sum\int_{\Omega_2}\left|D^{\beta}g(w)\right|^2\delta_2(w)^{-2j}\delta_2(w)^{-2n}dV(w). \tag{17.6}$$

Now if J is large enough, depending on the Sobolev imbedding theorem, then

$$\left|D^{\beta}g(w)\right|\le C\|g\|_{W_0^{j+J}}\cdot\delta_2(w)^{2n+2j}$$

(Remember that g is compactly supported in Ω_2.) Hence (17.6) is majorized by $C\|g\|_{W_0^{3+J}}$. □

Lemma 17.0.11 *10 For each $j\in\mathbb{N}$, there is an integer J so large that if $g\in W_0^{j+J}(\Omega_2)$, then $g\circ\phi\in W_0^{j}(\Omega_1)$. (Here ϕ is the biholomorphic mapping in 17.0.3.)*

Proof The Cauchy estimates give (since ϕ is bounded) that

$$\left|\frac{\partial^{\alpha}\phi_\ell}{\partial z^{\alpha}}(z)\right|\le C\cdot(\delta_1(z))^{-|\alpha|},\quad \ell=1,\ldots,n \tag{17.7}$$

and

$$\left|\nabla\phi^{-1}(w)\right| \le C\left(\delta_2(w)\right)^{-1}, \tag{17.8}$$

where $\phi = (\phi_1, \ldots, \phi_n)$. We will prove that

$$C \cdot \delta_1(z) \ge \delta_2(\phi(z)) \tag{17.9}$$

Then Lemma 17.0.10 gives the result.

To prove (17.9), let ρ be a smooth strictly plurisubharmonic defining function for Ω_1. Then $\rho \circ \phi^{-1}$ is a smooth plurisubharmonic function on Ω_2. Since ρ vanishes on $\partial\Omega_1$ and since ϕ^{-1} is proper, we conclude that $\rho \circ \phi^{-1}$ extends continuously to $\overline{\Omega}_2$. If $P \in \partial\Omega_2$ and ν_P is the unit outward normal to $\partial\Omega_2$ at P, then Hopf's lemma implies that the (lower) one-sided derivative $(\partial/\partial\nu_P)\left(\rho \circ \phi^{-1}\right)$ satisfies

$$\frac{\partial}{\partial\nu_P}\left(\rho \circ \phi^{-1}(P)\right) \ge C.$$

So, for $w = P - \epsilon\nu_P$, ϵ small, it holds that

$$-\rho \circ \phi^{-1}(w) \ge C \cdot \delta_2(w).$$

These estimates are uniform in $P \in \partial\Omega_2$. Using the comparability of $|\rho|$ and δ_1 yields

$$C\delta_1\left(\phi^{-1}(w)\right) \ge \delta_2(w).$$

Setting $z = \phi^{-1}(w)$ now gives

$$C'\delta_1(z) \ge \delta_2(\phi(z)),$$

which is (17.9). □

Exercise for the Reader
Let $\Omega \subset\subset \mathbb{C}^n$ be a smoothly bounded domain. Let $j \in \mathbb{N}$. There is an $N = N(j)$ so large that $g \in W_0^N$ implies that g vanishes to order j on $\partial\Omega$.

Lemma 17.0.12 *The function u is in $C^\infty\left(\overline{\Omega}_1\right)$.*

Proof It suffices to show that $u \in W^j(\Omega_1)$, every j. So fix j. Let $m = m(j)$ as in Condition R. According to (17.7), $|u(z)| \le C\delta_1(z)^{-2n}$. Then, by Lemma 17.0.11 and the exercise for the reader following it, there is a J so large that $g \in W_0^{m+J}(\Omega_2)$ implies $u \cdot (g \circ \phi) \in W^m(\Omega_1)$. Choose, by Lemma 17.0.8, a $g \in W_0^{m+J}(\Omega_2)$ such that $P_2 g \equiv 1$. Then Lemma 17.0.9 yields

$$P_1(u \cdot (g \circ \phi)) = u.$$

By Condition R, it follows that $u \in W^j(\Omega_1)$. □

Lemma 17.0.13 *The function u is bounded from 0 on* $\overline{\Omega}_1$.

Proof By symmetry, we may apply Lemma 17.0.12 to ϕ^{-1} and $\det J_{\mathbb{C}}\left(\phi^{-1}\right) = 1/u$. We conclude that $1/u \in C^\infty\left(\overline{\Omega}_2\right)$. Thus, u is nonvanishing on $\overline{\Omega}$. □

Proof of Fefferman's Theorem (Theorem 17.0.3) Use the notation of the proof of Lemma 17.0.12. Choose $g_1, \ldots, g_n \in W_0^{m+J}(\Omega_2)$ such that $P_2 g_i(w) = w_i$ (here w_i is the ith coordinate function). Then Lemma 17.0.9 yields that $u \cdot \phi_i \in W^j(\Omega_1)$, $i = 1, \ldots, n$. By Lemma 17.0.13, $\phi_i \in W^j(\Omega_1)$, $i = 1, \ldots, n$. By symmetry, $\phi^{-1} \in W^j(\Omega_2)$. Since j is arbitrary, the Sobolev imbedding theorem finishes the proof. □

It is important to understand the central role of Condition R in this proof. With some emendations, the proof we have presented shows that if $\Omega_1, \Omega_2 \subset\subset \mathbb{C}^n$ are smoothly bounded, pseudoconvex, and both satisfy Condition R, then a biholomorphic mapping from Ω_1 to Ω_2 extends smoothly to the boundary (in fact, S. Bell [54] has shown that it suffices for just one of the domains to satisfy Condition R). Condition R is known to hold on domains that have real analytic boundaries (see K. Diederich and J.E. Fornaess [134]), and more generally on domains of finite type (see [321, Section 11.5]). There are a number of interesting examples of nonpseudoconvex domains on which Condition R fails (see D. Barrett [43] and C. Kiselman [286]). F.M. Christ [113] has shown that Condition R fails on the Diederich-Fornaess worm domain.

L. Lempert [359] has derived a sharp boundary regularity result for biholomorphic mappings of strongly pseudoconvex domains with C^k boundary. The correct conclusion turns out to be that there is a loss of smoothness in some directions. So the sharp regularity result is formulated in terms of nonisotropic spaces. It is too technical to describe here.

Chapter 18
Solution of the $\overline{\partial}$ Problem

18.1 The Pompeiu Formula for Solutions of $\frac{\partial f}{\partial \bar{z}} = g$ in $\mathbb{C}$

A smooth function in a closed domain $\overline{\Omega} \subset \mathbb{C}$ can be expressed using the *Cauchy-Green-Pompéiu formula* (see, for example [178, Chapter 2, Section 2.1, 29]).

$$f(z) = \frac{1}{2\pi i} \int_{\partial\Omega} \frac{f(\zeta)}{\zeta - z} d\zeta + \frac{1}{2\pi i} \int_{\Omega} \frac{\partial f}{\partial \bar{\zeta}}(\zeta) \frac{1}{\zeta - z} d\bar{\zeta} \wedge d\zeta \,. \tag{18.1}$$

The first term on the right side is the formula which reproduces a holomorphic function in a domain in terms of its boundary values. The second term isolates the non-holomorphic part of f and yields a solution to the $\overline{\partial}$-equation $\frac{\partial f}{\partial \bar{z}} = g$.

The one-dimensional $\overline{\partial}$-problem is contained in the following: for a fixed function $g(z)$, find a function $v(z)$ such that $\frac{\partial v}{\partial \bar{z}} = g$.

One of the basic merits of the Cauchy-Green formula (18.1) is that it permits one to solve the one-dimensional $\overline{\partial}$-problem in explicit form:

$$f(z) = \frac{1}{2\pi i} \int_{\Omega} \frac{\partial f}{\partial \bar{\zeta}}(\zeta) \frac{1}{\zeta - z} d\bar{\zeta} \wedge d\zeta.$$

For functions of several variables, there does not exist such a simple and universal formula, and hence it is suitable to consider the *problem of integral formulas* for holomorphic functions and the *solvability of the $\overline{\partial}$-equation* separately.

P. V. Dovbush, S. G. Krantz, *The Geometric Theory of Complex Variables*,
https://doi.org/10.1007/978-3-031-77204-7_18

18.2 The $\overline{\partial}$-Problem in $\mathbb{C}^n$

The $\overline{\partial}$-problem is concerned with the solution of the inhomogeneous Cauchy-Riemann equation: given a (0, 1)-form

$$f = \sum_{j=1}^{n} f_j d\overline{z}_j \tag{18.2}$$

the problem is to find a function u such that

$$\overline{\partial} u = f\,. \tag{18.3}$$

Since $\overline{\partial}^2 = 0$, a necessary condition for solvability is

$$\overline{\partial} f = 0. \tag{18.4}$$

If f is given by (18.2), then

$$\overline{\partial} f = \sum_{k<j} \left(\overline{\partial}_k f_j - \overline{\partial}_j f_k\right) d\overline{z}_k \wedge d\overline{z}_j, \tag{18.5}$$

since $d\overline{z}_j \wedge d\overline{z}_k = -d\overline{z}_k \wedge d\overline{z}_j$. Thus, (18.4) is a brief way of saying that the components of f have to satisfy the system of differential equations

$$\overline{\partial}_k f_j = \overline{\partial}_j f_k \quad (j, k = 1, \ldots, n). \tag{18.6}$$

Of course, one is not only interested in the mere existence of solutions, but also one often wants to find solutions that satisfy good growth conditions or that have good smoothness properties.

18.2.1 The Leray (Cauchy-Fantappiè) Integral Representation

We start by noting that the Cauchy-Fantappiè formula, which was obtained by Leray [360, 361], can be derived from the Bochner-Martinelli formula [331, Chapter 1, Theorem 1.1.4, p. 22].

Let Ω be a bounded domain with a piecewise-smooth boundary, and suppose that for a fixed point $z \in \Omega$, there is defined on $\partial\Omega$ a continuously differentiable vector-valued function

$$\eta(\zeta) = (\eta_1(\zeta), \ldots, \eta_n(\zeta))$$

such that

$$\sum_{k=1}^{n} (\zeta_k - z_k)\, \eta_k(\zeta) = 1, \quad \zeta \in \partial\Omega.$$

Definition 18.2.1 The space $\mathcal{A}(\Omega)$ is the class of all continuous complex functions on the closure $\overline{\Omega}$ of Ω whose restriction to Ω is holomorphic there.

Theorem 18.2.2 (Leray) *Every function $f \in \mathcal{A}(\Omega)$ satisfies the equation*

$$f(z) = \frac{(n-1)!}{(2\pi i)^n} \int_{\partial D} f(\zeta)\omega'(\eta) \wedge d\zeta, \quad z \in \Omega \tag{18.7}$$

where

$$\omega'(\eta) = \sum_{k=1}^{n} (-1)^{k-1} \eta_k d\eta[k] \tag{18.8}$$

Proof The proof given here belongs to G.M. Khenkin. In the space $\mathbb{C}^{2n}$ of the variables

$$(\eta, \zeta) = (\eta_1, \ldots, \eta_n, \zeta_1, \ldots, \zeta_n)$$

consider the analytic hypersurface

$$L_z = \left\{ (\eta, \zeta) \in \mathbb{C}^{2n} : \sum_{k=1}^{n} (\zeta_k - z_k)\, \eta_k(\zeta) = 1 \right\}$$

on which the form $\omega'(\eta) \wedge d\zeta$ is closed. The two cycles

$$\Gamma_1 = \left\{ (\eta, \zeta) : \zeta \in \partial\Omega, \eta_j = \left(\overline{\zeta}_j - \overline{z}_j\right) |\zeta - z|^{-2}, j = 1, \ldots, n \right\}$$

and

$$\Gamma_2 = \left\{ (\eta, \zeta) : \zeta \in \partial\Omega, \eta_j = \eta_j(\zeta), j = 1, \ldots, n \right\}$$

in L_z are homotopic in L_z, the homotopy being given by the formula

$$\widetilde{\eta}_j = t \frac{\overline{\zeta}_j - \overline{z}_j}{|\zeta - z|^2} + (1-t)\eta_j(\zeta), \quad 0 \le t \le 1$$

That is, these are homologous cycles. Consequently,

$$\int_{\Gamma_1} f(\zeta)\omega'(\eta) \wedge d\zeta = \int_{\Gamma_2} f(\zeta)\omega'(\eta) \wedge d\zeta$$

when f is a holomorphic function. But

$$\omega'\left(\frac{\overline{\zeta}_1 - \overline{z}_1}{|\zeta - z|^2}, \ldots, \frac{\overline{\zeta}_n - \overline{z}_n}{|\zeta - z|^2}\right) = \frac{(2\pi i)^n}{(n-1)!} U(\zeta, z)$$

Hence (18.7) follows. □

The differential form (18.8) is called the *Leray form*. The Cauchy-Fantappié representation has turned out to be very useful, and it has many applications in multidimensional complex analysis.

18.3 The Khenkin-Ramirez Integral Representation

The bounded domain Ω is called a strongly pseudoconvex domain if there exists a neighborhood $U \supset \overline{\Omega}$ and a real-valued function $\rho \in \mathcal{C}^2(U)$ such that

$$\Omega = \{z \in U : \rho(z) < 0\},$$

where $\operatorname{grad} \rho \neq 0$ on $\partial\Omega$ and the function ρ is a strongly plurisubharmonic function in U, i.e., the inequality

$$\sum_{j,k=1}^{n} \frac{\partial^2 \rho(z)}{\partial z_j \partial \overline{z}_k} w_j \overline{w}_k > 0$$

holds for $z \in U$ and all $w \in \mathbb{C}^n$, $w \neq 0$.

Strongly pseudoconvex domains play an important role in multidimensional complex analysis. Any domain of holomorphy can be approximated from inside by these domains [24, Sect. 25].

Here are some auxiliary notations and statements we will need in the future. We denote

$$\Omega_\delta = \{z \in U : \rho(z) < \delta\}, \quad V_\delta = \{z \in U : |\rho(z)| < \delta\},$$
$$U_{\varepsilon,\delta} = \{(\zeta, z) : \zeta \in V_\delta, z \in \Omega_\delta, |\zeta - z| < \varepsilon\},$$

where $\varepsilon > 0$, $\delta > 0$. Let $C^1(V_\delta, \mathcal{H})$ be the space of functions of class C^1 in V_δ with values in space $\mathcal{H}$.

Lemma 18.3.1 ([24, Lemma 10.1]) *For every strongly pseudoconvex domain Ω, there exist positive constants ε, δ and functions $F \in C^1\left(U_{\varepsilon,\delta}\right), G \in C^1\left(U_{\varepsilon,\delta}\right), \Phi \in C^1\left(V_\delta, \mathcal{O}\left(\Omega_\delta\right)\right)$ such that:*

1. *$\Phi = FG$ on $U_{\varepsilon,\delta}$; $F(z,z) = 0$; $|G| > \delta$ on $U_{\varepsilon,\delta}$; $|\Phi| > \delta$ outside $U_{\varepsilon,\delta}$.*
2. *We have the inequality*

$$2\operatorname{Re} F(\zeta, z) \geq \rho(\zeta) - \rho(z) + \gamma|\zeta - z|^2, \quad \gamma > 0,$$

on $U_{\varepsilon,\delta}$.

3. *$\left|d_\zeta F(\zeta, z)\right|_{\zeta=z} = \left|d_z F(\zeta, z)\right|_{\zeta=z} = \partial\rho$.*

Lemma 18.3.2 ([24]) *Let $D = \{z \in U : \rho(z) < 0\}$ be a strongly pseudoconvex domain. For every point $\widetilde{\zeta} \in \partial D$, there is a biholomorphic map of the neighborhood $\widetilde{U_{\widetilde{\zeta}}}$ of the point $\widetilde{\zeta}$ to a neighborhood W of zero in space $\mathbb{C}^n_w$, such that the mapping domain $D \cap \widetilde{U_{\widetilde{\zeta}}}$ is biholomorphically equivalent to a convex domain in $\mathbb{C}^n_w$, while the inverse mapping turns the strongly plurisubharmonic function ρ into a strongly convex function.*

Proof Parallel translation can be arranged so that the point $\widetilde{\zeta}$ is zero; then in some neighborhood of zero, the Taylor expansion

$$\rho(z) = 2\operatorname{Re}\sum_{j=1}^n \frac{\partial\rho(0)}{\partial z_j} z_j + \frac{1}{2}\sum_{j,k=1}^n \frac{\partial^2\rho(0)}{\partial z_j \partial \overline{z}_k} z_j \overline{z}_k + \operatorname{Re}\sum_{j,k=1}^n \frac{\partial^2\rho(0)}{\partial z_j \partial z_k} z_j z_k + o\left(|z|^2\right)$$

is valid as $z \to 0$. By the implicit function theorem, the local coordinates in a smaller neighborhood of zero can be assumed to be given by the function

$$w_1(z) = 2\sum_{j=1}^n \frac{\partial\rho(0)}{\partial z_j} z_j + \sum_{j,k=1}^n \frac{\partial^2\rho(0)}{\partial z_j \partial z_k} z_j z_k$$

and any linear functions $w_2, \ldots, w_n$ being the coordinates of the complex hyperplane

$$\left\{z \in \mathbb{C}^n : \frac{\partial\rho(0)}{\partial z_1} z_1 + \cdots + \frac{\partial\rho(0)}{\partial z_n} z_n = 0\right\}.$$

Strong pseudoconvexity ρ at 0 now means that

$$\sum_{j,k=1}^n a_{jk}\xi_j\overline{\xi}_k \geq \gamma|\xi|^2$$

where $\gamma > 0$ and $a_{jk} = \frac{\partial^2\rho(0)}{\partial w_j \partial \overline{w}_k}$. In the new local coordinates of w, the function ρ is of the form

$$\rho(w) = \operatorname{Re} w_1 + \sum_{j,k=1}^{n} a_{jk} w_j \overline{w}_k + o\left(|w|^2\right) \tag{18.9}$$

Function (18.9), as is easily seen, is strongly convex in the coordinates of w. If $\varepsilon > 0$ is sufficiently small, then $W \cap D$ is a convex domain in the coordinates of w, where

$$W = \left\{w \in \mathbb{C}^n : |w| < \varepsilon\right\}$$

□

Remark 18.3.3 If we first perform a unitary transformation and make a shift so that the plane $\operatorname{Re} z_1 = 0$ is now a tangent plane at the point 0, and then repeat the whole procedure described in the previous lemma, we will find that locally the domain D is given by the function

$$\rho(w) = \operatorname{Re} w_1 + \sum_{j,k=2}^{n} a_{jk} w_j \overline{w}_k + o\left(|w|^2\right) . \tag{18.10}$$

Applying a unitary transformation and a stretching to the last equation, we can get a function ρ that will have the form

$$\rho(w) = \operatorname{Re} w_1 + \sum_{k=2}^{n} |w_k|^2 + o\left(|w|^2\right) .$$

Lemma 18.3.4 ([24]) *Under the conditions of Lemma 18.3.1,*

$$\Phi(\zeta, z) = \sum_{k=1}^{n} P_k(\zeta, z)\,(\zeta_k - z_k) \ , \tag{18.11}$$

where $P_k(\zeta, z) \in \mathcal{C}^1\left(V_\delta, \mathcal{O}\left(D_\delta\right)\right)$, $k = 1, \ldots, n$, $\zeta \in V_\delta$, $z \in D_\delta$.

Proof We denote

$$C(\zeta, z, w) = \Phi(\zeta, z) - \Phi(\zeta, w)$$

then

$$C(\zeta, z, w) \in C^1\left(V_\delta, \mathcal{O}\left(D_\delta \times D_\delta\right)\right)$$

For a fixed $\zeta \in V_\delta$, the function C belongs to the ideal J of holomorphic functions equal to zero on the set

$$\{(z, w) : z \in D_\delta,\ w \in D_\delta,\ z = w\}$$

By the Hefer theorem (see, for example, [24, Sect. 2.5]), the ideal J has generators

$$w_1 - z_1, \ldots, w_n - z_n$$

and decomposition of $C(\zeta, z, w)$ for these generators can be made continuously differentiable to the parameter ζ (see [24, Sect. 25.2']), i.e., there are functions

$$Q_k(\zeta, z, w) \in \mathcal{C}^1\left(V_\delta, \mathcal{O}\left(D_\delta \times D_\delta\right)\right), \quad k = 1, \ldots, n$$

such that

$$C(\zeta, z, w) = \sum_{k=1}^{n} Q_k(\zeta, z, w)\,(w_k - z_k)$$

Note that $F(\zeta, \zeta) = 0$; hence $\Phi(\zeta, \zeta) = 0$, therefore

$$\Phi(\zeta, z) = C(\zeta, z, \zeta) = \sum_{k=1}^{n} P_k(\zeta, z)\,(\zeta_k - z_k)$$

where $P_k(\zeta, z) = Q_k(\zeta, z, \zeta), k = 1, \ldots, n$. So formula (18.11) holds. Consider the Leray form (18.8)

$$\omega'(\eta) = \sum_{j=1}^{n}(-1)^{j-1}\eta_j d\eta[j]$$

for a given smooth function $\eta = \eta(\zeta, z, \lambda)$ with values in $\mathbb{C}^n$, where

$$(\zeta, z, \lambda) \in \mathbb{C}^n \times \mathbb{C}^n \times \mathbb{R}$$

Then we can write

$$\left(\omega'(\eta)\right) \wedge d\zeta \wedge dz = \left(\sum_{q=0}^{n} \omega'_q(\eta)\right) \wedge d\zeta \wedge dz$$

where $\omega'_q(\eta)$ is the differential form of type $(n - q - 1)$ by $d\overline{\zeta}$ and $d\lambda$ and type q by $d\overline{z}$. In particular, the form ω'_0 is a form of type 0 by $d\overline{z}$ and type $(n - 1)$ by $d\overline{\zeta}$ and $d\lambda$. In what follows, we will assume that $\omega'_{-1} = 0$. We note the obvious identity satisfied by the form ω'_q, [276, Sect. 8.2]:

$$d_\lambda \omega'_q + \overline{\partial}_\zeta \omega'_q + \overline{\partial}_z \omega'_{q-1} = 0. \tag{18.12}$$

□

From Theorem 18.2.2 and Lemma 18.3.4, we get the Khenkin-Ramirez integral formula (see [276, Sect. 4.2]).

Theorem 18.3.5 (Khenkin-Ramirez) *For any function $f \in A(D)$, the following integral representation is true*

$$f(z) = \frac{(n-1)!}{(2\pi i)^n} \int_{\partial D} f(\zeta) \frac{\omega_0'(P(\zeta, z)) \wedge d\zeta}{[\Phi(\zeta, z)]^n}, \quad z \in D \tag{18.13}$$

where the vector function $P(\zeta, z) = (P_1(\zeta, z), \ldots, P_n(\zeta, z))$.

Formula (18.13) is one of the most successful realizations of the general Cauchy-Fantappiè formula in multidimensional complex analysis (see [276, Sect. 4.2], [451]).

18.4 Formula for the Solution of the $\overline{\partial}$-Problem

Let us consider in the domain Ω the system of differential equations (18.4) where $\{f_k\}$ are infinitely differentiable functions on $\overline{\Omega}$ satisfying in Ω the compatibility conditions (18.6).

Theorem 18.4.1 *The function $v(z)$ defined in Ω by the formula*

$$v(z) = \frac{(n-1)!}{(2\pi i)^n} \left\{ \int_{\partial\Omega \times [0,1]} \langle f \cdot d\overline{\zeta} \rangle \wedge \omega'(\eta) \wedge \omega(\zeta) - \int_{\Omega} \frac{\langle f \cdot (\overline{\zeta} - \overline{z}) \rangle}{|\zeta - z|^2} \omega(\overline{\zeta}) \wedge \omega(\zeta) \right\}, \tag{18.14}$$

where

$$\eta_k = \lambda \frac{\overline{\zeta}_k - \overline{z}_k}{|\zeta - z|^2} + (1 - \lambda) \frac{p_k(z, \zeta)}{\Phi(z, \zeta)} \quad (k = 1, 2, \ldots, n),$$

$$\zeta \in \partial\Omega, \lambda \in [0, 1],$$

is an infinitely differentiable solution of the $\overline{\partial}$-problem (18.4).

Proof (Exercise: check this.) □

Chapter 19
Harmonic Measure

Prologue It is unfortunate that most graduate complex analysis courses do not treat harmonic measure. Indeed, most complex analysis texts do not cover the topic.

The term "*harmonic measure*" was introduced by R. Nevanlinna in the 1920s. But the ideas were anticipated in the work of Riesz brothers and others. It is still studied intensively today.

Harmonic measure is a natural outgrowth of the study of the Dirichlet problem. It is a decisive tool for measuring the growth of functions, and to seeing how they bend. It has been used in diverse applications such as the corona problem and in mapping problems. Harmonic measure has particularly interesting applications in probability theory—especially in the consideration of diffustions and Brownian motion.

It requires a little sophistication to appreciate harmonic measure. And calculating the measure—working through the examples—requires some work. But the effort is worth it, and the results are substantial.

This chapter gives a basic introduction to the idea of harmonic measure.

19.1 The Idea of Harmonic Measure

Capsule Harmonic measure is a way of manipulating the data that comes from the Dirichlet problem. It is a way of measuring the geometry of a domain and can be used to control the growth of harmonic functions.

Let $\Omega \subseteq \mathbb{C}$ be a domain with boundary consisting of finitely many Jordan curves (we call such a domain a *Jordan domain*). Let E be a finite union of arcs in $\partial\Omega$ (we allow the possibility of an entire connected boundary component—a simple, closed curve—to be one of these arcs). Then the harmonic measure of E at the point $z \in \Omega$ with respect to Ω is the value at z of the bounded harmonic function ω on Ω with boundary limit 1 at points of E and boundary limit 0 at points of $\partial\Omega \setminus E$ (except

P. V. Dovbush, S. G. Krantz, *The Geometric Theory of Complex Variables*,
https://doi.org/10.1007/978-3-031-77204-7_19

possibly at the endpoints of the arcs that make up E). We denote the harmonic measure by $\omega(z, \Omega, E)$.

The first question to ask about harmonic measure is that of existence and uniqueness: does harmonic measure always exist? If it does exist, is it unique? The answer to both these queries is "yes." Let us see why.

Fix a Jordan domain Ω. It is a classical result (see [15]) that there is a conformal mapping $\Phi : \Omega \to U$ of Ω to a domain U bounded by finitely many circles. If $F \subseteq \partial U$ consists of finitely many circular arcs, then it is obvious that the harmonic measure of F exists. It would just be the Poisson integral of the characteristic function of F, and that Poisson integral can be shown to exist by elementary conformal mapping arguments (i.e., mapping to the disc). Now a classical result of Carathéodory guarantees that Φ and its inverse extend continuously to the respective boundaries, so that the extended function is a homeomorphism of the closures. Thus, we may pull the harmonic measure back from U to Ω via Φ.

As for uniqueness, if Ω is bounded, then uniqueness follows from the standard maximum principle. For the general case, we need an extended maximum principle that is due to Lindelöf:

Proposition 19.1.1 *Let Ω be a domain whose boundary is not a finite set. Let u be a real-valued harmonic function on Ω, and assume that there is a real constant $M > 0$ such that*

$$u(z) \leq M, \qquad \textit{for } z \in \Omega.$$

Suppose that m is a real constant and that

$$\limsup_{z \to \zeta} u(z) \leq m \tag{19.1}$$

for all except possibly finitely many points $\zeta \in \partial\Omega$. Then $u(z) \leq m$ for all $z \in \Omega$.

Remark 19.1.2 Observe that any Jordan domain certainly satisfies the hypotheses of the proposition. Furthermore, by classical results coming from potential theory (see [15]), any domain on which the Dirichlet problem is solvable will satisfy the hypotheses of the theorem. Note, in particular, that Ω certainly need not be bounded.

We conclude by noting that this last result is a variant of the famous Phragmen-Lindelöf theorem (see [447]).

Proof of Proposition 19.1.1 First assume that Ω is bounded. We will remove this extra hypothesis later. Let the diameter of Ω be d. Let the exceptional boundary points (at which (19.1) does not hold) be called $\zeta_1, \ldots, \zeta_k$. Let $\epsilon > 0$. We then may apply the ordinary textbook maximum principle (see [202]) to the auxiliary function

$$h(z) = u(z) + \epsilon \sum_{j=1}^{k} \log \frac{|z - \zeta_j|}{d}.$$

Notice that h (instead of u) satisfies the hypotheses of the proposition at every boundary point. So $h \leq m$ on all of Ω. Then we let $\epsilon \to 0$ and the result follows. Now consider the case of Ω unbounded. If Ω has an exterior point (i.e., a point in the interior of the complement of the closure of Ω), then we may apply an inversion and reduce to the case in the preceding paragraph. Finally, if Ω has no exterior point, then let $R > |\zeta_j|$ for all $j = 1, \ldots, k$. Let

$$\Omega_1 = \{z \in \Omega : |z| < R\}$$

and

$$\Omega_2 = \{z \in \Omega : |z| > R\}.$$

These may not be domains (i.e., they could be disconnected), but they are certainly open sets. Let

$$S = \{z \in \Omega : |z| = R\}.$$

If $u \leq m$ on S, then we apply the result of the first paragraph on Ω_1 and the result of the second paragraph on Ω_2. If it is not the case that $u \leq m$ on S, then u will have a maximum N on S with $N > m$. But this will be a maximum for u on all of Ω. Since $u \leq m$ at the ends of the arcs of S, it follows that u actually achieves the maximum N on S. Therefore, by the usual maximum principle, u is identically equal to the constant N. But then the boundary condition (19.1) cannot be obtained. That is a contradiction. □

It is important to realize that if Ω has reasonably nice boundary, then $\omega(z, \Omega, E)$ is nothing other than the Poisson integral of the characteristic function of E. (We used this fact in our discussion of the existence of harmonic measure.) In any event, by the maximum principle, the function ω takes real values between 0 and 1. Therefore, $0 < \omega(z, \Omega, E) < 1$.

One of the reasons that harmonic measure is important is that it is a conformal invariant. Essential to this fact is Carathéodory's theorem that a conformal map of Jordan domains will extend continuously to the closures. We now formulate the invariance idea precisely.

Proposition 19.1.3 *Let Ω_1, Ω_2 be domains with boundaries consisting of Jordan curves in $\mathbb{C}$ and $\varphi : \Omega_1 \to \Omega_2$ a conformal map. If $E_1 \subseteq \partial\Omega_1$ is an arc and $z \in \Omega_1$, then*

$$\omega(z, \Omega_1, E_1) = \omega(\varphi(z), \Omega_2, \varphi(E_1)).$$

Here $\varphi(E_1)$ is well defined by Carathéodory's theorem.

Proof Let h denote the harmonic function $\omega(\varphi(z), \Omega_2, \varphi(E_1))$. Then $h \circ \varphi$ equals harmonic measure for Ω_1 (by Carathéodory's theorem). □

Remark 19.1.4 It is worth recording here an important result of F. and M. Riesz (see [314, p. 72]). Let $\Omega \subseteq \mathbb{C}$ be a Jordan domain bounded by a rectifiable boundary curve. Let $\varphi : \mathbb{D} \to \Omega$ be a conformal mapping. If $E \subseteq \partial\mathbb{D}$ has Lebesgue linear measure zero, then $\varphi(E)$ also has Lebesgue linear measure zero.

19.1.1 Some Examples

Capsule An idea like harmonic measure is a bit hollow without some concrete examples. In this section, we produce several very concrete instances.

Harmonic measure helps us understand the growth and value distribution of harmonic and holomorphic functions on Ω. It has become a powerful analytic tool. We begin to understand harmonic measure by first calculating some examples.

Example 19.1.5 Let $\mathcal{U}$ be the upper-half plane. Let E be an interval $[-T, T]$ on the real axis, centered at 0. Let us calculate $\omega(z, \mathcal{U}, E)$ for $z = x + iy \cong (x, y) \in \Omega$. It is a standard fact (see [202]) that the Poisson kernel for $\mathcal{U}$ is

$$P(x, y) = \frac{1}{\pi} \cdot \frac{y}{x^2 + y^2}.$$

Here we use the traditional real notation for the kernel. In other words, the harmonic function Ω that we seek is given by

$$\begin{aligned}\omega(x + iy) &= \frac{1}{\pi} \int_{-\infty}^{\infty} \chi_{[-T,T]}(t) \cdot \frac{y}{(x - t)^2 + y^2} dt \\ &= \frac{1}{\pi} \int_{-T}^{T} \frac{y}{(x - t)^2 + y^2} dt.\end{aligned}$$

Now it is a simple matter to rewrite the integral as

$$\begin{aligned}\omega(x + iy, \Omega, E) &= \frac{1}{\pi} \frac{1}{y} \int_{-T}^{T} \frac{1}{\left(\frac{x-t}{y}\right)^2 + 1} dt \\ &= -\frac{1}{\pi} \operatorname{Tan}^{-1} \left(\frac{x - t}{y}\right)\Big|_{-T}^{T} \\ &= \frac{1}{\pi} \operatorname{Tan}^{-1} \left(\frac{x + T}{y}\right) - \frac{1}{\pi} \operatorname{Tan}^{-1} \left(\frac{x - T}{y}\right).\end{aligned}$$

The reader may check that this function is harmonic, tends to 1 as (x, y) approaches $E \subseteq \partial\mathcal{U}$, and tends to 0 as (x, y) approaches $\partial\mathcal{U} \setminus E$. Glancing at Fig. 19.1, we see that the value of $\omega(z, \Omega, E)$ is simply α/π, where α is the angle subtended at the point (x, y) by the interval E.

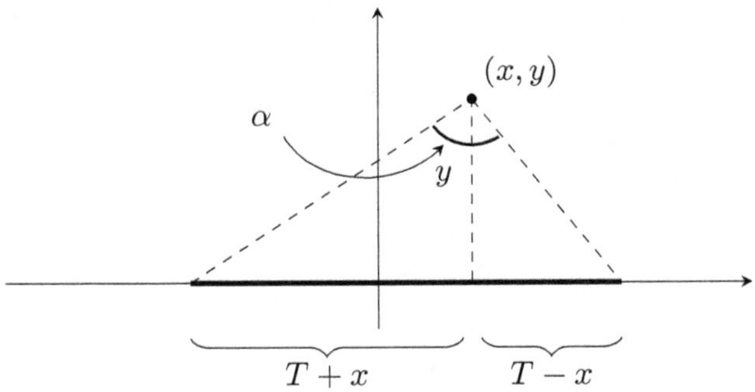

Fig. 19.1 The value of $\omega(z, \Omega, E)$

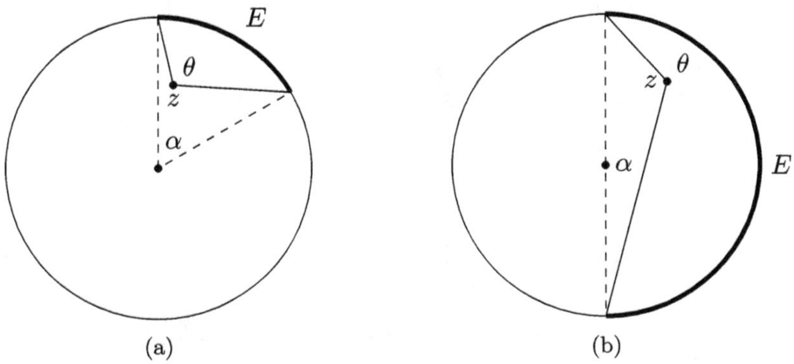

Fig. 19.2 (**a**) The angle subtended by E at z. (**b**) The case E is the arc from $-i$ to i

More generally, if E is any bounded, closed interval in the real line, then the harmonic measure $\omega(z, \mathcal{U}, E)$ will be $1/\pi$ times the angle subtended at the point (x, y) by the interval E. But we can say more. If now E is the finite disjoint union of closed, bounded intervals,

$$E = I_1 \cup \cdots \cup I_k,$$

then the harmonic measure of E is just the sum of the harmonic measures $\omega\left(z, \Omega, I_j\right)$ for each of the individual intervals. So the harmonic measure at z is just $1/\pi$ times the sum of the angles subtended at (x, y) by each of the intervals I_j.

Example 19.1.6 Let $\mathbb{D}$ be the unit disc. Let E be an arc of the circle with central angle α. Then, for $z \in \mathbb{D}$,

$$\omega(z, \mathbb{D}, E) = \frac{2\theta - \alpha}{2\pi}$$

where θ is the angle subtended by E at z. See Fig. 19.2a.

It turns out to be convenient to first treat the case where E is the arc from $-i$ to i. See Fig. 19.2b.

We may simply calculate the Poisson integral of $\chi(-\pi/2, \pi/2)$, where the argument of this characteristic function is the angle in radian measure on the circle. Thus, for $z = re^{i\lambda}$,

$$\begin{aligned}\omega(z, \mathbb{D}, E) &= \frac{1}{2\pi} \int_{-\pi/2}^{\pi/2} \frac{1-r^2}{1-2r\cos(\lambda - t) + r^2} dt \\ &= \frac{1-r^2}{2\pi} \int_{-\pi/2}^{\pi/2} \frac{dt}{(1+r^2) - 2r\cos(\lambda - t)} .\end{aligned}$$

At this point, it is useful to recall from calculus (and this may be calculated using the Weierstrass $w = \tan x/2$ substitution) that

$$\int \frac{dx}{a + b\cos x} = \frac{2}{\sqrt{a^2 - b^2}} \operatorname{Tan}^{-1}\left(\sqrt{\frac{a-b}{a+b}} \tan \frac{x}{2}\right) .$$

Therefore

$$\begin{aligned}\omega(z, \mathbb{D}, E) = \frac{1}{\pi} \Bigg[&\operatorname{Tan}^{-1}\left(\frac{1+r}{1-r} \tan \frac{\lambda + \frac{\pi}{2}}{2}\right) \\ &- \operatorname{Tan}^{-1}\left(\frac{1+r}{1-r} \tan \frac{\lambda - \frac{\pi}{2}}{2}\right)\Bigg] .\end{aligned}$$

Some easy but tedious manipulations now allow us to rewrite this last expression as

$$\begin{aligned}\omega(z, \mathbb{D}, E) = \frac{1}{\pi} \Bigg[&\operatorname{Tan}^{-1}\left(\left(\frac{1+r}{1-r}\right) \cdot \left(\frac{\sin\lambda + 1}{\cos\lambda}\right)\right) \\ &- \operatorname{Tan}^{-1}\left(\left(\frac{1+r}{1-r}\right) \cdot \left(\frac{\sin\lambda - 1}{\cos\lambda}\right)\right)\Bigg] .\end{aligned} \tag{19.2}$$

Recall, however, that

$$\tan(\alpha - \beta) = \frac{\tan\alpha - \tan\beta}{1 + \tan\alpha \tan\beta}$$

hence

$$\operatorname{Tan}^{-1} a - \operatorname{Tan}^{-1} b = \operatorname{Tan}^{-1}\left(\frac{a-b}{1+ab}\right) . \tag{19.3}$$

Applying this simple idea to (19.2) yields

$$\omega(z, \mathbb{D}, E) = \frac{1}{\pi}\,\mathrm{Tan}^{-1}\left(\frac{\left(\frac{1+r}{1-r}\right)\left(\frac{\sin\lambda+1}{\cos\lambda}\right) - \left(\frac{1+r}{1-r}\right)\left(\frac{\sin\lambda-1}{\cos\lambda}\right)}{1+\left(\frac{1+r}{1-r}\right)\left(\frac{\sin\lambda+1}{\cos\lambda}\right)\cdot\left(\frac{1+r}{1-r}\right)\left(\frac{\sin\lambda-1}{\cos\lambda}\right)}\right).$$

Elementary simplifications finally lead to

$$\omega(z, \mathbb{D}, E) = \frac{1}{\pi}\,\mathrm{Tan}^{-1}\left(\frac{1-r^2}{2r\cos\lambda}\right).$$

Now it is helpful to further rewrite the right-hand side of this last expression as

$$\frac{2\left[\mathrm{Tan}^{-1}\left(\frac{1-r\sin\lambda}{r\cos\lambda}\right) + \mathrm{Tan}^{-1}\left(\frac{1+r\sin\lambda}{r\cos\lambda}\right)\right] - \pi}{2\pi}$$

(remembering, of course, that $\pi/2 - \mathrm{Tan}^{-1}(\gamma) = \mathrm{Tan}^{-1}(1/\gamma)$). But the preceding expression is just the formula

$$\frac{2\theta - \alpha}{2\pi},$$

for the special angle $\alpha = \pi$, which was enunciated at the start of the example. For the case of general E, we may first suppose that E is an arc centered at the point $1 \in \partial\mathbb{D}$. Second, we may reduce the general case to the one just calculated by using a Möbius transformation.

Example 19.1.7 Suppose that U is an annulus with radii $0 < r_1 < r_2 < \infty$. Then one may verify by inspection that if E is the outer boundary circle of the annulus, then

$$\omega(z, U, E) = \frac{\log(|z|/r_1)}{\log(r_2/r_1)}.$$

In general, it is a tricky business to calculate exactly the harmonic measure for a given domain U and a given E. But one may often obtain useful estimates. The ensuing discussion will bear out this point.

19.1.2 Hadamard's Three-Circles Theorem

Capsule One of the big ideas in modern harmonic analysis is interpolation of operators. If a linear operator T maps a Banach space X to itself and another Banach space Y to itself, then is there automatically a range of "intermediate" spaces on which T is bounded? (This question goes back to M. Riesz's proof of the L^p boundedness of the Hilbert transform in 1926.)

The first genuine interpolation theorem (which treats the concept of intermediate space) is the Riesz-Thorin theorem. Its proof uses Hadamard's three-circles theorem.

Hadamard's three-circles theorem (sometimes called the "*three-lines theorem*") is an important sharpening of the classical maximum principle. It has proved useful in various parts of analysis, notably in proving the Riesz-Thorin interpolation theorem for linear operators. Here we shall give a thorough treatment of the three circles theorem from the point of view of harmonic measure. Afterward we shall discuss the result of Riesz and Thorin.

We begin by treating some general comparison principles regarding harmonic measure. It will facilitate our discussion to first introduce a slightly more general concept of harmonic measure.

Let $U \subseteq \mathbb{C}$ be, as usual, a domain. Let A be a closed set in the extended plane $\widehat{\mathbb{C}}$. Let E denote that part of $\partial(U \setminus A)$ that lies in A. [In what follows, we will speak of pairs (U, A).] Then $\omega(z, U \setminus A, E)$ will be called the harmonic measure of A with respect to U, assuming that the geometry is simple enough that we can compute this number component by component (of $U \setminus A$). Generally speaking, in the present chapter, we assume that all boundaries that arise consist of finitely many Jordan curves. This standing hypothesis is made so that we can readily apply Lindelöf's Proposition 19.1.1. Often this standing hypothesis will go unspoken.

The first new comparison tool is called the majorization principle.

Prelude We consider here a transformation property for harmonic measure.

Theorem 19.1.8 *Consider two pairs (U, A) and $(\widetilde{U}, \widetilde{A})$. Let*

$$f : U \setminus A \to \widetilde{U}$$

be holomorphic. Assume furthermore that if $U \ni z \to A$, then $\widetilde{U} \ni f(z) \to \widetilde{A}$. Then

$$\omega(z, U, A) \leq \omega(f(z), \widetilde{U}, \widetilde{A})$$

for $z \in f^{-1}(\widetilde{U} \setminus \widetilde{A})$.

Remark 19.1.9 Of course, we consider only holomorphic f because we want a mapping that preserves harmonic functions under composition.

Proof of Theorem 19.1.8 Let us abbreviate

$$\omega = \omega(z, U, A) \quad \text{and} \quad \widetilde{\omega} = (f(z), \widetilde{U}, \widetilde{A}).$$

Now we apply the maximum principle to $\omega - \widetilde{\omega}$ on a connected component V of $f^{-1}(\widetilde{U} \setminus \widetilde{A})$. As z approaches the boundary of V, either z tends to a boundary point of U that is not on A or else $f(z)$ tends to $\widetilde{A}$. In either of these circumstances,

$\limsup_{z \to \partial U}(\omega - \widetilde{\omega}) \leq 0$ except when $f(z)$ tends to an endpoint of the boundary arcs of $\widetilde{U} \setminus \widetilde{A}$ that lie on $\widetilde{A}$. Since there are only finitely many such points, Lindelöf's Proposition 19.1.1 tells us that the maximum principle still remains valid. We therefore conclude that $\omega \leq \widetilde{\omega}$ on all of $f^{-1}(\widetilde{U} \mid \widetilde{A})$. □

Corollary 19.1.10 *The function $\omega(z, U, A)$ increases if either U increases or A increases. That is to say, if $U \subseteq U^*$ and $A \subseteq A^*$, then*

$$\omega(z, U, A) \leq \omega\left(z, U^*, A\right) \quad \text{and} \quad \omega(z, U, A) \leq \omega\left(z, U, A^*\right).$$

Proof We prove the first statement and leave the second for the reader.

We apply the theorem with $f : U \setminus A \to U^*$ being the identity mapping. The result is now immediate. □

Now let U^* be an open disc of radius $R > 0$, and let A^* be a smaller closed disc of radius $0 < r < R$. One may see by inspection that

$$\omega\left(f(z), U^*, A^*\right) = \frac{\log(R/|f(z)|)}{\log R/r}.$$

We do not yet say what f is, and there is considerable latitude in practice. Nonetheless, we see immediately that the function ω is identically equal to 1 when $f(z) \in \partial A^*$ and identically equal to 0 when $f(z) \in \partial U^*$. Certainly it is harmonic.

Now we have our first result where the quantitative properties of the harmonic measure play a decisive role.

Prelude This next result is reminiscent of our arguments (below) using the three-circles theorem to establish the Riesz-Thorin theorem.

Theorem 19.1.11 *Let f be a holomorphic function on a domain U. Let A be a closed set. If $|f(z)| \leq M$ in U and $|f(z)| \leq m < M$ on A, then, for $0 \leq \theta \leq 1$,*

$$|f(z)| \leq m^{\theta} M^{1-\theta}$$

at points z where $\omega(z, U, A) \geq \theta$.

Remark 19.1.12 This result is of fundamental philosophical importance. It shows how the harmonic measure is a device for interpolating information about the function f.

Proof Of course, $f : U \to \mathbb{D}(0, M)$. We take $\widetilde{A} = \mathbb{D}(0, m)$, $A = f^{-1}(\widetilde{A})$, and $\widetilde{U} = f(U)$. Then we apply the previous theorem and the remark following. So

$$\omega(z, U, A) \leq \omega(f(z), \widetilde{U}, \widetilde{A}) = \frac{\log(M/|f(z)|)}{\log M/m}.$$

At a point z for which $\omega(z, U, A) \geq \theta$, we have

$$\theta \le \frac{\log(M/|f(z)|)}{\log M/m}.$$

This inequality is equivalent to the desired conclusion. □

If we take U and A in this last theorem to be annuli, then we can draw an important and precise conclusion. Thus, we have the following version of the three-circles theorem.

Prelude We now have the celebrated Hadamard's three-circles theorem, whose uses in analysis are many and varied.

Theorem 19.1.13 (Hadamard) *Let f be a holomorphic function on an annulus $\mathcal{A} = \{z \in \mathbb{C} : c < |z| < C\}$. For $c < \tau < C$, we let*

$$M(\tau) \equiv \max_{|z|=\tau} |f(z)|.$$

If $c < r < \rho < R < C$, then we have

$$\log M(\rho) \le \frac{\log R - \log \rho}{\log R - \log r} \cdot \log M(r) + \frac{\log \rho - \log r}{\log R - \log r} \cdot \log M(R)$$

(This inequality says quite plainly that $\log M(s)$ is a convex function of $\log s$.)

Proof Let

$$U = \{z \in \mathbb{C} : r < |z| < R\},$$

and let

$$A = \{z \in \mathbb{C} : r \le |z| \le r + \epsilon\}, \text{ some } \epsilon > 0.$$

We assume that f is holomorphic on U, $|f| \le M$ on U, and $|f| \le m$ on A. The preceding theorem tells us that

$$|f(z)| \le m^\theta M^{1-\theta} \tag{19.4}$$

on the set where $\omega(z, U, A) \ge \theta$, that is, on the set where

$$\frac{\log(|z|/R)}{\log((r+\epsilon)/R)} \ge \theta.$$

(Note that the roles of $r + \epsilon$ and R are reversed from their occurrence in Example 19.1.7, just because now we are creating harmonic measure for the inside circle of the annulus.) In particular, inequality (19.4) holds when $|z| = (r+\epsilon)^\theta R^{1-\theta}$. We take

$$\theta = \frac{\log R - \log \rho}{\log R - \log(r + \epsilon)}$$

and

$$|z| = (r + \epsilon)^{\theta} R^{1-\theta} \equiv \rho_{\epsilon}$$

Then (19.4) translates to

$$M(\rho_{\epsilon}) \leq M(r + \epsilon)^{\theta} \cdot M(R)^{1-\theta}.$$

Letting $\epsilon \to 0^{+}$ and taking the logarithm of both sides yields the desired result. □

We note that Ahlfors [15], from which our exposition derives, likes to express the conclusion of this last result as

$$\det \begin{pmatrix} 1 & 1 & 1 \\ \log r & \log \rho & \log R \\ \log M(r) & \log M(\rho) & \log M(R) \end{pmatrix} \geq 0.$$

We close this section by formulating the three-lines version of Hadamard's theorem. It is proved from Theorem 19.1.13 just by using conformal mapping, and we leave the details to the interested reader.

Prelude We have the "three-lines" version of the Hadamard's theorem.

Theorem 19.1.14 *Let f be a holomorphic function on the strip $S = \{z \in \mathbb{C} : 0 < \operatorname{Re} z < 1\}$. For $0 < x < 1$, we let*

$$M(x) \equiv \max_{\operatorname{Re} z = x} |f(z)|.$$

If $0 < a < \rho < b < 1$, then we have

$$\log M(\rho) \leq \frac{\log b - \log \rho}{\log b - \log a} \cdot \log M(a) + \frac{\log \rho - \log a}{\log b - \log a} \cdot \log M(b).$$

(This inequality says quite plainly that $\log M(s)$ is a convex function of $\log s$.)

19.2 A Discussion of Interpolation of Linear Operators

Capsule As predicted in the last **Capsule**, we now apply the three-lines theorem to prove a result about interpolation of operators. This will be a new idea, and a profound one.

Marcel Riesz discovered the idea of interpolating linear operators in the following context: Perhaps the most important linear operator in all of analysis is the Hilbert transform

$$H : f \to \int_{\mathbb{R}} \frac{f(t)}{x - t} dt.$$

This operator arises naturally in the study of the boundary behavior of conjugates of harmonic functions on the disc, in the existence and regularity theory for the Laplacian, in the summability theory of Fourier series, and, more generally, in the theory of singular integral operators. It had been an open problem for some time to show that H is a bounded operator on L^p, $1 < p < \infty$.

It turns out that the boundedness on L^2 is easy to prove and follows from Plancherel's theorem in Fourier analysis. Riesz cooked up some extremely clever tricks to derive boundedness on L^p when p is an even integer. (These ideas are explained in detail in [330].) He needed to find some way to derive therefrom the boundedness on the "intermediate" L^p spaces. Now we introduce enough language to explain precisely what the concept of "intermediate space" means.

Let X_0, X_1, Y_0, Y_1 be Banach spaces. Intuitively, we will think of a linear operator T such that

$$T : X_0 \to Y_0$$

continuously and

$$T : X_1 \to Y_1$$

continuously. We wish to posit the existence of spaces X_θ and Y_θ, $0 \leq \theta \leq 1$ such that these new spaces are natural "intermediaries" of X_0, X_1 and Y_0, Y_1, respectively. Furthermore, we want that

$$T : X_\theta \to Y_\theta$$

continuously. There is interest in knowing how the norm of T acting on X_θ depends on the norms of T acting on X_0 and X_1. One rigorous method for approaching this situation is as follows (see [73, 486], and [264] for our inspiration):

Let X_0, Y_0, X_1, Y_1 be given as in the last paragraph. Suppose that

$$T : X_0 \cap X_1 \longrightarrow Y_0 \cup Y_1$$

is a linear operator with the properties that

(i) $\|T_x\|_{Y_0} \leq C_0 \|x\|_{X_0}$ for all $x \in X_0 \cap X_1$;
(ii) $\|Tx\|_{Y_1} \leq C_0 \|x\|_{X_1}$ for all $x \in X_0 \cap X_1$.

Then we want to show that there is a collection of norms $\|\cdot\|X_\theta$ and $\|\cdot\|Y_\theta$, $0 < \theta < 1$, such that

$$\|Tx\|_{Y_\theta} \leq C_\theta \|x\|_{X_\theta}.$$

So the problem comes down to how to construct the "intermediate norms" $\|\cdot\|_{X_\theta}$ and $\|\cdot\|_{Y_\theta}$ from the given norms $\|\cdot\|_{X_0}$, $\|\cdot\|_{Y_0}$, $\|\cdot\|_{X_1}$, and $\|\cdot\|_{Y_1}$.

There are in fact a number of paradigms for effecting the indicated construction. Most prominent among these are the "real method" (usually attributed to Lyons and Peetre—see [363]) and the "complex method" (usually attributed to Calderón-see [103]). In this text, we shall concentrate on the complex method, which was inspired by the ideas of Riesz and Hadamard that were described at the beginning of this section.

Instead of considering the general paradigm for complex interpolation, we shall concentrate on the special case that was of interest to Riesz and Thorin. We now formulate our main theorem.

Prelude The Riesz-Thorin theorem was the very first theorem in the study of interpolation of operators. This is now a very well-developed part of analysis, and there are entire monographs devoted to the subject.

Theorem 19.2.1 *Let $1 \leq p_0 < p_1 \leq \infty$ and $1 \leq q_0 < q_1 \leq \infty$. Suppose that*

$$T : L^{p_0}\left(\mathbb{R}^2\right) \cap L^{p_1}\left(\mathbb{R}^2\right) \longrightarrow L^{q_0}\left(\mathbb{R}^2\right) \cup L^{q_1}\left(\mathbb{R}^2\right)$$

is a linear operator on Lebesgue spaces satisfying

$$\|Tf\|_{L^{q_0}} \leq C_0 \|f\|_{L^{p_0}}$$

and

$$\|Tf\|_{L^{q_1}} \leq C_0 \|f\|_{L^{p_1}}.$$

Define p_θ and q_θ by

$$\frac{1}{p_\theta} = (1-\theta)\cdot\frac{1}{p_0} + \theta\cdot\frac{1}{p_1}$$

and

$$\frac{1}{q_\theta} = (1-\theta)\cdot\frac{1}{q_0} + \theta\cdot\frac{1}{q_1},$$

with $0 \leq \theta \leq 1$. Then T is a bounded operator from the L^{p_θ} norm to the L^{q_θ} norm, and

$$\|Tf\|_{L^{q_\theta}} \le C_0^{1-\theta} C_1^\theta \|f\|_{L^{p_\theta}}$$

for all $f \in L^{p_0} \cap L^{p_1}$.

Proof Fix a nonzero function f that is continuous and with compact support in $\mathbb{R}^2$. We will prove an a priori inequality for this f, and then extend to general f at the end. Now certainly $f \in L^{p_0} \cap L^{p_1}$. We consider the holomorphic function

$$H : \zeta \mapsto \frac{\left(\int_{\mathbb{R}^2} |Tf(x)|^{(1-\zeta)q_0+\zeta q_1} dx\right)^{(1-\zeta)/q_0+\zeta/q_1}}{\left(\int_{\mathbb{R}^2} |f(x)|^{(1-\zeta)p_0+\zeta p_1} dx\right)^{(1-\zeta)/p_0+\zeta/p_1}}$$

on the strip $S = \{\zeta \in \mathbb{C} : 0 < \operatorname{Re} \zeta < 1\}$. We define

$$M(s) = \sup_{\operatorname{Re}\zeta = s} |H(\zeta)|, 0 < s < 1.$$

Notice that

$$\begin{aligned} M(0) &= \sup_{t\in\mathbb{R}} \left| \frac{\int_{\mathbb{R}^2} |Tf(x)|^{(1-it)q_0+itq_1} dx^{(1-it)/q_0+it/q_1}}{\int_{\mathbb{R}^2} |f(x)||^{(1-it)p_0+itp_1} dx^{(1-it)/p_0+it/p_1}} \right| \\ &\le \sup_{t\in\mathbb{R}} \frac{\int_{\mathbb{R}^2} |Tf(x)|^{q_0} dx^{1/q_0}}{\int_{\mathbb{R}^2} |f(x)|^{p_0} dx^{1/p_0}} \\ &\le C_0. \end{aligned}$$

A similar calculation shows that

$$M(1) \le C_1.$$

Obviously, this is grist for the three-lines theorem. We may conclude that $M(s) \le C_0^{1-s} \cdot C_1^s$, $0 < s < 1$. Now let p_ζ, q_ζ be given by

$$\frac{1}{p_\zeta} = (1-\zeta) \cdot \frac{1}{p_0} + \zeta \cdot \frac{1}{p_1}$$

and

$$\frac{1}{q_\zeta} = (1-\zeta) \cdot \frac{1}{q_0} + \zeta \cdot \frac{1}{q_1}.$$

Let $\zeta = s + it$. Then

$$\frac{\|Tf\|_{L^{q_\zeta}}}{\|f\|_{L^{p_\zeta}}} \le M(s) \le C_0^{1-s} C_1^s.$$

In other words,

$$\|Tf\|_{L^{q_s}} \leq C_0^{1-s} C_1^s \|f\|_{L^{p_s}}.$$

This is the desired conclusion for a function f that is continuous with compact support. The result for general $f \in L^{p_\theta}$ can be achieved with a simple approximation argument. □

19.3 The F. and M. Riesz Theorem

Capsule One of the big ideas of twentieth-century function theory is due to the brothers F. and M. Riesz. The result has many interpretations—in terms of the absolute continuity of harmonic measure, in terms of the vanishing of negative Fourier coefficients, and many other ideas as well.

The theorem of F. and M. Riesz makes an important statement about the absolute continuity of harmonic measure. We begin with a preliminary result that captures the essence of the theorem. Our exposition here follows the lead of [172, Ch. 6].

For $0 < p < \infty$, we define the Hardy space

$$H^p(\mathbb{D}) = \left\{ f \text{ holomorphic on } \mathbb{D} : \sup_{0<r<1} \left(\int_0^{2\pi} \left| f\left(re^{i\theta}\right) \right|^p d\theta \right)^{1/p} \right.$$
$$\left. \equiv \|f\|_{H^p} < \infty \right\}.$$

Hardy spaces are generalizations of the more classical space $H^\infty(\mathbb{D})$ of bounded, holomorphic functions.

The most important fact about a function in a Hardy space is that it has a boundary function. We enunciate that result here, and refer the reader to [331, Chapter 8] for the details.

Prelude The theory of boundary limits for functions in Hardy spaces goes back to the work of Fatou in 1906. Later on, Riesz and Hardy played a decisive role in fleshing out the theory and pointing in the directions that we study today.

Theorem 19.3.1 *Let $0 < p \leq \infty$ and let $f \in H^p(\mathbb{D})$. Then there is a function f^* $\in L^p(\partial\mathbb{D})$ such that:*

$$\lim_{r\to 1^-} f\left(re^{i\theta}\right) = f^*\left(e^{i\theta}\right) \text{ for almost every } \theta \in [0, 2\pi). \tag{19.5}$$

$$\text{Let } f_r\left(e^{i\theta}\right) = f\left(re^{i\theta}\right) \text{ for } 0 < r < 1. \tag{19.6}$$

Assume that $0 < p < \infty$. Then $\lim_{r\to 1^-} \|f - f_r\|_{L^p(\partial\mathbb{D})} = 0$.

In the following discussion, we shall consider rectifiable curves. Basically, a rectifiable curve is a curve with finite length. The technical definition is that the curve is locally the Lipschitz image of the unit interval (see [149]). But context will make it clear that the intuitive notion of rectifiability will suffice for our purposes.

Prelude Here it is important that we relate rectifiability for a curve with the theory of Hardy spaces. Hausdorff measure also comes into play.

Theorem 19.3.2 *Let U be a domain such that $\gamma = \partial U$ is a Jordan curve. Let*

$$\varphi : \mathbb{D} \to U$$

be a conformal map. Then the curve γ is rectifiable if and only if $\varphi' \in H^1$. In case $\varphi' \in H^1$, then we have

$$\|\varphi'\|_{H^1} = \text{length}(\gamma) = \mathcal{H}^1(\gamma). \tag{19.7}$$

Here $\mathcal{H}^1$ is a one-dimensional Hausdorff measure.

Remark 19.3.3 It is known (see [331, Chapter 8]) that a function in H^1 has a boundary limit function that is in $L^1(\partial\mathbb{D})$. Thus, the hypothesis of the theorem says, essentially, that φ is absolutely continuous on the boundary of $\mathbb{D}$. It makes sense, then, that φ would preserve length on the boundary. Theorem 19.3.2 is essentially equivalent to the result of F. and M. Riesz that we described in Remark 19.1.4.

Proof of Theorem 19.3.2 Once again, we invoke Carathéodory's theorem; thus, we know that φ and its inverse extend continuously and univalently to their respective boundaries. Let us assume that $\varphi' \in H^1$. Let

$$0 = \theta_0 < \theta_1 < \theta_2 < \cdots < \theta_k = 2\pi$$

be a partition of $[0, 2\pi]$. Then of course,

$$\begin{aligned} \sum_{j=1}^{k} \left|\varphi\left(e^{i\theta_j}\right) - \varphi\left(e^{i\theta_{j-1}}\right)\right| &= \lim_{r\to 1^-} \sum_{j=1}^{k} \left|\varphi\left(re^{i\theta_j}\right) - \varphi\left(re^{i\theta_{j-1}}\right)\right| \\ &= \lim_{r\to 1^-} \sum_{j=1}^{k} \left|\int_{\theta_{j-1}}^{\theta_j} \varphi'\left(re^{i\theta}\right) ire^{i\theta}\, d\theta\right| \\ &\le \|\varphi'\|_{H^1}\,. \end{aligned} \tag{19.8}$$

But plainly the length of γ is the supremum, over all partitions of $[0, 2\pi)$, of the left-hand side of (19.8). We conclude that γ is rectifiable and

$$\text{length}\,(\gamma) \le \|\varphi'\|_{H^1}\,.$$

For the converse, assume that γ is rectifiable. If $0 < r < 1$ is fixed, then let $\gamma_r = \varphi(\{z \in \mathbb{C} : |z| = r\})$. Let $\epsilon > 0$. Now choose a partition $\{\theta_0, \theta_1, \ldots, \theta_k\}$ of the interval $[0, 2\pi)$ as before so that

$$\sum_{j=1}^{k} \left|\varphi\left(re^{i\theta_j}\right) - \varphi\left(re^{i\theta_{j-1}}\right)\right| \geq \text{length}\,(\gamma_r) - \epsilon.$$

We write

$$\eta(z) = \sum_{j=1}^{k} \left|\varphi\left(ze^{i\theta_j}\right) - \varphi\left(ze^{i\theta_{j-1}}\right)\right|.$$

Then η, being the sum of absolute values of holomorphic functions, is subharmonic. By Carathéodory's theorem, η is continuous on $\overline{\mathbb{D}}$. Hence

$$\sup_{\mathbb{D}} \eta(z) = \sup_{\theta} \eta\left(e^{i\theta}\right) \leq \text{ length }(\gamma).$$

We conclude that

$$\int_0^{2\pi} \left|\varphi'\left(re^{i\theta}\right)\right| d\theta = \text{ length } (\gamma_r) \leq \eta(r) + \epsilon \leq \text{ length } (\gamma) + \epsilon.$$

It follows that $\phi' \in H^1$ and equality (19.7) is valid. We conclude by noting that, for a rectifiable curve, the ordinary notion of length and the one-dimensional Hausdorff measure are the same. □

The reason that we include Theorem 19.3.2 in the present chapter—apart from its general aesthetic interest—is that it can be interpreted in the language of harmonic measure. Let notation be as in Theorem 19.3.2. To wit, let $\gamma = \partial U$ be a rectifiable Jordan curve as above and let $F \subseteq \gamma$ be a subcurve. Let φ be a conformal mapping of U to the unit disc $\mathbb{D}$. Let $F = \varphi(E)$ and $\alpha = \varphi(0)$. Then Carathéodory's theorem, Example 19.1.6, and Hadamard's theorem tell us that

$$\omega(\alpha, U, F) = \omega(0, \mathbb{D}, E) = \frac{1}{2\pi}|E|,$$

where $|\cdot|$ denotes arc length. Since F is an arc, we see that Theorem 19.3.2 and the proof of (19.7) tell us that

$$\mathcal{H}^1(F) = \mathcal{H}^1(\varphi(E)) = \lim_{r\to 1^-} \int_E \left|\varphi'\left(re^{i\theta}\right)\right| d\theta = \int_E \left|\varphi'\left(e^{i\theta}\right)\right| d\theta. \tag{19.9}$$

Of course, if (19.9) holds for arcs, then, by passing to unions and intersections, we see that it holds for Borel sets F. Thus, we derive the important conclusion

$$\omega(\alpha, U, F) = 0 \Rightarrow \mathcal{H}^1(F) = 0.$$

Conversely, since (by a standard uniqueness theorem—see [314])

$$\left|\left\{\theta : \left|\varphi'\left(e^{1\theta}\right)\right| = 0\right\}\right| = 0$$

we see that

$$\mathcal{H}^1(F) = 0 \Rightarrow \omega(\alpha, U, F) = 0.$$

What we have proved, then, is that when $\gamma = \partial U$ is rectifiable, then harmonic measure for U and linear measure on γ are mutually absolutely continuous. We now summarize this result in a formally enunciated theorem.

Prelude The *F. and M. Riesz theorem* is one of the celebrated results of classical function theory. There are many different renditions of the result and many different proofs. We give here a geometric rendition of the theorem.

Theorem 19.3.4 (F. and M. Riesz, 1916) *Let U be a simply connected planar domain such that $\gamma = \partial U$ is a rectifiable Jordan curve. Assume that*

$$\varphi : \mathbb{D} \to U$$

is conformal. Then $\varphi' \in L^1(\partial\mathbb{D})$. For any Borel set $E \subseteq \partial\mathbb{D}$, it holds that

$$\mathcal{H}^1(\varphi(E)) = \int_E \mid \varphi' e^{i\theta} \mid d\theta.$$

and for any Borel set $F \subset \partial\Omega$,

$$\omega(F) = 0 \Longleftrightarrow \mathcal{H}^1(F) = 0$$

We conclude by noting that Theorems 19.3.2 and 19.3.4 are equivalent, just because formula (19.7) is valid precisely when $\varphi' \in H^1$.

When the domain Ω has a nice boundary, harmonic measure is particularly easy to understand.

Suppose that Ω is a domain on the sphere and that $p \in \Omega$. A function $g(z; p)$ is a *Green's function* for Ω with pole (or singularity) at p, $p \neq \infty$, if

$$g(z; p) \text{ is harmonic on } \Omega - \{p\} \tag{19.10}$$

$$g(z; p) + \log|z - p| \text{ is harmonic near } p \tag{19.11}$$

$$\lim\{g(z; p) : z \to \zeta\} = 0 \quad \text{for all } \zeta \in \partial\Omega \tag{19.12}$$

If $p = \infty$, then (19.11) is modified to

$$g(z; \infty) - \log |z| \text{ is harmonic near } \infty \tag{19.13}$$

Theorem 19.3.5 *Suppose that Ω is bounded by a finite number of disjoint analytic simple closed curves. For each $p \in \Omega$, we have*

$$d\omega_p = \frac{-1}{2\pi} \frac{\partial}{\partial n} g(\cdot; p) ds \tag{19.14}$$

where $g(\cdot; p)$ is the Green's function for Ω with pole at p, $\partial/\partial n$ is the derivative in the direction of the outward normal at Γ, and ds is arc length.

Proof The proof is an application of Green's theorem. Let h be a smooth function on Γ and for small $\delta > 0$, let

$$\Omega_\delta = \Omega \backslash \{z : |z - p| \leqslant \delta\}$$

Let u be the harmonic extension to Ω of h and let $v(z) = g(z; p)$. Then by Green's theorem on Ω_δ,

$$\iint_{\Omega_\delta} [u \Delta v - v \Delta u] dx dy = \int_{\partial \Omega_\delta} \left[u \frac{\partial v}{\partial n} - v \frac{\partial u}{\partial n} \right] ds \tag{19.15}$$

Now the left-hand side of (19.15) is zero since both u and v are harmonic on Ω_δ.

Further, $v = 0$ on Γ; so (19.15) simplifies to

$$\int_\Gamma h \frac{\partial v}{\partial n} ds = \int_{|z-p|=\delta} \left[u \frac{\partial v}{\partial n} - v \frac{\partial u}{\partial n} \right] ds$$

Let us set $z = p + re^{it}$, $0 \leqslant r \leqslant \delta$, and $0 \leqslant t \leqslant 2\pi$. Then the normal derivative is just the radial derivative and $g(z; p) = -\log r + G(z)$, where G is harmonic near p. Thus,

$$\frac{\partial g}{\partial n} = -\frac{1}{r} + \text{continuous term}$$

Since u is also continuous near p, we have

$$\begin{aligned} \int_{|z-p|=\delta} u \frac{\partial g}{\partial n} ds &= -\int_0^{2\pi} u \left(p + \delta e^{it} \right) dt + 0(\delta) \\ &\to -2\pi u(p) \quad \text{as } \delta \to 0 \end{aligned}$$

Finally, the other term

$$-\int_{|z-p|=\delta} g\frac{\partial u}{\partial n}ds$$

behaves like $-\delta\log\delta$ as $\delta\to 0$ and so goes to zero as $\delta\to 0$. □

It is convenient to put here two more facts related to harmonic measure. Let us assume that $\Gamma=\partial\Omega$ consists of $m+1$ disjoint analytic simple closed curves. Let $p\in\Omega$ and let $h(z)=h(z;p)$ be the harmonic conjugate of $g(z;p)$, the Green's function for Ω and p. Of course, h is not single-valued since it has periods about the holes in Ω (we will explore this type of difficulty later in some detail), but locally $g+ih$ is analytic and its derivative is single-valued on Ω. We write $Q=g+ih$ and Q' for its complex derivative.

Proposition 19.3.6 *We have* $d\omega_p(\zeta)=\frac{i}{2\pi}Q'(\zeta)d\zeta$.

Proof Let n be a unit outward normal at $\zeta\in\Gamma$, and let τ be a unit tangent at ζ. By the Cauchy-Riemann equations

$$\frac{\partial h}{\partial n}(\zeta)=\frac{\partial g}{\delta\tau}(\zeta)=0$$

since g vanishes on Γ. Next, note that $n=-i\tau=-id\zeta/|d\zeta|$ and

$$Q'(\zeta)=\lim_{t\downarrow 0}\frac{Q(\zeta+tn)-Q(\zeta)}{tn}$$

Further,

$$\frac{\partial g}{\partial n}=\lim_{t\downarrow 0}\frac{g(\zeta+tn)-g(\zeta)}{t}$$

Thus,

$$\begin{aligned}iQ'(\zeta)d\zeta&=i\lim_{t\downarrow 0}\frac{Q(\zeta+tn)-Q(\zeta)}{tn}d\zeta\\&=-\lim_{t\downarrow 0}\frac{g(\zeta+tn)-g(\zeta)}{t}|d\zeta|\\&=-\frac{\partial g}{\partial n}ds\end{aligned}$$

which establishes the result in light of Theorem 19.3.5. □

Proposition 19.3.7 *Let* $\Gamma=\partial\Omega$ *consist of* $m+1$ *disjoint analytic simple closed curves, let* $p\in\Omega$*, and let* $Q(z)=g(z;p)+ih(z;p)$*, where* h *is the (multiplevalued) harmonic conjugate of* g*. Then*

(a) Q' *does not vanish on* Γ.
(b) Q' *has precisely* m *zeros in* Ω, *counting multiplicities.*

Proof We know that Q' has a single pole of order one at p. Further, from Proposition 19.3.6, we also know that $i Q'(z)dz$ is a nonnegative measure on Γ. Thus, the total change in $1/2\pi$ arg $Q'(z)$ as z traverses Γ once is precisely $m - 1$. But this must be exactly the number of zeros of Q' minus the number of poles, and hence (b) is proven. To see (a), note that the Cauchy-Riemann equations imply that h is increasing locally on Γ, and thus Q must be one-to-one in a neighborhood of each $\zeta \in \Gamma$. □

Chapter 20
Quadrature

20.1 Introduction

We begin by saying a few words about the birth of this theory.

For the numerical integration of functions $f : [-1, 1] \to \mathbb{R}$, there are many classical quadrature rules available. Examples are the rectangle or midpoint rule, the trapezoidal rule, Simpson's rule, or the Gauss rule, which all have the general form

$$T_m(f) := \sum_{k=0}^{m} q_k f(t_k)$$

with nodes

$$t_0, t_1, \ldots, t_m \in [-1, 1]$$

and with integration weights

$$q_0, q_1, \ldots, q_m \in \mathbb{R}.$$

It was known already at the time of Newton that "quadrature," i.e., the evaluation of integrals, could sometimes be effected, approximately or exactly, by replacing $\int_a^b f(t)dt$ with a sum of the type $T_m(f)$.

For example

$$3\int_{-1}^{1} f(t)dt = f(-1) + 4f(0) + f(1) \tag{20.1}$$

P. V. Dovbush, S. G. Krantz, *The Geometric Theory of Complex Variables*,
https://doi.org/10.1007/978-3-031-77204-7_20

is valid when f is any cubic polynomial as well as for many other functions (e.g., all odd continuous functions on $[-1, 1]$). This formula can be scaled to an arbitrary interval; if we do so for a number of consecutive equal "small" intervals and add the resulting equations, we get "Simpson's rule."

The first person to systematically study general identities of the kind

$$\int_U f dA = \sum_{k=1}^{n} c_k f(a_k), (a_k \in U \subset \mathbb{C}, c_k \in \mathbb{C}) \tag{20.2}$$

was an American academic applied mathematician Philip J. Davis (see his excellent book [130] and references therein).

In *Encyclopædia Britannica*, one can read: " 'Quadrature' in mathematics, the process of determining the area of a plane geometric figure by dividing it into a collection of shapes of known area (usually rectangles) and then finding the limit (as the divisions become ever finer) of the sum of these areas. ...Because the definite integral of a function determines the area under its curve, integration is still sometimes referred to as quadrature."

From the above discussion, it is not hard to understand why Davis suggested the name "quadrature identity" for Eq. (20.1).

The "quadrature identities" we speak of in this chapter have formal similarities with (20.1) but also these differences: the domain of integration will be an open set in $\mathbb{R}^n$ for some $n \geq 2$, and the family of f for which the formula is to be exact will be some class of harmonic or analytic functions. Moreover, the point of view is different: we are not concerned here with quadrature identities as a method of numerically integrating harmonic (or other) functions over the domains involved. We are instead interested in more theoretical aspects.

With this chapter, the authors wish to give a general overview of the theory of quadrature domains, along with indications of some of its ramifications. There will be given some proofs and extensive bibliography. More details for most of the material known up to 1992 can be found in the book [468].

This chapter is not wholly self-contained. The reader will find that they should be familiar with the elementary portions of linear algebra and of the theory of functions of a complex variable.

20.2 Classical Quadrature Domains

Let $\mathbb{D}(a, r) = \{z \in \mathbb{C} : |z - a| < r\}$ be a disc centered at the point $a \in \mathbb{C}$ and of radius r, and let $A = \pi r^2$ its area. One of the most basic properties of a holomorphic (or harmonic) function f is its mean value property:

$$f(a) = A^{-1} \int_{\mathbb{D}(a,r)} f dA. \tag{20.3}$$

By $dA := \frac{i}{2}dz \wedge d\overline{z}$, here and throughout this chapter, we denote the area element on $\mathbb{R}^2 \cong \mathbb{C}$. We can also interpret (20.3) as a "one-point quadrature identity": to integrate a holomorphic function over a disc $\mathbb{D}(a, r)$, evaluate it at the center of the disc, then multiply by the area of the disc.

We shall call a bounded domain U in the complex plane a *(classical) quadrature domain* if there exist finitely many points $a_1, \ldots, a_m \in U$ and coefficients $c_{kj} \in \mathbb{C}$ so that

$$\int_U f dA = \sum_{k=1}^{m} \sum_{j=0}^{n_k-1} c_{kj} f^{(j)}(a_k) \tag{20.4}$$

for all integrable holomorphic functions f in U. The identity (20.4) is then called a *quadrature identity* and the integer $n = \sum_{k=1}^{m} n_k$ is the *order* of the quadrature identity (assuming $c_{k,n_k-1} \neq 0$).

We shall find it convenient in the sequel to write the quadrature identity (20.4) simply as

$$\int_U f dA = \sum_{k=1}^{n} c_k f(a_k) \tag{20.5}$$

where in the sequence $a_1, \ldots, a_n \in U$ repetitions are allowed, a repeated a_k being interpreted as the occurrence of derivatives of f at a_k.

In the simplest case, $n = 1$, it is known that discs $\mathbb{D}(a, r)$ are the only quadrature domains [6, 146, 147, 454, 468], and the quadrature identity then reduces to the ordinary mean value property for analytic functions.

Theorem 20.2.1 *Let U be a simply connected plane domain of finite area and p a point of U such that for every function u harmonic in U and integrable over U, the mean value of u over the area of U equals $u(p)$. Then U is a disc and p is its center.*

Proof Let $f(z)$ be any function holomorphic in U with finite quadratic integral $\iint_U |f(z)|^2 dA$; by the Cauchy-Schwarz inequality, the real and imaginary parts of $f(z)$ are each (absolutely) integrable over U, and so

$$f(p) = \iint_U f(z) \cdot A^{-1} dA \tag{20.6}$$

where A denotes the area of U. (It is at this point in the argument that for every function u harmonic in U and integrable over U, the mean value of u over the area of U equals $u(p)$ is employed.)

On the other hand, for every holomorphic quadratically integrable function, the equality

$$f(p) = \iint_U K(p, z) f(z) dA \tag{20.7}$$

holds, where $K(p,z)$ denotes the Bergman kernel associated to U.

Note that the function that is identically one is with finite quadratic integral because of the finite area assumption. We obtain from (20.6) and (20.7)

$$K(p,z) = A^{-1}. \tag{20.8}$$

Let $g : U \to \mathbb{D}(0,1)$ be a Riemann mapping function associated to the point p in U, i.e. $g(p) = 0$. Recall that the Bergman kernel for the unit disc is $\pi^{-1}/(1 - z\overline{w})^2$. Let $G = g^{-1}$ be the inverse of the Riemann map.

The transformation formula for the Bergman kernels yields

$$G'(z)K(p, G(z)) = \frac{1}{\pi(1 - z\overline{g(p)})^2}\overline{g'(p)}$$

and since $g(p) = 0$ and by (20.8) $K(p,\cdot)$ is constant, we conclude that $G'(z)$ is a constant. Consequently, $G(z)$ is complex linear and U is a disc. □

Thus, the quadrature identity (20.5) can be thought of as a generalized mean value property and a quadrature domain as a generalized disc (at least if all the c_k are positive).

20.3 The Schwarz Function

H.A. Schwarz showed us how to extend the notion of reflection in straight lines and circles to reflection in an arbitrary analytic arc. Reflection is an anti-holomorphic mapping. By taking its complex conjugate, Davis and Pollak arrive at an analytic function that they named the *Schwarz function* of the analytic arc [129].

We suppose that the arc C is written in rectangular form

$$f(x,y) = 0. \tag{20.9}$$

In conjugate coordinates, this becomes

$$f\left(\frac{z+\overline{z}}{2}, \frac{z-\overline{z}}{2i}\right) \equiv g(z,\overline{z}) = 0. \tag{20.10}$$

Suppose, for the moment, that the function $g(z,\overline{z})$ of (20.10) is an irreducible polynomial of a certain degree (i.e., g cannot be expressed in the form $g \equiv g_1 g_2$ where g_1 and g_2 are real polynomials and whose degree $\geqq 1$). Write

$$g(u,v) = \sum_{j,k=0}^{m,n} a_{jk} u^j v^k$$

and

$$\overline{g}(u,v)=\sum_{j,k=0}^{m,n}\overline{a}_{jk}u^{j}v^{k}\,.$$

Along C we have $g(z,\overline{z})=0$, and by conjugating this equation, we have $\overline{g(z,\overline{z})}\equiv\overline{g}(\overline{z},z)=0$. It follows by a well-known theorem of algebra that the two polynomials $g(z,\overline{z})$ and $\overline{g}(\overline{z},z)$ must be proportional (see, e.g., B1ôcher, [87, p. 211]). Thus,

$$\overline{g}(\overline{z},z)=\lambda g(z,\overline{z}),\quad \lambda=\text{ const }\neq 0. \tag{20.11}$$

Polynomials with the property (20.11) are called *self-conjugate*.

Self-conjugacy is therefore a necessary condition that a polynomial in conjugate coordinates $z,\overline{z}$ represent a real curve. It is not sufficient, as is seen from the example $g(z,\overline{z})=z\overline{z}+1$.

We now drop the polynomial hypothesis and assume only that g is a holomorphic function of z and $\overline{z}$. If at z_0, a point of the curve C, we have $\partial g/\partial\overline{z}|_{z_0}\neq 0$, then by the implicit function theorem, we may solve uniquely for $\overline{z}$ in terms of z to obtain

$$\overline{z}=S(z) \tag{20.12}$$

where $S(z)$ is a regular holomorphic function of z in some neighborhood of z_0 with $|z-z_0|\leqq\rho\,(z_0)$.

At z_0, we have $\overline{z}_0 = S\,(z_0)$ and the identity (20.12) persists along C in a neighborhood of z_0.

The condition $\partial g/\partial\overline{z}\neq 0$ is equivalent to

$$\frac{\partial}{\partial\overline{z}}f\left(\frac{z+\overline{z}}{2},\frac{z-\overline{z}}{2i}\right)=\frac{1}{2}\left(\frac{\partial f}{\partial x}+i\frac{\partial f}{\partial y}\right)\neq 0$$

or to asserting that $\partial f/\partial x=0$, $\partial f/\partial y=0$ do not occur simultaneously. A point of C at which $\partial g/\partial\overline{z}\neq 0$ is called a *regular point* of C.

If, along a portion of a simple arc, endpoints included, $\partial g/\partial\overline{z}\neq 0$, we may solve for $\overline{z}$ at each point and obtain a regular $S(z)$. Thus, by a standard argument of analytic continuation, $S(z)$ can be defined as the branch of a single-valued analytic function in a strip-like region containing the arc in its interior. If C intersects itself, then the strip-like region will intersect itself; that is, $S(z)$ can be defined in a strip-like region lying on a Riemann surface.

If C is a simple closed analytic curve, then $S(z)$ is a branch of a single-valued analytic function in an annulus-like region containing C in its interior. It is possible that $S(z)$ may be continuable analytically to other portions of the plane as a single- or a multivalued function. For example, if g is a polynomial, then $S(z)$ is generally a multivalued algebraic function of z.

If $z_0 = 0$ lies on C, then $g(0, 0) = 0$, and $S(z)$ has the following representation as a line integral:

$$S(z) = \frac{1}{2\pi i} \int_{\Gamma} \frac{w g_w(z, w)}{g(z, w)} dw$$

where Γ is a suitably chosen contour surrounding $w = 0$.

For the implicit function theorem for holomorphic functions leading to the above representation, see, e.g., Hille [235, p. 269].

We shall call $S(z)$ the *Schwarz function* of C, for reasons which will emerge subsequently. By the uniqueness theorem for analytic functions, an analytic function is determined uniquely by the values it takes along an arc.

The Schwarz function for C can, therefore, be defined alternatively as the unique analytic function $S(z)$ which at each point z along C takes on the value $\overline{z}$.

This tells us that the conjugate of the Schwarz function of an analytic arc yields the Schwarzian reflection of a point in that arc. Insofar as the Schwarz function is determined uniquely by the arc, we learn that the operation of reflection is independent of the parameterization of the arc. Schwarzian reflection is an anti-conformal transformation, i.e., one that reverses the sense of angles. The reflection of the reflection of a point z must be z itself. Hence, it follows that

$$\overline{S(\overline{S(z)})} \equiv z.$$

This is a functional equation satisfied by $S(z)$.

The Schwarz function $S(z)$, *qua* analytic function, was named and stressed in Davis and Pollak [129]. Applications were given to problems of complex moments and analytic continuation.

Despite the fact that the term "Schwarz function" is used in the theory of automorphic functions in a totally different sense, there is ample justification to allow the term in its present sense.

A domain is a quadrature domain if and only if the Schwarz function for the boundary curve extends meromorphically to the inside of the domain.

The conjugate of the Schwarz function is an antiholomorphic reflection function for the curve that maps one side to the other, fixes the curve, and is idempotent.

Theorem 20.3.1 ([130, p. 103]) *The Schwarz function of an analytic arc C is an entire function of z if and only if C is a straight line.*

When can the Schwarz function be rational? The answer is provided by the next theorem.

Theorem 20.3.2 ([130, p. 104]) *The Schwarz function of an analytic arc C is a rational function of z if and only if C is an arc of a circle or a straight line.*

Let U be a region of the complex plane. The class $\mathcal{A}(U)$ will designate the functions f continuous on $\overline{U}$ and holomorphic in U.

Definition 20.3.3 A linear functional L, defined on functions holomorphic in U, is said to be of class $\mathcal{D} = \mathcal{D}(U)$ if it can be expressed in the form

$$L(f) = \sum_{n=1}^{N} \sum_{k=0}^{n_k} c_{nk} f^{(k)}(a_n) \tag{20.13}$$

where $a_1, \ldots, a_N$ are distinct points in (the interior of) U and where c_{nk} are constants independent of f.

The class $\mathcal{D}$ may be described as the point differential functionals of finite order.

As preparation for the next result, we quote a theorem of J.L. Walsh, which will be used in the proof.

Theorem 20.3.4 *Let $U \subset \mathbb{C}$ be a bounded, simply connected region whose boundary C is an analytic Jordan curve. Let $t(z)$ be continuous on C. Then*

$$\int_C t(z) f(z) dz = 0 \tag{20.14}$$

for every $f \in \mathcal{A}(U)$ if and only if there exists a function $g \in \mathcal{A}(U)$ which coincides with t on C.

We can now prove:

Theorem 20.3.5 *Let U be a bounded, simply connected region with analytic boundary ∂U. Let the Schwarz function of ∂U be $S(z)$. Then, $S(z)$ is meromorphic in U if and only if there is an $L \in \mathcal{D}$ such that*

$$\iint_U f(z) dA = L(f) \tag{20.15}$$

for all functions $f \in \mathcal{A}(U)$.

Proof Suppose, first, that $S(z)$ is meromorphic. By Green's theorem in a complex analytic forms

$$\iint_U f(z) dA = (1/2i) \int_{\partial U} S(z) f(z) dz.$$

Now, since S is regular on ∂U and meromorphic in U, $S(z)$ has only a finite number of poles, say at $a_1, \ldots, a_N \in U$. Hence,

$$\frac{1}{2i} \int_{\partial U} S(z) f(z) dz = \sum_{n=1}^{N} \frac{\pi}{2\pi i} \int_{C_n} S(z) f(z) dz$$

where C_n is a circle of sufficiently small radius centered at a_n. Now, in C_n, $S(z)$ has an expansion of the form

$$S(z) = \zeta_n(z) + \frac{B_{1n}}{z - z_n} + \frac{B_{2n}}{(z - z_n)^2} + \cdots + \frac{B_{p_n n}}{(z - z_n)^{p_n}},$$

where $\zeta_n(z)$ is regular in the closure of C_n. Hence,

$$\frac{\pi}{2\pi i} \int_{C_n} S(z) f(z) dz = \pi \sum_{k=1}^{p_n} \frac{B_{kn} f^{(k-1)}(z_n)}{(k-1)!}$$

and therefore we can conclude that (20.15) is valid.

Conversely, suppose that (20.15) holds for some $L \in \mathcal{D}$ given by (20.13), and for all $f \in \mathcal{A}(U)$. Let $R(z)$ be the rational function

$$R(z) = \frac{1}{\pi} \sum_{n=1}^{N} \sum_{k=1}^{n_k} \frac{k! c_{nk}}{(z - z_n)^{k+1}}$$

so that for all $f \in \mathcal{A}(U)$, one has $L(f) = (1/2i) \int_{\partial U} R(z) f(z) dz$. Therefore, for all $f \in \mathcal{A}(U)$, we have

$$\int_{\partial U} (S(z) - R(z)) f(z) dz = 0 \text{ for all } f \in \mathcal{A}(U).$$

The function $S(z) - R(z)$ is certainly continuous on ∂U (in fact, it is regular in an annulus-like strip containing ∂U). Hence, by Walsh's Theorem, there exists a function $\zeta(z) \in \mathcal{A}(U)$ such that $S(z) - R(z) = \zeta(z)$, $z \in \partial U$. The function $\zeta(z)$, therefore, takes holomorphic values on ∂U, and hence it must be holomorphic in $U \cup \partial U$. It, therefore, has a common region of analyticity with $S(z) - R(z)$. By the Uniqueness Theorem, it follows that $S(z) - R(z) = \zeta(z)$ throughout U, so that $S(z) = R(z) + \zeta(z)$ and is therefore meromorphic in U. □

At this stage of the analysis, it is convenient to suppose that U is merely a bounded domain in the plane so that we can work with the Hilbert space $L^2(U)$ of functions that are single valued in U and are such that $\iint_U |f(z)|^2 dA < \infty$. The inner product is given by

$$(f, g)_U = \iint_U f(z) \overline{g(z)} dA.$$

The Bergman space, denoted $H^2(U)$, is the space of holomorphic functions on U which are square integrable on U with respect to area measure dA, i.e., h in $H^2(U)$ are holomorphic functions such that $\iint_U |h|^2 dA < \infty$. We may think of $H^2(U)$ as being a subset of $L^2(U)$ by adopting the standard convention that two functions that agree almost everywhere are the same function. Since $H^2(U)$ is a closed subspace of a Hilbert space $L^2(U)$, it too is a Hilbert space.

The Bergman kernel $K(z, \overline{w})$ has the characteristic property of reproducing functions of $L^2(U)$ according to the formula $f(w) = \langle f(z), K(z, \overline{w}) \rangle$. If z_0 is a point in the interior of U, then for each n the linear functional $L_n(f) = f^{(n)}(z_0)\,, n = 0, 1, \cdots$ is bounded over $L^2(U)$. If C is a rectifiable arc lying in the interior (endpoints included) of U and if $\int_C |w(z)||dz| < \infty$, then the linear functional $L(f) = \int_c w(z) f(z) dz$ is bounded over $L^2(U)$. If L is any bounded linear functional over $L^2(U)$, its representative $r(z)$ is given by

$$r(z) = \overline{L_v K(\overline{z}, w)}\,.$$

This means that

$$L(f) = (f, r) = \iint_U f(z)\overline{r(z)} dA, \quad f \in L^2(U).$$

If U is a simply connected region and if $w = M(z)$, $M(0) = 0$ performs a one-to-one conformal map of U onto the unit circle $|w| \leqq 1$, then the Bergman kernel function can be expressed in terms of $M(z)$ as follows:

$$K(\overline{z}, w) = \frac{1}{\pi} \frac{\overline{M'(z)} M'(w)}{(1 - \overline{M(z)} M(w))^2}.$$

If $L(f) = \iint_U f(z) dA$, then $r(z) \equiv 1$, and we have

$$\overline{L_\eta K(\overline{z}, \eta)} \equiv 1\,.$$

This is the functional equation for functionals whose representative is identically 1. If we assume that U is simply connected, then

$$1 = \overline{\frac{1}{\pi} L_\eta \left(\frac{\overline{M'(z)} M'(\eta)}{(1 - \overline{M(z)} M(\eta))^2} \right)}$$

or

$$1 \equiv \frac{1}{\pi} \sum_{n=0}^{\infty} (n+1)(M(z))^n M'(z) \overline{L\left((M(\eta))^n M'(\eta)\right)},$$

where the convergence is absolute and uniform on every compact subregion of U. Thus,

$$1 = \frac{1}{\pi} \frac{d}{dz} \sum_{n=0}^{\infty} (M(z))^{n+1} \overline{L\left(M^n(\eta) M'(\eta)\right)}.$$

Integrating from $z = 0$ to $z = z$ in U,

$$\pi z = M(z) \sum_{n=0}^{\infty} (M(z))^n \overline{L\,(M^n(\eta)M'(\eta))}.$$

From this, it follows that

$$\overline{z} = \overline{M(z)} L_\eta \left(\frac{M'(\eta)}{1 - M(\eta)\overline{M(z)}} \right). \tag{20.16}$$

If $z = m(w)$, $m(0) = 0$ designates the function inverse to $w = M(z)$, then (20.16) can be written as

$$\pi \overline{m(w)} = \overline{w} L_\eta \left(\frac{M'(\eta)}{1 - \overline{w} M(\eta)} \right). \tag{20.17}$$

Theorem 20.3.6 *Let U be a bounded simply connected region whose boundary ∂U is analytic and has Schwarz function $S(z)$. Assume that $z = 0$ is contained in U and that $z = m(w)$, $m(0) = 0$ maps the unit disc $|w| \leqq 1$ one-to-one conformally onto U. Then $S(z)$ is meromorphic in U if and only if $m(w)$ is a rational function of w.*

Proof Suppose that $m(w)$ is rational; we can write

$$m(w) = \frac{aw\,(1 - \beta_1 w) \cdots (1 - \beta_p w)}{(1 - \alpha_1 w) \cdots (1 - \alpha_n w)} = aw + \cdots$$

where the α's differ from the β's. We have

$$\begin{aligned} \iint_U f(z)dA &= \frac{1}{2i} \int_{\partial U} \overline{z} f(z) dz \\ &= \frac{1}{2i} \int_{|w|=1} \overline{m(w)} f(m(w)) m'(w) dw \\ &= \frac{1}{2i} \int_{|w|=1} \overline{m} \left(\frac{1}{w} \right) f(m(w)) m'(w) dw\,. \end{aligned}$$

Now

$$\overline{m} \left(\frac{1}{w} \right) = \frac{\overline{a} w^{n-p-1} \left(w - \overline{\beta}_1 \right) \cdots \left(w - \overline{\beta}_p \right)}{(w - \overline{\alpha}_1) \cdots (w - \overline{\alpha}_n)}.$$

If $n \geqq p + 1$, then $\overline{m}(1/w)$ has poles at $\overline{\alpha}_1, \ldots, \overline{\alpha}_n$ and no other singularities.
If $n < p + 1$, then $\overline{m}(1/w)$ also has a pole of order $p + 1 - n$ at 0.

If we first assume that $n \geqq p + 1$ and that the points α_i are distinct, then by the residue theorem,

$$\iint_U f(z)dA$$
$$= \pi \overline{a} \sum_{k=1}^{m} f(m(\overline{\alpha}_k)) m'(\overline{\alpha}_k) \frac{\overline{\alpha}_k^{n-p-1}(\overline{\alpha}_k - \overline{\beta}_1)\cdots(\overline{\alpha}_k - \overline{\beta}_p)}{\overline{P}'(\alpha_k)},$$

$$P(w) = (w - \alpha_1)\cdots(w - \alpha_n).$$

Thus, we have an identity of the form $\iint_U f(z)dA = \sum_{k=1}^{n} c_k f(z_k)$, where c_k and the abscissas $z_k = m(\overline{\alpha}_k)$ are independent of f.

If the α_i are not distinct, then each point of higher multiplicity τ_k contributes a differential operator of order $\tau_k - 1$ evaluated at $\overline{\alpha}_k$.

If $n < p+1$, then the point $\alpha = 0$ is a pole and hence $f(0)(m(0) = 0)$ is present by itself if $n - p - 1 = -1$ or with its higher derivatives if $n - p - 1 < -1$.

In any case, therefore, $\iint_U f(z)dA = L(f)$, $L \in \mathcal{D}$, for all $f \in \mathcal{A}(U)$. By the last theorem, therefore, $S(z)$ must be meromorphic in U.

Conversely, suppose that $S(z)$ is meromorphic in U. Then by the last theorem, for $f \in \mathcal{A}(U)$, $\iint_U f dA = L(f)$, with $L \in \mathcal{D}$. Now by (20.17),

$$\overline{\pi m(w)} = \overline{w} L_\eta \left(\frac{M'(\eta)}{1 - \overline{w}M(\eta)}\right).$$

Since L is a point differential functional, it follows that

$$L_\eta\left(\left(M'(\eta)\right)/(1 - \overline{w}M(\eta))\right)$$

and hence $\overline{m(w)}$ is a rational function of $\overline{w}$. □

The following theorem shows that the Schwarz function of domain U must have a singularity inside U.

Theorem 20.3.7 *Let f be holomorphic in $U \cup \partial U$, and let*

$$\overline{f}(S(z))\left(= \overline{f(\overline{S(z)})}\right),$$

be holomorphic in all of U. Then f is identically constant.

Proof Green's theorem in two variables tells us that

$$\int_{\partial U} \alpha dy - \beta dx = \iint_U \left(\alpha_x + \beta_y\right) dA \tag{20.18}$$

Quite general sufficient conditions on α, β, and U for this to be true will be found, e.g., in [449, 10.45 Green's theorem, p. 283].

Consider now the integral

$$\frac{1}{2i}\int_{\partial U}\overline{f(z)}g(z)dz = \frac{1}{2i}\int_{\partial U}\overline{f}g(dx+idy)$$

where f and g are analytic functions regular (for simplicity) in $U \cup \partial U$. Write

$$\alpha = \frac{1}{2}\overline{f}g, \quad \beta = \frac{i}{2}\overline{f}g.$$

Then $\alpha_x = \frac{1}{2}(\partial/\partial z + \partial/\partial\overline{z})\overline{f}g$, $\beta_y = -\frac{1}{2}(\partial/\partial z - \partial/\partial\overline{z})\overline{f}g$, so that $\alpha_x + \beta_y = \partial/\partial\overline{z}(\overline{f}g) = \partial/\partial\overline{z}(\overline{f}(\overline{z})g(z)) = \overline{f'(z)}g(z)$. From (20.18),

$$\begin{aligned}\iint_U \overline{f'(z)}g(z)dA &= \frac{1}{2i}\int_{\partial U}\overline{f(z)}g(z)dz\\ &= \frac{1}{2i}\int_{\partial U}\overline{f}(\overline{z})g(z)dz.\end{aligned}$$

This establishes Green's theorem in a complex analytic form. Suppose now that ∂U consists of a single analytic curve whose Schwarz function is $S(z)$. Then $\overline{z} = S(z)$ along ∂U so that

$$\iint_U \overline{f'(z)}g(z)dA = \frac{1}{2i}\int_{\partial U}\overline{f}(S(z))g(z)dz. \tag{20.19}$$

In particular, the selection $f(z) \equiv z$ yields

$$\iint_U g(z)dA = \frac{1}{2i}\int_{\partial U}S(z)g(z)dz. \tag{20.20}$$

(Note that for values of z interior to U and close to ∂U, $\overline{S(z)}$ is exterior to U and close to ∂U. Therefore, $\overline{f}(S(z))$ is initially known to be holomorphic near C without the further hypothesis.) We have from (20.19) and by Cauchy's theorem,

$$\int_U\int |f'(z)|^2 dA = \frac{1}{2i}\int_{\partial U}\tilde{f}(S(z))f'(z)dz = 0.$$

Therefore, $f'(z) \equiv 0$ and the conclusion follows. □

By selecting $f \equiv z$, we see that $S(z)$ must have a singularity inside U and furthermore that the singularities of S cannot be annihilated by being composed with a regular function.

The nature of the singularity of $S(z)$ inside U crucially governs the kind of application that can be made.

By choosing $f(\zeta) = \frac{1}{z-\zeta}$ for $z \in \mathbb{C}\setminus U$ in (20.5), one realizes that the Cauchy transform

$$\widehat{\chi}_U(z) = -\frac{1}{\pi}\int_U \frac{dA(\zeta)}{z-\zeta} \tag{20.21}$$

of the characteristic function of a quadrature domain U is a rational function outside U. To be precise, there exists a rational function $R(z)$ of the form

$$R(z) = \frac{Q(z)}{P(z)} \tag{20.22}$$

where $P(z) = \prod_{k=1}^{n}(z-a_k)$ and where $Q(z)$ is a polynomial of degree $n-1$, such that

$$\widehat{\chi}_U = R \quad \text{on } \mathbb{C}\backslash U. \tag{20.23}$$

Conversely, by an approximation theorem of L. Bers [80], the linear combinations of the above Cauchy kernels are dense in the integrable analytic functions, so (20.23) holding for some rational function R is actually equivalent to U being a quadrature domain.

The main tool for the approximation used in [80] is a system of ingeniously constructed mollifiers, nowadays often named the "Ahlfors-Bers mollifiers." These mollifiers play an important role in the theory of quadrature domains, also in higher dimensions. See, e.g., [220, 263, 454] for their use.

Assume that U is a quadrature domain (20.4). Since $\frac{\partial}{\partial \bar{z}}\widehat{\chi}_U = \chi_U$ in the sense of distributions, the function $S(z)$ defined on $\overline{U}$ by

$$S(z) = \bar{z} + R(z) - \widehat{\chi}_U(z) \tag{20.24}$$

is meromorphic in U, and by (20.23),

$$S(z) = \bar{z} \quad \text{for } z \in \partial U\,. \tag{20.25}$$

Thus, $S(z)$ is a one-sided *Schwarz function* of ∂U. To be a two-sided Schwarz function, one requires that $S(z)$ is defined and analytic in a full neighborhood of ∂U. The use of such a function associated to an analytic curve can be traced back to G. Herglotz [233, pp. 299-355]. It was named after H.A. Schwarz's reflection principle in [129]. Full accounts are given in [130, 468].

A two-sided Schwarz function, defined in some neighborhood of ∂U, exists if and only if ∂U is analytic, while a one-sided Schwarz function allows for certain singular points of ∂U. A complete characterization of those boundaries allowing a one-sided Schwarz function was given by M. Sakai in [455].

The previous argument also goes the other way around, so a bounded domain U is a quadrature domain if and only if there exists a meromorphic function $S(z)$ in U, continuous up to ∂U, so that (20.25) holds. This result was obtained in [130] under some smoothness assumptions on ∂U and in full generality in [6].

Indeed, one of the major achievements in [6] is that the authors prove, without any regularity assumptions whatsoever, that ∂U is a subset of an algebraic curve: there exists a nontrivial polynomial $Q(z, w)$ such that

$$\partial U \subset \{z \in \mathbb{C} : Q(z, \overline{z}) = 0\}. \tag{20.26}$$

From this, additional regularity of ∂U follows easily.

Differentiating (20.25) gives

$$s(z)dz = d\overline{z} \quad \text{along } \partial U , \tag{20.27}$$

where $s(z) = S'(z)$. The relation (20.27) holds for some meromorphic function $s(z)$ (not necessarily having a single-valued primitive) if and only if a quadrature identity of the kind

$$\int_U f dA = \sum_{k=1}^{m} \sum_{j=0}^{n_k-1} c_{kj} f^{(j)}(a_k) + \sum_{i=1}^{r} b_i \int_{\gamma_i} f dz \tag{20.28}$$

holds, where γ_i are (smooth) closed or nonclosed curves, compactly contained in U, and $b_i \in \mathbb{C}$. See [6, 35, 219, 262].

20.4 Area Quadrature Domains

We now turn to the study of area quadrature domains. We will follow [64, Chapter 22] for this section. In this section, we will show that area quadrature domains share many of the properties of the unit disc and that they turn out to be very abundant. We will also explain how to view our results as an improvement upon the Riemann mapping theorem.

Assume that U is an area quadrature domain. Note that the function that is identically 1 is in the Bergman space because of the finite area assumption, and if h is another function in the Bergman space, then

$$\iint_U h dA = (h, 1)_U .$$

Thus, we see that h is in $L^1(U)$ by the Cauchy-Schwarz inequality, and so the integral in the quadrature identity for U is well defined.

Let θ be a real-valued radially symmetric C^∞ function compactly supported in the unit disc such that $\iint_{\mathbb{D}} \theta dA = 1$. Fix a point a_0 in U and let $\epsilon > 0$ be less than the distance from a_0 to the boundary of U. For a near a_0, set

$$\varphi_a(z) = \frac{1}{\epsilon^2}\theta\left(\frac{z-a}{\epsilon}\right)$$

Then

$$h(a) = (h, \varphi_a) \tag{20.29}$$

for all $h \in H^2(U)$ and that is why we were able to conclude that $K_a = B\varphi_a$. Note that this function is in $C_0^\infty(U)$ and that it is the conjugate of $(\partial^m/\partial a^m)\,\varphi_a$ because φ_a is real valued. If a is close to a_0, we may differentiate (20.29) m times with respect to a to obtain

$$h^{(m)}(a) = (h, \varphi_a^m)\,.$$

Let $K_a^m = B\varphi_a^m$. This function represents the functional $h \mapsto h^{(m)}(a)$ in the sense that

$$h^{(m)}(a) = (h, K_a^m).$$

Let K_a^0 denote the Bergman kernel K_a. The Bergman span of U is defined to be the space of all linear combinations of functions of the form K_a^m as a ranges over U and m ranges over nonnegative integers. We may equate (K_a^m, K_z) with the conjugate of (K_z, K_a^m) to deduce that

$$K_a^m(z) = \frac{\partial^m}{\partial \bar{a}^m} K(z, a).$$

We remark here that equating (K_U^n, K_a^m) with the conjugate of (K_a^m, K_U^n) allows one to express all the derivatives of the Bergman kernel in both variables in terms of integrals that are easily bounded on compact sets. In this way, it can be deduced that $K(z, w)$ is C^∞ smooth in z and w in $U \times U$, holomorphic in z, and antiholomorphic in w.

Let Q denote the following linear combination of functions $K_{z_j}^k$ in the Bergman span inspired by the quadrature identity (20.4) that we assume holds:

$$Q(z) = \sum_{j=1}^{N}\sum_{k=0}^{m_j} \bar{c}_{jk} K_{z_j}^k\,.$$

Notice that

$$(h, 1) = (h, Q)$$

for $h \in H^2(U)$ because both sides yield the quadrature identity in h. Since the functions 1 and Q are both in the Bergman space, and since h is arbitrary, it must be that $Q \equiv 1$. The converse of this fact is also clearly true, namely that if the function 1 is in the Bergman span, then U is an area quadrature domain.

Theorem 20.4.1 *A domain of finite area is an area quadrature domain if and only if the function that is identically equal to one is in the Bergman span.*

We will now combine this theorem with the transformation properties of the Bergman kernel functions under biholomorphic mappings to obtain the following theorem.

Theorem 20.4.2 *Suppose that $f : U_1 \to U_2$ is a biholomorphic mapping between domains of finite area. Then U_2 is an area quadrature domain if and only if f' belongs to the Bergman span of U_1.*

Proof Suppose f is a biholomorphic mapping as in the statement of the theorem. Because $|f'|^2$ is equal to the absolute value of the real Jacobian determinate of the mapping as a mapping from $\mathbb{R}^2$ into itself, it follows from the classical change of variables formula that

$$\iint_{U_2} h dA = (h, 1) = \iint_{U_1} |f'|^2 (h \circ f) dA = (f'(h \circ f), f').$$

If f' is in the Bergman span, then the last inner product yields a linear combination of values and derivatives of $f'(h \circ f)$ at finitely many points in U_1, which is a fixed finite linear combination of values and derivatives of h at finitely many points in U_2. Consequently, U_2 is an area quadrature domain.

Conversely, suppose that U_2 is an area quadrature domain. Let $F = f^{-1}$. If h is a function in $H^2(U_1)$, then

$$(h, f') = (F'(h \circ F), 1) = \iint_{U_2} F'(h \circ F) dA$$

and this last integral is equal to a linear combination of values and derivatives of $F'(h \circ F)$ by virtue of the quadrature identity for U_2. Such a linear combination is a linear combination of values and derivatives of h at finitely many points in U_1. There is a function Q in the Bergman span of U_1 that yields the same linear combination when paired with h. Since $\langle h, f' \rangle_{U_1} = \langle h, Q \rangle_{U_1}$ holds for all h in the Bergman space, it follows that $f' = Q$, i.e., that f' is in the Bergman span. □

When Theorem 20.4.2 is combined with the Riemann mapping theorem and the known formula for the Bergman kernel on the disc, the following result is obtained.

Theorem 20.4.3 *A simply connected domain of finite area is an area quadrature domain if and only if the inverse of a Riemann mapping function is rational and without poles on the boundary.*

Proof Suppose that U is a simply connected domain of finite area and let $f : U \to \mathbb{D}$ be a Riemann mapping function associated to a point in U. Recall that the Bergman kernel for the unit disc is $\pi^{-1}/(1 - z\overline{w})^2$. Let $F = f^{-1}$ be the inverse of the Riemann map. Theorem 20.4.2 yields that U is an area quadrature domain if and only if F' is in the Bergman span of the disc. It is easy to verify that the Bergman span of the disc is exactly the set of rational functions with residue free poles outside the closed unit disc. (Note that the complex polynomials belong to this space.) Such rational functions have rational antiderivatives, and the set of all such antiderivatives is the set of rational functions with poles outside the closed unit disc. □

For arclength quadrature domains, see [64, Chapter 23]. Steve Bell has done quite a lot of work on quadrature domains. Bell proves the density of quadrature domains among smooth domains with real analytic boundary. Further, he constructs conformal mappings to nearby arclength quadrature domains and also to domains that are like the disc in that they are simultaneously area and arclength quadrature domains. See [51, 58–65, 67–69, 213].

20.5 Riemann Surfaces (the Schottky Double)

One way to understand the algebraicity (20.26) of ∂U is as follows: The relation (20.25), with $S(z)$ meromorphic in U, may be interpreted as saying that the pair of functions $(S(z), \overline{z})$ constitutes a meromorphic function on the *Schottky double* [460, 461] $\widehat{U}$ of U, namely the compact Riemann surface obtained by completing U with a backside $\widetilde{U}$ having the opposite conformal structure, and gluing the two together along ∂U. (We are here assuming a priori some mild regularity of ∂U.) Then $S(z)$ represents the values of the function on U and $\overline{z}$ the values on $\widetilde{U}$.

The opposite pair $(z, \overline{S(z)})$ will also represent a meromorphic function on $\widehat{U}$, and since any two meromorphic functions on a compact Riemann surface are related by a polynomial equation, it follows that there exists a polynomial $Q(z, w)$ such that

$$Q(z, S(z)) = 0 \quad (z \in U)\,.$$

In particular, $Q(z, \overline{z}) = 0$ on ∂U, i.e., we obtain (20.26) again.

More detailed investigations [219] show that $Q(z, w)$ (if chosen without extraneous factors) has degree exactly n in each of z and w (total degree $2n$) and that the difference set in (20.26) consists of at most finitely many points. Additional information on the structure of $Q(z, w)$ can be derived from properties of the exponential transform; see [221, 222, 427, 428].

If U is any simply connected bounded domain, then U is a quadrature domain if and only if any conformal map $g : \mathbb{D} \to U$ is a rational function. This was proved in [130] under some regularity assumptions on ∂U and in [6] without such assumptions. The idea for the nontrivial direction is to use the fact that $z \mapsto \overline{S(z)} \quad (z \in U)$ is the anti-conformal reflection in U to extend the mapping function g from $\mathbb{D}$ to the entire Riemann sphere.

For multiple connected domains, there are analogous results [219]: Let U be a finitely connected domain representing a certain conformal type and let $(g, \overline{h})$ represent a meromorphic function on the Schottky double $\widehat{U}$ so that g and h are meromorphic in U and $g = \overline{h}$ on ∂U. If g is holomorphic and univalent on U, then it will map U onto a quadrature domain, because the relation $g = \overline{h}$ on ∂U will become a relation of the form (20.25) in the image domain.

Other techniques to construct quadrature domains, based on Riemann surface ideas (automorphic functions, Schottky groups, Poincaré series, etc.), have been developed by Avci [35], Richardson, [432], and Crowdy and Marshall [122, 123].

Chapter 21
Teichmüller Theory

We have certainly benefitted from two sources in writing this chapter.

First, we originally profited from Tromba's *Teichmüller Theory in Riemannian Geometry* (based on lecture notes by Jochen Denzler) [511], which gives a detailed presentation of a differential geometric approach to Teichmüller's moduli space of closed Riemann surfaces which had been developed by the same author and A. Fisher.

Second, the text of the lectures is presented by Tromba [510] in the international course on "Topics in Calculus of Variations" which was held at Montecatini, Italy, July 20–28, 1987.

Wienhard [529, p. 1033] writes: "Let S be a closed connected oriented topological surface of negative Euler characteristic $\chi(S) = 2 - 2g < 0$, where g is the genus of S. The Teichmüller space $\mathcal{T}(S)$ of S is the space of marked conformal classes of Riemannian metrics on S. It has been well studied using the theory of quasi-conformal maps as well as methods from hyperbolic geometry. By the uniformization theorem, there is a unique hyperbolic, i.e., constant curvature -1, metric in each conformal class. This identifies $\mathcal{T}(S)$ with the moduli space of marked hyperbolic structures. A marked hyperbolic structure is a pair (X, f_X), where X is a hyperbolic surface and $f_X : S \to X$ is an orientation preserving homeomorphism. Two marked hyperbolic structures (X, f_X) and (Y, f_Y) are equivalent if there exists an isometry $g : X \to Y$ such that $g \circ f_X$ is isotopic to f_Y. The mapping class group of S acts naturally on $\mathcal{T}(S)$ by changing the marking. This action is properly discontinuous, and the quotient of $\mathcal{T}(S)$ by this action is the moduli space $\mathfrak{M}(S)$ of Riemann surfaces of topological type given by S. Teichmüller space is homeomorphic to $\mathbb{R}^{6g-6}$ and the universal cover of $\mathfrak{M}(S)$."

In this chapter, we use only methods of global analysis and Riemannian geometry to develop "a priori" Teichmüller theory. We shall give several descriptions of Teichmüller space as a finite dimensional differentiable manifold, and we prove that all these manifolds are diffeomorphic. A consequence of these characterizations is a

P. V. Dovbush, S. G. Krantz, *The Geometric Theory of Complex Variables*,
https://doi.org/10.1007/978-3-031-77204-7_21

straightforward proof of Teichmüller's famous result on the structure of the moduli space for compact Riemann surfaces that now bears his name.

We have attempted to make the chapter as self-contained as possible. Some preliminary definitions and results are contained in Chap. 7.

21.1 What Is Teichmüller Space?

Homeomorphic smooth manifolds are not necessarily diffeomorphic. It is not always possible to find a diffeomorphism between M and S^n! Examples of smooth manifolds homeomorphic, but not diffeomorphic, to S^7 were first constructed by J. Milnor in [379]. It is also true that homeomorphic manifolds have equal dimensions, but the proof requires a bit of algebraic topology.

From now on, a surface is a two-dimensional manifold. Every closed surface is classified up to diffeomorphism by its genus, the number of holes (or equivalently its Euler characteristic), and whether or not it is orientable. The classical statement says that the genus of a surface characterizes its topology (that is, two surfaces M and N are homeomorphic if and only if they have the same genus g.)

Theorem 21.1.1 (Classification of Compact Orientable Surfaces) *([240, Theorem 3.5, p. 204]) Let M be a compact connected orientable surface without boundary. Then there is a unique integer* $g = g(M) \geqslant 0$ *such that M is an orientable surface of genus g (a "sphere with g handles"). The Euler characteristic* $\chi(M)$ *of M determines g by the formula* $\chi(M) = 2 - 2g$*. In particular,* $\chi(M)$ *is even and* $\leqslant 2$*.*

Induction on g shows that an orientable surface of genus g is compact, connected, and reversible.[1] It has Euler characteristic $2 - 2g$. Therefore, orientable surfaces of different genus are not diffeomorphic. On the other hand, induction on g shows that two orientable surfaces of the same genus are diffeomorphic (see, for example, [240, 1.9 Theorem, p. 193]).

Already in dimension two, the classification of non-compact surfaces becomes difficult, due to the existence of exotic spaces such as Jacob's ladder (i.e., the cylinder with infinitely many handles attached in the direction of both ends, [175, Théorème A]) (Fig. 21.1).

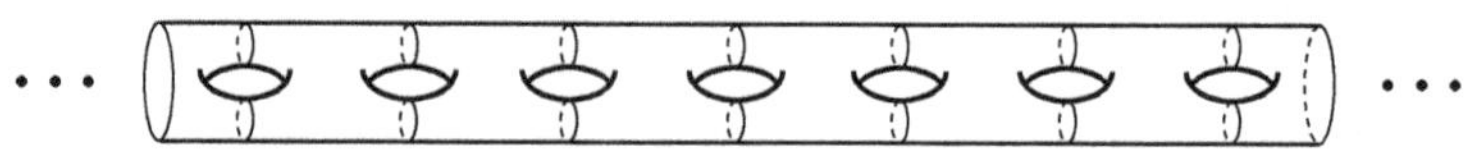

Fig. 21.1 The Jacob's ladder surface

[1] A connected manifold is called *reversible* if it is orientable and admits an orientation-reversing diffeomorphism.

Before we turn to working out our program, let us recall some classical facts from the theory of compact two-dimensional manifolds, to stress the existence of an abyss between $\mathcal{C}^\infty$ structures and the complex ones. In the two-dimensional case, this abyss is much deeper than between topological and differentiable structures.

Moreover, a classical result in surface theory [183] yields

Theorem 21.1.2 *If $\mathcal{M}$, N are compact, oriented $\mathcal{C}^\infty$ 2-manifolds (without boundary) which are homeomorphic, then they are diffeomorphic.*

This shows that the classification of compact, connected 2-manifolds up to homeomorphism, is the same as the classification up to diffeomorphism, and thus the $\mathcal{C}^\infty$ property does not provide us with any finer classification.

Riemann knew that the situation changes dramatically once complex structure is introduced; this was the source of the problem of moduli.

Let us recall the notion of complex structure on M. It is merely an atlas $c := \{\mathcal{U}_j, \varphi_j\}$ such that $\varphi_i \circ \varphi_j^{-1}$ are holomorphic maps.

When we think of M having a fixed complex structure c, we shall designate this by writing the pair (M, c). However, a given M may have many complex structures.

For example, let $f : M \longrightarrow M$ be a $\mathcal{C}^\infty$ self-mapping of M with a $\mathcal{C}^\infty$ inverse f^{-1} (such an f is a diffeomorphism). Given one such complex structure, we can construct a new complex structure f^*c, with $f^*c := \left\{\left(f^{-1}(U_i), \varphi_i \circ f\right)\right\}$ where $c = \{(U_i, \varphi_i)\}$.

Now f^*c will be a different complex structure (different due to the charts, which are not only different from the former ones, but even incompatible with the former ones).

Clearly,

$$\left(\varphi_j \circ f\right) \circ \left(\varphi_j \circ f\right)^{-1} = \varphi_i \circ f \circ f^{-1} \circ \varphi_j^{-1} = \varphi_j \circ \varphi_j^{-1}.$$

It follows easily that $f^\star c$ is a complex structure if and only if c is a complex structure. Note that (M, c) and (M, f^*c) are holomorphically equivalent, whereas (in general) $f : (M, c) \to (M, c)$ is not a holomorphic equivalence.

This was already known to Riemann.

Let c, c' be two complex structures on an oriented, compact 2-manifold $\mathcal{M}$. Then the Riemann surfaces (M, c) and $\left(M, c'\right)$ need not be holomorphically equivalent.

A. The existence of several distinct complex structures can be easily seen in the case of tori (genus 1). It is classical knowledge of course: A holomorphic equivalence $f : T = \mathbb{C}/\Omega \to T' = \mathbb{C}/\Omega'$ can be lifted to a holomorphic equivalence $\widetilde{f} : \mathbb{C} \to \mathbb{C}$ which commutes with the translations by lattice vectors in Ω and Ω', respectively:

$$\widetilde{f}(z + \omega_i) = \widetilde{f}(z) + \omega_i' \text{ for basis vectors } \omega_1, \omega_2 \in \Omega, \omega_1', \omega_2' \in \Omega'.$$

This implies that $|\widetilde{f}(z)| \leq A + B|z|$, and, using Cauchy's integral formula, holomorphy forces $\widetilde{f}(z) = a + bz$. In other words, Ω maps to Ω' under a similarity transformation. But there is a continuum of parallelograms, which are not similar to each other and give rise to tori which are not biholomorphically equivalent: in particular, two parallelograms P and P' with sides given by complex vectors ω_1, ω_2 and $\omega_1', \omega_2' \in \mathbb{C}$ will be similar iff $\omega_1/\omega_2 = \omega_1'/\omega_2'$.

As an aside, we note that the question of holomorphic (conformal) equivalence of Riemann surfaces is also equivalent to the question of isometry (up to scaling by a constant) of the corresponding Riemannian (real) 2-manifolds. In the case of tori, these are all equipped with the "flat" metric (which comes from Euclidean $\mathbb{R}^2$); all are locally, but not in general globally, isometric.

B. An analogous though less transparent situation holds for genus ≥ 2. Here again, holomorphic equivalence classes can be viewed as equivalence classes of isometries, but this time of hyperbolic metrics (metrics of constant negative curvature) as opposed to flat metrics (metrics of zero curvature) as in the case of a torus. We can argue this point heuristically as follows: The universal conformal covering space of a surface with a hyperbolic metric will be the hyperbolic plane. As the Euclidean plane can be tiled in many different ways by 4-gons, the hyperbolic plane can be tiled in many different ways by $4g$-gons ($4g$ of them meeting at each corner). If $g \geq 2$, these tilings give rise to globally non-isometric surfaces of genus g which are, however, locally isometric to the hyperbolic plane, their universal cover. Since the automorphisms (= holomorphic equivalences) of the upper half plane $\mathbb{H}$, considered as a Riemann surface, are the same as the isometry transformations of $\mathbb{H}$, considered as a Riemannian 2-manifold with the hyperbolic metric, we obtain many different holomorphic equivalence classes of a Riemann surface.

Hence, the classification of Riemann surfaces should be such that (M, c) and (M, f^*c) are in the same equivalence class.

Denote by $\mathcal{D} = \text{Diff}^+(M)$ the group of orientation-preserving diffeomorphisms of M and by $\mathcal{C}$ the *set of all complex structures* on M. Then we have a natural action of $\mathcal{D}$ on $\mathcal{C}$ (from the right) by pullback:

$$\mathcal{C}(M) \times \text{Diff}^+(M) \to \mathcal{C}(M), \quad (c, f) \to f^*c$$

Definition 21.1.3 The *Riemann moduli space* of M denoted by $\mathcal{M}$ or $\mathcal{R}(M)$ is the space of $\mathcal{D}$ orbits

$$\mathcal{R}(M) := \mathcal{C}(M)/\mathcal{D}(M).$$

Thus, $\mathcal{M}_p$ is the set of Riemann surfaces of genus p that are not holomorphically equivalent.

The following examples were known to Riemann:

Example 21.1.4 $\mathcal{M}_0 = \{\text{point}\}$, since any Riemann surface homeomorphic to the sphere is isomorphic to the Riemann sphere $\mathbb{P}^1(\mathbb{C})$.

The second example is much more interesting (elliptic case) and requires some knowledge of elliptic functions.

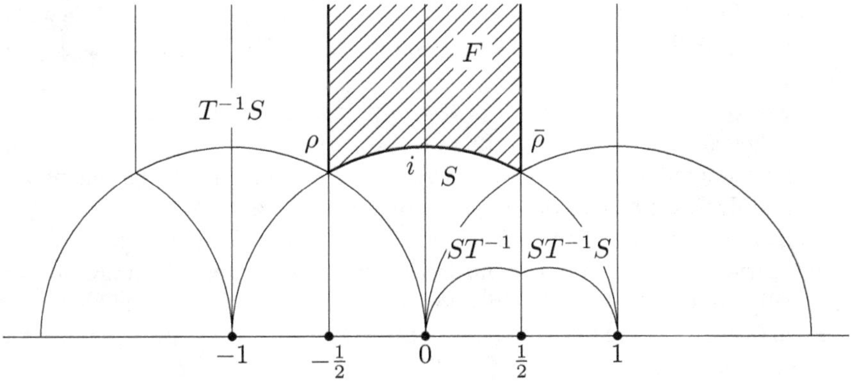

Fig. 21.2 The modular group

Example 21.1.5 $\mathcal{M}_1 = \mathbb{C}$. Any Riemann surface of genus 1 is homeomorphic to the torus $T^1 = \mathbb{C}/\Lambda$. Two such tori E_Λ and $E_{\Lambda'}$ are biholomorphic if and only if there exists an automorphism $z \to az + b$ of $\mathbb{C}$ carrying the lattice Λ to Λ'. If the lattice Λ is generated by $\lambda_1, \lambda_2 \in \mathbb{C}$ and Λ' by λ_1' and λ_2', this will be the case if and only if $\tau = \lambda_1/\lambda_2$ and $\tau' = \lambda_1'/\lambda_2'$ are related by

$$\tau' = \frac{\alpha\tau + \beta}{\gamma\tau + \delta} \quad \begin{pmatrix} \alpha & \beta \\ \gamma & \delta \end{pmatrix} \in SL(2, \mathbb{Z})\,.$$

Since we can take τ lying in the upper half plane $\mathbb{H} = \{z \in \mathbb{C} : \operatorname{Im} z > 0\}$, $\mathcal{M}_1$ is the set of isomorphism classes of elliptic curves and can be identified with $\mathbb{H}/\mathrm{SL}(2, \mathbb{Z})$. The set of moduli is the fundamental domain F of the group $\mathrm{SL}(2, \mathbb{Z})$, called the *modular group* (see Fig. 21.2):

$$F = \left\{z \in \mathbb{H} : -\frac{1}{2} \leqslant \operatorname{Re} z < \frac{1}{2}, |z| \geqslant 1 \text{ if } \operatorname{Re} z \leqslant 0, |z| > 1 \text{ if } \operatorname{Re} z > 0\right\}.$$

Thus, $\mathcal{M}_1 \simeq F \simeq \mathbb{C}$.

We have, therefore, the Riemann observation (and conjecture):

$$\dim_{\mathbb{C}} \mathcal{M}_p = \begin{cases} 0, & p = 0 \\ 1, & p = 1 \\ 3(p-1), & p > 1 \end{cases}$$

This last example is of great interest as it is a sort of pilot case: we encountered the basic features of the moduli problems (this holds as well for the case of objects different from Riemann surfaces, for example, instantons—the self-dual connections.)

From now on, M will be fixed as a $\mathcal{C}^\infty$ smooth compact oriented surface without boundary. We shall henceforth always assume that M has a fixed genus g greater than one.

Surfaces of a fixed genus are all diffeomorphic. Therefore, if one has a complex structure they all have a complex structure.

Riemann wanted to identify (M, c) with (M, f^*c) and thus (under this relation) consider equivalence classes of complex structures on a fixed M.

When classifying Riemann surfaces, one does not attempt to distinguish different complex structures, but one distinguishes complex structures which are not holomorphically equivalent. The classification of Riemann surfaces should be such that (M, c) and (M, f^*c) fall into the same equivalence class. We want to identify these two complex structures.

What we have here is an infinite-dimensional space $\mathcal{C}$ of geometrical structures (in our case, conformal structures) on M (torus in our case) and the infinite-dimensional group $\mathcal{D}$ acting on the space $\mathcal{C}$, but the space of orbits $\mathcal{C}/\mathcal{D}$ is finite-dimensional and its points (orbits) are called *moduli*.

The term "moduli" is due to Riemann; the phrase "a curve of genus $g > 2$ depends on $3g - 3$ moduli (numbers)" occurs in a paper by Riemann [434]. The upshot of this work is that one should create "universal families" of whatever one is trying to parametrize.

The structure of the moduli space $\mathcal{M}_p$ for $p > 1$ is mysterious; it is still not completely understood.

Denote by $\mathcal{R}(M)$ the quotient space $\mathcal{C}/\mathcal{D}$.

This is known as the *Riemann space of moduli*.

The word "moduli" is due to Riemann; the phrase "a curve of genus genus $g > 2$ depends on $3g-3$ numbers moduli" occurs in a paper by Riemann [434]. The upshot of this work is that one should create "universal families" of whatever one is trying to parametrize.

In our case, we will fit all Riemann surfaces of an appropriate type into a family. What "appropriate" means is straightforward for compact Riemann surfaces; even for Riemann surfaces of finite type, things are good.

The famous Riemann moduli problem is the following: Characterize the set of isomorphism classes of Riemann surfaces of a given topological type.

Riemann solved this problem for simply connected Riemann surfaces. The uniformization theorem states that there are only three different types of simply connected Riemann surfaces; namely the Riemann sphere, the complex plane, and the unit disc.

Moduli spaces arise naturally in classification problems of geometry. Typically, one has a collection of objects and an equivalence relation, and the problem is to describe the set of equivalence classes. Usually, there are discrete invariants that partition this collection into a countable number of subsets. In most cases, there exist continuous families of objects, and one would like to give the set of equivalence classes some geometric structure to reflect this fact. This is the object of the theory of moduli spaces.

Riemann conjectured that $\mathcal{C}/\mathcal{D} = \mathcal{R}(M)$ is a space of dimension 6(genus M)-6. The space $\mathcal{R}(M)$ is now called the *Riemann space of moduli.*

21.1.1 Inner Automorphisms

Definition 21.1.6 Let G be a group. An isomorphism from G onto itself is called an *automorphism* of G. The set of all automorphisms of G is denoted by Aut(G).

A simple verification shows that Aut(G), the automorphism group of G, is a group under composition of automorphisms (composition of automorphisms is defined since the domain and range of each automorphism is the same).

One of the most important examples of an automorphism of a group G is provided by conjugation by a fixed element in G. More specifically, the action of G on H by conjugation is defined for each $g \in G$ by

$$h \mapsto ghg^{-1} \quad \text{for each } h \in H.$$

For each $g \in G$, conjugation by g is an automorphism of H.

An *inner automorphism* of a group G is an automorphism of the form $\phi(g) = h^{-1}gh$, where h is a fixed element of G. An inner automorphism is any automorphism that arises from conjugation [139, Chapter 4.4 AUTOMORPHISMS, p. 132]

The composition of two inner automorphisms is again an inner automorphism, and with this operation, the collection of all inner automorphisms of G is a group, the inner automorphism group of G denoted by Inn(G). Inn(G) is a normal subgroup of the full automorphism group Aut(G) of G. The outer automorphism group, Out(G), is the quotient group

$$\mathrm{Out}(G) = \mathrm{Aut}(G)/\,\mathrm{Inn}(G).$$

The outer automorphism group measures, in a sense, how many automorphisms of G are not inner. Every non-inner automorphism yields a nontrivial element of Out(G), but different non-inner automorphisms may yield the same element of Out(G).

Saying that conjugation of x by a leaves X unchanged is equivalent to saying that a and X commute:

$$a^{-1}xa = x \Longleftrightarrow xa = ax.$$

Therefore, the existence and number of inner automorphisms that are not the identity mapping is a kind of measure of the failure of the commutative law in the group (or ring).

An automorphism of a group G is inner if and only if it extends to every group containing G (see [463]).

By associating the element $a \in G$ with the inner automorphism $f(x) = a^{-1}xa = x^a$ in $\mathrm{Inn}(G)$ as above, one obtains an isomorphism between the quotient group $G/\mathrm{Z}(G)$ (where $\mathrm{Z}(G)$ is the center of G) and the inner automorphism group:

$$G/\mathrm{Z}(G) \cong \mathrm{Inn}(G)$$

This is a consequence of the first isomorphism theorem, because $\mathrm{Z}(G)$ is precisely the set of those elements of G that give the identity mapping as corresponding inner automorphism (conjugation changes nothing).

21.1.2 The Fundamental Group

Let P and Q be two points on a Riemann surface S and γ_{PQ} a curve, i.e., a continuous map $\gamma : [0, 1] \to S$, connecting them: $\gamma_{PQ}(0) = P$, $\gamma_{PQ}(1) = Q$.

Definition 21.1.7 Two curves γ^1_{PQ} and γ^2_{PQ} on S with the initial point P and the terminal point Q are called *homotopic* if they can be continuously deformed one to another, i.e., provided there is a continuous map $\gamma : [0, 1] \times [0, 1] \to S$ such that $\gamma(t, 0) = \gamma^1_{PQ}(t)$, $\gamma(t, 1) = \gamma^2_{PQ}(t)$, $\gamma(0, \lambda) = P$, $\gamma(1, \lambda) = Q$. The set of homotopic curves forms a *homotopy class*, which we denote by $\Gamma_{PQ} = \left[\gamma_{PQ}\right]$.

If the terminal point of γ_1 coincides with the initial point of γ_2, then the curves can be multiplied:

$$\gamma_1 \cdot \gamma_2(t) = \begin{cases} \gamma_1(2t), & 0 \le t \le \dfrac{1}{2} \\ \gamma_2(2t-1), & \dfrac{1}{2} \le t \le 1 \end{cases}$$

This multiplication is well defined also for the corresponding homotopy classes

$$\Gamma_1 \cdot \Gamma_2 = [\gamma_1 \cdot \gamma_2]$$

Any two closed curves through P can be multiplied. The set of homotopy classes of these curves forms a group $\pi_1(S, P)$ with the multiplication defined above. The curves, which can be contracted to a point, correspond to the identity element of the group. It is easy to see that the groups $\pi_1(S, P)$ and $\pi_1(S, Q)$ based at different points are isomorphic as groups. Considering this group, one can omit the second argument in the notation

$$\pi_1(S, P) \approx \pi_1(S, Q) \approx \pi_1(S)$$

Definition 21.1.8 The group $\pi_1(S)$ is called the *first fundamental group* of S.

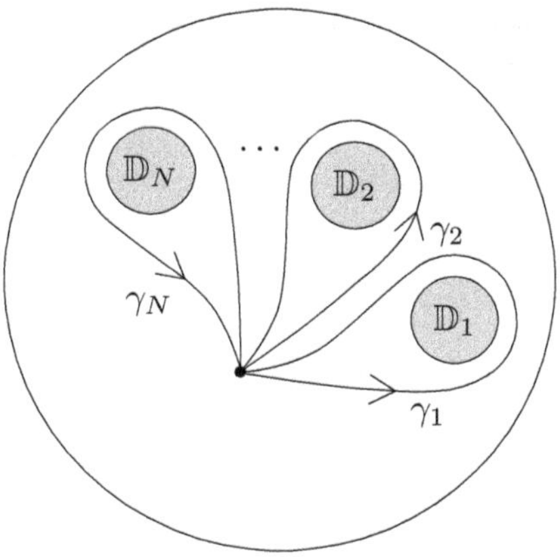

Fig. 21.3 Fundamental group of a sphere with N holes

Definition 21.1.9 We say that a path-connected Riemann surface is *simply connected* if its first fundamental group is trivial.

If a Riemann surface S is path connected, then, the group $\pi_1(S, P_0)$ does not depend on P_0; thus, we can speak of the fundamental group $\pi_1(S)$ without specifying the point P_0.

Example 21.1.10 *Sphere with N holes*
$S = \mathbb{S}^2 \setminus \left\{\bigcup_{n=1}^N H_n\right\}$.

The fundamental group is generated by the homotopy classes of the closed curves $\gamma_1, \ldots, \gamma_N$ each going around one of the holes (Fig. 21.3). The curve $\gamma_1\gamma_2\ldots\gamma_N$ can be contracted to a point, which implies the relation

$$\Gamma_1\Gamma_2\ldots\Gamma_N = 1$$

in $\pi_1\left(\mathbb{S}^2 \setminus \left\{\bigcup_{n=1}^N H_n\right\}\right)$.

Example 21.1.11 (Compact Riemann Surface of Genus g)
It is convenient to consider the $4g$-gon model F_g (Fig. 21.4). The curves $a_1, b_1, \ldots, a_g, b_g$ are closed on $\mathcal{R}$. Their homotopy classes, which we denote by $A_1, B_1, \ldots, A_g, B_g$, generate $\pi_1(\mathcal{R})$. The curve

$$a_1b_1a_1^{-1}b_1^{-1}\ldots a_gb_ga_g^{-1}b_g^{-1}$$

is the oriented boundary of F_g. This implies the relation

$$A_1B_1A_1^{-1}B_1^{-1}\ldots A_gB_gA_g^{-1}B_g^{-1} = 1 \qquad (21.1)$$

in the fundamental group. There are no other independent relations. Indeed, such a relation would mean that some product p of the curves $a_1, \ldots, b_g$ can be contracted

Fig. 21.4 Fundamental group of a compact surface of genus g

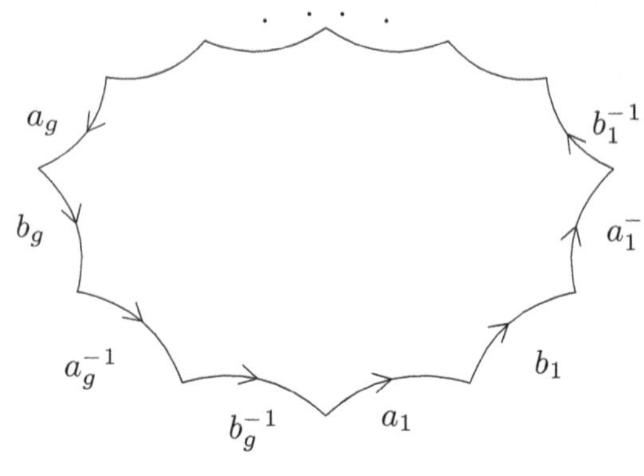

to a point. Since all the points of $\mathcal{R}$ are equivalent, this point can be chosen inside F_g. This proves that $[p]$ is a multiple of (21.1).

For more on this topic, see [95, 189, 318, 443, 480].

In 1939, Oswald Teichmüller in a series of brillant papers [503] broke up the problem into two parts as follows: Let $\mathcal{D}_0 \subset \mathcal{D}$ be those diffeomorphisms which are isotopic to the identity (i.e., homotopic through diffeomorphisms). Define the *Teichmüller space* $\mathcal{T} = \mathcal{T}(M)$ to be given by $\mathcal{C}/\mathcal{D}_0$. Here $\mathcal{C}$ is the set of all complex structures on M.

By a fundamental theorem of Nielsen [403] and Baer [36], $\mathcal{D}/\mathcal{D}_0$ is isomorphic to $\operatorname{Aut}\pi_1(M)/\operatorname{Inn}\operatorname{Aut}\pi_1(M)$,

the group of outer automorphisms of the fundamental group of M, and this is, therefore, an infinite discrete group. Thus,

$$\mathcal{R}(M) = \mathcal{C}/\mathcal{D} = (\mathcal{C}/\mathcal{D}_0)\,/\,(\mathcal{D}/\mathcal{D}_0) = \mathcal{T}(M)/\,(\mathcal{D}/\mathcal{D}_0)$$

(to our present knowledge, at least as sets).

We will not go into a study of these types of questions here. Since the publication in 1966 of *Lectures on quasiconformal mappings* by Lars Ahlfors, a number of books on Teichmüller theory have appeared, including *The Real Analytic Theory of Teichmüller Space* by W. Abikoff [2], *The Complex Analytic Theory of Teichmüller Spaces* by Subhashis Nag [390], *An Introduction to Teichmüller Spaces* by Y. Imayoshi and M. Taniguchi [253], *Quasiconformal Teichmüller Theory* by Frederick Gardiner and Nikola Lakic [169], *Teichmüller theory in Riemannian Geometry* by Anthony Tromba [511], and *Teichmüller Theory and Applications to Geometry, Topology, and Dynamics* by John H. Hubbard [248]. The books [508, 511], where Teichmüller theory is explained only by using basic ideas in geometry and analysis, stand apart from the others. These books are all excellent, and we recommend them highly, but these references are sometimes quite difficult to read. Our main justification for writing the present chapter is to make that task easier.

We can attempt to attain an understanding of $\mathcal{R}(M)$ by first understanding $\mathcal{T}(M)$ and then the action of $\Gamma = \mathcal{D}/\mathcal{D}_0$ on $\mathcal{T}(M)$. It is this second question whose answer is not yet at hand. Teichmüller was able to put a topology (in fact a metric) on $\mathcal{T}(M)$ and to show that $\mathcal{T}(M)$ is homeomorphic to Euclidean space $\mathbb{R}^{6\,\text{genus}(M)-6}$.

21.2 The Space $\mathcal{A}$ of Almost Complex Structures

As we shall later, see the space of almost complex structures $\mathcal{A}$ on M is in one-to-one correspondence with the space of complex structures $\mathcal{C}$.

Recall that each complex linear structure on $\mathbf{R}^{2n}$ is determined by an automorphism J on $\mathbf{R}^{2n}$ such that $J^2 = -id$, which corresponds to multiplication by $\sqrt{-1}$. Next, an almost complex structure on a manifold M is given by an automorphism of $T(M)$, denoted by J or by $\sqrt{-1}$, such that $J^2 = -\text{id}$. Almost complex manifolds (M, J) with $\dim M = 2$ are called Riemann surfaces.

A $\mathcal{C}^1$-map between two almost complex manifolds, say $f : (M_1, J_1) \to (M_2, J_2)$, is called holomorphic if the differential D_f is a complex linear map between the fibers $T_v(M_1) \to T_{f(v)}(M_2)$ for all $v \in M$. This is equivalent to the identity $D_f \circ J_1 = J_2 \circ D_f$.

A $\mathcal{C}^\infty(1, 1)$ tensor H is a mapping such that for each $x \in \mathcal{M}$, $H_x : T_x\mathcal{M} \longrightarrow T_x\mathcal{M}$ is a linear mapping from the tangent space $T_x\mathcal{M}$ to itself and such that $x \to H_x$ is $\mathcal{C}^\infty$ smooth.

The space of all $\mathcal{C}^\infty(1, 1)$ tensors forms an infinite dimensional Fréchet space.

Let $\phi : E \to F$ be a continuous map between two Fréchet spaces E and F, satisfying

$$\phi(0) = 0.$$

Does there exist a local inverse map ψ satisfying $\phi \circ \psi = \text{id}$ in an open neighborhood V of $0 \in F$?

If E and F are Banach spaces and if ϕ is of class $\mathcal{C}^1$, then ϕ is a local $\mathcal{C}^1$-diffeomorphism in some neighborhood of 0, provided $\phi'(0)$ is an isomorphism of E onto F [365, pp. 240–242].

By contrast, the situation in Fréchet spaces is quite different. Take for example, $\phi : f \to \exp(f)$ from $\mathcal{C}^\infty(\mathbb{R})$ into itself, or from the Fréchet space of entire functions into itself. This map is smooth and injective, its derivative at every point is an isomorphism of the Fréchet space, but the range of ϕ is clearly nowhere dense; hence there is no inverse map on an open set.

The space of almost complex structures $\mathcal{A}$ is the subset of all $\mathcal{C}^\infty(1, 1)$ tensors J such that for each $x \in \mathcal{M}$, $J_x^2 = -\text{id}_x$, $\text{id}_x : T_x\mathcal{M} \longrightarrow T_x\mathcal{M}$ is the identity map, and second for each tangent vector $X_x \in T_x\mathcal{M}$, the pair X_x, $J_x X_x$ is an oriented basis for $T_x\mathcal{M}$.

One can naturally ask what sort of structure $\mathcal{A}$ has. Is it a manifold? A variety? The answer to this question is provided by the following theorem:

Theorem 21.2.1 *Let $\mathcal{A}$ be a $\mathcal{C}^\infty$ smooth Fréchet manifold. The tangent space to $\mathcal{A}$ at a $J_0 \in \mathcal{A}$ is the subspace of all $\mathcal{C}^\infty(1,1)$ tensors H satisfying the algebraic relation $HJ_0 = -J_0H$.*

Proof Let $\mathcal{C}^\infty\left(T_1^1(M)\right)$ denote the space of $\mathcal{C}^\infty(1,1)$ tensors. Define a mapping $\Phi : \mathcal{C}^\infty\left(T_1^1(M)\right) \longrightarrow \mathcal{C}^\infty\left(T_1^1(M)\right)$ by

$$\Phi(H) = H^2.$$

Then $\mathcal{A} = \Phi^{-1}(-id)$, where $-id_x : T_xM \to T_x\mathcal{M}$ is the identity mapping.

If (i) $\mathcal{A}$ were modeled on a Banach space (C^r) or Sobolev space of tensors H^S, where the inverse function theorem applied and (ii) $-id$ were a regular value for Φ then by the implicit function theorem, we would be able to conclude that $\mathcal{A}$ was a manifold (in fact a submanifold of $\mathcal{C}^\infty\left(T_1^1(M)\right)$); moreover, the tangent space $T_{J_0}\mathcal{A}$ is given by the kernel of the differential of Φ, namely

$$D\Phi_J(H) = HJ + JH$$

or

$$T_{J_0}\mathcal{A} = \{H \mid HJ = -JH\}. \tag{21.2}$$

Although neither (i) nor (ii) is valid, the tangent space to $\mathcal{A}$ at J_0 is given by (21.2). In fact, we can write down explicit global coordinates for $\mathcal{A}$ discovered by Uwe Abresch and Arthur Fischer. For fixed J_0, define

$$\varphi : T_{J_0}\mathcal{A} \longrightarrow \mathcal{A}$$

by

$$\varphi(H) = (I + H)J_0(I + H)^{-1}. \tag{21.3}$$

Formula (21.3) makes it clear that φ maps $T_{J_0}A$ to $\mathcal{A}$. Since J_0 anticommutes with H, we can rewrite φ symbolically as

$$\varphi(H) = J_0 \cdot \frac{I - H}{I + H},$$

which shows that φ is a linear fractional transformation. What is more surprising is that φ has an "algebraic inverse." We leave it to the reader to verify that if $\psi = \varphi^{-1}$,

$$\begin{aligned}\psi(J) &= (J - J_0)\,(J + J_0)^{-1}\\ &= \frac{J - J_0}{J + J_0}.\end{aligned}$$

In two dimensions, it follows that $J^2 = -id$, $J_0^2 = -id$ implies that $J + J_0$ fails to be invertible only when $J = -J_0$. However, $J = -J_0$ (because of the orientation condition) is not in $\mathcal{A}$. From this, we can conclude that φ is a global chart for $\mathcal{A}$ which concludes Theorem 21.2.1. □

Remark 21.2.2 The relation $HJ = -JH$ implies that each such H is trace-free. To see this note that

$$-\operatorname{tr}(H) = \operatorname{tr}(JJH) = -\operatorname{tr}(JHJ) = \operatorname{tr}(H).$$

The group $\mathcal{D}$ and therefore $\mathcal{D}_0$ acts on $\mathcal{A}$ as follows: If $f \in \mathcal{D}$, then

$$\left(f^\star J\right)_x = df_x^{-1} \cdot J_{f(x)} \cdot df_x .$$

Clearly, $(f^\star J)^2 = -id$ if $J^2 = -id$.

The bijective correspondence between $\mathcal{C}$ and $\mathcal{A}$ is $\mathcal{D}$-equivariant so that if $c \rightsquigarrow J$, then $f^*c \rightsquigarrow f^*J$. Therefore, this correspondence induces a bijective correspondence between $\mathcal{C}/\mathcal{D}_0$ and $\mathcal{A}/\mathcal{D}_0$.

Before going further in developing Teichmüller's theory, we wish to make one further remark about the manifold $\mathcal{A}$. One can ask whether this manifold of almost complex structures is, in fact, a complex manifold.

We can begin by asking the simpler question of whether $\mathcal{A}$ is an almost complex manifold. The answer is "yes," in a very natural sense.

Define, for each $J \in \mathcal{A}$, a linear map $\Xi_J : T_J\mathcal{A} \longrightarrow T_J\mathcal{A}$ by

$$\Xi_J(H) = JH. \tag{21.4}$$

It follows immediately that

$$\Xi_J^2 = -I,$$

and $I_J : T_J\mathcal{A} \longrightarrow T_J\mathcal{A}$ is the identity mapping. Thus, $\mathcal{A}$ is an almost complex manifold with almost complex structure Ξ. The space $\mathcal{A}$ will be a complex manifold if there is a collection of coordinate charts (φ_i, U_i), U_i open in $T_{J_0}\mathcal{A}$ for some fixed J_0, such that $\varphi_i\,(U_i)$ cover $\mathcal{A}$ and $\varphi_i^*(\Xi) = \Xi_{J_0}$, the fixed almost complex structure on $T_{J_0}\mathcal{A}$. The surprising observation of Abresch-Fischer is that our global coordinate chart φ given in (21.3) satisfies

$$\varphi^\star(\Xi) = \Xi_{J_0}$$

and thus $\mathcal{A}$ is a complex manifold.

We now return to the relation between $\mathcal{A}$ and the study of Teichmüller's moduli space.

Let $c \in \mathcal{C}$ be a complex structure. Then to each such c, we can associate an almost complex structure $J \in \mathcal{A}$ as follows: Let $x_0 \in \mathcal{M}$ and $\psi : U \to \mathbb{R}^2$, $U \subset \mathcal{M}$ an open neighborhood of x_0 in $\mathcal{M}$, ψ a complex coordinate chart for $\mathcal{M}$. Now $\mathbb{R}^2$ has a natural almost complex structure $\widehat{J}$ given by the matrix $\begin{pmatrix} 0 & -1 \\ 1 & 0 \end{pmatrix}$. For $x \in U$, define $J_x : T_x M \to T_x M$ by

$$J_x = d\varphi_x^{-1} \widehat{J} d\varphi_x .$$

On the surface, it would appear that J_x depends on the choice of coordinate mapping φ. However, we leave it as an easy exercise for the reader to verify that if ψ is any other complex coordinate mapping for a neighborhood of x_0, then

$$J_x = d\psi_x^{-1} \widehat{J} d\psi_x .$$

Thus, $\{J_x\}_{x \in M}$ is a globally defined almost complex structure on M. This defines a mapping

$$\mathcal{C} \longrightarrow \mathcal{A}, \quad c \longrightarrow J . \tag{21.5}$$

Our next observation is that the diffeomorphism group $\mathcal{D}$ acts on $\mathcal{A}$ in a natural way, namely, if $f \in \mathcal{D}$ then

$$(f, J) \longrightarrow f^{\star} J = df_x^{-1} J_{f(x)} df_x .$$

Moreover, $c \to J$ has the property that it is $\mathcal{D}$ equivariant, namely if $c \longrightarrow J$, then $f^{\star} c \longrightarrow f^{\star} J$.

Later we shall see that this map has a natural inverse and is therefore bijective. Thus, in order to study $T(M) = \mathcal{C}/\mathcal{D}_0$, it suffices to study $\mathcal{A}/\mathcal{D}_0$.

Finally, we remark that the almost complex structure Ξ is $\mathcal{D}$-invariant; we leave this verification to the reader. This last observation is important in realizing the quotient space, Teichmüller's space $T(M) \approx A/\mathcal{D}_0$ as a "complex manifold."

Thus, we now restrict our attention to the study of the space $\mathcal{A}/\mathcal{D}_0$.

21.3 The Space of Riemannian Metrics $\mathcal{M}$

Let S_2 be the space of all $(0, 2)$ tensors on M. Thus, $h \in S_2$ iff for each $x \in M$, $h(x) : T_x M \times T_x M \to R$ is a symmetric bilinear form. Let $\mathcal{M} = \{g \in S_2 \mid g(x) > 0\}$ for each $x \in M$. Thus, for $v \in T_x M$, $g(x)(v, v) > 0$ if $v \neq 0$. Clearly, $\mathcal{M} \subset S_2$ is open and thus a submanifold. The diffeomorphism group ν also acts on M, namely $(f, g) \longrightarrow f^{\star} g$, where

$$\left(f^*g\right)(x)(u, v) = g\left(f(x)\right)\left(df_x u, df_x v\right)$$

for $u, v \in T_x M$.

There is a very natural mapping from $\mathcal{M}$ to $\mathcal{A}$ described as follows: For $g \in \mathcal{M}$, $v \in T_x M$, define $J_x : T_x M \to T_x M$ by $J_x v$ is the rotation of v by 90° in the "counter-clockwise direction." Given $g \in \mathcal{M}$ and the orientation on M, this is well defined. This gives us a map $\varphi : \mathcal{M} \longrightarrow \mathcal{A}$.

The problem with this definition is that it is implicit as opposed to explicit. As analysts, we would like to be able to differentiate mappings, even in the infinite-dimensional context. Fortunately, we can describe the mapping Ψ explicitly, namely

$$\psi(g) = -g^{-1}\mu_g\,, \tag{21.6}$$

where $\mu_g(x) : T_x M \times T_x M \longrightarrow R$ is the unique antisymmetric two form, the volume element associated to g and the given orientation of M. More precisely, by $\psi(g) = J$, we mean

$$-\,g(x)\left(J_x u, v\right) = \mu_g(x)(u, v) \tag{21.7}$$

for all $u, v \in T_x M$.

It is not too hard to check that $\Psi : \mathcal{M} \longrightarrow \mathcal{A}$ and is $\mathcal{D}$-equivariant, namely

$$\Psi(f^*g) = f^*\Psi(g)\,.$$

However, ψ is not a bijection, and this is not difficult to see. Let $p : M \longrightarrow \mathrm{R}^+$ be a strictly positive $\mathcal{C}^\infty$ function. Then $p \cdot g \in \mathcal{M}$ if $g \in \mathcal{M}$ $\{(p \cdot g)(x)(u, v) = p(x)g(x)(u, v)\}$ since $\mu_{p \cdot g} = p \cdot \mu_g$ and one sees immediately from (21.6) that

$$\Psi(pg) = \Psi(g)\,.$$

Let $\mathcal{P}$ be the space of all $\mathcal{C}^\infty$ positive functions. One can check that $\mathcal{P}$ acts freely and properly on $\mathcal{M}$, the action given by the map

$$(p, g) \longmapsto p \cdot q\,.$$

If we were in the $\mathcal{C}^\infty$ or Sobolev H^S class of metrics and functions, it would follow immediately that the quotient space, say $\mathcal{M}^S/\mathcal{P}^S$, is a manifold. However, in the $\mathcal{C}^\infty$ class of metrics $\mathcal{M}$ and $\mathcal{C}^\infty$ positive functions $\mathcal{P}$, this result is still true, namely that $\mathcal{M}/\mathcal{P}$ is a $\mathcal{C}^\infty$ Fréchet manifold. Before we state this as part of the main theorem of this section, let us see what the tangent space of this manifold must be. First, if $h \in S_2$ and $g \in \mathcal{M}$, we can define the trace of h with respect to g by

$$\operatorname{trace}_g h = \operatorname{tr}_g h = \operatorname{trace} H\,, \tag{21.8}$$

where H is the $(1, 1)$ tensor associated to h via the metric g by way of the relation

$$g(x)\,(H_x u, v) = h(x)(u, v) \tag{21.9}$$

where $x \longmapsto (\mathrm{tr}_g h)_x$ is a $\mathcal{C}^\infty$ function on M. Now let $h \in S_2$. Then

$$\begin{aligned} h &= \left(h - \frac{1}{2}(\mathrm{tr}_g h)\cdot g\right) + \left(\frac{1}{2}\mathrm{tr}_g h\right)\cdot g \\ &= h^T + \lambda \cdot g \end{aligned}$$

where $\mathrm{tr}_g h^T = 0$.

Thus, every $(0, 2)$ tensor h can be decomposed as a trace-free tensor plus a multiple of the metric g.

Consider now the map

$$p \longmapsto p \cdot g$$

for fixed g. The image of this mapping is the orbit of the group $\mathcal{P}$ through the point g. The tangent space of this orbit is clearly the set of $\lambda \cdot g$ where $\lambda : M \longrightarrow R$ is a $\mathcal{C}^\infty$ function. From this observation and the previous remarks, we can state the following:

Theorem 21.3.1 *The quotient space $\mathcal{M}/\mathcal{P}$ is a $\mathcal{C}^\infty$ Fréchet manifold whose tangent space at an equivalence class $[g] \in \mathcal{M}/\mathcal{P}$ can be identified with those $h \in S_2$ such that* $\mathrm{tr}_g h = 0$.

Moreover, we have

Theorem 21.3.2 *The map $\Psi : \mathcal{M} \longrightarrow \mathcal{A}$ induces a $\mathcal{C}^\infty$ $\mathcal{D}$-equivariant diffeomorphism $\Psi : \mathcal{M}/\mathcal{P} \longrightarrow \mathcal{A}$.*

The conclusion of this section is that in order to understand the structure of the space $T(M) \approx \mathcal{A}/\mathcal{D}_0$, it suffices to understand the structure of the double quotient space $T(M) \approx \mathcal{M}/\mathcal{P}/\mathcal{D}_0$.

21.4 The Manifold $\mathcal{M}_{-1}$ of Metrics of Negative Constant Curvature Minus One

The quotient space $\mathcal{M}/\mathcal{P}$ is still a little too cumbersome to work with. Let us now introduce another model for $\mathcal{M}/\mathcal{P}$. Given a metric g, there is a $\mathcal{C}^\infty$ function on M called the *scalar curvature* of g at x. We denote this function by $R(g)$. If $\mathcal{C}^\infty(M)$ denotes the space of $\mathcal{C}^\infty$ functions on M, then $R : \mathcal{M}/\mathcal{P} \to \mathcal{C}^\infty(M)$.

Definition 21.4.1 We have that $\mathcal{M}_{-1} = R^{-1}(-1)$. That is, $\mathcal{M}_{-1}$ consists of all those $\mathcal{C}^\infty$ metrics whose scalar curvature is the constant function -1.

In this section and later, we build upon a classical and important result of Poincaré:

Theorem 21.4.2 *Let M be an oriented surface of (genus M) > 1. Then given any metric $g \in \mathcal{M}$, there exists a unique $\lambda : M \rightarrow R$ such that the Gauss or scalar curvature of λg, $R(\lambda g)$, is the constant function -1.*

This establishes the bijective correspondence between $\mathcal{M}_{-1}$ and $\mathcal{M}/\mathcal{P}$, and hence $\mathcal{C}$.

We begin by asking whether $\mathcal{M}_{-1}$ is a manifold. This is answered by:

Theorem 21.4.3 *The space $\mathcal{M}_{-1}$ of metrics of constant negative Gauss curvature is a smooth C^{∞} Fréchet submanifold of $\mathcal{M}$.*

Proof We sketch a proof in the Sobolev class $\mathcal{M}^S$ of metrics. This result follows immediately from the implicit function theorem. The result in the C^{∞} case involves slightly more work. That $\mathcal{M}^S_{-1}$ is a manifold would follow if we could show that -1 is a regular value for the map

$$R : \mathcal{M}^S \longrightarrow F^{S-2}.$$

Now the derivative formula for

$$DR(g) : S_2^S \longrightarrow F^{S-2}$$

with $S_2^S \approx T_g\mathcal{M}^S$, the H^S smooth (0, 2) tensors on $\mathcal{M}$, is well known. In the case $R(g) \equiv -1$ it is given by the formula

$$DR(g)h = -\Delta\left(t_g h\right) + \delta_g\delta_g h + \frac{1}{2}\left(tr_g h\right) \tag{21.10}$$

where Δ is the Laplace-Betrami operator on functions, $\mathrm{tr}_g\, h$ is the trace of h with respect to g, and $\delta_g g_g h$ is the function on M defining the double covariant divergence of h with respect to g [see Eisenhart [144]].

We may conclude that $\mathcal{M}^S_{-1}$ is a submanifold of $\mathcal{M}^S$ with tangent space Kernel $DR(g)$ if we can show that the operator $H \longmapsto DR(g)H$ is onto F^{S-2}. The simplest way to see this surjectivity is not by considering all variations h but instead by considering the subspace of those h of the form $\rho \cdot g$, where $\rho \in F^S$. Then $\mathrm{tr}_g(\rho g) = 2\rho$ and $\delta_g\delta_g(\rho g) = \Delta p$, where again Δ is the Laplace-Beltrami operator. Thus, if $h = \rho g$, then

$$DR(g)h = -2\Delta\rho + \Delta\rho + \rho = -\Delta\rho + \rho. \tag{21.11}$$

Now it is completely standard in elliptic theory that $\rho \longmapsto -\Delta\rho + \rho$ as a map from $F^S \longrightarrow F^{S-2}$ is surjective. This concludes the proof of Theorem 21.4.3 . □

But we get more.

Theorem 21.4.4 *The tangent space to the submanifold $\mathcal{M}_{-1} \subset S_2$ consists of those $(0, 2)$ tensors h such that*

$$- \Delta \left(\mathrm{tr}_g\, h\right) + \delta_g \delta_g h + \frac{1}{2}\left(tr_g h\right) = 0\,. \tag{21.12}$$

Consider now the natural quotient map $\pi : \mathcal{M} \longrightarrow \mathcal{M}/\mathcal{P}$. Let π_{-1} be the restriction of π to the submanifold $\mathcal{M}_{-1}$.

We can now prove that:

Theorem 21.4.5 *The map $\pi_{-1} : \mathcal{M}_{-1} \longrightarrow \mathcal{M}/\mathcal{P}$ is a $\mathcal{D}$-equivariant diffeomorphism of Fréchet manifolds.*

Proof First, we should remark that $\mathcal{D}$ acts on $\mathcal{M}_{-1}$; i.e., the $\mathcal{D}$ action on $\mathcal{M}$ preserves $\mathcal{M}_{-1}$. This follows from the relation

$$R(f^* g) = f^* R(g) = R(g) \circ f\,. \tag{21.13}$$

Thus, if $R(g) \equiv -1$, then $R\left(f^*\right) \equiv -1$. That π_{-1} is $\mathcal{D}$-equivariant follows from the fact that $\pi : \mathcal{M} \to \mathcal{M}/\mathcal{P}$ is naturally $\mathcal{D}$-equivariant.

From Poincaré's Theorem 21.4.2, it follows that π_{-1} is a bijection. We must show that π_{-1} is a diffeomorphism, or that $\mathcal{D}\pi_{-1} : T_g\mathcal{M}_{-1} \longrightarrow T_{[g]}\mathcal{M}/\mathcal{P}$ is an isomorphism of Fréchet spaces. From Theorem 21.3.1, we know that $T_{[g]}\mathcal{M}/\mathcal{P}$ can be identified with those elements of S_2 which are trace-free with respect to g. Moreover, $D\pi_g(h) = h^T = \left\{h - \left(\frac{1}{2}\,\mathrm{tr}_g\, h\right) g\right\}$, h^T the trace-free part of h. From this, it immediately follows that

$$D\pi_{-1}(h) = h^T\,. \tag{21.14}$$

Now suppose $D\pi_{-1}(h) = 0$. This means that $h = \rho g$, $\rho \in F$. But $DR(g)h = 0$ which implies as in (21.11) that

$$-\Delta\rho + \rho = 0$$

or that $\rho = 0$. Thus, $D\pi_{-1}$ is injective. Now let $v \in F$. To show $D\pi_{-1}$ is surjective, we must produce an $h \in T_g\mathcal{M}_1$ such that $D\pi_{-1}(h) = v$. Let $h = \rho g$. Then it suffices to find a ρ such that

$$-\Delta\rho + \rho = v.$$

But $\rho \longrightarrow -\Delta\rho + \rho$ is surjective which completes the proof of Theorem 21.4.5. □

The conclusion of this section is that in order to understand the structures of $\mathcal{M}/P/\mathcal{D}_0 \cong A/\mathcal{D}_0 \cong C/\mathcal{D}_0$, it suffices to understand the structure of $\mathcal{M}_1/\mathcal{D}_0$. Our intuition is now very clear. We have an infinite-dimensional Lie group ∇_0 acting on a smooth $\mathcal{C}^\infty$ manifold $\mathcal{M}_1$, and we wish to understand the quotient space $\mathcal{M}_1/\mathcal{D}_0$. Our first objective is to identify the tangent space with the quotient. This we do in Sect. 21.6.

21.5 Conformal Coordinates and the $\mathcal{D}$-equivariant Equivalence Between $\mathcal{M}_{-1}$ and C, and C and $\mathcal{A}$

Let $x \in M$ and $g \in \mathcal{M}$ be arbitrary. Then it is a standard fact that we can find a diffeomorphism $\psi : V \longrightarrow U$, where V is a neighborhood of 0 in $\mathbb{R}^2$ and U is a neighborhood of x in M, such that the pullback of the metric g satisfies

$$\left(\psi^* g\right)_{ij} = \lambda\delta_{ij}\,, \tag{21.15}$$

where δ_{ij} is the Kronecker delta; i.e., the pullback of the metric g to a metric on $V \subset \mathbb{R}^2$ is given by a positive function times the standard Euclidean metric. The set of all $\left\{\psi^{-1}, U\right\}$ naturally forms a complex structure for M. What this means is the following: the element $g \in \mathcal{M}_{-1}$, a conformal coordinate system, yields a complex structure $c \in \mathcal{C}$.

One can readily see that this is a $\mathcal{D}$-equivariant mapping which is the inverse of the composition of the maps

$$c \longrightarrow \tau \longrightarrow g \in \mathcal{M}_{-1}$$

developed in the last sections. Thus, the existence of conformal coordinate systems provides an inverse for the map $c \longrightarrow \tau$, $\mathcal{C} \to \mathcal{A}$ introduced in Sect. 21.2.

21.6 The L_2-Decomposition of the Tangent Space to $\mathcal{M}_{-1}$

We have already observed that $\mathcal{D}$ acts on $\mathcal{A}$. What is the tangent space to this action?

Let $g \in \mathcal{M}_{-1}$ be fixed and consider the orbit of the diffeomorphism group $\mathcal{D}$; thus, look at the manifold

$$\mathcal{O}_\mathcal{D}(g) = \left\{f^\alpha g | f \in \mathcal{D}\right\}.$$

What is the tangent space to this orbit? Let f_t be a smooth one-parameter group of diffeomorphisms with $-\varepsilon < t < \varepsilon$, $\varepsilon > 0$, $f_0(x) = x$, $\frac{df}{dt}|_{t=0} = X$, a vector field on M. A tangent vector $\beta \in T_g\mathcal{O}_\mathcal{D}(g)$ will be given by

$$\frac{d}{dt}\left\{f_t^* g\right\}_{t=0} = \beta\,. \tag{21.16}$$

But (21.16) is a well-known object in differential geometry—the Lie derivative of g with respect to the vector field X. In a local coordinate system where g is given by g_{ij}, we can write this Lie derivative $L_X g$ as

$$(L_X)_{ij} = x^k \frac{\partial g_{ij}}{\partial x^k} + g_{kj} \frac{\partial x^k}{\partial x^i} + g_{ik} \frac{\partial x^k}{\partial x^j} \,. \tag{21.17}$$

Summarizing we have:

Theorem 21.6.1 *The tangent space to the orbit $\mathcal{O}_{\mathcal{D}}(g)$ is the set of all Lie derivatives $L_X g$, X a vector field on M.*

If we are to have a natural candidate for the quotient space $\mathcal{M}_{-1}/\mathcal{D}_0$, then we should produce a natural complement, say an orthogonal complement to the tangent space $T_g\mathcal{O}_{\mathcal{D}}(g)$. But "orthogonal complement" necessitates the existence of a metric. Therefore we ask:

Is there a "natural" metric on the space of all metrics $\mathcal{M}$, where by natural, we mean that the diffeomorphism group should act as a group of isometries? The positive answer was well known to the physicists and general relativists for a long time, namely the so-called L_2-metric on the space of metrics which is defined as follows: Let $h, k \in T_g\mathcal{M}_{-1}$ and let H, K be the $(1, 1)$ tensors $\{g(x)\,(H_x u, v) = h(x)(u, v)\}$ corresponding to h, k via the metric g. Define

$$\ll h, k \gg_g = \int_M \text{trace}(HK) d\mu_g. \tag{21.18}$$

Here $x \longmapsto \text{trace}\,(H_x K_x)$ is a $\mathcal{C}^\infty$ function on M integrated with respect to the volume element μ_g. Clearly,

$$\ll, \gg : T_g\mathcal{M}_{-1} \times T_g\mathcal{M}_{-1} \longrightarrow \mathrm{R}.$$

Theorem 21.6.2 *The space $\mathcal{D}$ acts as a group of isometries for $\ll, \gg$.*

Proof For $f \in \mathcal{D}$, consider the map

$$\theta_f(g) = f^* g \tag{21.19}$$

Since $g \longmapsto f^* g$ is "linear" in g, it follows that the derivative

$$D\theta_f(g) : S_2 \longrightarrow S_2$$

$$D\theta_f(g)(h) = f^* h. \tag{21.20}$$

θ_f is an isometry if

$$\ll D\theta_f(g)(h), D\theta_f(g)k \gg_{\theta_f(g)} = \ll h, k \gg_g \,. \tag{21.21}$$

But the left-hand side of (21.21) is equal to

$$\int_M \operatorname{trace}\left(f^\star H f^\star K\right)_x d\mu_{f^\star g}$$
$$= \int_M \operatorname{trace}(HK)_{f(x)} d\mu_{f^\alpha g}$$

which by the ordinary change of variables formula is equal to

$$\int_M \operatorname{trace}(HK)_x d\mu_g$$

which proves Theorem 21.6.2. □

The next question is whether we can identify the orthogonal complement of those $L_X g \in T_g\mathcal{M}_{-1}$ with respect to this metric. The first step in this direction is given by the L_2-decomposition of symmetric tensors given in a famous splitting theorem whose first complete proof was given by Berger and Ebin.

Theorem 21.6.3 (Berger-Ebin) *Let $h \in S_2$ and $g \in \mathcal{M}$. Then h can be uniquely decomposed as*

$$h = h^o + L_X g \tag{21.22}$$

where h° is a divergence-free $(0, 2)$ tensor; i.e.,

$$\delta_g h^o = 0$$

where in a local coordinate system

$$\left(\delta_g h\right)_i = \frac{1}{\sqrt{g}} \frac{\partial}{\partial x^j}\left(h_i^j \sqrt{g}\right) - \frac{1}{2} g^{kl} h_1^i \frac{\partial g_{jk}}{\partial x^i}. \tag{21.23}$$

Here $\sqrt{g} = \det\left(g_{ij}\right)$ is the determinant of the g_{ij}, g^{kl} is the matrix of the inverse of $\left\{g_{ij}\right\}$, and finally $h_i^j = g^{jk} h_{ki}$. And in local coordinates, h is given by $\{h_{ki}\}$. Moreover, the decomposition (21.22) is L_2-orthogonal.

We are now in a position to determine the orthogonal complement in $T_g\mathcal{M}_{-1}$ of the space of Lie derivatives $L_X g$, the tangent space to the orbit of $\mathcal{D}$. Let $g \in \mathcal{M}_{-1}$ and suppose that

$$h \in T_g\mathcal{M}_{-1}\left\{-\Delta\left(\operatorname{tr}_g h + \delta_g\delta_g h + \frac{1}{2}\left(\operatorname{tr}_g h\right)\right) = 0\right\}.$$

Then, by Berger-Ebin,

$$h = h^o + L_X g\,.$$

But, since $L_X g \in T_g\mathcal{M}_{-1}$, it follows that

$$0 = -\Delta\left(\operatorname{tr}_g\left(L_X g\right)\right) + \delta_g\delta_g\left(L_X g\right) + \frac{1}{2}\left(\operatorname{tr}_g L_X g\right).$$

Thus, if $h \in T_g\mathcal{M}_{-1}$, $h = h^o + L_X g$, then necessarily

$$-\Delta\left(\operatorname{tr}_g h^\circ\right) + \delta_g\delta_g h^\circ + \frac{1}{2}\operatorname{tr}_g h^\circ = 0.$$

Since $\delta_g h^\circ = 0$, $\delta_g\delta_g h^0 = 0$, we see that

$$-\Delta\left(\operatorname{tr}_g h^o\right) + \frac{1}{2}\operatorname{tr}_g h^o = 0\,.$$

If $\rho = \operatorname{tr}_g h^o$, then

$$-\Delta\rho + \rho = 0\,,$$

which immediately implies (just integrate by parts) that $\rho = 0$.

Our conclusion is the following: Let $h \in T_g\mathcal{M}_{-1}$. Then

$$h = h^o + L_X g\,,$$

where h^o is both divergence-free and trace-free. For clarity, it would be better to denote such a trace-free, divergence-free two-tensor by h^{TT} and the set of all such trace-free and divergence-free two-tensors with respect to g by $S_2^{TT}(g)$. Paraphrasing our result, we see that

Theorem 21.6.4 *Any $h \in T_g\mathcal{M}_{-1}$ can be uniquely decomposed as*

$$h = h^{TT} + L_X g$$

where $h^{TT} \in S_2^{TT}(g)$, and this decomposition is L_2-orthogonal.

We should remark that in general, although $L_X g$ is unique, X need not be uniquely determined. It will be determined only up to a Killing field Y, $L_Y g = 0$. If $g \in \mathcal{M}_{-1}$ there are no nonzero Killing fields ($L_Y g = 0$ implies $Y = 0$).

Back to Teichmüller space! If $\mathcal{M}_{-1}/\mathcal{D}_0$ is a manifold, then the dimension of the space of $S_2^{TT}(g)$ must be finite and the same dimension for all $g \in \mathcal{M}_{-1}$. How can we compute this dimension? Let us see what it means for h^{TT} to belong to $S_2^{TT}(g)$ in a local conformal coordinate system for M (recall that such a conformal coordinate system is also a complex coordinate system). In such a system, we can write h^{TT} in classical notation

$$h^{TT} = udx^2 - udy^2 - 2vdxdy\,,$$

with u, v local functions on the coordinate neighborhood. We have already used the fact that h^{TT} is trace-free by writing the coefficient of dy^2 as the negative of that of dx^2. What does it mean for H^{TT} to be divergence-free? Using formula (21.23), the reader may verify that $\delta_g h^{TT} = 0$ reduces to the system of equations

$$\frac{\partial u}{\partial x} = \frac{\partial v}{\partial y}\,,$$
$$\frac{\partial u}{\partial y} = -\frac{\partial v}{\partial x}\,,$$

that is, the Cauchy-Riemann equations! Thus, formally

$$\begin{aligned} h^{\mathrm{TT}} &= \mathrm{Re}\left\{(u+iv)(dx+idy)^2\right\} \\ &= \mathrm{Re}\left\{\xi(z)dz^2\right\}, \end{aligned} \tag{21.24}$$

where $\xi(z)$ is a local holomorphic function.

Thus, h^{TT} is the real part of a complex-valued symmetric tensor which in conformal coordinates is given by a holomorphic function $\xi(z)$. This is the definition of a holomorphic quadratic differential on $\mathcal{M}$ with respect to the complex structures $c(g)$ arising from the metric g.

We are now back to Teichmüller's original observation that as a consequence of the theorem of Riemann-Roch, one can compute the dimension of the space of holomorphic quadratic differentials to be 6(genus $\mathcal{M}$) $-$ 6, exactly the number that Riemann had conjectured was the dimension of $R(M)$, the Riemann space of moduli (cf. Sect. 21.2). As a corollary, we obtain

Corollary 21.6.5 *For any* $g \in \mathcal{M}_{-1}$

$$\dim S_2^{TT}(g) = 6(genus\ M) - 6.$$

Paraphrasing Theorem 21.6.4, we have

Theorem 21.6.6 *Any* $h \in T_g\mathcal{M}_{-1}$ *can be uniquely decomposed as*

$$h = \mathrm{Re}\left(\xi(z)dz^2\right) + L_X g$$

where $\xi(z)dz^2$ *is a holomorphic quadratic differential.*

Thus, if Teichmüller space is a manifold, it must be of dimension 6(genus M)$-$6! In the next section, we construct a coordinate chart for Teichmüller space.

21.7 Teichmüler's Moduli Space Is a $\mathcal{C}^\infty$ Manifold of Dimension 6(genus $\mathcal{M}$) − 6

Given $g \in \mathcal{M}_{-1}$, we would like to construct a submanifold $S_g \subset \mathcal{M}_{-1}$ through g transverse to the orbit $\mathcal{O}_{\mathcal{D}}(g)$ of $\mathcal{D}$ through g. Let $H^{TT} \in S_2^{TT}(g)$ be "small." Then $g + H^{TT} \in \mathcal{M}$. By Poincaré's Theorem 21.4.2, we can find a unique function $\lambda\left(H^{\mathrm{TT}}\right) : M \longrightarrow \mathrm{R}^+$ such that the Gauss curvature $R\left(\lambda\left(H^{TT}\right)\left\{g + H^{TT}\right\}\right) \equiv -1$. Moreover, $H^{TT} \longmapsto \lambda\left(h^{TT}\right)$ is smooth and $\lambda(0) = 1$, since $G(g) \equiv -1$. Consider the map

$$h^{TT} \overset{\Theta}{\longmapsto} \lambda\left(h^{TT}\right)\left\{g + H^{TT}\right\}.$$

One sees that $\mathcal{D}\theta(0)h^{TT} = h^{TT}$. By the standard finite-dimensional implicit function theorem, this implies that the image under θ of a neighborhood of 0 in $S_2^{TT}(g)$ is a submanifold S_g of $\mathcal{M}_{-1}$ of dimension 6 genus $M - 6$. This "local slice" S_g is our candidate for a coordinate chart for $T(M) = \mathcal{M}_{-1}/\mathcal{D}_0$ (see Fig. 21.5).

Up till now, we have not distinguished between the actions of $\mathcal{D}_o$ and $\mathcal{D}$. Through each point $\widehat{g} \in S_g$, there is an orbit of $\mathcal{D}$. It is possible that no matter how small S_g is chosen different points of S_g may correspond to the same orbits of $\mathcal{D}$. This is exactly what happens. This is the reason why the Riemann space of moduli is not a manifold and Teichmüller space is, for if S_g is small enough, each point $\widehat{g} \in S_g$ corresponds to a unique orbit of D_o, and thus locally, points in the quotient space may be uniquely identified with points in the slice S_g. This gives the $\mathcal{C}^\infty$ manifold structure to $T(M)$. The fact that each point $\widehat{g} \in L_g$ corresponds to a unique orbit of $\mathcal{D}_o$ follows from two theorems. The first due to Ebin-Palais says that the action of $\mathcal{D}$ on M is proper. We state this as:

Theorem 21.7.1 (Ebin-Palais) *The space $\mathcal{D}$ acts properly on $\mathcal{M}$, i.e., if $f_n \in \mathcal{D}$ and $g_n \in \mathcal{M}$ are sequences such that $f_n^* g_n \longrightarrow g_1$ and $g_n \longrightarrow g_0$, then there is a subsequence of f_n which converges ($f_n \longrightarrow f$) to some $f \in \mathcal{D}$.*

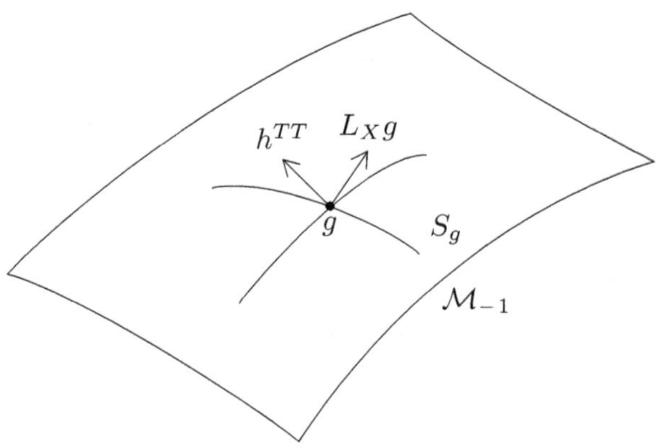

Fig. 21.5 A submanifold S_g of manifold $\mathcal{M}_{-1}$

For a proof, see [143].

Theorem 21.7.2 *The space $\mathcal{D}_o$ acts freely on $\mathcal{M}_{-1}$.*

Remark 21.7.3 This theorem is definitely not true if $\mathcal{D}_o$ is replaced by $\mathcal{D}$.

Proof Suppose $f \in \mathcal{D}_o$ and $f^*g = g$ i.e., f is an isometry for g. If $c(g)$ is the unique complex structure associated to g, it follows that $f^\star c(g) = c(g)$. This says that $f : (M, C(g)) \longrightarrow (M, c(g))$ is a holomorphic self-map. Let us suppose that f is not the identity. Then since f is holomorphic, it follows that the fixed point set of f is isolated and therefore finite in number. Since f is holomorphic, the Lefschetz fixed point index $\Lambda(f) > 0$. Since f is homotopic to the identity

$$\Lambda(f) = 2(1 - \text{ genus } M) < 0$$

the Euler characteristic of M. This is a contradiction. Therefore, f must be the identity, which finishes Theorem 21.7.2. □

Theorem 21.7.4 *If S_g is small enough, each point $\widehat{g} \in S_g$ corresponds to a unique orbit of $\mathcal{D}_o$.*

Proof Suppose not. Then there exists a sequence $f_n \in \mathcal{D}_0$ and $g_n \in S_g$ with $g_n \longrightarrow g$ and $f_n^\star g_n \longrightarrow g$ with all f_n outside a fixed neighborhood of the identity map of M. By the Ebin-Palais lemma, we may conclude that there exists a subsequence of the f_n, say f_n such that $f_n \longrightarrow f$. Then necessarily $f^*g = g$, $f \neq id$ and f homotopic to the identity contradicting Theorem 21.7.2. □

Thus, a neighborhood of $[g]$ in $\mathcal{M}_{-1}/\mathcal{D}_o$ can be identified with S_g which provides a coordinate chart for the quotient $T(M) = \mathcal{M}_{-1}/\mathcal{D}_o$. It is a standard exercise that the transition maps for these coordinate mappings are $\mathcal{C}^\infty$ smooth. Thus, we may conclude this section with:

Theorem 21.7.5 *Teichmüller's space $T(M) \cong \mathcal{M}_{-1}/\mathcal{D}_o \cong \mathcal{M}/\mathcal{P}/\mathcal{D}_o \cong \mathcal{A}/\mathcal{D}_o \cong \mathcal{C}/\mathcal{D}_o$ has the natural structure of a smooth $\mathcal{C}^\infty$ manifold of dimension $\mathbb{R}^{6genus\ M-6}$.*

In the next section, we shall see why Teichmüller's space must be diffeomorphic to Euclidean space.

21.8 Teichmüller's Space Is Diffeomorphic to $\mathbb{R}^{6\text{genus } M-6}$

Our main objective in this section is to indicate that Dirichlet's energy on Teichmüller space is a proper smooth $\mathcal{C}^\infty$ function with one nondegenerate minimum point.

Let $g_0 \in \mathcal{M}_{-1}$ be a fixed metric and let $g \in \mathcal{M}_{-1}$ be arbitrary. Furthermore, for simplicity of exposition, we assume (without loss of generality) that (M, g_0) is isometrically embedded in some Euclidean space of dimension K.

Let $S : (M, g) \longrightarrow (M, g_o) \subset \mathbb{R}^K$ be a smooth map. Then $S = \left(S^1, \ldots, S^K\right)$. Define Dirichlet's energy of S by

$$E(g, S) = \frac{1}{2} \sum_{j=1}^{K} \int_M g(x) \left(\nabla_g S^j, \nabla_g S^i\right) d\mu_g \tag{21.25}$$

with $\nabla_g S^i$ the gradient of the jth component function S^j with respect to the metric g.

We now have a basic result toward which many people have contributed including Eells, Sampson, Schoen, Yau, and Jost.

Theorem 21.8.1 *For each fixed $g \in \mathcal{M}_{-1}$, there exists a unique critical map (a minimum map) $S(g) : M \longrightarrow M$ of E which is homotopic to the identity.*

Moreover, $S(g) \in \mathcal{D}_o$ and depends smoothly on g. $S(g)$ is a harmonic map from (M, g) to (M, g_o).

Dirichlet's energy has an invariance property under the action of $\mathcal{D}$, namely

$$E(f^* g, f^* S) = E(g, S)\,, \tag{21.26}$$

where $f * (S) = S \circ f$. Thus, E as a function of two variables is $\mathcal{D}$-invariant. Define the function $\widetilde{E} : \mathcal{M}_{-1} \longrightarrow R$ by

$$\widetilde{E}(g) = E(g, S(g)). \tag{21.27}$$

Theorem 21.8.2 *The function $\widetilde{E}$ is $\mathcal{D}_0$ invariant (but not $\mathcal{D}$ invariant!).*

Proof $\widetilde{E}\,(f^\star g) = E(f^\star g, S(f^\star g))$.

Now $S(f^* g)$ is the unique harmonic map from $(M, f^* g)$ to (M, g_o) homotopic to the identity. But $f^* S(g) = S(g) \circ f$ is also a harmonic map from $(M, f^* g)$ to (M, g_o). If $f \in \mathcal{D}_0$, $S(g) \circ f \in \mathcal{D}_0$, and therefore by uniqueness

$$S\left(f^* g\right) = f^* S(g).$$

By (21.26), we may conclude that

$$\widetilde{E}\left(f^* g\right) = \widetilde{E}(g),\, f \in \mathcal{D}_0$$

concluding Theorem 21.8.2. □

Thus, the smooth map $\widetilde{E}$ passes to a smooth map

$$\widetilde{E} : T(M) \longrightarrow R.$$

Based on the results of Schoen-Yau and Jost, the authors observed that:

Theorem 21.8.3 *The map* $\widetilde{E} : T(M) \longrightarrow R$ *is a smooth proper map; i.e., the inverse image of a compact set is compact.*

This result is the analytical heart of Teichmüller theory.

To state our final result, we would like to have a minor diversion and discuss a natural Riemannian metric on $T(M)$, the Weil-Petersson metric.

Classical Teichmüller theory realizes the holomorphic quadratic differentials as the co-tangent space to Teichmüller's space. Around 1952, André Weil suggested to Lars Ahlfors a metric on Teichmüller space. Petersson had earlier introduced a bilinear pairing on modular forms of weight k. Holomorphic quadratic differentials are a particular case of modular forms of weight 2. If ξ and n are holomorphic quadratic differentials on $(M, c(g))$, $g \in \mathcal{M}_{-1}$ and $c(g)$ the associated complex structure, then consider the expression

$$\left(\xi(z)dz^2\right) \cdot \left(\overline{\eta(z)dz^2}\right) = \xi(z)\overline{\eta}(z)|dz|^4 .$$

In conformal coordinates, the metric g can be written as $\lambda|dz|^2$. Thus, the expression

$$\frac{\xi(z)\overline{\eta}(z)|dz|^4}{\lambda^2|dz|^4} = \frac{\xi(z)\overline{\eta}(z)}{\lambda^2}$$

is invariant, i.e., independent of the choice of coordinate system and is thus a function on M. The metric on $T(M)$ suggested by Weil is the Petersson pairing, for $[g] \in T(M)$:

$$\begin{aligned}\langle \xi, \eta\rangle_{[g]} &= \mathrm{Re} \int_M \frac{\xi(z)\overline{\eta(z)}}{\lambda^2} d\mu_g \\ &= \mathrm{Re} \int_M \frac{\xi(z)\overline{\eta(z)}}{\lambda} |dz|^2.\end{aligned}$$

There is another way of obtaining a nice metric on $T(M) = \mathcal{M}_{-1}/\mathcal{D}_0$. We have already seen that D acts as a group of isometries with respect to the L_2-metric $\ll, \gg$ on $\mathcal{M}_{-1}$. Thus, this metric passes to a metric $<, >$ on $\mathcal{M}_{-1}/\mathcal{D}_0$ in the following way: Let $h, k \in T_{[g]}\mathcal{M}_{-1}/\mathcal{D}_0$. Then there exists unique $\widetilde{h}, \widetilde{k} \in T_g\mathcal{M}_{-1}, \widetilde{h}, \widetilde{k} \in S_2^{TT}(g)$ with $\mathcal{D}\pi_{-1}(\widetilde{h}) = h$, $\mathcal{D}\pi_{-1}(\widetilde{k}) = k$. Define

$$< h, k >_{[g]} = \ll \widetilde{h}, \widetilde{k} \gg_g .$$

In fact that $\mathcal{D}$ acts as a group of isometries implies that $<, >$ is well defined on $T(M)$.

Theorem 21.8.4 *Suppose* $\widetilde{H} = \mathrm{Re}\, \xi(z)dz^2, \widetilde{k} = \mathrm{Re}\, n(z)dz^2$. *Then*

$$\langle \xi, n\rangle_{[g]} = \langle h, k\rangle_{[g]} ,$$

i.e., the Weil-Petersson metric is the quotient metric on $\mathcal{M}_{-1}/\mathcal{D}_0$.

Let us return to Dirichlet's energy map $\widetilde{E} : T(M) \longrightarrow R$. A nontrivial calculation now yields:

Theorem 21.8.5 *The map* $\widetilde{E}$ *has only one critical point, namely* $[g_0]$, *which is a minimum. The Hessian of* $\widetilde{E}$ *at* $[g_0]$ *is given by the formula*

$$D^2\widetilde{E}\,[g_o]\,(h,k) = \langle h,k\rangle_{[g]}\,,$$

i.e., the Weil-Petersson metric on $T(M)$. *Thus,* $[g_o]$ *is a nondegenerate minimum point for* $\widetilde{E}$.

As an immediate consequence of Theorems 21.8.3 and 21.8.5, we have

Theorem 21.8.6 *Teichmüller's space is diffeomorphic to* $\mathbb{R}^{6(genus\ M)-6}$.

Proof Applying a simple case of the gradient deformation techniques of Morse theory, we see that any n-manifold which admits a proper function which is bounded below and has only one nondegenerate minimum is diffeomorphic to Euclidean space. This ends the proof of Teichmüller's theorem. □

21.9 Proof of Poincaré Theorem

Poincaré probably proved his theorem using the uniformization theorem for Riemann surfaces; today, it is possible to give a straightforward proof using the calculus of variations. Tomi and Tromba give a proof of a more general version of Poincaré's theorem in [507, Theorem 1.6, p. 23].

Proof By the theorem of Korn-Lichtenstein (see [317, 362]), there exist local isothermal parameters x, y so that locally the metric can be written

$$g = \lambda\left(dx^2 + dy^2\right), \quad \lambda > 0.$$

In general, computing Gauss curvature in terms of a metric is complicated, but in these coordinates, there is a very nice formula

$$R(g) = -\frac{1}{2\lambda}\Delta\log\lambda \quad \left(\Delta = \frac{\partial^2}{\partial x^2} + \frac{\partial^2}{\partial y^2}\right)$$

and $\Delta_g = \frac{1}{\lambda}\Delta$ (see, for example, [249, Theorem 2.2.1, p. 31]).

We have to find $\rho > 0$ such that $R(\rho\cdot g) \equiv -1$; since $\rho > 0$, we may write $\rho = e^v$. Now,

$$R\left(e^{v}g\right) = -\frac{1}{2\lambda e^{v}}\Delta \log\left(e^{v}\lambda\right) = \frac{1}{e^{v}}\left(-\frac{1}{2\lambda}\Delta \log \lambda - \frac{1}{2\lambda}\Delta v\right).$$

So now abandoning conformal coordinates and requiring $R(\rho g) = -1$, we have to solve the global intrinsically defined PDE

$$-\Delta_g v + 2R(g) + 2e^{v} = 0. \tag{21.28}$$

respectively, where Δ_g denotes the Laplace-Beltrami operator of (M, g).

The uniqueness of R follows immediately from (21.28) by means of the maximum principle for elliptic equations. We remark that (21.28) is the Euler equation of the functional

$$I(v) = E(v, g) + 2\int_M \left(e^{v} + R(g)v\right) d\mu_g$$

where $E(v, g) = \int_M g(\nabla v, \nabla v)d\mu_g, and \nabla v$ is the gradient of v with respect to g. This is, in fact, Dirichlet's functional evaluated on the pair (v, g).

We shall solve (21.28) by minimizing I in an appropriate set of function C. We observe that if a solution v of (21.28) attains its maximal value in a point $z \in M$, then necessarily $-\Delta_g v(z) \geqq 0$ and hence

$$0 \leqq e^{v(z)} \leqq -R(g)(z),$$

i.e., $v(z) \leqq \log|\min \mathrm{R}(g)| =: x$.

We, therefore, choose

$$C = \left\{u \in H_2^1(M) \mid u(x) \leqq 1 + x\right\}.$$

Since the first eigenvalue of $-\Delta_g$ on M is 0, the corresponding eigenspace consisting of the constant functions, we have the estimate

$$E(u, g) \geqq \lambda_2 \int u^2 d\mu_g \tag{21.29}$$

for all $u \in H_2^1(M)$ with $\int u d\mu_g = 0$, where λ_2 is the second eigenvalue of $-\Delta_g$. It follows that

$$\sqrt{E(u, g)} + \frac{1}{A}\left|\int u d\mu_g\right|, \quad A := \int d\mu_g \tag{21.30}$$

is a norm on $H_2^1(m)$. Let us now show that I is bounded from below on C and any minimizing sequence bounded in the norm (21.30). Let $u \in C$ and decompose $u = u_0 + m$ with $\int u_0 d\mu_g = 0$ and $m = \frac{1}{A}\int u d\mu_g$. Using (21.29), we obtain

$$
\begin{aligned}
I(u) &\geqq E(u,g) + 2\int R(g)u_0 d\mu_g + 2m\int R(g)d\mu_g \\
&\geqq E(u,g) - \frac{1}{2}\lambda_2\int u_0^2 d\mu_g - 2\lambda_2^{-1}\int R(g)^2 d\mu_g + 4\pi m\chi(M) \\
&\geqq \frac{1}{2}E(u,g) + 4\pi m\chi(M) - 2\lambda_2^{-1}\int R(g)^2 d\mu_g.
\end{aligned}
$$

In case $m \geqq 0$, we have $|m| = m \leqq 1 + x$ and therefore

$$
I(u) \geqq \frac{1}{2}E(u,g) - 4\pi(1+x)|x(M)| - 2\lambda_2^{-1}\int R(g)^2 d\mu_g,
$$

if, however, $m < 0$, we obtain

$$
I(u) \geqq \frac{1}{2}E(u,g) + 4\pi|\chi(M)||m| - 2\lambda_2^{-1}\int R(g)^2 d\mu_g.
$$

We conclude immediately from the last two inequalities that I is bounded from below on C and that any minimizing sequence in C is bounded in $H_2^1(M)$. Standard compactness and lower semicontinuity arguments [384] then give the existence of a minimizer $v \in C$.

Let us show that $v \leqq x$. Let $w \in C$ be arbitrary. Then

$$
I(v) \leqq I((1-t)v + tw) = I(v + t(w-v)).
$$

Then using the fact that $\frac{d}{dt}I(v+t(w-v))\big|_{t=0} \geqq 0$, we clearly obtain the variational inequality

$$
\int \left(g(\nabla v, \nabla(w-v)) + 2(e^v + R(g)(w-v))\right) d\mu_g \geqq 0
$$

for all $w \in C$, where we again use ∇ as an abbreviation for the gradient with respect to the metric g.

Inserting $w = \min\{x, v\}$, we obtain

$$
\int_{v>x} \left(g(\nabla v, \nabla v) + 2\left(\mathrm{e}^v + \mathrm{R}(g)\right)(v-x)\right) d\mu_g \leqq 0
$$

By the choice of x, we have $e^v + \mathrm{R}(g) \geqq 0$ for $v > x$, and we conclude that $v = 0$ almost everywhere on the set $v(x) > x$. We thus have shown that $v \leqq x$ a.e. and, consequently, v lies in the interior of the set C with respect to the L_∞ norm. The Euler equations for v can then be derived in the usual way, giving that v is a weak solution of (21.28). Since the nonlinear term e^v is already bounded regularity of v follows in a standard way.

The theorem is now completely proved. □

21.10 The Classical Schwarzian Derivative

Given a meromorphic function f on a Riemann surface M and a holomorphic coordinate z, the Schwarzian derivative of f with respect to z is given by the expression

$$S_z(f) = \left(\frac{f''}{f'}\right)' - \frac{1}{2}\left(\frac{f''}{f'}\right)^2 . \tag{21.31}$$

Here $f(z)$ is a function in one complex variable and $f'(z)$, $f''(z), \dots$ are its derivatives.

We recall that the Schwarzian derivative measures by how much a holomorphic map distorts cross-ratios of four points (see [169, Section 6.1]). In particular, the Schwarzian derivative of a Möbius map is zero.

The Schwarzian derivative, or the Schwarzian, for short, was discovered by Lagrange in his treatise "Sur la construction des cartes géographiques" (1779) (see [346, XXXVIII, pp. 637–664]); the Schwarzian also appeared in a paper by Kummer (1836) [344], and it was named after Schwarz by Cayley [108].

Expression (21.31) is ubiquitous and tends to appear in seemingly unrelated fields of mathematics: classical complex analysis, differential equations, and one-dimensional dynamics, as well as Teichmüller theory, integrable systems, and conformal field theory. Leaving these numerous applications aside, we focus on the basic properties of the Schwarzian itself.

One can quickly check that the Schwarzian $S_z(f)$ vanishes if and only if $f = \frac{az+b}{cz+d}$ is fractional linear in z. Moreover, two meromorphic functions f and g are related by a fractional linear transformation

$$g = \frac{af+b}{cf+d}$$

if and only if their Schwarzians $S_z(f) = S_z(g)$ agree. The last property indicates that the Schwarzian is an invariant for maps into $\mathbb{CP}^1$. We will develop this viewpoint further.

Let $f : M \to \mathbb{CCP}^1$ be a holomorphic immersion, i.e., $df_p \neq 0$ for all $p \in M$, and fix a holomorphic coordinate z on M. A *lift* of f is a holomorphic map $\psi : M \to \mathbb{C}^2\backslash\{0\}$ so that $f(p)$ is the line spanned by $\psi(p)$. In other words, ψ is a holomorphic section of the pullback via f of the tautological line bundle over $\mathbb{CP}^1$. Our assumption that f is an immersion is equivalent to ψ and ψ_z being linearly independent at each point $p \in M$. Using a fixed complex volume form det on $\mathbb{C}^2$, we can find, up to sign, a unique normalized lift ψ of f by demanding that

$$\det(\psi, \psi_z) = 1 . \tag{21.32}$$

Indeed, two lifts $\widetilde{\psi} = \psi\lambda$ are related by a nowhere vanishing scalar $\lambda : M \to \mathbb{C}$ and

$$\det\left(\widetilde{\psi}, \widetilde{\psi}_z\right) = \lambda^2 \det(\psi, \psi_z)$$

which proves our assertion. Taking the derivative of (21.32), we obtain

$$\det(\psi, \psi_{zz}) = 0$$

so that ψ satisfies Hill's equation

$$\psi_{zz} + \frac{c}{2}\psi = 0 \tag{21.33}$$

for some holomorphic function $c : M \to \mathbb{C}$. From our construction, it is clear that c depends only on f and the coordinate z. If the holomorphic map f into $\mathbb{CP}^1$ is written in homogeneous coordinates $[f : 1]$, then one easily checks that

$$c = S_z(f) \tag{21.34}$$

is the Schwarzian derivative. The abovementioned properties of the Schwarzian now follow immediately from (21.33). Moreover, given c, we solve Hill's equation for

$$\psi = \begin{pmatrix} \psi_1 \\ \psi_2 \end{pmatrix}$$

and then $f = [\psi_1 : \psi_2]$ gives a map into $\mathbb{CP}^1$, unique up to the action of $Sl(2, \mathbb{C})$, with Schwarzian $S_z(f) = c$. Thus, the Schwarzian is an infinitesimal invariant, with respect to the Möbius group $PSl(2, \mathbb{C})$, for maps into $\mathbb{CP}^1$.

A surprise hidden in formula (21.31) is that the Schwarzian is actually not a function. The difference between a function and a more complicated tensor field is in its behavior under coordinate changes. Choose another coordinate w. What is the formula for $S_w(f)$? For a function, the answer is simply $f(w) = f(z(w))$, but for its derivative, due to the Chain Rule, it is different: $f'(w) = f'(z(w))z'(w)$; this is why the invariant (geometric) quantity is not the derivative but the differential $df = f'(z)dw$!

To obtain a better understanding of what type of object the Schwarzian is, we differentiate with respect to another holomorphic coordinate w on M. An elementary calculation using (21.33) reveals the transformation formula

$$S_w(f)dw^2 = (S_z(f) - S_z(w))\, dz^2\,. \tag{21.35}$$

Recall that a projective structure on M is given by a holomorphic atlas $\mathcal{A}$ whose coordinate transition functions are fractional linear. Given two projective structures $\mathcal{A}$ and $\widetilde{\mathcal{A}}$, we have the holomorphic quadratic differential

$$Q := S_z(\widetilde{z})dz^2.$$

where z and $\widetilde{z}$ are charts in $\mathcal{A}$ and $\widetilde{\mathcal{A}}$. This is well defined by (21.35). Therefore, the space of all projective structures on a Riemann surface is an affine space modeled on the vector space of holomorphic quadratic differentials. If we have fixed a projective structure on M, the transformation rule (21.35) implies that the Schwarzian of a function may be regarded as a holomorphic quadratic differential.

21.11 The Schwarzian Derivative

A Riemann surface carries an analytic structure—an atlas of local coordinates such that the change of coordinate maps are analytic. One can refine this notion by requiring that the local coordinates take their values in the Riemann sphere $\mathbb{CP}^1$ and that the changes of coordinates be restrictions of automorphisms of $\mathbb{CP}^1$, i.e., restrictions of Möbius transformations. Such an atlas is said to endow the Riemann surface with a *projective structure*.

The generalized Schwarzian derivative $S\{f, g\}$, also called the *Schwarzian*, measures the difference between the projective structures induced by f and g: it measures how much $g \circ f^{-1}$ differs from a Möbius transformation. Let U be a Riemann surface and let $f, g : U \to \mathbb{CP}^1$ be analytic maps with nonvanishing derivatives. For each $z \in U$, there exists a unique Möbius transformation A such that the Taylor series of f and of $A \circ g$ coincide to second order: they have the same value, and the same first and second derivatives. Then the leading term $\left[D^3(f - A \circ g)(z)\right]$ of $f - (A \circ g)$ at z is naturally a cubic map

$$T_zU \to T_{f(z)}\mathbb{CP}^1 . \tag{21.36}$$

Let us compose this map with the inverse of the isomorphism $Df(z) : T_zU \to T_{f(z)}\mathbb{CP}^1$ to get a cubic map $T_zU \to T_zU$:

$$(Df(z))^{-1} \circ D^3(f - (A \circ g))(z) . \tag{21.37}$$

Now invoke the following trivial result from linear algebra:
let E be a one-dimensional vector space over any field $\mathbb{K}$. Then the map

$$\alpha \mapsto (w \mapsto \alpha(w)w) \tag{21.38}$$

is an isomorphism from the one-dimensional vector space of quadratic maps $E \to \mathbb{K}$ to the one-dimensional vector space of cubic maps $E \to E$.

Thus, the composition

$$(Df(z))^{-1} \circ D^3(f - (A \circ g))(z) \tag{21.39}$$

is naturally a quadratic form on T_zU. We denote this field of quadratic forms, i.e., this quadratic differential, by $\frac{1}{6}S\{f, g\}$. The 6 in the denominator makes our definition of the Schwarzian derivative agree with the standard definition.

Definition 21.11.1 (Generalized Schwarzian Derivative) Let U be a Riemann surface. Let $f, g : U \to \mathbb{CP}^1$ be maps with nonvanishing derivatives and let A be the unique Möbius transformation such that the Taylor series of f and of $A \circ g$ coincide to second order at a point $z \in U$. Then the Schwarzian derivative $S\{f, g\}$ at z is given by

$$S\{f, g\}(z) = 6\left((Df(z))^{-1} \circ D^3(f - A \circ g)(z)\right).$$

The fundamental example is the following, which is the classical definition of Schwarzian derivative:

Proposition 21.11.2 (Computing the Schwarzian Derivative) *Let $U \subset \mathbb{C}$. If $f : U \to \mathbb{C}$ is an analytic function with nonvanishing derivative, and $g(z) := z$, then*

$$S\{f, z\} = \left(\frac{f'''}{f'} - \frac{3}{2}\left(\frac{f''}{f'}\right)^2\right) dz^2 = S_z(f)dz^2. \tag{21.40}$$

Proof This is a straightforward computation. We may assume that $z = 0$ and $f(0) = 0$, so that

$$f(z) = a_1 z + \frac{a_2}{2}z^2 + \frac{a_3}{6}z^3 + \cdots$$

The Möbius transformation that best approximates f is

$$A := \frac{\alpha z}{1 + \beta z} = \alpha z - \alpha\beta z^2 + \alpha\beta^2 z^3 - \cdots,$$

where

$$\alpha = a_1 \quad \text{and} \quad \beta = -\frac{a_2}{2a_1}.$$

Thus, the cubic term of $f(z) - A(z)$ is

$$\frac{a_3}{6} - \frac{a_2^2}{4a_1}.$$

We now compose this cubic term with $(Df(z))^{-1}$, i.e., divide by a_1, to find that

$$\frac{a_3}{6a_1} - \frac{a_2^2}{4a_1^2} = \frac{1}{6}\left(\frac{f'''(0)}{f'(0)} - \frac{3}{2}\left(\frac{f''(0)}{f'(0)}\right)^2\right).$$

□

The following properties of the Schwarzian derivative can be computed from Eq. (21.40), but it is much more instructive to derive them from the definition.

Proposition 21.11.3 (Properties of the Schwarzian Derivative) *The Schwarzian derivative satisfies the following properties, where α is a Möbius transformation:*

1. $S\{f, g\} = -S\{g, f\}$.
2. $S\{f, g\} = 0$ *if and only if* $f = \alpha \circ g$.
3. $S\{f, g\} = S\{\alpha \circ f, g\} = S\{f, \alpha \circ g\}$.

Kummer [344] (see also [343]) observed the following: if y_1 and y_2 are independent solutions of a general second-order differential equation

$$y'' + py' + qy = 0, \tag{21.41}$$

then their quotient $f = y_1/y_2$ satisfies

$$S_z(f) = 2q - \frac{1}{2}p^2 - p'. \tag{21.42}$$

This identity has the honor of surviving as the last exercise in Ahlfors' complex analysis book [18, Excercise 4, p. 318].

Kummer's identity (21.42) when $p = 0$ says that the ratio $f = y_1/y_2$ of two independent solutions of $y'' + qy = 0$ satisfies $S_z(f) = 2q$. This fact is what one most often sees in treatments of the Schwarzian.

Another way of linking the Schwarzian to second-order differential equations is to treat $S_z(f) = 2q$ as a Riccati equation

$$w' - \frac{1}{2}w^2 = 2q$$

for

$$w = \frac{f''}{f'}.$$

If we now let v be an antiderivative of w (in this case $v = \log f'$) and

$$u = e^{-v/2} = \left(f'\right)^{-1/2},$$

then u satisfies $u'' + qu = 0$ as before.

The projective geometry of a second-order equation is discussed in Arnold's book [32]. The solutions to a Riccati equation also have a "projective structure" (rather than forming a vector space) since any solution is a Möbius transformation of three given solutions; see, for instance, [237].

Lemma 21.11.4 *For any holomorphic function h in a simply connected domain $\Omega \subset \mathbb{C}$, there is a locally univalent meromorphic function f in Ω such that $S_z(f) = h$, and any other meromorphic function with this property has the form $A \circ f$, where A is a linear fractional map.*

Proof We can assume that $0 \in \Omega$. As above, we set $y = f''/f'$ and reduce the equation $S_z(f) = h$ to the Ricatti equation $y' - y^2/2 = h$. By standard substitution $y = -2w'/w$, we obtain the second-order linear equation $w'' + \frac{1}{2}hw = 0$. We solve it in the neighborhood of 0 with the normalization $w(z) = z + o(z)$. Then $w(z) = z - \frac{1}{2}\int_0^z \int_0^\zeta (hw)(\eta)d\eta d\zeta$, and this integral equation is uniquely solvable in the circle $\mathbb{D}_r := \{z \in \mathbb{C} : |z| < r\}$ if $r \cdot \max_{\mathbb{D}_r} |h| < 1$ (using the method of successive approximations, an operator with a double integral is contractive).

Since Ω is simply connected, this solution continues analytically along any path in Ω (in small circles the solutions exists) and gives a solution w_0 holomorphic in Ω, unique with normalization $w_0(0) = 0$, $w_0'(0) = 1$.

Similarly, solving the equation $w(z) = 1 - \frac{1}{2}\int_0^z \int_0^\zeta (hw)(\eta)d\eta d\zeta$, we obtain the solution w_1 with normalization $w_1(0) = 1$, $w_1'(0) = 0$. By direct calculation, we find that $\left(w_0'w_1 - w_1'w_0\right)' \equiv 0$; therefore $w_0'w_1 - w_1'w_0 \equiv 1$.

Let us put $f = w_0/w_1$. Then $f' = 1/w_1^2$, $f'' = -2w_1'/w_1^3$ and $f''/f' = -2w_1'/w_1$ therefore $S_z(f) = h$. From the relation $w_0'w_1 - w_1'w_0 = 1$, it is clear that all zeros of w_1 in $\mathbb{D}$ simple hence it follows that f is meromorphic locally univalent in Ω.

Any other such function F must satisfy the equation $F''/F' = -2w'/w$ for some solution w of the equation $w'' + \frac{1}{2}hw$. Replacing F with its fractional linear transformation, we can assume that $F(z) = z + o\left(|z|^2\right)$, like f, and therefore $F''/F' = o(1)$. Since any solution w is a linear combination of w_0, w_1, then $w = Cw_1$ and $F''/F' = f''/f'$ due to the same normalizations at 0, $F'/F = f'/f$ and $F = f$. □

Recall that the hyperbolic metric on $\widehat{\mathbb{C}}\backslash\overline{\mathbb{D}}$ is $\rho = \frac{|dz|}{(|z|^2-1)}$.

For any f which is locally univalent in $\widehat{\mathbb{C}}\backslash\overline{\mathbb{D}}$, the function $|S\{f, z\}|/\rho^2$ on $\widehat{\mathbb{C}}\backslash\overline{\mathbb{D}}$ is bounded at ∞, and one can define the "sup-norm"

$$\|S\{f, g\}\|_\infty = \sup\{|S_z(f)|(|z|^2 - 1)^2 : z \in \widehat{\mathbb{C}}\backslash\overline{\mathbb{D}}\}$$

(quotes because it can equal ∞).

It is important that in the ratio $\frac{|S\{f,z\}|}{\rho^2}$, the $\left|dz^2\right|$ coming from Eq. (21.40) cancels with the $\left|dz^2\right|$ coming from $\rho^2 = \frac{|dz|^2}{(|z|^2-1)^2}$. Thus, the ratio is a real-valued function, and it makes sense to compare it with a number.

Although solutions of the Schwarzian differential equation $S_f(z) = q$ tend not to be injective, solutions of $S_f(z) = 0$ are Möbius transformations, hence certainly injective. Thus, one might expect some connection between injective solutions and small Schwarzians. Indeed, Theorem 21.11.5, due to Zeev Nehari, says that if an analytic mapping is injective, then its Schwarzian is small, when measured in the sup-norm $\|\cdot\|_\infty$. Thus, the sup-norm is the right norm to consider in this context.

It is because of inequality (21.43) that the space of bounded quadratic differentials plays a leading role in Teichmüller theory.

Theorem 21.11.5 (Nehari's Theorem) *If f is univalent in $\widehat{\mathbb{C}}\backslash\overline{\mathbb{D}}$, then*

$$\|S\{f, g\}\|_{\infty} \leqslant 6. \tag{21.43}$$

Proof According to the invariance of $S\{f, g\}$ under automorphisms of the domain (see property **3** of Proposition 21.11.3), it is sufficient to prove the inequality at one point, for example, at ∞. For $R > 1$, there exists the diffeomorphism f_1 of the Riemann sphere coinciding with f for $|z| \geqslant R$. That is why

$$\frac{1}{2i}\int_{|z|=R} \overline{f} df = \int_{|z|<R} \frac{i}{2} df_1 \wedge d\overline{f}_1 > 0.$$

By property **3** of Proposition 21.11.3, $S\{A \circ f, z\} = S\{f, z\}$ for all linear fractional A. Therefore, we can assume that $f(z) = z + o(1)$ in a neighborhood of ∞. Then the Laurent series for f has the form $f(z) = z + b_1/z + \cdots$. This means that we can write

$$\left(\overline{f} f'\right)(z) = \left(\overline{z} + \overline{b}_1/\overline{z} + \cdots\right) \cdot \left(1 - b_1/z^2 - \cdots\right).$$

Since $\int_{|z|=R} \overline{z}^k z^l dz = 2\pi i R^{2k}$ only for $k - l = 1$, and in other cases it is equal to 0, we have

$$\frac{1}{2i}\int_{|z|=R} \overline{f} df = \pi\left(R^2 - |b_1|^2/R^2 - \cdots\right) \leqslant \pi\left(R^2 - |b_1|^2/R^2\right).$$

Since $R > 1$ is arbitrary, it follows $|b_1| \leqslant 1$.

The Schwarzian derivative is easy to compute since

$$f' = 1 - b_1/z^2 - \cdots,$$

$$f'' = 2/z^3 + \cdots,$$

$$f''' = -6/z^4 - \cdots,$$

the Schwarzian is

$$S\{f, z\} = \left(-6b_1 z^{-4} + \cdots\right) dz^2.$$

Dividing by ρ^2 and taking the limit as $z \to \infty$ gives

$$\lim_{z\to\infty} \left(|z|^2 - 1\right)^2 |S\{f, z\}| = 6|b_1| \leq 6.$$

□

The Koebe function

$$k(z) = \frac{z}{(1-z)^2} = z + 2z^2 + 3z^3 + \cdots$$

has

$$S_z(k) = \frac{-6}{\left(1 - z^2\right)^2}$$

which shows that the necessary condition for univalence is sharp.

The estimate in Theorem 21.11.5 is exact. The maximum is achieved by the Koebe function $k(z) = \frac{z}{(1-z)^2}$, which is univalent in $\mathbb{D}$ and which in $\widehat{\mathbb{C}} \backslash \overline{\mathbb{D}}$ has the Schwartzian derivative equal $k(z) = \frac{-6}{(1-z)^2}$ with the norm $S_z(k) = 6$ (in both domains).

The following theorem, in a sense the converse of Theorem 21.11.5, was proved by Ahlfors.

Theorem 21.11.6 *If h is holomorphic in $\mathbb{D}$ and $\|S\{f, z\}\|_\infty < 2$ then the equation $h = S_z(F)$ has the solution F, where F is univalent in $\mathbb{D}$ and extends to a quasiconformal homeomorphism of $\widehat{\mathbb{C}}$ with the Beltrami coefficient μ, such that $\mu(z) = -\frac{1}{2\overline{z}^4}\left(|z|^2 - 1\right)^2 h(1/\overline{z})$, $|z| > 1$, and $\mu = 0$ in $\mathbb{D}$.*

Proof Let $f = w_0/w_1$ be the function from Lemma 21.11.4 and $S_z(f) = h$. Replacing h with $h(rz)$ with $0 < r < 1$ arbitrarily close to 1, we first assume that h is holomorphic in a neighborhood of $\overline{\mathbb{D}}$, and w_1 has no zeros on $\partial\mathbb{D}$. Let us put

$$g(\zeta) = \frac{w_0(\zeta) - (\zeta - 1/\overline{\zeta})w_0'(\zeta)}{w_1(\zeta) - (\zeta - 1/\overline{\zeta})w_1'(\zeta)}.$$

From the relation $w_0'w_1 - w_1'w_0 \equiv 1$, it follows that the numerator and denominator g do not have common zeros. It is clear that g is continuously differentiable everywhere in $\overline{\mathbb{D}} \backslash \left\{0 \cup g^{-1}(\infty)\right\}$. A simple (but tedious) calculation gives the expressions at these points

$$g_{\overline{\zeta}} = \frac{1}{\overline{\zeta}^2 \left(w_1 - (\zeta - 1/\overline{\zeta})w_1'\right)^2}$$

$$g_\zeta = -\frac{(\zeta - 1/\overline{\zeta})^2 h}{2\left(w_1 - (\zeta - 1/\overline{zeta})w_1'\right)^2}.$$

It follows $\left|g_\zeta/g_{\overline{\zeta}}\right| = \frac{1}{2}|h|\left(|\zeta|^2 - 1\right)^2 < 1$ therefore g is locally one-to-one in the neighborhood of each such point (but changes orientation). Applying this reasoning to $1/g$, we find that g is locally one-to-one in $\overline{\mathbb{D}}\backslash 0$. From $w_1 = 1+a\zeta^2(1+o(1))$ and $w_1''(0) = \frac{1}{2}h(0)$, we obtain that $|a| < 1/2$ and $g(\zeta) = (1+o(1))/(\overline{\zeta} - 2a\zeta + o(|\zeta|))$ locally one-to-one in a neighborhood of 0 is also ($g(0) = \infty$).

Let us put

$$F(z) = \begin{cases} f(z), & |z| \leqslant 1 \\ g(1/\overline{z}), & |z| > 1 \end{cases}$$

Then $F : \widehat{\mathbb{C}} \to \widehat{\mathbb{C}}$ is continuous, $F(\infty) = \infty$; F preserves orientation and is locally one-to-one in $\widehat{\mathbb{C}}\backslash\partial\mathbb{D}$. Moreover, outside $\overline{\mathbb{D}}$

$$F_z(z) = -g_{\overline{\zeta}}(1/\overline{z})/z^2, \quad F_{\overline{z}}(z) = -g_\zeta(1/\overline{z})/\overline{z}^2.$$

Then $F_{\overline{z}} \to 0$ and $F_z \to -1/w_1^2 = f'$ for $|z| \to 1$. In fact, the function F is continuously differentiable everywhere outside the poles $f(z)$ and $g(1/\overline{z})$, in the neighborhood of which it is one-to-one and hence $F : \widehat{\mathbb{C}} \to \widehat{\mathbb{C}}$ is covering it follows that F is a homeomorphism. Because

$$\begin{aligned}\mu(z) &:= \frac{F_{\overline{z}}}{F_z}(z) \\ &= \frac{z^2}{\overline{z}^2}\frac{g_\zeta(1/\overline{z})}{g_{\overline{\zeta}}(1/\overline{z})} \\ &= -\frac{1}{2\overline{z}^4}\left(|z|^2 - 1\right)^2 h(1/\overline{z})\end{aligned}$$

then $\|F_{\overline{z}}/F_z\| < 1$; hence F is a quasiconformal homeomorphism of $\widehat{\mathbb{C}}$.

In the general case (h is holomorphic only in $\mathbb{D}$), replacing $h(z)$ with $h(rz)$ with suitable $r < 1$ and then passing to the limit at $r \to 1$, we obtain, as in the general case the mapping constructed above F is also a quasiconformal homeomorphism $\widehat{\mathbb{C}}$ with the indicated Beltrami differential. □

The estimate in Ahlfors's theorem is also exact. The function $f(z) = \log\frac{1+z}{1-z}$ in $\mathbb{D}$ is univalent, the sup-norm of its Schwartz derivative $S\{f, z\}$ is equal to 2, but f does not extend continuously even to the closure of $\mathbb{D}$ (at the points ± 1).

Corollary 21.11.7 *If h is holomorphic in $\widehat{\mathbb{C}}\backslash\overline{\mathbb{D}}$, and $\|S\{h, z\}\|_\infty < 2$, then $h = S_z(F_\mu)$, where F_μ is the canonical solution of the Beltrami equation with the coefficient μ concentrated in $\overline{\mathbb{D}}$ and $\|\mu\|_\infty = \frac{1}{2}\|S\{h, z\}\|$.*

Proof Let us apply Lemma 21.11.6 to the function $h(1/z)/z^4$, which is holomorphic in $\mathbb{D}$ since h has a zero of order $\geqslant 4$ in ∞, and $\sup(1 - |z|^2)^2\left|h(1/z)/z^4\right| =$

$\|S\{h, z\}\|_\infty < 2$. Let F be the quasiconformal extension of $h(1/z)/z^4$ constructed there to the entire $\widehat{\mathbb{C}}$, $S_z(F) = h(1/z)/z^4$ in $\mathbb{D}$. Then $f(z) := S_{1/z}(F)$ is a quasiconformal homeomorphism of $\widehat{\mathbb{C}}$ with the Beltrami coefficient μ concentrated in $\overline{\mathbb{D}}$ and $S_z(f) = S_{1/z}(F)/z^4 = h(z)$ in $\mathbb{C}\backslash\overline{\mathbb{D}}$. The canonical solution F_μ is a fractional linear function of f and therefore also $S_z(F_\mu) = h$ outside $\mathbb{D}$. □

Notes to This Chapter

In the literature, there are many proofs of Teichmüller's theorem. We refer, for example, to [2, 168], or [248] and the bibliography therein.

Standard references for the classical theory of the Teichmüller space are Abikoff [2], Ahlfors [20], Bers [77], Gardiner [168], Grothendieck [208], Hubbard [248], Imayoshi-Taniguchi [253], Nag [390], and Tromba [511]. See also [407].

Our treatment of Teichmüller's theorem is an expansion of the point of view taken by Tromba in [510] and [511].

Chapter 22
Appendix on the Structure Equations and Curvature

22.1 Introduction

Here we give a brief presentation of the connection between the calculus notion of curvature (see [504]) and the more abstract notion of curvature which leads to the definition of κ in Sect. 12.2. It is a pleasure to acknowledge our debt to the clear and compelling exposition in [404].

All necessary background on differential forms may be found in [449] or in [404]

22.2 Expressing Curvature Intrinsically

22.2.1 Coordinates on a Surface

Surfaces in $\mathbb{R}^3$ are given in three different ways:

1. (The simplest) as the graph of a function of two variables

$$z = f(x, y)$$

2. More generally as the graph of the equation

$$F(x, y, z) = 0$$

3. By parametric equations

$$r = r(u, v)$$

or, in more detail,

P. V. Dovbush, S. G. Krantz, *The Geometric Theory of Complex Variables*,
https://doi.org/10.1007/978-3-031-77204-7_22

$$x = x(u, v), \quad y = y(u, v), \quad z = z(u, v)$$

where the pair (u, v) ranges over some region in the plane, the domain of values of the parameters u, v. We shall often refer suggestively to the parameters u, v as "coordinates" for the surface.

All functions will be assumed to be continuously differentiable.

Definition 22.2.1 We say that the surface given by the equation $F(x, y, z) = 0$ is *non-singular* at the point $P = (x_0, y_0, z_0)$ on it (i.e., satisfying $F(x_0, y_0, z_0) = 0$), if the gradient of F at P is nonzero:

$$\frac{\partial F}{\partial x} e_1 + \frac{\partial F}{dy} e_2 + \frac{\partial F}{\partial z} e_3 \neq 0 \quad \text{at } x = x_0, \quad y = y_0, \quad z = z_0$$

By the Implicit Function Theorem, if say $\partial F / \partial z|_{x_0, y_0, z_0} \neq 0$, then near the point $P(x_0, y_0, z_0)$ on the surface, the equation $F(x, y, z) = 0$ can be solved for z as a (once again continuously differentiable) function of x and y; i.e., there exists a (continuously differentiable) function $z = f(x, y)$, such that $z_0 = f(x_0, y_0)$, and, in a neighborhood of (i.e., region containing), the point (x_0, y_0), $F(x, y, f(x, y)) = 0$. Formulas for the partial derivatives of f are obtained as follows: On differentiating the identity $F(x, y, f(x, y)) \equiv 0$ implicitly with respect to x, we obtain (using the "chain rule," and bearing in mind that $z = f(x, y)$):

$$\frac{\partial F}{\partial z}\frac{\partial f}{\partial x} + \frac{\partial F}{\partial x}\frac{\partial x}{\partial x} = 0$$

whence

$$\frac{\partial f}{\partial x} = -\frac{\partial F / \partial x}{\partial F / \partial z}. \tag{22.1}$$

Similarly

$$\frac{\partial f}{\partial y} = -\frac{\partial F / \partial y}{\partial F / \partial z}. \tag{22.2}$$

From Definition 22.2.1 and the Implicit Function Theorem, it follows that around each non-singular point of a surface given by $F(x, y, z) = 0$, there is a neighborhood such that the piece of the surface enclosed by that neighborhood is the graph of a function (where if not z, then either x or y may serve as the dependent variable). Hence, locally, i.e., in a neighborhood of a non-singular point, the surface is given by the parametric equations $z = f(u, v)$, $x = u$, $y = v$ (around the point $x_0 = u_0$, $y_0 = v_0$). This is often expressed as follows:

In a neighborhood of a non-singular point of a surface, there exist local coordinates u, v.

Suppose now that a surface is given parametrically:

$$r = r(u, v); \quad x = x(u, v), \quad y = y(u, v), \quad z = z(u, v). \tag{22.3}$$

Definition 22.2.2 A point

$$P = (x_0, y_0, z_0) = (x(u_0, v_0), y(u_0, v_0), z(u_0, v_0))$$

on the surface given by (22.3) is said to be *non-singular* if the matrix

$$A = \begin{pmatrix} \frac{\partial x}{\partial u} & \frac{\partial y}{\partial u} & \frac{\partial z}{\partial u} \\ \frac{\partial x}{\partial v} & \frac{\partial y}{\partial v} & \frac{\partial z}{\partial v} \end{pmatrix}_{u_0, v_0}$$

has rank 2.

The next theorem tells us that Definitions 22.2.1 and 22.2.2 are "locally" equivalent, but, more significantly, we can conclude from it that locally, i.e., in a neighborhood of a non-singular point of a surface, the above three ways of presenting a surface equationally are equivalent.

Theorem 22.2.3 *If a surface is given parametrically (as in (22.3)) and the point $P(x_0, y_0, z_0)$ corresponding to (u_0, v_0) is non-singular, then there is an equation $F(x, y, z) = 0$, which defines the surface in a neighborhood of P, and has the property that* $(\text{grad } F)_{x_0, y_0, z_0} \neq 0$.

Proof Since P is non-singular, by 22.2.2 the matrix A has rank 2; hence, by renaming x, y, z if necessary, we may suppose that the determinant

$$\frac{\partial x}{\partial u}\frac{\partial y}{\partial v} - \frac{\partial x}{\partial v}\frac{\partial y}{\partial u} \neq 0$$

However, this determinant is the Jacobian of the transformation $(u, v) \mapsto (x, y)$, evaluated at the point (u_0, v_0). Since it is nonzero, we deduce from the Inverse Function Theorem that there is a neighborhood of the point (x_0, y_0) (where $x_0 = x(u_0, v_0), y_0 = y(u_0, v_0)$) on which the restriction of the transformation $(x, y) \mapsto (u, v)$ has an inverse and on which, in addition, the matrix $\begin{pmatrix} u_x & v_x \\ u_y & v_y \end{pmatrix}$ is the inverse of $\begin{pmatrix} x_u & y_u \\ x_v & y_v \end{pmatrix}$. If we define $f(x, y)$ by

$$f(x, y) = z(u(x, y), v(x, y))$$

then the equation $z = f(x, y)$ (or rather $z - f(x, y) = 0$) will fulfill the claims of the theorem. □

22.2.2 Tangent Planes

If a surface in $\mathbb{R}^3$ is given in parametric form $r = r(u, v)$, $r = (x, y, z)$, where u, v are the parameters or, in other terminology, coordinates on the surface. A curve $u = u(t)$, $v = v(t)$ in the (u, v)-region defines a curve $r(t) = r(u(t), v(t))$ lying on our surface in $\mathbb{R}^3$. The tangent vector $\dot{r}(t)$ to the curve has the form

$$\dot{r}(t) = r_u\dot{u} + r_v\dot{v}, \quad r_u = \frac{\partial r}{\partial u}, \quad r_v = \frac{\partial r}{\partial v}. \tag{22.4}$$

By Definition 22.2.2, a non-singular point of the surface is one at which the vectors $r_u = (x_u, y_u, z_u)$ and $r_v = (x_v, y_v, z_v)$ are linearly independent. Putting this together with (22.4), which tells us that every vector tangent to our surface is a linear combination of the vectors r_u and r_v, we deduce that the totality of vectors tangent to our surface at a given non-singular point P forms a two-dimensional subspace having the pair r_u, r_v as a basis. We call this subspace the tangent plane to the surface at the given point. (Note that in terms of the basis r_u, r_v, the components of the tangent vector $\dot{r}$ are the quantities $\dot{u}$, $\dot{v}$.)

Suppose now that a surface in $\mathbb{R}^3$ is given to us instead via the equation $F(x, y, z) = 0$. If $r = r(t) = (x(t), y(t), z(t))$ is a curve on this surface, i.e., if $F(x(t), y(t), z(t)) = 0$ for the specified interval of values of t, then $(d/dt)F(x(t), y(t), z(t)) = 0$ for those t, whence, applying the chain rule yet again, we get

$$F_x\dot{x} + F_y\dot{y} + F_z\dot{z} = 0. \tag{22.5}$$

At a non-singular point $P\,(x_0, y_0, z_0)$ of the surface we have (by definition), $\operatorname{grad} F = \left(F_x, F_y, F_z\right) \neq 0$. Hence writing A, B, C for F_x, F_y, F_z, respectively, evaluated at P, we see from (22.5) that every vector $(\dot{x}, \dot{y}, \dot{z})$ tangent to the surface at P is in the plane $Ax + By + Cz = 0$. Hence, there is a tangent plane to the surface at the non-singular point P: it is the two-dimensional subspace consisting of all vectors (x, y, z) satisfying $Ax + By + Cz = 0$.

Suppose now that a surface M in $\mathbb{R}$ is given parametrically:

$$x = x(u, v), \quad y = y(u, v), \quad z = z(u, v),$$

where the pair (u, v) ranges over some region in the plane, the domain of values of the parameters u, v. We shall often refer suggestively to the parameters u, v as "coordinates" for the surface.

In what follows any points of the surface which come up for consideration will be tacitly assumed to be non-singular (see Definition 22.2.2). We see from (22.5) that the vectors given by the rows of the matrix A span the tangent plane to M at each point. Applying the Gram-Schmidt orthonormalization procedure to the row vectors of matrix A, and shrinking U, M if necessary, we may create vector fields

$$E_1 : M \longrightarrow \mathbb{R}^3$$

$$E_2 : M \longrightarrow \mathbb{R}^3$$

such that $E_1\,(x, y, z)$ and $E_2\,(x, y, z)$ are orthonormal and tangent to M at each $P = (x, y, z) \in M$. Denote by $T_P(M)$ the collection of linear combinations

$$aE_1\,(x, y, z) + bE_2\,(x, y, z)\,, \qquad a, b \in \mathbb{R}\,.$$

We call $T_P(M)$ the tangent space to M at P.

Let $E_3(P)$ be the unit normal to M at P given by $E_1(P)\times E_2(P)$. The functions E_1, E_2, E_3 are smooth vector fields on M: they assign to each $P \in M$ a triple of orthonormal vectors.

Let e_1, e_2, e_3 denote the standard basis for vectors in $\mathbb{R}^3$:

$$e_1 = (1, 0, 0),$$

$$e_2 = (0, 1, 0),$$

$$e_3 = (0, 0, 1).$$

(Many calculus books call these vectors **i**, **j**, and **k**.) Then we may write

$$E_i = \sum_j a_{i,j}\,(x, y, z)\,e_j, \qquad i = 1, 2, 3$$

The matrix

$$\mathcal{A} \equiv \left(a_{i,j}\right)_{i,j=1}^{3}$$

where the $a_{i,j}$ are functions of the space variables, is called the *attitude matrix* of the frame (or basis) E_1, E_2, E_3. Since $\mathcal{A}$ transforms one orthonormal frame to another, $\mathcal{A}$ is an orthogonal matrix. Hence

$$\mathcal{A}^{-1} = {}^t\mathcal{A}$$

Definition 22.2.4 If $P \in M$, $v \in T_P(M)$, and f is a smooth function on M, then define

$$D_v f(P) = \frac{d}{dt} f \circ \phi(t)\Big|_{t=0}$$

where ϕ is any smooth curve in M such that $\phi(0) = P$ and $\phi'(0) = v$. One checks that this definition is independent of the choice of ϕ.

Definition 22.2.5 If

$$\alpha : M \longrightarrow \mathbb{R}^3$$

is a vector field on M,

$$\alpha(P) = \alpha_1(P)e_1 + \alpha_2(P)e_2 + \alpha_3(P)e_3,$$

and $v \in T_P(M)$ then define

$$\nabla_v \alpha(P) = (D_v\alpha_1)(P)e_1 + (D_v\alpha_2)(P)e_2 + (D_v\alpha_3)(P)e_3$$

The operation ∇_v is called *covariant differentiation* of the vector field α.

Definition 22.2.6 If $P \in M, v \in T_P(M)$, we define the *shape operator* (or Weingarten map) for M at P to be

$$S_P(v) = -\nabla_v E_3(P).$$

Lemma 22.2.7 *We have that $S_P(v) \in T_P(M)$.*

Proof Now $E_3 \cdot E_3 \equiv 1$, hence

$$\begin{aligned} 0 &= D_v (E_3 \cdot E_3)|_P \\ &= (2\nabla v E_3) \cdot E_3|_P \\ &= -2S_P(v) \cdot E_3(P) \end{aligned}$$

Hence, $S_P(v) \perp E_3(P)$ so $S_P(v) \in T_P(M)$. □

Notice that the shape operator assigns to each $P \in M$ a linear operator S_P on the two-dimensional tangent space $T_P(M)$.

We can express this linear operator as a matrix with respect to the basis $E_1(P), E_2(P)$. So S assigns to each $P \in M$ a 2×2 matrix $\mathcal{M}_P$.

The linear operator S_P measures the rate of change of E_3 in any tangent direction v. It can be shown that $\mathcal{M}_P$ is diagonalizable. The (real) eigenvectors of $\mathcal{M}_P$ correspond to the principal curvatures of M at P (these are the directions of greatest and least curvature), and the corresponding eigenvalues measure the amount of curvature in those directions.

The preceding observations about $\mathcal{M}_P$ motivate the following definition:

Definition 22.2.8 5. The *Gaussian curvature* $\kappa(P)$ of M at a point $P \in M$ is the determinant of $\mathcal{M}_P$—the product of the two eigenvalues.

Our aim is to express $\kappa(P)$ in terms of the intrinsic geometry of M, without reference to E_3—that is, without reference to the way that M is situated in space.

To this end, we define covector fields θ_i which are dual to the vector fields E_i:

$$\theta_i E_j(P) = \delta_{ij}$$

Then the θ_i may be expressed as linear combinations of the standard basis covectors dx_1, dx_2, dx_3. Indeed, if $\mathcal{A} = \left(a_{i,j}\right)$ is the attitude matrix then

$$\theta_i = \sum a_{i,j} dx_j$$

(remember that $\mathcal{A}^{-1} = {}^t\mathcal{A}$). Thus, for each i, θ_i is a differential form, and the standard calculus of differential forms—including exterior differentiation—applies.

From now on, we restrict attention to 1- and 2-forms acting on tangent vectors to M. Thus any 1-form α may be expressed as

$$\alpha = \alpha\,(E_1)\,\theta_1 + \alpha\,(E_2)\,\theta_2$$

and any 2-form β may be expressed as

$$\beta = \beta\,(E_1, E_2)\,\theta_1 \wedge \theta_2 .$$

Now we define covector fields $\omega_{i,j}$, $i, j \in \{1, 2, 3\}$ by the formula

$$\omega_{i,j}(v) = (\nabla_v E_i) \cdot E_j(P) .$$

Here "$\cdot$" is the Euclidean dot product. We think of $\omega_{i,j}$ as a differential 1-form. Notice that since $E_i \cdot E_j \equiv \delta_{i,j}$, we have for $v \in T_P(M)$ that

$$\begin{aligned} 0 &= D_v\left(E_i \cdot E_j\right) \\ &= (\nabla_v E_i) \cdot E_j + E_i \cdot \left(\nabla_v E_j\right) \\ &= \omega_{i,j}(v) + \omega_{j,i}(v) . \end{aligned}$$

Thus,

$$\omega_{i,j} = -\omega_{j,i} .$$

In particular,

$$\omega_{i,i} = 0 .$$

If $v \in T_P(M)$, then it is easy to check, just using linear algebra, that

$$\nabla_v E_i = \sum_j \omega_{i,j}(v) E_j, \quad 1 \le i \le 3 .$$

We call the $\omega_{i,j}$ the *connection forms* for M. We can now express the shape operator in terms of these connection forms.

Proposition 22.2.9 *Let $P \in M$ and $v \in T_P(M)$. Then*

$$S_P(v) = \omega_{1,3}(v)E_1(P) + \omega_{2,3}(v)E_2(P)\,.$$

Proof We have that

$$\begin{aligned}
S_P(v) &= -\nabla_v E_3 \\
&= -\sum_{j=1}^{3} \left(\nabla_v E_3 \cdot E_j\right) E_j \\
&= -\sum_{j=1}^{3} \omega_{3,j}(v) E_j \\
&= \omega_{1,3}(v)E_1 + \omega_{2,3}(v)E_2
\end{aligned}$$

since $\omega_{3,3} = 0$. Now we can express Gaussian curvature in terms of the $\omega_{i,j}$. □

Proposition 22.2.10 *We have that*

$$\omega_{1,3} \wedge \omega_{2,3} = \kappa\theta_1 \wedge \theta_2\,.$$

Proof We need to calculate $\mathcal{M}_P$ in terms of the $\omega_{i,j}$. Proposition 22.2.9 gives that

$$S_P\left(E_1\right) = \omega_{1,3}\left(E_1\right)E_1 + \omega_{2,3}\left(E_1\right)E_2$$

and

$$S_P\left(E_2\right) = \omega_{1,3}\left(E_2\right)E_1 + \omega_{2,3}\left(E_2\right)E_2$$

so the matrix of S_P, in terms of the basis E_1, E_2, is

$$\mathcal{M}_P = \begin{pmatrix} \omega_{1,3}\left(E_1\right) & \omega_{2,3}\left(E_1\right) \\ \omega_{1,3}\left(E_2\right) & \omega_{2,3}\left(E_2\right) \end{pmatrix}.$$

We know that $\omega_{1,3} \wedge \omega_{2,3}$, being a 2-form, can be written as $\lambda \cdot \theta_1 \wedge \theta_2$. On the other hand,

$$\begin{aligned}
\kappa &= \det \mathcal{M}_P \\
&= \omega_{1,3}\left(E_1\right)\omega_{2,3}\left(E_2\right) - \omega_{1,3}\left(E_2\right)\omega_{2,3}\left(E_1\right) \\
&= \left(\omega_{1,3} \wedge \omega_{2,3}\right)\left(E_1, E_2\right) \\
&= \lambda\,.
\end{aligned}$$

Therefore,

$$\omega_{1,3} \wedge \omega_{2,3} = \lambda \theta_1 \wedge \theta_2 = \kappa \theta_1 \wedge \theta_2 .$$

□

Our goal now is to express κ using only those $\omega_{i,j}$ with $i \neq 3$, $j \neq 3$. For this, we require a technical lemma about the attitude matrix.

Lemma 22.2.11 *We have that*

$$\omega_{i,j} = \sum_k a_{j,k} da_{i,k}, \quad 1 \leq i, j \leq 3 .$$

Proof If $v \in T_P(M)$ then

$$\omega_{i,j}(v) = \nabla_v E_i \cdot E_j(P) .$$

But

$$E_i = \sum_k a_{i,k} \delta_k .$$

Therefore,

$$\nabla_v E_i = \sum_k \left(D_v a_{i,k}\right) \delta_k .$$

Then

$$\begin{aligned} \omega_{i,j}(v) &\equiv \nabla_v E_i \cdot E_j \\ &= \left(\sum_k \left(D_v a_{i,k}\right) \delta_k\right) \cdot \left(\sum_k a_{j,k} \delta_k\right) \\ &= \sum_k \left(D_v a_{i,k}\right) a_{j,k} \\ &= \sum_k da_{i,k}(v) a_{j,k} . \end{aligned}$$

As a result,

$$\omega_{i,j} = \sum_k a_{j,k} da_{i,k} .$$

□

Now we have reached a milestone. We can derive the important Cartan structural equations, which are the key to our intrinsic formulas for curvature.

Theorem 22.2.12 *We have*

$$d\theta_i = \sum_j \omega_{i,j} \wedge \theta_j , \tag{22.6}$$

$$d\omega_{i,j} = \sum_k \omega_{i,k} \wedge \omega_{k,j} . \tag{22.7}$$

Proof We have that

$$\theta_i = \sum_j a_{i,j} dx_j$$

hence

$$d\theta_i = \sum_j da_{i,j} \wedge dx_j .$$

Since the attitude matrix $\mathcal{A}$ is orthogonal, we may solve the equations in Lemma 22.2.11 for $da_{i,k}$ in terms of $\omega_{i,j}$. Thus,

$$da_{i,j} = \sum_k \omega_{i,k} a_{k,j} .$$

Substituting this into our last formula gives

$$\begin{aligned} d\theta_i &= \sum_j \left[\left(\sum_k \omega_{i,k} a_{k,j} \right) \wedge dx_j \right] \\ &= \sum_k \left[\omega_{i,k} \wedge \sum_j a_{k,j} dx_j \right] \\ &= \sum_k \omega_{i,k} \wedge \theta_k . \end{aligned}$$

This is the first structural equation. For the second equation, notice that the formula

$$\omega_{i,j} = \sum_k a_{j,k} da_{i,k}$$

implies

$$d\omega_{i,j} = - \sum_k da_{i,k} \wedge da_{j,k} .$$

On the other hand,

$$\begin{aligned}
\sum_k \omega_{i,k} \wedge \omega_{k,j} &= \sum_k \left(\sum_\ell a_{k,\ell} da_{i,\ell} \right) \wedge \left(\sum_m a_{j,m} da_{k,m} \right) \\
&= \sum_k \left(\sum_\ell a_{k,\ell} da_{i,\ell} \right) \wedge \left(- \sum_m a_{k,m} da_{j,m} \right) \\
&= - \left(\sum_{\ell,m} \left\{ \sum_k a_{k,\ell} a_{k,m} \right\} \cdot da_{i,\ell} \wedge da_{j,m} \right) \\
&= - \sum_m da_{i,m} \wedge da_{j,m} \\
&= d\omega_{i,j} \,.
\end{aligned}$$

In the antepenultimate line, we have used the fact that $\mathcal{A}^{-1} = {}^t\mathcal{A}$. The result now follows: □

The following corollary will prove critical:

Corollary 22.2.13 *We have*

$$d\omega_{1,2} = -\kappa \theta_1 \wedge \theta_2 \,.$$

Proof The second structural equation gives

$$\begin{aligned}
d\omega_{1,2} &= \sum_k \omega_{1,k} \wedge \omega_{k,2} \\
&= \omega_{1,1} \wedge \omega_{1,2} + \omega_{1,2} \wedge \omega_{2,2} + \omega_{1,3} \wedge \omega_{3,2} \,.
\end{aligned}$$

Only the third summand does not vanish. But we calculated in Proposition 22.2.10 that it equals $-\kappa\theta_1 \wedge \theta_2$. □

The Corollary has been our main goal in this subsection. It gives an intrinsic way to calculate Gaussian curvature in the classical setting, hence a way to define Gaussian curvature in more abstract settings. We now proceed to develop this more abstract point of view.

22.3 Curvature Calculations on Planar Domains

Let $\Omega \subseteq \mathbb{C}$ be a domain which is equipped with a metric ρ. Assume for simplicity that $\rho(z) > 0$ at all points of Ω. Define functions

$$E_1 \equiv \frac{(1,0)}{\rho} \quad \text{and} \quad E_2 \equiv \frac{(0,1)}{\rho}.$$

Then

$$\theta_1 = \rho dx \quad \text{and} \quad \theta_2 = \rho dy$$

are the dual covector fields. We define $\omega_{i,j}$ according to the first structure equation:

$$\begin{aligned} d\theta_1 &= \omega_{1,2} \wedge \theta_2\,, \\ d\theta_2 &= \omega_{2,1} \wedge \theta_1\,. \end{aligned}$$

We define Gaussian curvature according to the Corollary to Theorem 22.2.12 in the previous section:

$$d\omega_{1,2} = -\kappa\theta_1 \wedge \theta_2\,.$$

One can check that these definitions are independent of the choice of frame (or basis) E_1, E_2, but this is irrelevant for our purposes. We conclude this Appendix by proving that the definition of curvature which we just elicited from the structural equations coincides with the one given in Sect. 12.2.

First, by the way that we have defined θ_1 and θ_2, we have

$$\begin{aligned} d\theta_1 &= d\rho \wedge dx \\ &= \big(\rho_x dx + \rho_y dy\big) \wedge dx \\ &= \rho_y dy \wedge dx \\ &= -\frac{\rho_y}{\rho} dx \wedge \rho dy \\ &= -\frac{\rho_y}{\rho} dx \wedge \theta_2\,. \end{aligned}$$

Similarly,

$$\begin{aligned} d\theta_2 &= d\rho \wedge dy \\ &= \big(\rho_x dx + \rho_y dy\big) \wedge dy \\ &= \rho_x dx \wedge dy \\ &= -\frac{\rho_x}{\rho} dy \wedge \rho dx \\ &= -\frac{\rho_x}{\rho} dy \wedge \theta_1\,. \end{aligned}$$

Comparison with the first structural equation gives

$$\omega_{1,2} = -\frac{\rho_y}{\rho}dx + \tau dy$$

and

$$\omega_{1,2} = -\omega_{2,1} = -\left(-\frac{\rho_x}{\rho}dy\right) + \sigma dx$$

for some unknown functions τ and σ.

The only way these equations can be consistent is if

$$\omega_{1,2} = -\frac{\rho_y}{\rho}dx + \frac{\rho_x}{\rho}dy\,.$$

Thus,

$$\begin{aligned}
d\omega_{1,2} &= -\frac{\partial}{\partial y}\left(\frac{\rho_y}{\rho}\right) dy \wedge dx + \frac{\partial}{\partial x}\left(\frac{\rho_x}{\rho}\right) dx \wedge dy \\
&= \left(-\frac{\rho_{yy}}{\rho} + \frac{\rho_y\rho_y}{\rho^2}\right) dy \wedge dx + \left(\frac{\rho_{xx}}{\rho} - \frac{\rho_x\rho_x}{\rho^2}\right) dx \wedge dy \\
&= \frac{1}{\rho^2}\left(\rho\Delta\rho - (\rho_y)^2 - (\rho_x)^2\right) dx \wedge dy \\
&= \frac{1}{\rho^4}\left(\rho\Delta\rho - (\rho_y)^2 - (\rho_x)^2\right) \theta_1 \wedge \theta_2\,.
\end{aligned}$$

The second structural equation now implies that

$$\kappa = -\frac{1}{\rho^4}\left(\rho\Delta\rho - |\nabla\rho|^2\right)\,.$$

On the other hand, in Sect. 12.2, we defined

$$\kappa = -\frac{\Delta \log \rho}{\rho^2}\,.$$

We have

$$\frac{\partial}{\partial \bar{z}} \log \rho = \frac{1}{\rho}\frac{\partial \rho}{\partial \bar{z}}$$

and

$$\begin{aligned}\Delta \log \rho &= 4\frac{\partial^2}{\partial z \partial \bar{z}} \log \rho \\ &= 4\left(-\frac{1}{\rho^2}\frac{\partial \rho}{\partial z}\frac{\partial \rho}{\partial \bar{z}} + \frac{1}{\rho}\frac{\partial^2 \rho}{\partial z \partial \bar{z}}\right) \\ &= -\frac{1}{\rho^2} \cdot |\nabla \rho|^2 + \frac{1}{\rho} \cdot \Delta \rho .\end{aligned}$$

It follows that, according to the definition in Sect. 12.2,

$$\kappa = -\frac{1}{\rho^4}\left(\rho \Delta \rho - |\nabla \rho|^2\right) .$$

Thus, as claimed, the definition of curvature from Sect. 12.2 is equal to that which arises from the structural equations.

Chapter 23
Concluding Remarks

The geometric viewpoint represents an entirely new perception of the subject of complex analysis. As demonstrated in this book by uniformization, Teichmüller theory, hyperbolic spaces, and many other topics, geometry raises new questions and provides profound and powerful answers. The techniques treated here are powerful tools for answering a vast array of questions—both traditional analytic questions and modern geometric questions.

The world of geometry in the subject of complex variables is vast. Thus, our coverage in this book is perforce selective. But we hope to have given the reader a dynamic introduction to this important part of our subject. The reader is encouraged to engage in his or her own explorations and diversions. The result can only be a richer and more rewarding mathematical life.

P. V. Dovbush, S. G. Krantz, *The Geometric Theory of Complex Variables*,
https://doi.org/10.1007/978-3-031-77204-7_23

References

1. The correspondence of Isaac Newton, Vol. II: 1676–1687. Cambridge University Press, New York (1960)
2. Abikoff, W.: The real analytic theory of Teichmüller space, *Lecture Notes in Mathematics*, vol. 820. Springer, Berlin (1980)
3. Abraham, R., Marsden, J.E., Raţiu, T.S.: Manifolds, tensor analysis, and applications, *Global Analysis Pure and Applied: Series B*, vol. 2. Addison-Wesley Publishing Co., Reading, MA (1983)
4. Abramowitz, M., Stegun, I.A. (eds.): Handbook of mathematical functions, with formulas, graphs, and mathematical tables. Dover Publications, Inc., New York (1966)
5. Adams, R.A.: Sobolev spaces, *Pure and Applied Mathematics*, vol. Vol. 65. Academic Press [Harcourt Brace Jovanovich, Publishers], New York-London (1975)
6. Aharonov, D., Shapiro, H.S.: Domains on which analytic functions satisfy quadrature identities. J. Analyse Math. **30**, 39–73 (1976). https://doi.org/10.1007/BF02786704
7. Ahlfors, L.: Zur Theorie der überlagerungsflächen. Acta Math. **65**(1), 157–194 (1935). https://doi.org/10.1007/BF02420945
8. Ahlfors, L., Bers, L.: Riemann's mapping theorem for variable metrics. Ann. of Math. (2) **72**, 385–404 (1960). https://doi.org/10.2307/1970141
9. Ahlfors, L., Beurling, A.: Conformal invariants and function-theoretic null-sets. Acta Math. **83**, 101–129 (1950). https://doi.org/10.1007/BF02392634
10. Ahlfors, L.V.: An extension of Schwarz's lemma. Trans. Amer. Math. Soc. **43**(3), 359–364 (1938). https://doi.org/10.2307/1990065
11. Ahlfors, L.V.: Bounded analytic functions. Duke Math. J. **14**, 1–11 (1947). http://projecteuclid.org/euclid.dmj/1077473984
12. Ahlfors, L.V.: Complex analysis. An introduction to the theory of analytic functions of one complex variable. McGraw-Hill Book Co., Inc., New York-Toronto-London (1953)
13. Ahlfors, L.V.: Development of the theory of conformal mapping and Riemann surfaces through a century. In: Contributions to the theory of Riemann surfaces, *Ann. of Math. Stud.*, vol. no. 30, pp. 3–13. Princeton Univ. Press, Princeton, NJ (1953)
14. Ahlfors, L.V.: On quasiconformal mappings. J. Analyse Math. **3**, 1–58; correction, 207–208 (1954). https://doi.org/10.1007/BF02803585
15. Ahlfors, L.V.: Complex analysis: An introduction of the theory of analytic functions of one complex variable, second edn. McGraw-Hill Book Co., New York-Toronto-London (1966)
16. Ahlfors, L.V.: Lectures on quasiconformal mappings, *Van Nostrand Mathematical Studies*, vol. No. 10. D. Van Nostrand Co., Inc., Toronto, Ont.-New York-London (1966)

P. V. Dovbush, S. G. Krantz, *The Geometric Theory of Complex Variables*,
https://doi.org/10.1007/978-3-031-77204-7

17. Ahlfors, L.V.: Conformal invariants: topics in geometric function theory. McGraw-Hill Series in Higher Mathematics. McGraw-Hill Book Co., New York-Düsseldorf-Johannesburg (1973)
18. Ahlfors, L.V.: Complex analysis, third edn. International Series in Pure and Applied Mathematics. McGraw-Hill Book Co., New York (1978)
19. Ahlfors, L.V.: Commentary on: "Zur theorie der überlagerungsflächen" (1935) [reprinted from MR0688648 (84k:01066a)]. In: Fields Medallists' lectures, *World Sci. Ser. 20th Century Math.*, vol. 5, pp. 8–9. World Sci. Publ., River Edge, NJ (1997). https://doi.org/10.1142/9789812385215_0002
20. Ahlfors, L.V.: Lectures on quasiconformal mappings, *University Lecture Series*, vol. 38, second edn. American Mathematical Society, Providence, RI (2006). https://doi.org/10.1090/ulect/038
21. Ahlfors, L.V.: Conformal invariants. AMS Chelsea Publishing, Providence, RI (2010). https://doi.org/10.1090/chel/371
22. Ahlfors, L.V., Bers, L.: Prostranstva Rimanovykh poverkhnostei i kvazikonformnye otobrazheniya. Izdat. Inostr. Lit., Moscow (1961)
23. Ahlfors, L.V., Sario, L.: Riemann surfaces, *Princeton Mathematical Series*, vol. No. 26. Princeton University Press, Princeton, NJ (1960)
24. Aĭzenberg, I.A., Yuzhakov, A.P.: Integral representations and residues in multidimensional complex analysis, *Translations of Mathematical Monographs*, vol. 58. American Mathematical Society, Providence, RI (1983). https://doi.org/10.1090/mmono/058
25. Aizenberg, L.A.: Multidimensional analogues of the Carleman formula with integration over boundary sets of maximal dimension. In: Multidimensional complex analysis (Russian), pp. 12–22, 272. Akad. Nauk SSSR Sibirsk. Otdel., Inst. Fiz., Krasnoyarsk (1985)
26. Alexander, H.: Holomorphic mappings from the ball and polydisc. Math. Ann. **209**, 249–256 (1974). https://doi.org/10.1007/BF01351851
27. Andreev, E.M.: Convex polyhedra in Lobačevskiĭ spaces. Mat. Sb. (N.S.) **81(123)**, 445–478 (1970)
28. Andreev, E.M.: Convex polyhedra of finite volume in Lobačevskiĭ space. Mat. Sb. (N.S.) **83(125)**, 256–260 (1970)
29. Appell, P., Goursat, É.: Théorie des fonctions algébriques et de leurs intégrales. I: Étude des fonctions analytiques sur une surface de Riemann. 2. ed. revue et augmentee par P. Fatou. Paris: Gauthier-Villars. XXXV, 526 p (1929). (1929)
30. Ramirez de Arellano, E.: Ein Divisionsproblem und Randintegraldarstellungen in der komplexen Analysis. Math. Ann. **184**, 172–187 (1970). https://doi.org/10.1007/BF01351561
31. Armstrong, M.A.: Basic topology. Undergraduate Texts in Mathematics. Springer-Verlag, New York-Berlin (1983)
32. Arnold, V.I.: Geometrical methods in the theory of ordinary differential equations, *Grundlehren der Mathematischen Wissenschaften*, vol. 250. Springer-Verlag, New York-Berlin (1983)
33. Aronszajn, N.: Theory of reproducing kernels. Trans. Amer. Math. Soc. **68**, 337–404 (1950). https://doi.org/10.2307/1990404
34. Astala, K., Iwaniec, T., Martin, G.: Elliptic partial differential equations and quasiconformal mappings in the plane, *Princeton Mathematical Series*, vol. 48. Princeton University Press, Princeton, NJ (2009)
35. Avci, Y.: Quadrature identities and the Schwarz Function. Ph.D. thesis, Stanford (1977)
36. Baer, R.: Isotopie von Kurven auf orientierbaren, geschlossenen Flächen und ihr Zusammenhang mit der topologischen Deformation der Flächen. J. Reine Angew. Math. **159**, 101–116 (1928). https://doi.org/10.1515/crll.1928.159.101
37. Baernstein II, A.: Ahlfors and conformal invariants. Ann. Acad. Sci. Fenn. Ser. A I Math. **13**(3), 289–312 (1988). https://doi.org/10.5186/aasfm.1988.1323
38. Bagemihl, F., Seidel, W.: Some boundary properties of analytic functions. Math. Z. **61**, 186–199 (1954). https://doi.org/10.1007/BF01181342

39. Banjai, L., Trefethen, L.N.: A multipole method for Schwarz-Christoffel mapping of polygons with thousands of sides. SIAM J. Sci. Comput. **25**(3), 1042–1065 (2003). https://doi.org/10.1137/S1064827502411675
40. Baracco, L., Zaitsev, D., Zampieri, G.: A Burns-Krantz type theorem for domains with corners. Math. Ann. **336**(3), 491–504 (2006). https://doi.org/10.1007/s00208-005-0727-2
41. Barker, S.R.: Two theorems on boundary values of analytic functions. Proc. Amer. Math. Soc. **68**(1), 54–58 (1978). https://doi.org/10.2307/2040907
42. Barletta, E., Bedford, E.: Existence of proper mappings from domains in $\mathbb{C}^2$. Indiana Univ. Math. J. **39**(2), 315–338 (1990). https://doi.org/10.1512/iumj.1990.39.39018
43. Barrett, D.E.: Irregularity of the Bergman projection on a smooth bounded domain in C^2. Ann. of Math. (2) **119**(2), 431–436 (1984). https://doi.org/10.2307/2007045
44. Barrett, D.E.: Behavior of the Bergman projection on the Diederich-Fornæss worm. Acta Math. **168**(1-2), 1–10 (1992). https://doi.org/10.1007/BF02392975
45. Beardon, A.F.: The geometry of discrete groups, *Graduate Texts in Mathematics*, vol. 91. Springer-Verlag, New York (1983). https://doi.org/10.1007/978-1-4612-1146-4
46. Beardon, A.F.: A primer on Riemann surfaces, *London Mathematical Society Lecture Note Series*, vol. 78. Cambridge University Press, Cambridge (1984)
47. Beardon, A.F., Minda, D.: Normal families: a geometric perspective. Comput. Methods Funct. Theory **14**(2-3), 331–355 (2014). https://doi.org/10.1007/s40315-014-0054-2
48. Beck, A.: A theorem on maximum modulus. Proc. Amer. Math. Soc. **15**, 345–349 (1964). https://doi.org/10.2307/2034500
49. Bedford, E., Dadok, J.: Bounded domains with prescribed group of automorphisms. Comment. Math. Helv. **62**(4), 561–572 (1987). https://doi.org/10.1007/BF02564462
50. Behnke, H., Stein, K.: Modifikation komplexer Mannigfaltigkeiten und Riemannscher Gebiete. Math. Ann. **124**, 1–16 (1951). https://doi.org/10.1007/BF01343548
51. Bell, J., Blå sjö, V.: Pietro Mengoli's 1650 proof that the harmonic series diverges. Math. Mag. **91**(5), 341–347 (2018). https://doi.org/10.1080/0025570X.2018.1506656
52. Bell, S.: Numerical computation of the Ahlfors map of a multiply connected planar domain. J. Math. Anal. Appl. **120**(1), 211–217 (1986). https://doi.org/10.1016/0022-247X(86)90211-8
53. Bell, S., Ligocka, E.: A simplification and extension of Fefferman's theorem on biholomorphic mappings. Invent. Math. **57**(3), 283–289 (1980). https://doi.org/10.1007/BF01418930
54. Bell, S.R.: Biholomorphic mappings and the $\bar{\partial}$-problem. Ann. of Math. (2) **114**(1), 103–113 (1981). https://doi.org/10.2307/1971379
55. Bell, S.R.: The Cauchy transform, potential theory, and conformal mapping. Studies in Advanced Mathematics. CRC Press, Boca Raton, FL (1992)
56. Bell, S.R.: Finitely generated function fields and complexity in potential theory in the plane. Duke Math. J. **98**(1), 187–207 (1999). https://doi.org/10.1215/S0012-7094-99-09805-8
57. Bell, S.R.: The role of the Ahlfors mapping in the theory of kernel functions in the plane. In: Reproducing kernels and their applications (Newark, DE, 1997), *Int. Soc. Anal. Appl. Comput.*, vol. 3, pp. 33–42. Kluwer Acad. Publ., Dordrecht (1999). https://doi.org/10.1007/978-1-4757-2987-0_4
58. Bell, S.R.: The Bergman kernel and quadrature domains in the plane. In: Quadrature domains and their applications, *Oper. Theory Adv. Appl.*, vol. 156, pp. 61–78. Birkhäuser, Basel (2005). https://doi.org/10.1007/3-7643-7316-4_3
59. Bell, S.R.: Quadrature domains and kernel function zipping. Ark. Mat. **43**(2), 271–287 (2005). https://doi.org/10.1007/BF02384780
60. Bell, S.R.: Bergman coordinates. Studia Math. **176**(1), 69–83 (2006). https://doi.org/10.4064/sm176-1-5
61. Bell, S.R.: Density of quadrature domains in one and several complex variables. Complex Var. Elliptic Equ. **54**(3-4), 165–171 (2009). https://doi.org/10.1080/17476930902759361
62. Bell, S.R.: An improved Riemann mapping theorem and complexity in potential theory. Ark. Mat. **51**(2), 223–249 (2013). https://doi.org/10.1007/s11512-012-0168-6

63. Bell, S.R.: The Dirichlet and Neumann and Dirichlet-to-Neumann problems in quadrature, double quadrature, and non-quadrature domains. Anal. Math. Phys. **5**(2), 113–135 (2015). https://doi.org/10.1007/s13324-014-0089-6
64. Bell, S.R.: The Cauchy transform, potential theory and conformal mapping, second edn. Chapman & Hall/CRC, Boca Raton, FL (2016)
65. Bell, S.R.: The Cauchy integral formula, quadrature domains, and Riemann mapping theorems. Comput. Methods Funct. Theory **18**(4), 661–676 (2018). https://doi.org/10.1007/s40315-018-0246-2
66. Bell, S.R., Boas, H.P.: Regularity of the Bergman projection in weakly pseudoconvex domains. Math. Ann. **257**(1), 23–30 (1981). https://doi.org/10.1007/BF01450652
67. Bell, S.R., Ebenfelt, P., Khavinson, D., Shapiro, H.S.: On the classical Dirichlet problem in the plane with rational data. J. Anal. Math. **100**, 157–190 (2006). https://doi.org/10.1007/BF02916759
68. Bell, S.R., Ebenfelt, P., Khavinson, D., Shapiro, H.S.: Algebraicity in the Dirichlet problem in the plane with rational data. Complex Var. Elliptic Equ. **52**(2-3), 235–244 (2007). https://doi.org/10.1080/17476930601065615
69. Bell, S.R., Gustafsson, B., Sylvan, Z.A.: Szegő coordinates, quadrature domains, and double quadrature domains. Comput. Methods Funct. Theory **11**(1), 25–44 (2011). https://doi.org/10.1007/BF03321788
70. Bell, S.R., Krantz, S.G.: Smoothness to the boundary of conformal maps. Rocky Mountain J. Math. **17**(1), 23–40 (1987). https://doi.org/10.1216/RMJ-1987-17-1-23
71. Berg, G.: Bounded holomorphic functions of several variables. Ark. Mat. **20**(2), 249–270 (1982). https://doi.org/10.1007/BF02390511
72. Berger, M.S.: Riemannian structures of prescribed Gaussian curvature for compact 2-manifolds. J. Differential Geometry **5**, 325–332 (1971). http://projecteuclid.org/euclid.jdg/1214429996
73. Bergh, J., Löfström, J.: Interpolation spaces. An introduction, *Grundlehren der Mathematischen Wissenschaften*, vol. No. 223. Springer-Verlag, Berlin-New York (1976)
74. Bergman, S., Schiffer, M.: Kernel functions and elliptic differential equations in mathematical physics. Academic Press, Inc., New York (1953)
75. Bergmann, S.: Über die Entwicklung der harmonischen Funktionen der Ebene und des Raumes nach Orthogonalfunktionen. Math. Ann. **86**(3-4), 238–271 (1922). https://doi.org/10.1007/BF01457987
76. Bers, L.: Riemann surfaces: Lectures given at the Institute of Mathematical Sciences, New York University 1957–1958 (1958)
77. Bers, L.: Quasiconformal mappings and Teichmüller's theorem. In: Analytic functions, *Princeton Math. Ser.*, vol. No. 24, pp. 89–119. Princeton Univ. Press, Princeton, NJ (1960)
78. Bers, L.: Uniformization by Beltrami equations. Comm. Pure Appl. Math. **14**, 215–228 (1961). https://doi.org/10.1002/cpa.3160140304
79. Bers, L.: Introduction to Several Complex Variables: Lectures by Lipman Bers 1962–1963. Courant Institute of Mathematical Sciences, New York University (1964)
80. Bers, L.: An approximation theorem. J. Analyse Math. **14**, 1–4 (1965). https://doi.org/10.1007/BF02806376
81. Bers, L.: On Hilbert's 22nd problem. In: Mathematical developments arising from Hilbert problems (Proc. Sympos. Pure Math., Northern Illinois Univ., De Kalb, Ill., 1974), *Proc. Sympos. Pure Math.*, vol. Vol. XXVIII, pp. 559–609. Amer. Math. Soc., Providence, RI (1976)
82. Bers, L., of Mathematical Sciences, C.I.: Riemann Surfaces. Lecture notes. Courant Institute of Mathematical Sciences, New York University (1958). https://books.google.md/books?id=n28JtgEACAAJ
83. Bieberbach, L.: Über die Koeffizienten derjenigen Potenzreihen, welche eine schlichte Abbildung des Einheitskreises vermitteln. Berl. Ber. **1916**, 940–955 (1916)
84. Boas, H.P.: Counterexample to the Lu Qi-Keng conjecture. Proc. Amer. Math. Soc. **97**(2), 374–375 (1986). https://doi.org/10.2307/2046535

85. Boas, H.P.: The Lu Qi-Keng conjecture fails generically. Proc. Amer. Math. Soc. **124**(7), 2021–2027 (1996). https://doi.org/10.1090/S0002-9939-96-03259-5
86. Boas, R.P.: Invitation to complex analysis, second edn. MAA Textbooks. Mathematical Association of America, Washington, DC (2010)
87. Bocher, M.: Introduction to higher algebra. Dover Publications, Inc., New York (1964)
88. Bôchner, S.: Über orthogonale Systeme analytischer Funktionen. Math. Z. **14**(1), 180–207 (1922). https://doi.org/10.1007/BF01215898
89. Bochner, S., Montgomery, D.: Locally compact groups of differentiable transformations. Ann. of Math. (2) **47**, 639–653 (1946). https://doi.org/10.2307/1969226
90. Borel, A., Serre, J.P.: Groupes de Lie et puissances réduites de Steenrod. Amer. J. Math. **75**, 409–448 (1953). https://doi.org/10.2307/2372495
91. Bottazzini, U., Gray, J.: Hidden harmony—geometric fantasies. Sources and Studies in the History of Mathematics and Physical Sciences. Springer, New York (2013). https://doi.org/10.1007/978-1-4614-5725-1
92. Boyarskii, B.V.: Homeomorphic solutions of Beltrami systems. Dokl. Akad. Nauk SSSR (N.S.) **102**, 661–664 (1955)
93. de Branges, L.: A proof of the Bieberbach conjecture. Acta Math. **154**(1-2), 137–152 (1985). https://doi.org/10.1007/BF02392821
94. Bremermann, H.J.: Holomorphic continuation of the kernel function and the Bergman metric in several complex variables. In: Lectures on functions of a complex variable, pp. 349–383. Univ. Michigan Press, Ann Arbor, MI (1955)
95. Brown, R.: Topology and groupoids, third edn. BookSurge, LLC, Charleston, SC (2006)
96. Bungart, L.: Holomorphic functions with values in locally convex spaces and applications to integral formulas. Trans. Amer. Math. Soc. **111**, 317–344 (1964). https://doi.org/10.2307/1994246
97. Burckel, R.B.: Classical analysis in the complex plane. Cornerstones. Birkhäuser/Springer, New York (2021). https://doi.org/10.1007/978-1-0716-1965-0
98. Burkholder, D.L., Gundy, R.F., Silverstein, M.L.: A maximal function characterization of the class H^p. Trans. Amer. Math. Soc. **157**, 137–153 (1971). https://doi.org/10.2307/1995838
99. Burns Jr., D., Shnider, S., Wells Jr., R.O.: Deformations of strictly pseudoconvex domains. Invent. Math. **46**(3), 237–253 (1978). https://doi.org/10.1007/BF01390277
100. Burns, D.M., Krantz, S.G.: Rigidity of holomorphic mappings and a new Schwarz lemma at the boundary. J. Amer. Math. Soc. **7**(3), 661–676 (1994). https://doi.org/10.2307/2152787
101. Buser, P.: Geometry and spectra of compact Riemann surfaces, *Progress in Mathematics*, vol. 106. Birkhäuser Boston, Inc., Boston, MA (1992)
102. Byun, J.: On the boundary accumulation points for the holomorphic automorphism groups. Michigan Math. J. **51**(2), 379–386 (2003). https://doi.org/10.1307/mmj/1060013203
103. Calderón, A.P.: Intermediate spaces and interpolation, the complex method. Studia Math. **24**, 113–190 (1964). https://doi.org/10.4064/sm-24-2-113-190
104. Carathéodory, C.: Untersuchungen über die konformen Abbildungen von festen und veränderlichen Gebieten. Math. Ann. **72**(1), 107–144 (1912). https://doi.org/10.1007/BF01456892
105. Carathéodory, C.: Über das Schwarzsche Lemma bei analytischen Funktionen von zwei komplexen Veränderlichen. Math. Ann. **97**(1), 76–98 (1927). https://doi.org/10.1007/BF01447861
106. Carathéodory, C.: Über die geometrie der analytischen abbildungen, die durch analytische funktionen von zwei veränderlichen vermittelt werden. Abh. Math. Sem. Univ. Hamburg **6**(1), 96–145 (1928). https://doi.org/10.1007/BF02940606
107. Cartan, H.: Fonctions analytiques de plusieurs variables complexes. Séminaire Henri Cartan de l'Ecole Supérieure (1951/1952), Secrétariat mathématique, 11 Rue Pierre Curie, Paris (1952)
108. Cayley, A.: On the Schwarzian derivative and the polyhedral functions. Trans. Camb. Philos. Soc. **13** (1880)
109. Chang, D.C.E., Krantz, S.G.: Holomorphic Lipschitz functions and application to the $\overline{\partial}$-problem. Colloq. Math. **62**(2), 227–256 (1991). https://doi.org/10.4064/cm-62-2-227-256

110. Chern, S.s.: An elementary proof of the existence of isothermal parameters on a surface. Proc. Amer. Math. Soc. **6**, 771–782 (1955). https://doi.org/10.2307/2032933
111. Chern, S.S., Moser, J.K.: Real hypersurfaces in complex manifolds. Acta Math. **133**, 219–271 (1974). https://doi.org/10.1007/BF02392146
112. Chevalley, C.: Introduction to the Theory of Algebraic Functions of One Variable, *Mathematical Surveys*, vol. No. VI. American Mathematical Society, New York (1951)
113. Christ, M.: Global $\mathcal{C}^\infty$ irregularity of the $\overline{\partial}$-Neumann problem for worm domains. J. Amer. Math. Soc. **9**(4), 1171–1185 (1996). https://doi.org/10.1090/S0894-0347-96-00213-5
114. Churchill, R.V.: Complex variables and applications. McGraw-Hill Book Co., Inc., New York-Toronto-London (1960)
115. Cima, J.A., Krantz, S.G.: The Lindelöf principle and normal functions of several complex variables. Duke Math. J. **50**(1), 303–328 (1983). https://doi.org/10.1215/S0012-7094-83-05014-7
116. Čirka, E.M.: The Lindelöf and Fatou theorems in $\mathbb{C}^n$. Mat. Sb. (N.S.) **92(134)**, 622–644, 648 (1973)
117. Clemens, C.H.: A scrapbook of complex curve theory. University Series in Mathematics. Plenum Press, New York-London (1980)
118. Conway, J.B.: Functions of one complex variable, *Graduate Texts in Mathematics*, vol. 11, second edn. Springer-Verlag, New York-Berlin (1978)
119. Córdoba, A., Fefferman, R.: On differentiation of integrals. Proc. Nat. Acad. Sci. U.S.A. **74**(6), 2211–2213 (1977). https://doi.org/10.1073/pnas.74.6.2211
120. Courant, R.: Dirichlet's Principle, Conformal Mapping, and Minimal Surfaces. Interscience Publishers, Inc., New York (1950)
121. Courant, R., Hilbert, D.: Methods of mathematical physics. Vol. II: Partial differential equations. Interscience Publishers (a division of John Wiley & Sons, Inc.), New York-London (1962)
122. Crowdy, D.: Multipolar vortices and algebraic curves. R. Soc. Lond. Proc. Ser. A Math. Phys. Eng. Sci. **457**(2014), 2337–2359 (2001). https://doi.org/10.1098/rspa.2001.0815
123. Crowdy, D., Marshall, J.: Constructing multiply connected quadrature domains. SIAM J. Appl. Math. **64**(4), 1334–1359 (2004). https://doi.org/10.1137/S0036139903438545
124. Crowdy, D., Marshall, J.: Conformal mappings between canonical multiply connected domains. Comput. Methods Funct. Theory **6**(1), 59–76 (2006). https://doi.org/10.1007/BF03321118
125. D'Angelo, J.P.: Real hypersurfaces, orders of contact, and applications. Ann. of Math. (2) **115**(3), 615–637 (1982). https://doi.org/10.2307/2007015
126. D'Angelo, J.P.: Several complex variables and the geometry of real hypersurfaces. Studies in Advanced Mathematics. CRC Press, Boca Raton, FL (1993)
127. D'Angelo, J.P.: A gentle introduction to points of finite type on real hypersurfaces. In: Explorations in complex and Riemannian geometry, *Contemp. Math.*, vol. 332, pp. 19–36. Amer. Math. Soc., Providence, RI (2003). https://doi.org/10.1090/conm/332/05928
128. David, G.: Solutions de l'équation de Beltrami avec $\|\mu\|_\infty = 1$. Ann. Acad. Sci. Fenn. Ser. A I Math. **13**(1), 25–70 (1988). https://doi.org/10.5186/aasfm.1988.1303
129. Davis, P., Pollak, H.: On the analytic continuation of mapping functions. Trans. Amer. Math. Soc. **87**, 198–225 (1958). https://doi.org/10.2307/1993097
130. Davis, P.J.: The Schwarz function and its applications, *The Carus Mathematical Monographs*, vol. No. 17. Mathematical Association of America, Buffalo, NY (1974)
131. Deger, E.: A biholomorphism from the Bell representative domain onto an annulus and kernel functions. Ph.D. thesis, Purdue Univ. (2007)
132. Denneberg, H.: Konforme Abbildung einer Klasse unendlich-vielfach zusammenhängender schlichter Bereiche auf Kreisbereiche. Ber. Verh. Sächs. Akad. Leipzig 84, 331–352 (1932). (1932)
133. Di Biase, F., Krantz, S.G.: Foundations of Fatou theory and a tribute to the work of E. M. Stein on boundary behavior of holomorphic functions. J. Geom. Anal. **31**(7), 7184–7296 (2021). https://doi.org/10.1007/s12220-021-00618-z

134. Diederich, K., Fornaess, J.E.: Pseudoconvex domains with real-analytic boundary. Ann. of Math. (2) **107**(2), 371–384 (1978). https://doi.org/10.2307/1971120
135. Dittmar, B.: Bemerkungen zu einem Funktional von M. Schiffer und N. S. Hawley in der Theorie der konformen Abbildungen. Math. Nachr. **97**, 21–37 (1980). https://doi.org/10.1002/mana.19800970104
136. Donaldson, S.K., Sullivan, D.P.: Quasiconformal 4-manifolds. Acta Math. **163**(3-4), 181–252 (1989). https://doi.org/10.1007/BF02392736
137. Drasin, D.: The inverse problem of the Nevanlinna theory. Acta Math. **138**(1-2), 83–151 (1976). https://doi.org/10.1007/BF02392314
138. Driscoll, T.A., Trefethen, L.N.: Schwarz-Christoffel mapping, *Cambridge Monographs on Applied and Computational Mathematics*, vol. 8. Cambridge University Press, Cambridge (2002). https://doi.org/10.1017/CBO9780511546808
139. Dummit, D.S., Foote, R.M.: Abstract algebra, third edn. John Wiley & Sons, Inc., Hoboken, NJ (2004)
140. Duren, P.: F. W. Gehring: a biographical sketch. In: Quasiconformal mappings and analysis (Ann Arbor, MI, 1995), pp. 1–4. Springer, New York (1998)
141. Duren, P.L.: Theory of H^p spaces, *Pure and Applied Mathematics*, vol. Vol. 38. Academic Press, New York-London (1970)
142. Duren, P.L.: Univalent functions, *Grundlehren der mathematischen Wissenschaften [Fundamental Principles of Mathematical Sciences]*, vol. 259. Springer-Verlag, New York (1983)
143. Ebin, D.G.: The manifold of Riemannian metrics. In: Global Analysis (Proc. Sympos. Pure Math., Vols. XIV, XV, XVI, Berkeley, Calif., 1968), *Proc. Sympos. Pure Math.*, vol. XIV-XVI, pp. 11–40. Amer. Math. Soc., Providence, RI (1970)
144. Eisenhart, L.P.: Riemannian Geometry. Princeton University Press, Princeton, NJ (1949)
145. Eisenman, D.A.: Intrinsic measures on complex manifolds and holomorphic mappings, *Memoirs of the American Mathematical Society*, vol. No. 96. American Mathematical Society, Providence, RI (1970)
146. Epstein, B.: On the mean-value property of harmonic functions. Proc. Amer. Math. Soc. **13**, 830 (1962). https://doi.org/10.2307/2034188
147. Epstein, B., Schiffer, M.M.: On the mean-value property of harmonic functions. J. Analyse Math. **14**, 109–111 (1965). https://doi.org/10.1007/BF02806381
148. Farkas, H.M., Kra, I.: Riemann surfaces, *Graduate Texts in Mathematics*, vol. 71, second edn. Springer-Verlag, New York (1992). https://doi.org/10.1007/978-1-4612-2034-3
149. Federer, H.: Geometric measure theory, *Die Grundlehren der mathematischen Wissenschaften*, vol. Band 153. Springer-Verlag New York, Inc., New York (1969)
150. Fefferman, C.: The Bergman kernel and biholomorphic mappings of pseudoconvex domains. Invent. Math. **26**, 1–65 (1974). https://doi.org/10.1007/BF01406845
151. Fefferman, C.: Parabolic invariant theory in complex analysis. Adv. in Math. **31**(2), 131–262 (1979). https://doi.org/10.1016/0001-8708(79)90025-2
152. Fefferman, C., Stein, E.M.: H^p spaces of several variables. Acta Math. **129**(3-4), 137–193 (1972). https://doi.org/10.1007/BF02392215
153. Fisher, S.D.: Function theory on planar domains. Pure and Applied Mathematics (New York). John Wiley & Sons, Inc., New York (1983)
154. Folland, G.B., Kohn, J.J.: The Neumann problem for the Cauchy-Riemann complex, *Annals of Mathematics Studies*, vol. No. 75. Princeton University Press, Princeton, NJ; University of Tokyo Press, Tokyo (1972)
155. Ford, L.R.: Automorphic functions, second edn. Chelsea Publishing Co., New York (1951)
156. Fornaess, J.E.: Embedding strictly pseudoconvex domains in convex domains. Amer. J. Math. **98**(2), 529–569 (1976). https://doi.org/10.2307/2373900
157. Forster, O.: Riemannsche Flächen, *Heidelberger Taschenbücher [Heidelberg Paperbacks]*, vol. Band 184. Springer-Verlag, Berlin-New York (1977)
158. Forster, O.: Lectures on Riemann surfaces, *Graduate Texts in Mathematics*, vol. 81. Springer-Verlag, New York-Berlin (1981)

159. Freeman, M.: Local complex foliation of real submanifolds. Math. Ann. **209**, 1–30 (1974). https://doi.org/10.1007/BF01432883
160. Fricke, R., Klein, F.: Lectures on the theory of automorphic functions. Vol. 2, *CTM. Classical Topics in Mathematics*, vol. 4. Higher Education Press, Beijing (2017)
161. Fridman, B.L.: A universal exhausting domain. Proc. Amer. Math. Soc. **98**(2), 267–270 (1986). https://doi.org/10.2307/2045696
162. Fu, S., Wong, B.: On a domain in $\mathbb{C}^2$ with generic piecewise smooth Levi-flat boundary and non-compact automorphism group. Complex Variables Theory Appl. **42**(1), 25–40 (2000). https://doi.org/10.1080/17476930008815269
163. Gaier, D.: Konstruktive Methoden der konformen Abbildung, *Springer Tracts in Natural Philosophy*, vol. Vol. 3. Springer-Verlag, Berlin (1964)
164. Gaier, D.: Konforme Abbildung mehrfach zusammenhängender Gebiete. Jahresber. Deutsch. Math.-Verein. **81**(1), 25–44 (1978)
165. Gamelin, T.W.: Uniform algebras. Prentice-Hall, Inc., Englewood Cliffs, NJ (1969)
166. Gamelin, T.W.: Complex analysis. Undergraduate Texts in Mathematics. Springer-Verlag, New York (2001). https://doi.org/10.1007/978-0-387-21607-2
167. Gamelin, T.W., Greene, R.E.: Introduction to topology, second edn. Dover Publications, Inc., Mineola, NY (1999)
168. Gardiner, F.P.: Teichmüller theory and quadratic differentials. Pure and Applied Mathematics (New York). John Wiley & Sons, Inc., New York (1987)
169. Gardiner, F.P., Lakic, N.: Quasiconformal Teichmüller theory, *Mathematical Surveys and Monographs*, vol. 76. American Mathematical Society, Providence, RI (2000). https://doi.org/10.1090/surv/076
170. Garnett, J.B.: Bounded analytic functions, *Pure and Applied Mathematics*, vol. 96. Academic Press, Inc. [Harcourt Brace Jovanovich, Publishers], New York-London (1981)
171. Garnett, J.B., Latter, R.H.: The atomic decomposition for Hardy spaces in several complex variables. Duke Math. J. **45**(4), 815–845 (1978). http://projecteuclid.org/euclid.dmj/1077313101
172. Garnett, J.B., Marshall, D.E.: Harmonic measure, *New Mathematical Monographs*, vol. 2. Cambridge University Press, Cambridge (2005). https://doi.org/10.1017/CBO9780511546617
173. Gavosto, E.: Thesis. Ph.D. thesis (1990)
174. Gehring, F.W., Lehto, O.: On the total differentiability of functions of a complex variable. Ann. Acad. Sci. Fenn. Ser. A I **272**, 9 (1959)
175. Ghys, É.: Topologie des feuilles génériques. Ann. of Math. (2) **141**(2), 387–422 (1995). https://doi.org/10.2307/2118526
176. Gleason, A.M.: The abstract theorem of Cauchy-Weil. Pacific J. Math. **12**, 511–525 (1962). http://projecteuclid.org/euclid.pjm/1103036490
177. Godement, R.: Topologie algébrique et théorie des faisceaux, *Publ. Inst. Math. Univ. Strasbourg*, vol. 13. Hermann, Paris (1958)
178. Gohberg, I., Leiterer, J.: Holomorphic operator functions of one variable and applications, *Operator Theory: Advances and Applications*, vol. 192. Birkhäuser Verlag, Basel (2009). https://doi.org/10.1007/978-3-0346-0126-9
179. Goluzin, G.M.: Geometric theory of functions of a complex variable, *Translations of Mathematical Monographs*, vol. Vol. 26. American Mathematical Society, Providence, RI (1969)
180. Gong, S.: Singular integrals in several complex variables. Xiandai Shuxue Congshu (Series of Modern Mathematics). Shanghai: Shanghai Scientific & Technical Publishers. III, 278 p. (1982). (1982)
181. Gong, S.: Harmonic analysis on classical groups. Berlin etc.: Springer-Verlag; Beijing: Science Press (1991)
182. Graham, I.: Boundary behavior of the Carathéodory and Kobayashi metrics on strongly pseudoconvex domains in C^n with smooth boundary. Trans. Amer. Math. Soc. **207**, 219–240 (1975). https://doi.org/10.2307/1997175

183. Gramain, A.: Topologie des surfaces, *Collection SUP: "Le Mathématicien"*, vol. 7. Presses Universitaires de France, Paris (1971)
184. Grauert, H.: Analytische Faserungen über holomorph-vollständigen Räumen. Math. Ann. **135**, 263–273 (1958). https://doi.org/10.1007/BF01351803
185. Grauert, H., Lieb, I.: Das Ramirezsche Integral und die Lösung der Gleichung $\bar{\partial} f = \alpha$ im Bereich der beschränkten Formen. Rice Univ. Stud. **56**(2), 29–50 (1970)
186. Grauert, H., Remmert, R.: Komplexe Räume. Math. Ann. **136**, 245–318 (1958). https://doi.org/10.1007/BF01362011
187. Gray, J.: On the history of the Riemann mapping theorem. In: Studies in the history of modern mathematics, I, pp. 47–94. Palermo: Circolo Matemático di Palermo (1994)
188. Greenberg, M.J.: Lectures on algebraic topology. W. A. Benjamin, Inc., New York-Amsterdam (1967)
189. Greenberg, M.J., Harper, J.R.: Algebraic topology, *Mathematics Lecture Note Series*, vol. 58. Benjamin/Cummings Publishing Co., Inc., Advanced Book Program, Reading, MA (1981)
190. Greene, R.E., Kim, K.T., Krantz, S.G.: The geometry of complex domains, *Progress in Mathematics*, vol. 291. Birkhäuser Boston, Ltd., Boston, MA (2011). https://doi.org/10.1007/978-0-8176-4622-6
191. Greene, R.E., Krantz, S.G.: The stability of the Bergman kernel and the geometry of the Bergman metric. Bull. Amer. Math. Soc. (N.S.) **4**(1), 111–115 (1981). https://doi.org/10.1090/S0273-0979-1981-14874-6
192. Greene, R.E., Krantz, S.G.: Stability properties of the Bergman kernel and curvature properties of bounded domains. In: Recent developments in several complex variables (Proc. Conf., Princeton Univ., Princeton, N. J., 1979), *Ann. of Math. Stud.*, vol. No. 100, pp. 179–198. Princeton Univ. Press, Princeton, NJ (1981)
193. Greene, R.E., Krantz, S.G.: The automorphism groups of strongly pseudoconvex domains. Math. Ann. **261**(4), 425–446 (1982). https://doi.org/10.1007/BF01457445
194. Greene, R.E., Krantz, S.G.: Deformation of complex structures, estimates for the $\bar{\partial}$ equation, and stability of the Bergman kernel. Adv. in Math. **43**(1), 1–86 (1982). https://doi.org/10.1016/0001-8708(82)90028-7
195. Greene, R.E., Krantz, S.G.: Stability of the Carathéodory and Kobayashi metrics and applications to biholomorphic mappings. In: Complex analysis of several variables (Madison, Wis., 1982), *Proc. Sympos. Pure Math.*, vol. 41, pp. 77–93. Amer. Math. Soc., Providence, RI (1984). https://doi.org/10.1090/pspum/041/740874
196. Greene, R.E., Krantz, S.G.: Characterization of complex manifolds by the isotropy subgroups of their automorphism groups. Indiana Univ. Math. J. **34**(4), 865–879 (1985). https://doi.org/10.1512/iumj.1985.34.34048
197. Greene, R.E., Krantz, S.G.: Normal families and the semicontinuity of isometry and automorphism groups. Math. Z. **190**(4), 455–467 (1985). https://doi.org/10.1007/BF01214745
198. Greene, R.E., Krantz, S.G.: Biholomorphic self-maps of domains. In: Complex analysis, II (College Park, Md., 1985–86), *Lecture Notes in Math.*, vol. 1276, pp. 136–207. Springer, Berlin (1987). https://doi.org/10.1007/BFb0078959
199. Greene, R.E., Krantz, S.G.: Characterizations of certain weakly pseudoconvex domains with noncompact automorphism groups. In: Complex analysis (University Park, Pa., 1986), *Lecture Notes in Math.*, vol. 1268, pp. 121–157. Springer, Berlin (1987). https://doi.org/10.1007/BFb0097301
200. Greene, R.E., Krantz, S.G.: Invariants of Bergman geometry and the automorphism groups of domains in $\mathbb{C}^n$. In: Geometrical and algebraical aspects in several complex variables (Cetraro, 1989), *Sem. Conf.*, vol. 8, pp. 107–136. EditEl, Rende (1991)
201. Greene, R.E., Krantz, S.G.: Techniques for studying automorphisms of weakly pseudoconvex domains. In: Several complex variables (Stockholm, 1987/1988), *Math. Notes*, vol. 38, pp. 389–410. Princeton Univ. Press, Princeton, NJ (1993)
202. Greene, R.E., Krantz, S.G.: Function theory of one complex variable, *Graduate Studies in Mathematics*, vol. 40, third edn. American Mathematical Society, Providence, RI (2006). https://doi.org/10.1090/gsm/040

203. Greene, R.E., Wu, H.: Analysis on noncompact Kähler manifolds. In: Several complex variables (Proc. Sympos. Pure Math., Vol. XXX, Part 2, Williams Coll., Williamstown, Mass., 1975), *Proc. Sympos. Pure Math.*, vol. Vol. XXX, Part 2, pp. 69–100. Amer. Math. Soc., Providence, RI (1977)
204. Greiner, P.C., Stein, E.M.: Estimates for the $\overline{\partial}$-Neumann problem, *Mathematical Notes*, vol. No. 19. Princeton University Press, Princeton, NJ (1977)
205. Griffiths, P., Harris, J.: Principles of algebraic geometry. Pure and Applied Mathematics. Wiley-Interscience [John Wiley & Sons], New York (1978)
206. Griffiths, P., Harris, J.: Principles of algebraic geometry. Wiley Classics Library. John Wiley & Sons, Inc., New York (1994). https://doi.org/10.1002/9781118032527
207. Griffiths, P.A.: Introduction to algebraic curves, *Translations of Mathematical Monographs*, vol. 76. American Mathematical Society, Providence, RI (1989). https://doi.org/10.1090/mmono/076
208. Grothendieck, A.: Techniques de construction en géométrie analytique. III. Produits fibrés d'espaces analytiques. In: Séminaire Henri Cartan Volume 13 (1960–1961) no. 1, Talk no. 10, vol. 13, pp. 1–11 (1960). http://www.numdam.org/item/SHC_1960-1961__13_1_A6_0/
209. Grötzsch, H.: Über die Verzerrung bei schlichten nichtkonformen Abbildungen und über eine damit zusammenhängende Erweiterung des Picardschen Satzes. Berichte Leipzig 80, 503–507 (1928). (1928)
210. Grötzsch, H.: Über das Parallelschlitztheorem der konformen Abbildung schlichter Bereiche. Bereiche Leipzig 84, 15–36 (1932). (1932)
211. Grötzsch, H.: Über möglichst konforme Abbildungen von schlichten Bereichen. Berichte Leipzig 84, 114–120 (1932). (1932)
212. Grötzsch, H.: Eine Bemerkung zum Koebeschen Kreisnormierungsprinzip. Ber. Verh. Sächs. Akad. Leipzig 87, 319–324 (1935). (1935)
213. Grout, R.W., Kolla, H., Minion, M.L., Bell, J.B.: Achieving algorithmic resilience for temporal integration through spectral deferred corrections. Commun. Appl. Math. Comput. Sci. **12**(1), 25–50 (2017). https://doi.org/10.2140/camcos.2017.12.25
214. Grunsky, H.: Lectures on theory of functions in multiply connected domains, *Studia Mathematica: Skript*, vol. 4. Vandenhoeck & Ruprecht, Göttingen (1978)
215. Guenot, J., Narasimhan, R.: Introduction à la théorie des surfaces de Riemann, *Monographies de L'Enseignement Mathématique [Monographs of L'Enseignement Mathématique]*, vol. No. 23. Université de Genève, L'Enseignement Mathématique, Geneva (1976)
216. Gunning, R.C.: Lectures on Riemann surfaces, Jacobi varieties, *Mathematical Notes*, vol. No. 12. Princeton University Press, Princeton, NJ; University of Tokyo Press, Tokyo (1972)
217. Gunning, R.C.: Riemann surfaces and generalized theta functions, *Ergebnisse der Mathematik und ihrer Grenzgebiete [Results in Mathematics and Related Areas]*, vol. Band 91. Springer-Verlag, Berlin-New York (1976)
218. Gunning, R.C., Rossi, H.: Analytic functions of several complex variables. Prentice-Hall, Inc., Englewood Cliffs, NJ (1965)
219. Gustafsson, B.: Quadrature identities and the Schottky double. Acta Appl. Math. **1**(3), 209–240 (1983). https://doi.org/10.1007/BF00046600
220. Gustafsson, B.: On quadrature domains and an inverse problem in potential theory. J. Analyse Math. **55**, 172–216 (1990). https://doi.org/10.1007/BF02789201
221. Gustafsson, B., Putinar, M.: Linear analysis of quadrature domains. II. Israel J. Math. **119**, 187–216 (2000). https://doi.org/10.1007/BF02810668
222. Gustafsson, B., Putinar, M.: Linear analysis of quadrature domains. IV. In: Quadrature domains and their applications, *Oper. Theory Adv. Appl.*, vol. 156, pp. 173–194. Birkhäuser, Basel (2005). https://doi.org/10.1007/3-7643-7316-4_8
223. H., W.: Some open problems in the study of noncompact Kähler manifolds. Geometric theory of several complex variables. RIMS Kōkyuroku **340**, 12–25 (1978). http://hdl.handle.net/2433/104265
224. Haas, A.: Linearization and mappings onto pseudocircle domains. Trans. Amer. Math. Soc. **282**(1), 415–429 (1984). https://doi.org/10.2307/1999596

225. Hahn, K.T.: Equivalence of the classical theorems of Schottky, Landau, Picard and hyperbolicity. Proc. Amer. Math. Soc. **89**(4), 628–632 (1983). https://doi.org/10.2307/2044595
226. Hahn, K.T., Kim, K.T.: Hyperbolicity of a complex manifold and other equivalent properties. Proc. Amer. Math. Soc. **91**(1), 49–53 (1984). https://doi.org/10.2307/2045267
227. He, Z.X., Schramm, O.: Fixed points, Koebe uniformization and circle packings. Ann. of Math. (2) **137**(2), 369–406 (1993). https://doi.org/10.2307/2946541
228. Heins, M.: On the number of $1-1$ directly conformal maps which a multiply-connected plane region of finite connectivity $p(> 2)$ admits onto itself. Bull. Amer. Math. Soc. **52**, 454–457 (1946). https://doi.org/10.1090/S0002-9904-1946-08590-0
229. Helgason, S.u.: Differential geometry and symmetric spaces, *Pure and Applied Mathematics*, vol. Vol. XII. Academic Press, New York-London (1962)
230. Henkin, G.M.: Integral representation of functions which are holomorphic in strictly pseudoconvex regions, and some applications. Mat. Sb. (N.S.) **78(120)**, 611–632 (1969)
231. Henrici, P.: Applied and computational complex analysis. Pure and Applied Mathematics. Wiley-Interscience [John Wiley & Sons], New York-London-Sydney (1974)
232. Henrici, P.: Applied and computational complex analysis. Vol. 3. Pure and Applied Mathematics (New York). John Wiley & Sons, Inc., New York (1986)
233. Herglotz, G.: Gesammelte Schriften. Vandenhoeck & Ruprecht, Göttingen (1979)
234. Hilbert, D.: Über das Dirichletsche Prinzip. Math. Ann. **59**(1-2), 161–186 (1904). https://doi.org/10.1007/BF01444753
235. Hille, E.: Analytic function theory. Vol. 1. Introductions to Higher Mathematics. Ginn and Company, Boston, MA (1959)
236. Hille, E.: Analytic function theory. Vol. II. Introductions to Higher Mathematics. Ginn and Company, Boston, Mass.-New York-Toronto (1962)
237. Hille, E.: Ordinary differential equations in the complex domain. Dover Publications, Inc., Mineola, NY (1997)
238. Hironaka, H.: Resolution of singularities of an algebraic variety over a field of characteristic zero. I, II. Ann. of Math. (2) **79**, 109–203; 79 (1964), 205–326 (1964). https://doi.org/10.2307/1970547
239. Hironaka, H., Rossi, H.: On the equivalence of imbeddings of exceptional complex spaces. Math. Ann. **156**, 313–333 (1964). https://doi.org/10.1007/BF01361027
240. Hirsch, M.W.: Differential topology, *Graduate Texts in Mathematics*, vol. No. 33. Springer-Verlag, New York-Heidelberg (1976)
241. Hoffman, K.: Banach spaces of analytic functions. Prentice-Hall Series in Modern Analysis. Prentice-Hall, Inc., Englewood Cliffs, NJ (1962)
242. Hörmander, L.: An introduction to complex analysis in several variables. D. Van Nostrand Co., Inc., Princeton, N.J.-Toronto, Ont.-London (1966)
243. Horstmann, H.: Zur Theorie der Funktionen mehrerer komplexen Veränderlichen: Carathéodorysche Metrik und Regularitätshüllen. Math. Ann. **108**(1), 208–217 (1933). https://doi.org/10.1007/BF01452833
244. Hua, L.K.: Harmonic analysis of functions of several complex variables in the classical domains. American Mathematical Society, Providence, RI (1963)
245. Huang, X.J.: Some applications of Bell's theorem to weakly pseudoconvex domains. Pacific J. Math. **158**(2), 305–315 (1993). http://projecteuclid.org/euclid.pjm/1102634504
246. Huang, X.J.: A preservation principle of extremal mappings near a strongly pseudoconvex point and its applications. Illinois J. Math. **38**(2), 283–302 (1994). http://projecteuclid.org/euclid.ijm/1255986801
247. Huang, X.J.: A boundary rigidity problem for holomorphic mappings on some weakly pseudoconvex domains. Canad. J. Math. **47**(2), 405–420 (1995). https://doi.org/10.4153/CJM-1995-022-3
248. Hubbard, J.H.: Teichmüller theory and applications to geometry, topology, and dynamics. Vol. 1. Matrix Editions, Ithaca, NY (2006)
249. Hubbard, J.H.: Teichmüller theory and applications to geometry, topology, and dynamics. Vol. 2. Matrix Editions, Ithaca, NY (2016)

250. Huber, H.: Über analytische Abbildungen von Ringgebieten in Ringgebiete. Compositio Math. **9**, 161–168 (1951)
251. Hurwitz, A.: Ueber algebraische Gebilde mit eindeutigen Transformationen in sich. Math. Ann. **41**(3), 403–442 (1892). https://doi.org/10.1007/BF01443420
252. Illarionov, M.A.: Kobayashi pseudometric on a fiber space. Vestnik Moskov. Univ. Ser. I Mat. Mekh. **35**(2), 35–36 (1980)
253. Imayoshi, Y., Taniguchi, M.: An introduction to Teichmüller spaces. Springer-Verlag, Tokyo (1992). https://doi.org/10.1007/978-4-431-68174-8
254. Isaev, A.V., Krantz, S.G.: Domains with non-compact automorphism group: a survey. Adv. Math. **146**(1), 1–38 (1999). https://doi.org/10.1006/aima.1998.1821
255. Iwaniec, T., Martin, G.: Quasiregular mappings in even dimensions. Acta Math. **170**(1), 29–81 (1993). https://doi.org/10.1007/BF02392454
256. Iwaniec, T., Martin, G.: Riesz transforms and related singular integrals. J. Reine Angew. Math. **473**, 25–57 (1996)
257. Iwaniec, T., Martin, G.: Geometric function theory and non-linear analysis. Oxford Mathematical Monographs. The Clarendon Press, Oxford University Press, New York (2001)
258. Jenkins, J.A.: Some results related to extremal length. In: Contributions to the theory of Riemann surfaces, *Ann. of Math. Stud.*, vol. no. 30, pp. 87–94. Princeton Univ. Press, Princeton, NJ (1953). https://doi.org/10.1017/s0022481200127112
259. Jeong, M., Oh, J.W., Taniguchi, M.: Equivalence problem for annuli and Bell representations in the plane. J. Math. Anal. Appl. **325**(2), 1295–1305 (2007). https://doi.org/10.1016/j.jmaa.2006.02.001
260. Jeong, M., Taniguchi, M.: Bell representations of finitely connected planar domains. Proc. Amer. Math. Soc. **131**(8), 2325–2328 (2003). https://doi.org/10.1090/S0002-9939-02-06823-5
261. Jones, G.A., Singerman, D.: Complex functions. Cambridge University Press, Cambridge (1987). https://doi.org/10.1017/CBO9781139171915
262. Karp, L.: Multivalued analytic continuation of the Cauchy transform. Potential Anal. **24**(1), 1–13 (2006). https://doi.org/10.1007/s11118-005-8566-z
263. Karp, L., Margulis, A.S.: Newtonian potential theory for unbounded sources and applications to free boundary problems. J. Anal. Math. **70**, 1–63 (1996). https://doi.org/10.1007/BF02820440
264. Katznelson, Y.: An introduction to harmonic analysis. John Wiley & Sons, Inc., New York-London-Sydney (1968)
265. Katznelson, Y.: An introduction to harmonic analysis, corrected edn. Dover Publications, Inc., New York (1976)
266. Kazdan, J.L., Warner, F.W.: Curvature functions for compact 2-manifolds. Ann. of Math. (2) **99**, 14–47 (1974). https://doi.org/10.2307/1971012
267. Kelley, J.L.: General topology. D. Van Nostrand Co., Inc., Toronto-New York-London (1955)
268. Kellogg, O.D.: Harmonic functions and Green's integral. Trans. Amer. Math. Soc. **13**(1), 109–132 (1912). https://doi.org/10.2307/1988618
269. Kerzman, N.: Topics in Complex Analysis
270. Kerzman, N.: Hölder and L^p estimates for solutions of $\bar{\partial} u = f$ in strongly pseudoconvex domains. Comm. Pure Appl. Math. **24**, 301–379 (1971). https://doi.org/10.1002/cpa.3160240303
271. Kerzman, N., Rosay, J.P.: Fonctions plurisousharmoniques d'exhaustion bornées et domaines taut. Math. Ann. **257**(2), 171–184 (1981). https://doi.org/10.1007/BF01458282
272. Kerzman, N., Stein, E.M.: The Cauchy kernel, the Szegö kernel, and the Riemann mapping function. Math. Ann. **236**(1), 85–93 (1978). https://doi.org/10.1007/BF01420257
273. Kerzman, N., Stein, E.M.: The Szegö kernel in terms of Cauchy-Fantappiè kernels. Duke Math. J. **45**(2), 197–224 (1978). http://projecteuclid.org/euclid.dmj/1077312816
274. Kerzman, N., Trummer, M.R.: Numerical conformal mapping via the Szegő kernel. pp. 111–123 (1986). https://doi.org/10.1016/0377-0427(86)90133-0

275. Kerzman, N., Trummer, M.R.: Numerical conformal mapping via the Szegö kernel. J. Comput. Appl. Math. **14**, 111–123 (1986). https://doi.org/10.1016/0377-0427(86)90133-0
276. Khenkin, G.M. (ed.): Several complex variables. V, *Encyclopaedia of Mathematical Sciences*, vol. 54. Springer-Verlag, Berlin (1993). https://doi.org/10.1007/978-3-642-58011-6
277. Kiernan, P.: Hyperbolic submanifolds of complex projective space. Proc. Amer. Math. Soc. **22**, 603–606 (1969). https://doi.org/10.2307/2037441
278. Kiernan, P.: On the relations between taut, tight and hyperbolic manifolds. Bull. Amer. Math. Soc. **76**, 49–51 (1970). https://doi.org/10.1090/S0002-9904-1970-12363-1
279. Kiernan, P.: Some results concerning hyperbolic manifolds. Proc. Amer. Math. Soc. **25**, 588–592 (1970). https://doi.org/10.2307/2036649
280. Kim, K.T.: Domains in $\mathbb{C}^n$ with a piecewise Levi flat boundary which possess a noncompact automorphism group. Math. Ann. **292**(4), 575–586 (1992). https://doi.org/10.1007/BF01444637
281. Kim, K.T.: On a boundary point repelling automorphism orbits. J. Math. Anal. Appl. **179**(2), 463–482 (1993). https://doi.org/10.1006/jmaa.1993.1362
282. Kim, K.T., Krantz, S.G.: Complex scaling and domains with non-compact automorphism group. Illinois J. Math. **45**(4), 1273–1299 (2001). http://projecteuclid.org/euclid.ijm/1258138066
283. Kim, K.T., Krantz, S.G., Spiro, A.F.: Analytic polyhedra in $\mathbb{C}^2$ with a non-compact automorphism group. J. Reine Angew. Math. **579**, 1–12 (2005). https://doi.org/10.1515/crll.2005.2005.579.1
284. Kim, K.T., Pagano, A.: Normal analytic polyhedra in $\mathbb{C}^2$ with a noncompact automorphism group. J. Geom. Anal. **11**(2), 283–293 (2001). https://doi.org/10.1007/BF02921967
285. Kirwan, F.: Complex algebraic curves, *London Mathematical Society Student Texts*, vol. 23. Cambridge University Press, Cambridge (1992). https://doi.org/10.1017/CBO9780511623929
286. Kiselman, C.O.: A study of the Bergman projection in certain Hartogs domains. In: Several complex variables and complex geometry, Part 3 (Santa Cruz, CA, 1989), *Proc. Sympos. Pure Math.*, vol. 52, Part 3, pp. 219–231. Amer. Math. Soc., Providence, RI (1991). https://doi.org/10.1090/pspum/052.3/1128596
287. Klein, F.: Vergleichende Betrachtungen über neuere geometrische Forschungen. Math. Ann. **43**, 63–100 (1893). https://doi.org/10.1007/BF01446615
288. Klein, F.: Riemann surfaces. Autographirte Vorlesung, gehalten zu Göttingen. W. S. 1891/92 und S. S. 1892. Teil I. 301 S., Teil II. 287 S. (1894). (1894)
289. Klein, F.: Vorlesungen über die Theorie der elliptischen Modulfunktionen. Band I: Grundlegung der Theorie, *Bibliotheca Mathematica Teubneriana*, vol. Band 10. Johnson Reprint Corp., New York; B. G. Teubner Verlagsgesellschaft, Stuttgart (1966)
290. Klein, F.: Gesammelte mathematische Abhandlungen. Springer-Verlag, Berlin-New York (1973)
291. Kobayashi, S.: Geometry of bounded domains. Trans. Amer. Math. Soc. **92**, 267–290 (1959). https://doi.org/10.2307/1993156
292. Kobayashi, S.: On the automorphism group of a certain class of algebraic manifolds. Tohoku Math. J. (2) **11**, 184–190 (1959). https://doi.org/10.2748/tmj/1178244581
293. Kobayashi, S.: Invariant distances on complex manifolds and holomorphic mappings. J. Math. Soc. Japan **19**, 460–480 (1967). https://doi.org/10.2969/jmsj/01940460
294. Kobayashi, S.: Hyperbolic manifolds and holomorphic mappings, *Pure and Applied Mathematics*, vol. 2. Marcel Dekker, Inc., New York (1970)
295. Kobayashi, S.: Transformation groups in differential geometry, *Ergebnisse der Mathematik und ihrer Grenzgebiete [Results in Mathematics and Related Areas]*, vol. Band 70. Springer-Verlag, New York-Heidelberg (1972)
296. Kobayashi, S.: Intrinsic distances, measures and geometric function theory. Bull. Amer. Math. Soc. **82**(3), 357–416 (1976). https://doi.org/10.1090/S0002-9904-1976-14018-9

297. Kobayashi, S.: Hyperbolic complex spaces, *Grundlehren der mathematischen Wissenschaften [Fundamental Principles of Mathematical Sciences]*, vol. 318. Springer-Verlag, Berlin (1998). https://doi.org/10.1007/978-3-662-03582-5
298. Kobayashi, S.: Differential geometry of curves and surfaces, japanese edn. Springer Undergraduate Mathematics Series. Springer, Singapore (2021). https://doi.org/10.1007/978-981-15-1739-6
299. Kobayashi, S., Nomizu, K.: Foundations of differential geometry. Vol I. Interscience Publishers (a division of John Wiley & Sons, Inc.), New York-London (1963)
300. Koebe, P.: Über die Uniformisierung beliebiger analytischer Kurven. Nachr. Ges. Wiss. Göttingen, Math.-Phys. Kl. **1907**, 191–210 (1907)
301. Koebe, P.: Über die Uniformisierung beliebiger analytischer Kurven. Zweite Mitteilung. Nachr. Ges. Wiss. Göttingen, Math.-Phys. Kl. **1907**, 633–669 (1907)
302. Koebe, P.: Über die Uniformisierung beliebiger analytischer Kurven. (Dritte Mitteilung.). Nachr. Ges. Wiss. Göttingen, Math.-Phys. Kl. **1908**, 337–358 (1908)
303. Koebe, P.: Über die Uniformisierung der algebraischen Kurven. I. Math. Ann. **67**, 145–224 (1909). https://doi.org/10.1007/BF01450180
304. Koebe, P.: Über die Uniformisierung beliebiger analytischer Kurven. Erster Teil: Das allgemeine Uniformisierungsprinzip. J. Reine Angew. Math. **138**, 192–253 (1910). https://doi.org/10.1515/crll.1910.138.192
305. Koebe, P.: Über die Uniformisierung beliebiger analytischer Kurven. II. Teil. Die zentralen Uniformisierungsprobleme. J. Reine Angew. Math. **139**, 251–292 (1911). https://doi.org/10.1515/crll.1911.139.251
306. Koebe, P.: Über die Uniformisierung der algebraischen Kurven. III. Math. Ann. **72**, 437–516 (1912). https://doi.org/10.1007/BF01456673
307. Koebe, P.: Über die Uniformisierung der algebraischen Kurven. IV. Zweiter Existenzbeweis der allgemeinen kanonischen uniformisierenden Variablen: Kontinuitätsmethode. Math. Ann. **75**, 42–129 (1914). https://doi.org/10.1007/BF01564522
308. Koebe, P.: Abhandlungen zur Theorie der konformen Abbildung. VI. Abbildung mehrfach zusammenhängender schlichter Bereiche auf Kreisbereiche. Uniformisierung hyperelliptischer Kurven. (Iterationsmethoden). Math. Z. **7**, 235–301 (1920). https://doi.org/10.1007/BF01199400
309. Koebe, P.: Über die konforme Abbildung endlich- und unendlich-vielfach zusammenhängender symmetrischer Bereiche. Acta Math. **43**, 263–287 (1922). https://doi.org/10.1007/BF02401759
310. Koebe, P.: Methoden der konformen Abbildung und Uniformisierung. In: Atti del congresso internazionale dei matematici : Bologna, 3–10 settembre 1928., vol. Tomo 3, pp. 195–203 (1928). https://www.mathunion.org/fileadmin/ICM/Proceedings/ICM1928.3/ICM1928.3.ocr.pdf
311. Koebe, P.: Methoden der konformen Abbildung und Uniformisierung. Atti Congresso Bologna 3, 195–203 (1930). (1930)
312. Koebe, P.: Kontaktprobleme der konformen Abbildung. Ber. Sächs. Akad. Wiss. Leipzig, Math.-phys. Kl. 88, 141–164 (1936). (1936)
313. Koosis, P.: Introduction to H_p spaces, *London Mathematical Society Lecture Note Series*, vol. 40. Cambridge University Press, Cambridge-New York (1980)
314. Koosis, P.: Introduction to H_p spaces, *Cambridge Tracts in Mathematics*, vol. 115, second edn. Cambridge University Press, Cambridge (1998)
315. Korányi, A.: Boundary behavior of Poisson integrals on symmetric spaces. Trans. Amer. Math. Soc. **140**, 393–409 (1969). https://doi.org/10.2307/1995145
316. Korányi, A.: Harmonic functions on Hermitian hyperbolic space. Trans. Amer. Math. Soc. **135**, 507–516 (1969). https://doi.org/10.2307/1995029
317. Korn, A.: Zwei Anwendungen der Methode der sukzessiven Annäherungen. Schwarz-Festschr. 215–229 (1914). (1914)
318. Kosniowski, C.: A first course in algebraic topology. Cambridge University Press, Cambridge-New York (1980)

319. Krantz, S.G.: Intrinsic Lipschitz classes on manifolds with applications to complex function theory and estimates for the $\bar{\partial}$ and $\bar{\partial}_b$ equations. Manuscripta Math. **24**(4), 351–378 (1978). https://doi.org/10.1007/BF01168882
320. Krantz, S.G.: Estimates for integral kernels of mixed type, fractional integration operators, and optimal estimates for the $\bar{\partial}$ operator. Manuscripta Math. **30**(1), 21–52 (1980). https://doi.org/10.1007/BF01305989
321. Krantz, S.G.: Function theory of several complex variables. Pure and Applied Mathematics. John Wiley & Sons, Inc., New York (1982)
322. Krantz, S.G.: Characterization of smooth domains in $\mathbb{C}$ by their biholomorphic self-maps. Amer. Math. Monthly **90**(8), 555–557 (1983). https://doi.org/10.2307/2322791
323. Krantz, S.G.: Lipschitz spaces, smoothness of functions, and approximation theory. Exposition. Math. **1**(3), 193–260 (1983)
324. Krantz, S.G.: A compactness principle in complex analysis. Division de Matematicas, Univ. Autonoma de Madrid Seminarios **3**, 171–194 (1987)
325. Krantz, S.G.: Complex analysis: the geometric viewpoint, *Carus Mathematical Monographs*, vol. 23. Mathematical Association of America, Washington, DC (1990)
326. Krantz, S.G.: Convexity in complex analysis. In: Several complex variables and complex geometry, Part 1 (Santa Cruz, CA, 1989), *Proc. Sympos. Pure Math.*, vol. 52, Part 1, pp. 119–137. Amer. Math. Soc., Providence, RI (1991). https://doi.org/10.1090/pspum/052.1/1128519
327. Krantz, S.G.: Invariant metrics and the boundary behavior of holomorphic functions on domains in $\mathbb{C}^n$. J. Geom. Anal. **1**(2), 71–97 (1991). https://doi.org/10.1007/BF02938115
328. Krantz, S.G.: Function theory of several complex variables, second edn. The Wadsworth & Brooks/Cole Mathematics Series. Wadsworth & Brooks/Cole Advanced Books & Software, Pacific Grove, CA (1992)
329. Krantz, S.G.: Partial differential equations and complex analysis. Studies in Advanced Mathematics. CRC Press, Boca Raton, FL (1992)
330. Krantz, S.G.: A panorama of harmonic analysis, *Carus Mathematical Monographs*, vol. 27. Mathematical Association of America, Washington, DC (1999)
331. Krantz, S.G.: Function theory of several complex variables. AMS Chelsea Publishing, Providence, RI (2001). https://doi.org/10.1090/chel/340
332. Krantz, S.G.: A new proof and a generalization of Ramadanov's theorem. Complex Var. Elliptic Equ. **51**(12), 1125–1128 (2006). https://doi.org/10.1080/17476930600758855
333. Krantz, S.G.: Geometric analysis of the Bergman kernel and metric, *Graduate Texts in Mathematics*, vol. 268. Springer, New York (2013). https://doi.org/10.1007/978-1-4614-7924-6
334. Krantz, S.G.: A new, elementary proof of the Riemann mapping theorem. Complex Var. Elliptic Equ. **68**(5), 729–731 (2023). https://doi.org/10.1080/17476933.2021.2014460
335. Krantz, S.G., Ma, D.: Bloch functions on strongly pseudoconvex domains. Indiana Univ. Math. J. **37**(1), 145–163 (1988). https://doi.org/10.1512/iumj.1988.37.37007
336. Krantz, S.G., Parks, H.R.: The implicit function theorem. Birkhäuser Boston, Inc., Boston, MA (2002). https://doi.org/10.1007/978-1-4612-0059-8
337. Kriete, T.L., MacCluer, B.D.: A rigidity theorem for composition operators on certain Bergman spaces. Michigan Math. J. **42**(2), 379–386 (1995). https://doi.org/10.1307/mmj/1029005235
338. Kruschkal, S.L., Kühnau, R.: Quasikonforme Abbildungen—neue Methoden und Anwendungen, *Teubner-Texte zur Mathematik [Teubner Texts in Mathematics]*, vol. 54. BSB B. G. Teubner Verlagsgesellschaft, Leipzig (1983)
339. Kufner, A., John, O., Fučik, S.: Function spaces. Monographs and Textbooks on Mechanics of Solids and Fluids, Mechanics: Analysis. Noordhoff International Publishing, Leiden; Academia, Prague (1977)
340. Kühnau, R.: Variation of diametrically symmetric or elliptically schlicht conformal mappings. J. Anal. Math. **89**, 303–316 (2003). https://doi.org/10.1007/BF02893085

341. Kühnau, R.: Bibliography of geometric function theory. In: Handbook of complex analysis: geometric function theory. Vol. 2, pp. 809–828. Elsevier Sci. B. V., Amsterdam (2005). https://doi.org/10.1016/S1874-5709(05)80023-X
342. Kühnau, R.: Canonical conformal and quasiconformal mappings. Identities. Kernel functions. In: Handbook of complex analysis: geometric function theory. Vol. 2, pp. 131–163. Elsevier Sci. B. V., Amsterdam (2005). https://doi.org/10.1016/S1874-5709(05)80008-3
343. Kummer, E.E.: Über die hypergeometrische Reihe $1 + \frac{\alpha\cdot\beta}{1.\gamma}x + \frac{\alpha(\alpha+1)\beta(\beta+1)}{1.2.\gamma(\gamma+1)}x^2 + \frac{\alpha(\alpha+1)(\alpha+2)\beta(\beta+1)(\beta+2)}{1.2.3.\gamma(\gamma+1)(\gamma+2)}x^3 + \cdots$. J. Reine Angew. Math. **15**, 127–172 (1836). https://doi.org/10.1515/crll.1836.15.127
344. Kummer, E.E.: Über die hypergeometrische Reihe $1 + \frac{\alpha.\beta}{1.\gamma}x + \frac{\alpha(\alpha+1)\beta(\beta+1)}{1.2.\gamma.(\gamma+1)}x^2 + \frac{\alpha(\alpha+1)(\alpha+2)\beta(\beta+1)(\beta+2)}{1.2.3.\gamma(\gamma+1)(\gamma+2)}x^3 + \cdots$. J. Reine Angew. Math. **15**, 39–83 (1836). https://doi.org/10.1515/crll.1836.15.39
345. Kwack, M.H.: Generalization of the big Picard theorem. Ann. of Math. (2) **90**, 9–22 (1969). https://doi.org/10.2307/1970678
346. de Lagrange, J.L.: Sur la construction des cartes géographiques, vol. IV, chap. XXXVIII. Gauthier Villars (1869). https://gallica.bnf.fr/ark:/12148/bpt6k229223s/f639
347. Landau, H.J., Osserman, R.: Some distortion theorems for multivalent mappings. Proc. Amer. Math. Soc. **10**, 87–91 (1959). https://doi.org/10.2307/2032893
348. Landau, H.J., Osserman, R.: On analytic mappings of Riemann surfaces. Journal d'Analyse Mathématique **7**, 249–279 (1960). https://doi.org/10.1007/BF02787688
349. Lang, S.: Complex analysis, *Graduate Texts in Mathematics*, vol. 103, second edn. Springer-Verlag, New York (1985). https://doi.org/10.1007/978-1-4757-1871-3
350. Lavrent'ev, M.A.: A fundamental theorem of the theory of quasi-conformal mapping of plane regions. Izv. Akad. Nauk SSSR Ser. Mat. **12**, 513–554 (1948)
351. Lavrentieff, M.A.: Sur une méthode géométrique dans la représentation conforme. Atti Congresso Bologna 3, 241–242 (1930). (1930)
352. Lavrentieff, M.A.: Sur une classe de représentations continues. C. R. Acad. Sci., Paris **200**, 1010–1012 (1935)
353. Lehto, O.: Univalent functions and Teichmüller spaces, *Graduate Texts in Mathematics*, vol. 109. Springer-Verlag, New York (1987). https://doi.org/10.1007/978-1-4613-8652-0
354. Lehto, O., Virtanen, K.I.: Boundary behaviour and normal meromorphic functions. Acta Math. **97**, 47–65 (1957). https://doi.org/10.1007/BF02392392
355. Lehto, O., Virtanen, K.I.: Quasiconformal mappings in the plane, *Die Grundlehren der mathematischen Wissenschaften*, vol. Band 126, second edn. Springer-Verlag, New York-Heidelberg (1973)
356. Lempert, L.: Boundary behaviour of meromorphic functions of several variables. Acta Math. **144**(1-2), 1–25 (1980). https://doi.org/10.1007/BF02392119
357. Lempert, L.: La métrique de Kobayashi et la représentation des domaines sur la boule. Bull. Soc. Math. France **109**(4), 427–474 (1981). http://www.numdam.org/item?id=BSMF_1981__109__427_0
358. Lempert, L.: Holomorphic retracts and intrinsic metrics in convex domains. Anal. Math. **8**(4), 257–261 (1982). https://doi.org/10.1007/BF02201775
359. Lempert, L.: A precise result on the boundary regularity of biholomorphic mappings. Math. Z. **193**(4), 559–579 (1986). https://doi.org/10.1007/BF01160475
360. Leray, J.: Fonction de variables complexes: sa représentation comme somme de puissances négatives de fonctions linéaires. Atti Accad. Naz. Lincei Rend. Cl. Sci. Fis. Mat. Nat. (8) **20**, 589–590 (1956)
361. Leray, J.: Le calcul différentiel et intégral sur une variété analytique complexe. (Problème de Cauchy. III). Bull. Soc. Math. France **87**, 81–180 (1959). http://www.numdam.org/item?id=BSMF_1959__87__81_0
362. Lichtenstein, L.: Zur Theorie der konformen Abbildung nichtanalytischer, singularitätenfreier Flächenstücke auf ebene Gebiete. Krak. Anz. 1916, 192–217 (1916). (1916)

363. Lions, J.L., Peetre, J.: On a class of interpolation spaces. Publ. Math., Inst. Hautes Étud. Sci. **19**, 5–68 (1964). https://doi.org/10.1007/BF02684796
364. Lucchini, A., Tamburini, M.C., Wilson, J.S.: Hurwitz groups of large rank. J. London Math. Soc. (2) **61**(1), 81–92 (2000). https://doi.org/10.1112/S0024610799008467
365. Luenberger, D.G.: Optimization by vector space methods. John Wiley & Sons, Inc., New York-London-Sydney (1969)
366. Lumer, G.: Espaces de Hardy en plusieurs variables complexes. C. R. Acad. Sci. Paris Sér. A-B **273**, A151–A154 (1971)
367. Lvovskii, M.N.: Integral representations of hyperbolic volumes. Vestnik Moskov. Univ. Ser. I Mat. Meh. **31**(2), 21–27 (1976)
368. Mair, B.A., McCullough, S.: Invariance of extreme harmonic functions on an annulus; applications to theta functions. Houston J. Math. **20**(3), 453–473 (1994)
369. Marsden, J.E.: Basic complex analysis. W. H. Freeman and Co., San Francisco, CA (1973)
370. Marshall, D.E.: Complex analysis. Cambridge Mathematical Textbooks. Cambridge University Press, Cambridge (2019)
371. Massey, W.S.: Algebraic topology: An introduction. Harcourt, Brace & World, Inc., New York (1967)
372. Massey, W.S.: Singular homology theory, *Graduate Texts in Mathematics*, vol. 70. Springer-Verlag, New York-Berlin (1980)
373. Massey, W.S.: A basic course in algebraic topology, *Graduate Texts in Mathematics*, vol. 127. Springer-Verlag, New York (1991)
374. Maurin, K.: Analysis. Part II. D. Reidel Publishing Co., Dordrecht-Boston, Mass.-London; PWN—Polish Scientific Publishers, Warsaw (1980)
375. Maurin, K.: The Riemann legacy, *Mathematics and its Applications*, vol. 417. Kluwer Academic Publishers Group, Dordrecht (1997). https://doi.org/10.1007/978-94-015-8939-0
376. Maz'ja, V.G.: Sobolev spaces. Springer Series in Soviet Mathematics. Springer-Verlag, Berlin (1985). https://doi.org/10.1007/978-3-662-09922-3
377. Meschkowski, H.: Über die konforme Abbildung gewisser Bereiche von unendlich hohem Zusammenhang auf Vollkreisbereiche. I. Math. Ann. **123**, 392–405 (1951). https://doi.org/10.1007/BF02054962
378. Meschkowski, H.: Über die konforme Abbildung gewisser Bereiche von unendlich hohem Zusammenhang auf Vollkreisbereiche. II. Math. Ann. **124**, 178–181 (1952). https://doi.org/10.1007/BF01343559
379. Milnor, J.: On manifolds homeomorphic to the 7-sphere. Ann. of Math. (2) **64**, 399–405 (1956). https://doi.org/10.2307/1969983
380. Milnor, J.: Dynamics in one complex variable. Friedr. Vieweg & Sohn, Braunschweig (1999)
381. Minda, D., Schober, G.: Another elementary approach to the theorems of Landau, Montel, Picard and Schottky. Complex Variables Theory Appl. **2**(2), 157–164 (1983). https://doi.org/10.1080/17476938308814039
382. Boutet de Monvel, L., Sjöstrand, J.: Sur la singularité des noyaux de Bergman et de Szegö. In: Journées: Équations aux Dérivées Partielles de Rennes (1975), *Astérisque*, vol. No. 34–35, pp. 123–164. Soc. Math. France, Paris (1976)
383. Morrey Jr., C.B.: On the solutions of quasi-linear elliptic partial differential equations. Trans. Amer. Math. Soc. **43**(1), 126–166 (1938). https://doi.org/10.2307/1989904
384. Morrey Jr., C.B.: Multiple integrals in the calculus of variations. Classics in Mathematics. Springer-Verlag, Berlin (2008). https://doi.org/10.1007/978-3-540-69952-1
385. Morrow, J., Kodaira, K.: Complex manifolds. Holt, Rinehart and Winston, Inc., New York-Montreal, Que.-London (1971)
386. Mueller, C., Rudin, W.: Proper holomorphic self-maps of plane regions. Complex Variables Theory Appl. **17**(1-2), 113–121 (1991). https://doi.org/10.1080/17476939108814502
387. Munkres, J.R.: Elementary differential topology, *Annals of Mathematics Studies*, vol. No. 54, revised edn. Princeton University Press, Princeton, NJ (1966)
388. Munkres, J.R.: Elements of algebraic topology. Addison-Wesley Publishing Company, Menlo Park, CA (1984)

389. Munkres, J.R.: Analysis on manifolds. Addison-Wesley Publishing Company, Advanced Book Program, Redwood City, CA (1991)
390. Nag, S.: The complex analytic theory of Teichmüller spaces. Canadian Mathematical Society Series of Monographs and Advanced Texts. John Wiley & Sons, Inc., New York (1988)
391. Nagel, A., Rosay, J.P., Stein, E.M., Wainger, S.: Estimates for the Bergman and Szegö kernels in $\mathbb{C}^2$. Ann. of Math. (2) **129**(1), 113–149 (1989). https://doi.org/10.2307/1971487
392. Nagel, A., Stein, E.M., Wainger, S.: Boundary behavior of functions holomorphic in domains of finite type. Proc. Nat. Acad. Sci. U.S.A. **78**(11), 6596–6599 (1981). https://doi.org/10.1073/pnas.78.11.6596
393. Narasimhan, R.: Several complex variables. Chicago Lectures in Mathematics. University of Chicago Press, Chicago, Ill.-London (1971)
394. Narasimhan, R.: Analysis on real and complex manifolds, *North-Holland Mathematical Library*, vol. 35. North-Holland Publishing Co., Amsterdam (1985)
395. Narasimhan, R.: Compact Riemann surfaces. Lectures in Mathematics ETH Zürich. Birkhäuser Verlag, Basel (1992). https://doi.org/10.1007/978-3-0348-8617-8
396. Needham, T.: Visual complex analysis, anniversary edn. Oxford University Press, Oxford (2023). https://doi.org/10.1093/oso/9780192868916.001.0001
397. Nehari, Z.: Conformal mapping. McGraw-Hill Book Co., Inc., New York-Toronto-London (1952)
398. Nehari, Z.: Conformal mapping. Dover Publications, Inc., New York (1975)
399. Nemirovskii, S.Y., Shafikov, R.G.: Uniformization of strictly pseudoconvex domains. I. Izv. Ross. Akad. Nauk Ser. Mat. **69**(6), 115–130 (2005). https://doi.org/10.1070/IM2005v069n06ABEH002295
400. Nemirovskii, S.Y., Shafikov, R.G.: Uniformization of strictly pseudoconvex domains. II. Izv. Ross. Akad. Nauk Ser. Mat. **69**(6), 131–138 (2005). https://doi.org/10.1070/IM2005v069n06ABEH002296
401. Nevanlinna, R.: Uniformisierung, *Die Grundlehren der mathematischen Wissenschaften in Einzeldarstellungen mit besonderer Berücksichtigung der Anwendungsgebiete*, vol. Band LXIV. Springer-Verlag, Berlin-Göttingen-Heidelberg (1953)
402. Neville, E.H.: Jacobian Elliptic Functions. Oxford University Press (1944)
403. Nielsen, J.: Untersuchungen zur Topologie der geschlossenen zweiseitigen Flächen. Acta Math. **50**(1), 189–358 (1927). https://doi.org/10.1007/BF02421324
404. O'Neill, B.: Elementary differential geometry. Academic Press, New York-London (1966)
405. Osgood, W.F.: On the existence of the Green's function for the most general simply connected plane region. Trans. Amer. Math. Soc. **1**(3), 310–314 (1900). https://doi.org/10.2307/1986285
406. Painlevé, P.: Sur les lignes singulières des fonctions analytiques. Ann. Fac. Sci. Toulouse Sci. Math. Sci. Phys. **2**, B1–B130 (1888). http://www.numdam.org/item?id=AFST_1888_1_2__B1_0
407. Papadopoulos, A. (ed.): Handbook of Teichmüller theory. Vol. V, *IRMA Lectures in Mathematics and Theoretical Physics*, vol. 26. European Mathematical Society (EMS), Zürich (2016)
408. Parreau, M.: Sur les moyennes des fonctions harmoniques et analytiques et la classification des surfaces de Riemann. Ann. Inst. Fourier (Grenoble) **3**, 103–197 (1951). http://www.numdam.org/item?id=AIF_1951__3__103_0
409. Pennisi, L.L.: Elements of complex variables. Holt, Rinehart and Winston, New York-Toronto-London (1966)
410. Petersen, P.: Riemannian geometry, *Graduate Texts in Mathematics*, vol. 171, second edn. Springer, New York (2006)
411. Pfluger, A.: Sur une propriété de l'application quasi conforme d'une surface de Riemann ouverte. C. R. Acad. Sci. Paris **227**, 25–26 (1948)
412. Pfluger, A.: Une propriété métrique de la représentation quasi conforme. C. R. Acad. Sci. Paris **226**, 623–625 (1948)
413. Pfluger, A.: Theorie der Riemannschen Flächen. Springer-Verlag, Berlin-Göttingen-Heidelberg (1957)

414. Pinčuk, S.I.: The analytic continuation of holomorphic mappings. Mat. Sb. (N.S.) **98(140)**(3(11)), 416–435, 495–496 (1975)
415. Pinčuk, S.I.: Holomorphic mappings of real-analytic hypersurfaces. Mat. Sb. (N.S.) **105(147)**(4), 574–593, 640 (1978)
416. Poincaré, H.: Mémoire. Acta Math. **3**(1), 49–92 (1883). https://doi.org/10.1007/BF02422441
417. Poincaré, H.: Sur un théorème de M. Fuchs. Acta Math. **7**(1), 1–32 (1885). https://doi.org/10.1007/BF02402192
418. Poincaré, H.: Sur l'uniformisation des fonctions analytiques. Acta Math. **31**, 1–64 (1907). https://doi.org/10.1007/BF02415442
419. Poincaré, H.: Sur l'uniformisation des fonctions analytiques. Acta Math. **31**(1), 1–63 (1908). https://doi.org/10.1007/BF02415442
420. Poincaré, H.: Œuvres. Tome II: Fonctions fuchsiennes. Publié avec la collaboration de N. E. Nørlund et d'Ernest Lebon. Préface: Éloge historique d'Henri Poincaré par G. Darboux, reprint of the 1916 original edn. Sceaux: Éditions Jacques Gabay (1995)
421. Pommerenke, C.: Univalent functions, *Studia Mathematica/Mathematische Lehrbücher [Studia Mathematica/Mathematical Textbooks]*, vol. Band XXV. Vandenhoeck & Ruprecht, Göttingen (1975)
422. Pommerenke, C.: Boundary behaviour of conformal maps, *Grundlehren der mathematischen Wissenschaften [Fundamental Principles of Mathematical Sciences]*, vol. 299. Springer-Verlag, Berlin (1992). https://doi.org/10.1007/978-3-662-02770-7
423. Pommerenke, C.: Conformal maps at the boundary. In: Handbook of complex analysis: geometric function theory, Vol. 1, pp. 37–74. North-Holland, Amsterdam (2002). https://doi.org/10.1016/S1874-5709(02)80004-X
424. de Possel, R.: Zum Parallelschlitztheorem unendlich-vielfach zusammenhängender Gebiete. Nachr. Ges. Wiss. Göttingen, Math.-Phys. Kl. **1931**, 199–202 (1931)
425. Preissmann, A.: Quelques propriétés globales des espaces de Riemann. Comment. Math. Helv. **15**, 175–216 (1943). https://doi.org/10.1007/BF02565638
426. Puiseux, V.: Recherches sur les fonctions algébriques. Journal de Mathématiques Pures et Appliquées **1e série, 15** (1850). http://www.numdam.org/item/JMPA_1850_1_15__365_0/
427. Putinar, M.: Linear analysis of quadrature domains. Ark. Mat. **33**(2), 357–376 (1995). https://doi.org/10.1007/BF02559714
428. Putinar, M.: Linear analysis of quadrature domains. III. J. Math. Anal. Appl. **239**(1), 101–117 (1999). https://doi.org/10.1006/jmaa.1999.6556
429. Reiffen, H.J.: Die differentialgeometrischen Eigenschaften der invarianten Distanzfunktion von Carathéodory. Schr. Math. Inst. Univ. Münster **26**, iii+66 (1963)
430. Reiffen, H.J.: Die Carathéodorysche Distanz und ihre zugehörige Differentialmetrik. Math. Ann. **161**, 315–324 (1965). https://doi.org/10.1007/BF01359970
431. Reyssat, É.: Quelques aspects des surfaces de Riemann, *Progress in Mathematics*, vol. 77. Birkhäuser Boston, Inc., Boston, MA (1989)
432. Richardson, S.: Hele-Shaw flows with time-dependent free boundaries involving a multiply-connected fluid region. European J. Appl. Math. **12**(5), 571–599 (2001). https://doi.org/10.1017/S0956792501004569
433. Rickman, S.: Quasiregular mappings, *Ergebnisse der Mathematik und ihrer Grenzgebiete (3) [Results in Mathematics and Related Areas (3)]*, vol. 26. Springer-Verlag, Berlin (1993). https://doi.org/10.1007/978-3-642-78201-5
434. Riemann, B.: Theorie der Abel'schen Functionen. J. Reine Angew. Math. **54**, 115–155 (1857). https://doi.org/10.1515/crll.1857.54.115
435. Riemann, B.: Über die Hypothesen, welche der Geometrie zu Grunde liegen. Neu herausgegeben und erläutert von H. Weyl. Berlin: J. Springer, VI u. 47 S. 8° (1919). (1919)
436. Riemann, B.: Gesammelte mathematische Werke, wissenschaftlicher Nachlass und Nachträge, *Teubner-Archiv zur Mathematik [Teubner Archive on Mathematics]*, vol. Suppl. 1, the edn. BSB B. G. Teubner Verlagsgesellschaft, Leipzig; Springer-Verlag, Berlin (1990). https://doi.org/10.1007/978-3-663-10149-9

437. Robert, A.: Elliptic curves. Notes from postgraduate lectures given in Lausanne 1971/72, *Lect. Notes Math.*, vol. 326. Springer, Cham (1973)
438. Rodin, B., Sullivan, D.: The convergence of circle packings to the Riemann mapping. J. Differential Geom. **26**(2), 349–360 (1987). http://projecteuclid.org/euclid.jdg/1214441375
439. Rodin, B., Warschawski, S.E.: Estimates of the Riemann mapping function near a boundary point. In: Romanian-Finnish Seminar on Complex Analysis (Proc., Bucharest, 1976), *Lecture Notes in Math.*, vol. 743, pp. 349–366. Springer, Berlin (1979)
440. Roman, S.: The formula of Faà di Bruno. Amer. Math. Monthly **87**(10), 805–809 (1980). https://doi.org/10.2307/2320788
441. Rosay, J.P.: Sur une caractérisation de la boule parmi les domaines de $\mathbb{C}^n$ par son groupe d'automorphismes. Ann. Inst. Fourier (Grenoble) **29**(4), ix, 91–97 (1979). http://www.numdam.org/item?id=AIF_1979__29_4_91_0
442. Rosenthal, P.: On the zeros of the Bergman function in doubly-connected domains. Proc. Amer. Math. Soc. **21**, 33–35 (1969). https://doi.org/10.2307/2036852
443. Rotman, J.J.: An introduction to algebraic topology, *Graduate Texts in Mathematics*, vol. 119. Springer-Verlag, New York (1988). https://doi.org/10.1007/978-1-4612-4576-6
444. Royden, H.L.: Remarks on the Kobayashi metric. In: Several complex variables, II (Proc. Internat. Conf., Univ. Maryland, College Park, Md., 1970), *Lecture Notes in Math.*, vol. Vol. 185, pp. 125–137. Springer, Berlin-New York (1971)
445. Royden, H.L.: Hyperbolicity in complex analysis. Ann. Acad. Sci. Fenn. Ser. A I Math. **13**(3), 387–400 (1988). https://doi.org/10.5186/aasfm.1988.1329
446. Rudin, W.: Analytic functions of class H_p. Trans. Amer. Math. Soc. **78**, 46–66 (1955). https://doi.org/10.2307/1992948
447. Rudin, W.: Real and complex analysis. McGraw-Hill Book Co., New York-Toronto-London (1966)
448. Rudin, W.: Functional analysis. McGraw-Hill Series in Higher Mathematics. McGraw-Hill Book Co., New York-Düsseldorf-Johannesburg (1973)
449. Rudin, W.: Principles of mathematical analysis, third edn. International Series in Pure and Applied Mathematics. McGraw-Hill Book Co., New York-Auckland-Düsseldorf (1976)
450. Rudin, W.: Holomorphic Lipschitz functions in balls. Comment. Math. Helv. **53**(1), 143–147 (1978). https://doi.org/10.1007/BF02566070
451. Rudin, W.: Function theory in the unit ball of $\mathbb{C}^n$, *Grundlehren der Mathematischen Wissenschaften*, vol. 241. Springer-Verlag, New York-Berlin (1980)
452. Rudin, W.: Holomorphic maps that extend to automorphisms of a ball. Proc. Amer. Math. Soc. **81**(3), 429–432 (1981). https://doi.org/10.2307/2043480
453. Saerens, R., Zame, W.R.: The isometry groups of manifolds and the automorphism groups of domains. Trans. Amer. Math. Soc. **301**(1), 413–429 (1987). https://doi.org/10.2307/2000347
454. Sakai, M.: Quadrature domains, *Lecture Notes in Mathematics*, vol. 934. Springer-Verlag, Berlin-New York (1982)
455. Sakai, M.: Regularity of a boundary having a Schwarz function. Acta Math. **166**(3-4), 263–297 (1991). https://doi.org/10.1007/BF02398888
456. Sario, L.: Über Riemannsche Flächen mit hebbarem Rand. Ann. Acad. Sci. Fennicae Ser. A. I. Math.-Phys. **1948**(50), 79 (1948)
457. Sario, L., Nakai, M.: Classification theory of Riemann surfaces, *Die Grundlehren der mathematischen Wissenschaften*, vol. Band 164. Springer-Verlag, New York-Berlin (1970)
458. Schiff, J.L.: Normal families. Universitext. Springer-Verlag, New York (1993). https://doi.org/10.1007/978-1-4612-0907-2
459. Schiffer, M.: On the modulus of doubly-connected domains. Quart. J. Math. Oxford Ser. **17**, 197–213 (1946). https://doi.org/10.1093/qmath/os-17.1.197
460. Schiffer, M., Spencer, D.C.: Functionals of finite Riemann surfaces. Princeton University Press, Princeton, NJ (1954)
461. Schottky, F.: On the conformal mapping of multiply connected planar surfaces. J. Reine Angew. Math. **83**, 300–351 (1877)

462. Schramm, O.: Transboundary extremal length. J. Anal. Math. **66**, 307–329 (1995). https://doi.org/10.1007/BF02788827
463. Schupp, P.E.: A characterization of inner automorphisms. Proc. Amer. Math. Soc. **101**(2), 226–228 (1987). https://doi.org/10.2307/2045986
464. Schwartz, L.: Théorie des distributions, *Publications de l'Institut de Mathématique de l'Université de Strasbourg*, vol. IX-X. Hermann, Paris (1966)
465. Seifert, H., Threlfall, W.: Lehrbuch der Topologie. Leipzig und Berlin: B. G. Teubner. VII, 353 S., 132 Fig. (1934). (1934)
466. Semmes, S.: A generalization of Riemann mappings and geometric structures on a space of domains in $\mathbb{C}^n$. Mem. Amer. Math. Soc. **98**(472), vi+98 (1992). https://doi.org/10.1090/memo/0472
467. Serre, J.P.: Géométrie algébrique et géométrie analytique. Ann. Inst. Fourier (Grenoble) **6**, 1–42 (1956). http://aif.cedram.org/item?id=AIF_1955__6__1_0
468. Shapiro, H.S.: The Schwarz function and its generalization to higher dimensions, *University of Arkansas Lecture Notes in the Mathematical Sciences*, vol. 9. John Wiley & Sons, Inc., New York (1992)
469. Shokurov, V.V.: Riemann surfaces and algebraic curves. In: Algebraic geometry, I, *Encyclopaedia Math. Sci.*, vol. 23, pp. 1–166. Springer, Berlin (1994)
470. Sibner, R.J.: Uniformization of symmetric Riemann surfaces by Schottky groups. Trans. Amer. Math. Soc. **116**, 79–85 (1965). https://doi.org/10.2307/1994105
471. Sibner, R.J.: Remarks on the Koebe Kreisnormierungsproblem. Comment. Math. Helv. **43**, 289–295 (1968). https://doi.org/10.1007/BF02564397
472. Sibony, N.: A class of hyperbolic manifolds. In: Recent developments in several complex variables (Proc. Conf., Princeton Univ., Princeton, N. J., 1979), *Ann. of Math. Stud.*, vol. No. 100, pp. 357–372. Princeton Univ. Press, Princeton, NJ (1981)
473. Sibony, N.: Unpublished note (1998)
474. Siegel, C.L.: Topics in complex function theory. Vol. I: Elliptic functions and uniformization theory, *Interscience Tracts in Pure and Applied Mathematics*, vol. No. 25. Wiley-Interscience [A division of John Wiley & Sons, Inc.], New York-London-Sydney (1969)
475. Sieradski, A.J.: An introduction to topology and homotopy. The Prindle, Weber & Schmidt Series in Advanced Mathematics. PWS-KENT Publishing Co., Boston, MA (1992)
476. Skwarczyński, M.: The invariant distance in the theory of pseudoconformal transformations and the Lu Qi-keng conjecture. Proc. Amer. Math. Soc. **22**, 305–310 (1969). https://doi.org/10.2307/2037043
477. Smith, D.E.: A Source Book in the History of Mathematics. Amer. Math. Monthly **35**(6), 280–282 (1928). https://doi.org/10.2307/2298668
478. Smith, K.T.: A generalization of an inequality of Hardy and Littlewood. Canadian J. Math. **8**, 157–170 (1956). https://doi.org/10.4153/CJM-1956-019-5
479. Spanier, E.H.: Algebraic topology. McGraw-Hill Book Co., New York-Toronto-London (1966)
480. Spanier, E.H.: Algebraic topology. Springer-Verlag, New York-Berlin (1981)
481. Springer, G.: Introduction to Riemann surfaces. Addison-Wesley Publishing Co., Inc., Reading, MA (1957)
482. Steenrod, N.: The Topology of Fibre Bundles, *Princeton Mathematical Series*, vol. vol. 14. Princeton University Press, Princeton, NJ (1951)
483. Stein, E.M.: Singular integrals and differentiability properties of functions, *Princeton Mathematical Series*, vol. No. 30. Princeton University Press, Princeton, NJ (1970)
484. Stein, E.M.: Boundary behavior of holomorphic functions of several complex variables, *Mathematical Notes*, vol. No. 11. Princeton University Press, Princeton, NJ; University of Tokyo Press, Tokyo (1972)
485. Stein, E.M.: Singular integrals and estimates for the Cauchy-Riemann equations. Bull. Amer. Math. Soc. **79**, 440–445 (1973). https://doi.org/10.1090/S0002-9904-1973-13205-7
486. Stein, E.M., Weiss, G.: Introduction to Fourier analysis on Euclidean spaces, *Princeton Mathematical Series*, vol. No. 32. Princeton University Press, Princeton, NJ (1971)

487. Stephenson, K.: Circle packing and discrete analytic function theory. In: Handbook of complex analysis: geometric function theory, Vol. 1, pp. 333–370. North-Holland, Amsterdam (2002). https://doi.org/10.1016/S1874-5709(02)80013-0
488. Stephenson, K.: Introduction to circle packing. Cambridge University Press, Cambridge (2005)
489. Stillwell, J.C.: Classical topology and combinatorial group theory, *Graduate Texts in Mathematics*, vol. 72. Springer-Verlag, New York-Berlin (1980)
490. Stout, E.L.: H^p-functions on strictly pseudoconvex domains. Amer. J. Math. **98**(3), 821–852 (1976). https://doi.org/10.2307/2373817
491. Strebel, K.: Über das Kreisnormierungsproblem der konformen Abbildung. Ann. Acad. Sci. Fennicae Ser. A. I. Math.-Phys. **1951**(101), 22 (1951)
492. Strebel, K.: Über die konforme Abbildung von Gebieten unendlich hohen Zusammenhangs. I. Comment. Math. Helv. **27**, 101–127 (1953). https://doi.org/10.1007/BF02564556
493. Suita, N., Yamada, A.: On the Lu Qi-keng conjecture. Proc. Amer. Math. Soc. **59**(2), 222–224 (1976). https://doi.org/10.2307/2041472
494. Szegö, G.: Über Orthogonalsysteme von Polynomen. Math. Z. **4**(1-2), 139–151 (1919). https://doi.org/10.1007/BF01203393
495. Tanaka, N.: On generalized graded Lie algebras and geometric structures. I. J. Math. Soc. Japan **19**, 215–254 (1967). https://doi.org/10.2969/jmsj/01920215
496. Tauraso, R., Vlacci, F.: Rigidity at the boundary for holomorphic self-maps of the unit disc. Complex Variables Theory Appl. **45**(2), 151–165 (2001). https://doi.org/10.1080/17476930108815374
497. Tegtmeyer, T.: The Ahlfors map and the Szegö kernel in multiply connected domains. Ph.D. thesis, Purdue Univ, PhD thesis (1998)
498. Tegtmeyer, T.J., Thomas, A.D.: The Ahlfors map and Szegő kernel for an annulus. Rocky Mountain J. Math. **29**(2), 709–723 (1999). https://doi.org/10.1216/rmjm/1181071660
499. Teichmüller, O.: Untersuchungen über konforme und quasikonforme Abbildung. Deutsche Math. **3**, 621–678 (1938)
500. Teichmüller, O.: Extremale quasikonforme Abbildungen und quadratische Differentiale. Abh. Preuss. Akad. Wiss. Math.-Nat. Kl. **1939**(22), 197 (1940)
501. Teichmüller, O.: Ein Verschiebungssatz der quasikonformen Abbildung. Deutsche Math. **7**, 336–343 (1944)
502. Teichmüller, O.: Veränderliche Riemannsche Flächen. Deutsche Math. **7**, 344–359 (1944)
503. Teichmüller, O.: Gesammelte Abhandlungen. Springer-Verlag, Berlin-New York (1982)
504. Thomas, G.B.j., Finney, R.L.: Calculus and analytic geometry., 8. ed. edn. Reading, MA: Addison-Wesley (1992)
505. Thurston, W.: The geometry and topology of 3-manifolds, Princeton University Notes, Preprint.
506. Thurston, W.P.: Three-dimensional geometry and topology. Vol. 1, *Princeton Mathematical Series*, vol. 35. Princeton University Press, Princeton, NJ (1997)
507. Tomi, F., Tromba, A.J.: Existence theorems for minimal surfaces of nonzero genus spanning a contour. Mem. Amer. Math. Soc. **71**(382), iv+83 (1988). https://doi.org/10.1090/memo/0382
508. Tomi, F., Tromba, A.J.: The index theorem for minimal surfaces of higher genus. Mem. Amer. Math. Soc. **117**(560), vi+78 (1995). https://doi.org/10.1090/memo/0560
509. Trèves, F.: Topological vector spaces, distributions and kernels. Academic Press, New York-London (1967)
510. Tromba, A.J.: A classical variational approach to Teichmüller theory. In: Topics in calculus of variations (Montecatini Terme, 1987), *Lecture Notes in Math.*, vol. 1365, pp. 155–185. Springer, Berlin (1989). https://doi.org/10.1007/BFb0089181
511. Tromba, A.J.: Teichmüller theory in Riemannian geometry. Lectures in Mathematics ETH Zürich. Birkhäuser Verlag, Basel (1992). https://doi.org/10.1007/978-3-0348-8613-0
512. Trummer, M.R.: An efficient implementation of a conformal mapping method based on the Szegő kernel. SIAM J. Numer. Anal. **23**(4), 853–872 (1986). https://doi.org/10.1137/0723055

513. Väisälä, J.: Lectures on n-dimensional quasiconformal mappings, *Lecture Notes in Mathematics*, vol. Vol. 229. Springer-Verlag, Berlin-New York (1971)
514. Valiron, G.: Théorie des Fonctions. Masson et Cie, Paris (1942)
515. Vekua, I.N.: Generalized analytic functions. Pergamon Press, London-Paris-Frankfurt; Addison-Wesley Publishing Company, Inc., Reading, MA (1962)
516. Velling, J.: Thesis. Ph.D. thesis, Stanford (1985)
517. Warschawski, S.E.: On the boundary behavior of conformal maps. Nagoya Math. J. **30**, 83–101 (1967). http://projecteuclid.org/euclid.nmj/1118796802
518. Warschawski, S.E.: On boundary derivations in conformal mapping. Ann. Acad. Sci. Fenn. Ser. A I **420**, 22 (1968)
519. Warschawski, S.E.: On Hölder continuity at the boundary in conformal maps. J. Math. Mech. **18**, 423–427 (1969). https://doi.org/10.1512/iumj.1969.18.18032
520. Wegmann, R.: Methods for numerical conformal mapping. In: Handbook of complex analysis: geometric function theory. Vol. 2, pp. 351–477. Elsevier Sci. B. V., Amsterdam (2005). https://doi.org/10.1016/S1874-5709(05)80013-7
521. Weierstraß, K.: Mathematische Werke: Vorlesungen über die Theorie der elliptischen Funktionen. v. 5. G. Olms (1967). https://books.google.md/books?id=aUUhnwEACAAJ
522. Weil, A.: Elliptic functions according to Eisenstein and Kronecker, *Ergeb. Math. Grenzgeb.*, vol. 88. Springer-Verlag, Berlin (1976)
523. Wells Jr., R.O.: Differential analysis on complex manifolds, *Graduate Texts in Mathematics*, vol. 65, second edn. Springer-Verlag, New York-Berlin (1980)
524. Weyl, H.: Die Idee der Riemannschen Fläche. B. G. Teubner Verlagsgesellschaft, Stuttgart (1955)
525. Weyl, H.: The concept of a Riemann surface, German edn. ADIWES International Series in Mathematics. Addison-Wesley Publishing Co., Inc., Reading, Mass.-London (1964)
526. Weyl, H.: Die Idee der Riemannschen Fläche. B. G. Teubner, Stuttgart (1974)
527. Widman, K.O.: On the boundary behavior of solutions to a class of elliptic partial differential equations. Ark. Mat. **6**, 485–533 (1967). https://doi.org/10.1007/BF02591926
528. Wiegerinck, J.J.O.O.: Domains with finite-dimensional Bergman space. Math. Z. **187**(4), 559–562 (1984). https://doi.org/10.1007/BF01174190
529. Wienhard, A.: An invitation to higher Teichmüller theory. In: Proceedings of the International Congress of Mathematicians—Rio de Janeiro 2018. Vol. II. Invited lectures, pp. 1013–1039. World Sci. Publ., Hackensack, NJ (2018)
530. Wilson, R.A.: The Monster is a Hurwitz group. J. Group Theory **4**(4), 367–374 (2001). https://doi.org/10.1515/jgth.2001.027
531. Wogen, W.R.: Composition operators acting on spaces of holomorphic functions on domains in $\mathbb{C}^n$. In: Operator theory: operator algebras and applications, Part 2 (Durham, NH, 1988), *Proc. Sympos. Pure Math.*, vol. 51, Part 2, pp. 361–366. Amer. Math. Soc., Providence, RI (1990). https://doi.org/10.1090/pspum/051.2/1077457
532. Wolf, J.A.: Spaces of constant curvature, sixth edn. AMS Chelsea Publishing, Providence, RI (2011). https://doi.org/10.1090/chel/372
533. Wong, B.: Characterization of the unit ball in $\mathbb{C}^n$ by its automorphism group. Invent. Math. **41**(3), 253–257 (1977). https://doi.org/10.1007/BF01403050
534. Wong, B.: On complex manifolds with noncompact automorphism groups. In: Explorations in complex and Riemannian geometry, *Contemp. Math.*, vol. 332, pp. 287–304. Amer. Math. Soc., Providence, RI (2003). https://doi.org/10.1090/conm/332/05943
535. Wu, H.: Normal families of holomorphic mappings. Acta Math. **119**, 193–233 (1967). https://doi.org/10.1007/BF02392083
536. Yang, K.: Complex algebraic geometry, *Monographs and Textbooks in Pure and Applied Mathematics*, vol. 149. Marcel Dekker, Inc., New York (1991)
537. Yau, S.T.: A general Schwarz lemma for Kähler manifolds. Amer. J. Math. **100**(1), 197–203 (1978). https://doi.org/10.2307/2373880
538. Zeidler, E.: Nonlinear functional analysis and its applications. I. Springer-Verlag, New York (1986). https://doi.org/10.1007/978-1-4612-4838-5

539. Zeilberger, D.: A combinatorial proof of Newton's identities. Discrete Math. **49**(3), 319 (1984). https://doi.org/10.1016/0012-365X(84)90171-7

540. Zorich, V.A.: Quasiconformal mappings and the asymptotic geometry of manifolds. Uspekhi Mat. Nauk **57**(3(345)), 3–28 (2002). https://doi.org/10.1070/RM2002v057n03ABEH000510

Index

P. V. Dovbush, S. G. Krantz, *The Geometric Theory of Complex Variables*,
https://doi.org/10.1007/978-3-031-77204-7

www.ingramcontent.com/pod-product-compliance
Lightning Source LLC
Chambersburg PA
CBHW071636030225
21296CB00011B/192
9783031772030